Catalysis for a Sustainable Environment

Catalysis for a Sustainable Environment

Reactions, Processes and Applied Technologies

Volume 3

Edited by

Professor Armando J. L. Pombeiro
Instituto Superior técnico
Lisboa, Portugal

Dr. Manas Sutradhar
Universidade Lusófona de Humanidades e Tecnologias
Faculdade de Engenharia
Lisboa, Portugal

Professor Elisabete C. B. A. Alegria
Instituto Politécnico de Lisboa
Departamento de Engenharia Química
Lisboa, Portugal

WILEY

This edition first published 2024
© 2024 John Wiley and Sons Ltd

All rights reserved. No part of this publication may be reproduced, stored in a retrieval system, or transmitted, in any form or by any means, electronic, mechanical, photocopying, recording or otherwise, except as permitted by law. Advice on how to obtain permission to reuse material from this title is available at http://www.wiley.com/go/permissions.

The right of Armando J.L. Pombeiro, Manas Sutradhar, and Elisabete C.B.A. Alegria to be identified as the author of the editorial material in this work has been asserted in accordance with law.

Registered Offices
John Wiley & Sons, Inc., 111 River Street, Hoboken, NJ 07030, USA
John Wiley & Sons Ltd, The Atrium, Southern Gate, Chichester, West Sussex, PO19 8SQ, UK

Editorial Office
The Atrium, Southern Gate, Chichester, West Sussex, PO19 8SQ, UK

For details of our global editorial offices, customer services, and more information about Wiley products visit us at www.wiley.com.

Wiley also publishes its books in a variety of electronic formats and by print-on-demand. Some content that appears in standard print versions of this book may not be available in other formats.

Trademarks: Wiley and the Wiley logo are trademarks or registered trademarks of John Wiley & Sons, Inc. and/or its affiliates in the United States and other countries and may not be used without written permission. All other trademarks are the property of their respective owners. John Wiley & Sons, Inc. is not associated with any product or vendor mentioned in this book.

Limit of Liability/Disclaimer of Warranty
In view of ongoing research, equipment modifications, changes in governmental regulations, and the constant flow of information relating to the use of experimental reagents, equipment, and devices, the reader is urged to review and evaluate the information provided in the package insert or instructions for each chemical, piece of equipment, reagent, or device for, among other things, any changes in the instructions or indication of usage and for added warnings and precautions. While the publisher and authors have used their best efforts in preparing this work, they make no representations or warranties with respect to the accuracy or completeness of the contents of this work and specifically disclaim all warranties, including without limitation any implied warranties of merchantability or fitness for a particular purpose. No warranty may be created or extended by sales representatives, written sales materials or promotional statements for this work. The fact that an organization, website, or product is referred to in this work as a citation and/or potential source of further information does not mean that the publisher and authors endorse the information or services the organization, website, or product may provide or recommendations it may make. This work is sold with the understanding that the publisher is not engaged in rendering professional services. The advice and strategies contained herein may not be suitable for your situation. You should consult with a specialist where appropriate. Further, readers should be aware that websites listed in this work may have changed or disappeared between when this work was written and when it is read. Neither the publisher nor authors shall be liable for any loss of profit or any other commercial damages, including but not limited to special, incidental, consequential, or other damages.

A catalogue record for this book is available from the Library of Congress

Hardback ISBN: 9781119870524; ePub ISBN: 9781119870630; ePDF ISBN: 9781119870623;
oBook ISBN: 9781119870647

Cover image: © Sasha Fenix/Shutterstock
Cover design by Wiley

Set in 9.5/12.5pt STIXTwoText by Integra Software Services Pvt. Ltd, Pondicherry, India
Printed and bound by CPI Group (UK) Ltd, Croydon, CR0 4YY

Contents

VOLUME 1

About the Editors *xiii*
Preface *xv*

1 **Introduction** *1*
Armando J.L. Pombeiro, Manas Sutradhar, and Elisabete C.B.A. Alegria

Part I Carbon Dioxide Utilization *5*

2 **Transition from Fossil-C to Renewable-C (Biomass and CO_2) Driven by Hybrid Catalysis** *7*
Michele Aresta and Angela Dibenedetto

3 **Synthesis of Acetic Acid Using Carbon Dioxide** *25*
Philippe Kalck

4 **New Sustainable Chemicals and Materials Derived from CO_2 and Bio-based Resources: A New Catalytic Challenge** *35*
Ana B. Paninho, Malgorzata E. Zakrzewska, Leticia R.C. Correa, Fátima Guedes da Silva, Luís C. Branco, and Ana V.M. Nunes

5 **Sustainable Technologies in CO_2 Utilization: The Production of Synthetic Natural Gas** *55*
M. Carmen Bacariza, José M. Lopes, and Carlos Henriques

6 **Catalysis for Sustainable Aviation Fuels: Focus on Fischer-Tropsch Catalysis** *73*
Denzil Moodley, Thys Botha, Renier Crous, Jana Potgieter, Jacobus Visagie, Ryan Walmsley, and Cathy Dwyer

7 **Sustainable Catalytic Conversion of CO_2 into Urea and Its Derivatives** *117*
Maurizio Peruzzini, Fabrizio Mani, and Francesco Barzagli

Part II Transformation of Volatile Organic Compounds (VOCs) *139*

8   Catalysis Abatement of NO_x/VOCs Assisted by Ozone *141*
Zhihua Wang and Fawei Lin

9   Catalytic Oxidation of VOCs to Value-added Compounds Under Mild Conditions *161*
Elisabete C.B.A. Alegria, Manas Sutradhar, and Tannistha R. Barman

10   Catalytic Cyclohexane Oxyfunctionalization *181*
Manas Sutradhar, Elisabete C.B.A. Alegria, M. Fátima C. Guedes da Silva, and Armando J.L. Pombeiro

Part III Carbon-based Catalysis *207*

11   Carbon-based Catalysts for Sustainable Chemical Processes *209*
Katarzyna Morawa Eblagon, Raquel P. Rocha, M. Fernando R. Pereira, and José Luís Figueiredo

12   Carbon-based Catalysts as a Sustainable and Metal-free Tool for Gas-phase Industrial Oxidation Processes *225*
Giulia Tuci, Andrea Rossin, Matteo Pugliesi, Housseinou Ba, Cuong Duong-Viet, Yuefeng Liu, Cuong Pham-Huu, and Giuliano Giambastiani

13   Hybrid Carbon-Metal Oxide Catalysts for Electrocatalysis, Biomass Valorization and, Wastewater Treatment: Cutting-Edge Solutions for a Sustainable World *247*
Clara Pereira, Diana M. Fernandes, Andreia F. Peixoto, Marta Nunes, Bruno Jarrais, Iwona Kuźniarska-Biernacka, and Cristina Freire

VOLUME 2

About the Editors *xiii*
Preface *xv*

Part IV Coordination, Inorganic, and Bioinspired Catalysis *299*

14   Hydroformylation Catalysts for the Synthesis of Fine Chemicals *301*
Mariette M. Pereira, Rui M.B. Carrilho, Fábio M.S. Rodrigues, Lucas D. Dias, and Mário J.F. Calvete

15   Synthesis of New Polyolefins by Incorporation of New Comonomers *323*
Kotohiro Nomura and Suphitchaya Kitphaitun

16	**Catalytic Depolymerization of Plastic Waste** *339* Noel Angel Espinosa-Jalapa and Amit Kumar	
17	**Bioinspired Selective Catalytic C-H Oxygenation, Halogenation, and Azidation of Steroids** *369* Konstantin P. Bryliakov	
18	**Catalysis by Pincer Compounds and Their Contribution to Environmental and Sustainable Processes** *389* Hugo Valdés and David Morales-Morales	
19	**Heterometallic Complexes: Novel Catalysts for Sophisticated Chemical Synthesis** *409* Franco Scalambra, Ismael Francisco Díaz-Ortega, and Antonio Romerosa	
20	**Metal-Organic Frameworks in Tandem Catalysis** *429* Anirban Karmakar and Armando J.L. Pombeiro	
21	**(Tetracarboxylate)bridged-di-transition Metal Complexes and Factors Impacting Their Carbene Transfer Reactivity** *445* LiPing Xu, Adrian Varela-Alvarez, and Djamaladdin G. Musaev	
22	**Sustainable Cu-based Methods for Valuable Organic Scaffolds** *461* Argyro Dolla, Dimitrios Andreou, Ethan Essenfeld, Jonathan Farhi, Ioannis N. Lykakis, and George E. Kostakis	
23	**Environmental Catalysis by Gold Nanoparticles** *481* Sónia Alexandra Correia Carabineiro	
24	**Platinum Complexes for Selective Oxidations in Water** *515* Alessandro Scarso, Paolo Sgarbossa, Roberta Bertani, and Giorgio Strukul	
25	**The Role of Water in Reactions Catalyzed by Transition Metals** *537* A.W. Augustyniak and A.M. Trzeciak	
26	**Using Speciation to Gain Insight into Sustainable Coupling Reactions and Their Catalysts** *559* Skyler Markham, Debbie C. Crans, and Bruce Atwater	
27	**Hierarchical Zeolites for Environmentally Friendly Friedel Crafts Acylation Reactions** *577* Ana P. Carvalho, Angela Martins, Filomena Martins, Nelson Nunes, and Rúben Elvas-Leitão	

VOLUME 3

About the Editors *xiii*
Preface *xv*

Part V   Organocatalysis   609

28   Sustainable Drug Substance Processes Enabled by Catalysis: Case Studies from the Roche Pipeline   *611*
Kurt Püntener, Stefan Hildbrand, Helmut Stahr, Andreas Schuster, Hans Iding, and Stephan Bachmann
28.1   Introduction   *611*
28.2   Case Studies   *612*
28.2.1   Aleglitazar   *612*
28.2.2   Idasanutlin   *619*
28.2.3   Danoprevir   *623*
28.2.4   Ipatasertib   *626*
28.3   Conclusions   *635*
References   *636*

29   Supported Chiral Organocatalysts for Accessing Fine Chemicals   *639*
Ana C. Amorim and Anthony J. Burke
29.1   Introduction   *639*
29.2   Organocatalyst Immobilizations   *640*
29.2.1   Proline Immobilizations   *640*
29.2.2   Diphenylprolinol Silyl Ether (Jørgensen-Hayashi Organocatalyst) Immobilizations   *643*
29.2.3   Organocatalysts Based on Immobilized Pyrrolidines   *645*
29.2.4   Organocatalysts Based on Immobilized Imidazolidinones   *647*
29.2.5   Other Amino Acid and Peptide Type Catalysts   *648*
29.2.5.1   Supported-primary Amino Acid Catalysts   *649*
29.2.5.2   Supported-peptide Derivative Catalysts   *650*
29.2.6   Immobilized Amino-Cinchona Based Organocatalysts   *651*
29.2.6.1   Cinchona Picolinamide Derivatives   *651*
29.2.6.2   Cinchona Squaramide Derivatives   *652*
29.2.7   Other Organocatalysts   *653*
29.2.7.1   Phosphoric Acid Catalysts   *654*
29.2.7.2   Isothiourea Catalysts   *655*
29.3   Conclusions   *657*
References   *657*

30   Synthesis of Bio-based Aliphatic Polyesters from Plant Oils by Efficient Molecular Catalysis   *659*
Kotohiro Nomura and Nor Wahida Binti Awang
30.1   Introduction   *659*
30.2   Synthesis of Bio-Based Aliphatic Polyesters by Condensation Polymerization   *660*
30.2.1   Synthesis of Bio-Based Aliphatic Polyesters by Condensation Polymerization and Dehydrogenative Condensation   *661*

30.2.2	Synthesis of BioBasd Aliphatic Polyesters by Acyclic Diene Metathesis (ADMET) Polymerization and Subsequent Hydrogenation *663*	
30.2.3	One Pot Synthesis of Bio-Based Long Chain Aliphatic Polyesters by Tandem ADMET Polymerization and Hydrogenation. Depolymerization by Reaction with Ethylene *670*	
30.3	Concluding Remarks and Outlook *671*	
	References *672*	

31 Modern Strategies for Electron Injection by Means of Organic Photocatalysts: Beyond Metallic Reagents *675*
Takashi Koike

31.1	Introduction *675*
31.2	Basic and Advanced Concepts for $1e^-$ Injection by Organic Photoredox Catalysis *675*
31.3	Triarylamine-based Highly Reducing Organic Photocatalysts *677*
31.4	Consecutive Photoinduced Electron Transfer (*conPET*) by Organic Photoredox Catalysis *681*
31.5	Consecutive Photoinduced Electron Transfer (*conPET*) by the Combination of Organic Photocatalysis and Electrolysis *683*
31.6	Summary and Outlook *685*
	References *685*

32 Visible Light as an Alternative Energy Source in Enantioselective Catalysis *687*
Ana Maria Faisca Phillips and Armando J.L. Pombeiro

32.1	Introduction *687*
32.2	Dual Chiral Organocatalysis and Photoredox Catalysis *690*
32.2.1	Chiral Amines as Catalysts *690*
32.2.2	N-Heterocyclic Carbenes (NHCs) as Catalysts *695*
32.2.3	Chiral Phosphoric Acids as Catalysts *696*
32.2.4	Miscellaneous *700*
32.3	Metal Catalyzed Processes *702*
32.3.1	Dual Transition Metal/Photoredox Catalysis *702*
32.3.2	Dual Chiral Lewis Acid/Photoredox Catalysis *707*
32.4	Chiral Photocatalysts *708*
32.4.1	Chiral-at-Metal Photocatalysts *710*
32.4.2	Organic Photocatalysts *711*
32.5	Conclusions *712*
	Acknowledgements *712*
	References *712*

Part VI Catalysis for the Purification of Water and Liquid Fuels *717*

33 Heterogeneous Photocatalysis for Wastewater Treatment: A Major Step Towards Environmental Sustainability *719*
Shima Rahim Pouran and Aziz Habibi-Yangjeh

33.1	Introduction *719*
33.2	Heterogeneous Photocatalysis *720*
33.3	Sustainable Photocatalysts *721*

33.3.1	Metal Oxide-based Photocatalysts	722
33.3.1.1	Magnetic Metal Oxide Semiconductors	725
33.3.1.2	Green Synthesis Routes	730
33.3.2	Carbonaceous Photocatalysts	732
33.4	Remarks and Future Perspectives	737
	Acknowledgments	737
	References	737

34 Sustainable Homogeneous Catalytic Oxidative Processes for the Desulfurization of Fuels 743

Federica Sabuzi, Giuseppe Pomarico, Pierluca Galloni, and Valeria Conte

34.1	Introduction	743
34.2	Vanadium	743
34.3	Manganese	746
34.4	Iron	746
34.5	Cobalt	748
34.6	Molybdenum	749
34.7	Tungsten	750
34.8	Polyoxometalates	750
34.9	Ionic Liquids	751
34.10	Conclusions	753
	References	753

35 Heterogeneous Catalytic Desulfurization of Liquid Fuels: The Present and the Future 757

Rui G. Faria, Alexandre Viana, Carlos M. Granadeiro, Luís Cunha-Silva, and Salete S. Balula

35.1	Introduction	757
35.2	Hydrodesulfurization	758
35.3	Adsorptive Desulfurization	761
35.3.1	ADS with Carbon-based Materials	762
35.3.2	ADS with Zeolites	763
35.3.3	ADS with Mesoporous Silica	765
35.3.4	ADS with Metal-organic Frameworks	765
35.4	Oxidative Desulfurization	767
35.4.1	Oxidants for (EC)ODS	768
35.4.2	Heterogeneous Catalysts for (EC)ODS	768
35.4.2.1	(EC)ODS with Zeolites	768
35.4.2.2	(EC)ODS with Metal-organic Frameworks	769
35.4.2.3	(EC)ODS with Carbon-based Materials	771
35.4.2.4	(EC)ODS with Mesoporous Silicas	773
35.4.2.5	(EC)ODS with Titanate Nanotubes	773
35.4.3	(EC)ODS Catalyzed by Heterogeneous Polyoxometalates	774
35.4.3.1	Carbonaceous Composites	774
35.4.3.2	MOF Composites	775
35.4.3.3	Zeolite Composites	775
35.4.3.4	Mesoporous Silica Composites	776

35.5	(EC)ODS Catalyzed by Membranes *777*	
35.6	Future Perspectives *778*	
	Acknowledgments *779*	
	References *779*	

Part VII Hydrogen Formation, Storage, and Utilization *783*

36 Paraformaldehyde: Opportunities as a C1-Building Block and H_2 Source for Sustainable Organic Synthesis *785*

Ana Maria Faísca Phillips, Maximilian N. Kopylovich, Leandro Helgueira de Andrade, and Martin H.G. Prechtl

36.1	Introduction *785*
36.2	Carbonylation and Related Reactions *787*
36.2.1	Alkoxycarbonylation of Olefins *788*
36.2.2	Carbonylation of Aryl Halides *790*
36.2.3	Cascade C–H Activation/carbonylation/cyclization Reactions and Related Processes: The Synthesis of Heterocycles *793*
36.2.4	Hydroformylation of Alkenes *794*
36.2.5	*N*-formylation *798*
36.3	Methylation and Related Reactions *799*
36.4	Hydrogen Generation and Transfer-hydrogenation Reactions *810*
36.5	Summary and Outlook *815*
	Acknowledgement *815*
	References *815*

37 Hydrogen Storage and Recovery with the Use of Chemical Batteries *819*

Henrietta Horváth, Gábor Papp, Ágnes Kathó, and Ferenc Joó

37.1	Introduction *819*
37.2	Hydrogen as an Energy Storage Material *820*
37.3	Chemical Hydrogen Storage *821*
37.4	Liquid Organic Hydrogen Carriers *822*
37.5	Definitions and Fundamental Questions *824*
37.6	Catalysts Applied in Hydrogen Batteries *826*
37.7	Formic Acid and Formate Salts as Storage Materials in Hydrogen Batteries *827*
37.7.1	Formic Acid as a Hydrogen Storage Material *828*
37.7.2	Formate Salts as Hydrogen Storage Materials *829*
37.8	Catalysts and Reaction Conditions Potentially Applicable in Hydrogen Batteries Based on the Formate-bicarbonate Equilibrium *831*
37.9	Functional Hydrogen Batteries *832*
37.9.1	Hydrogen Batteries Based on CO_2–formic Acid Cycles *833*
37.9.2	Hydrogen Batteries Based on Formate–bicarbonate Cycles *834*
37.9.3	Hydrogen Batteries Based on N-heterocyclic Compounds *836*
37.9.4	Hydrogen Batteries Based on Alcohols *837*
37.9.5	Hydrogen Batteries Based on Whole-cell Biocatalysis *839*
37.10	Summary and Conclusions *840*
	Acknowledgements *840*
	References *841*

38	**Low-cost Co and Ni MOFs/CPs as Electrocatalysts for Water Splitting Toward Clean Energy-Technology** *847*	
	Anup Paul, Biljana Šljukić, and Armando J.L. Pombeiro	
38.1	Introduction *847*	
38.2	Fundamentals of Water Splitting Reactions *849*	
38.3	MOFs/CPs as Electrocatalysts for Water Splitting Reactions *852*	
38.3.1	Co MOFs and Derived Electrocatalysts for OER and HER *852*	
38.3.1.1	Co MOFs and Derived Electrocatalysts for OER *852*	
38.3.1.2	Co MOFs and Derived Composites for HER *857*	
38.3.2	Ni MOFs and Derived Composites for OER and HER *859*	
38.3.2.1	Ni MOFs and Derived Composites for the OER *859*	
38.3.2.2	Ni MOFs and Derived Composites for HER *863*	
38.3.3	Polyhomo and Heterometallic MOFs of Co(II) or Ni(II) for OER and HER *864*	
38.4	Conclusions *868*	
	Acknowledgements *868*	
	References *868*	

Index *871*

About the Editors

Armando Pombeiro is a Full Professor Jubilado at Instituto Superior Técnico, Universidade de Lisboa (ULisboa), former Distant Director at the People's Friendship University of Russia (RUDN University), a Full Member of the Academy of Sciences of Lisbon (ASL), the President of the Scientific Council of the ASL, a Fellow of the European Academy of Sciences (EURASC), a Member of the Academia Europaea, founding President of the College of Chemistry of ULisboa, a former Coordinator of the Centro de Química Estrutural at ULisboa, Coordinator of the Coordination Chemistry and Catalysis group at ULisboa, and the founding Director of the doctoral Program in Catalysis and Sustainability at ULisboa. He has chaired major international conferences. His research addresses the activation of small molecules with industrial, environmental, or biological significance (including alkane functionalization, oxidation catalysis, and catalysis in unconventional conditions) as well as crystal engineering of coordination compounds, polynuclear and supramolecular structures (including MOFs), non-covalent interactions in synthesis, coordination compounds with bioactivity, molecular electrochemistry, and theoretical studies.

He has authored or edited 10 books, (co-)authored *ca.* 1000 research publications, and registered *ca.* 40 patents. His work received *over.* 30,000 citations (over 12,000 citing articles), h-index *ca.* 80 (Web of Science).

Among his honors, he was awarded an Honorary Professorship by St. Petersburg State University (Institute of Chemistry), an Invited Chair Professorship by National Taiwan University of Science & Technology, the inaugural SCF French-Portuguese Prize by the French Chemical Society, the Madinabeitia-Lourenço Prize by the Spanish Royal Chemical Society, and the Prizes of the Portuguese Chemical and Electrochemical Societies, the Scientific Prizes of ULisboa and Technical ULisboa, and the Vanadis Prize. Special issues of Coordination Chemistry Reviews and the Journal of Organometallic Chemistry were published in his honor.

https://fenix.tecnico.ulisboa.pt/homepage/ist10897

Manas Sutradhar is an Assistant Professor at the Universidade Lusófona, Lisbon, Portugal and an integrated member at the Centro de Química Estrutural, Instituto Superior Técnico, Universidade de Lisboa, Portugal. He was a post-doctoral fellow at the Institute of Inorganic and Analytical Chemistry of Johannes Gutenberg University of Mainz, Germany and a researcher at the Centro de Química Estrutural, Instituto Superior Técnico, Universidade de Lisboa. He has published 72 papers in international peer review journals (including three reviews + 1 reference module), giving him an h-index 28 (ISI Web of Knowledge) and more than 2250 citations. In addition, he has 11 book chapters in books with international circulation and one patent. He is one of the editors of the book *Vanadium Catalysis*, published by the Royal Society of Chemistry. His main areas of work include metal complexes with aroylhydrazones, oxidation catalysis of industrial importance and sustainable environmental significance, magnetic properties of metal complexes, and bio-active molecules. The major contributions of his research work are in the areas of vanadium chemistry and oxidation catalysis. He received the 2006 Young Scientist Award from the Indian Chemical Society, India and the Sir P. C. Ray Research Award (2006) from the University of Calcutta, India.

https://orcid.org/0000-0003-3349-9154

Elisabete C.B.A. Alegria is an Adjunct Professor at the Chemical Engineering Department of the Instituto Superior de Engenharia de Lisboa (ISEL) of the Polytechnic Institute of Lisbon, Portugal. She is a researcher (Core Member) at the Centro de Química Estrutural (Coordination Chemistry and Catalysis Group). She has authored 86 papers in international peered review journals and has an h-index of 23 with over 1600 citations, four patents, five book chapters, and over 180 presentations at national and international scientific meetings. She was awarded an Honorary Distinction (2017–2020) for the Areas of Technology and Engineering (Scientific Prize IPL-CGD). She is an editorial board member, and has acted as a guest editor and reviewer for several scientific journals. Her main research interests include coordination and sustainable chemistry, homogeneous and supported catalysis, stimuli-responsive catalytic systems, green synthesis of metallic nanoparticles for catalysis, and biomedical applications. She is also interested in mechanochemistry (synthesis and catalysis) and molecular electrochemistry.

https://orcid.org/0000-0003-4060-1057

Preface

Aiming to change the world for the better, 17 Sustainable Development Goals (SDGs) were adopted by the United Nations (UN) Member States in 2015, as part of the UN 2030 Agenda for Sustainable Development that concerns social, economic, and *environmental sustainability*. Hence, a 15-year plan was set up to achieve these Goals and it is already into its second half.

However, the world does not seem to be on a good track to reach those aims as it is immersed in the Covid-19 pandemic crisis and climate emergency, as well as economic and political uncertainties. Enormous efforts must be pursued to overcome these obstacles and chemical sciences should play a pivotal role. *Catalysis* is of particular importance as it constitutes the most relevant contribution of chemistry towards sustainable development. This is true even though the SDGs are integrated and action in one can affect others.

For example, the importance of chemistry and particularly catalysis is evident in several SDGs. Goal 12, addresses "Responsible Consumption and Production Patterns" and is aligned with the circularity concept with sustainable loops or cycles (e.g., in recycle and reuse processes that are relevant within the UN Environmental Program). Goal 7 addresses "Affordable and Clean Energy" and relates to efforts to improve energy conversion processes, such as hydrogen evolution and oxygen evolution from water, that have a high environmental impact. Other SDGs in which chemistry and catalysis play an evident role with environmental significance include Goal 6 ("Clean Water and Sanitation"), Goal 9 ("Industry, Innovation and Infrastructure") 13 ("Climate Action"), Goal 14 ("Life Below Water"), and Goal 15 ("Life on Land").

The book is aligned with these SDGs by covering recent developments in various *catalytic processes* that are designed for a *sustainable environment*. It gathers skilful researchers from around the world to address the use of catalysis in various approaches, including homogeneous, supported, and heterogeneous catalyses as well as photo- and electrocatalysis by searching for innovative green chemistry routes from a sustainable environmental angle. It illustrates, in an authoritative way, state-of-the-art knowledge in relevant areas, presented from modern perspectives and viewpoints topics in coordination, inorganic, organic, organometallic, bioinorganic, pharmacological, and analytical chemistries as well as chemical engineering and materials science.

The chapters are spread over seven main sections focused on Carbon Dioxide Utilization, Transformation of Volatile Organic Compound (VOCs), Carbon-based Catalysts, Coordination, Inorganic, and Bioinspired Catalysis, Organocatalysis, Catalysis for the Purification of Water and Liquid Fuels,and Hydrogen Formation, Storage, and Utilization. These sections are gathered together as a contribution towards the development of the challenging topic.

The book addresses topics in (i) activation of relevant small molecules with strong environmental impacts, (ii) catalytic synthesis of important added value organic compounds, and (iii) development of systems operating under environmentally benign and mild conditions toward the establishment of sustainable energy processes.

This work is expected to be a reference for academic and research staff of universities and research institutions, including industrial laboratories. It is also addressed to post-doctoral, post-graduate, and undergraduate students (in the latter case as a supplemental text) working in chemical, chemical engineering, and related sciences. It should also provide inspiration for research topics for PhD and MSc theses, projects, and research lines, in addition to acting as an encouragement for the development of the overall field.

The topic Catalysis for Sustainable Environment is very relevant in the context of modern research and is often implicit, although in a non-systematic and disconnected way, in many publications and in a number of initiatives such as international conferences. These include the XXII International Symposium on Homogeneous Catalysis (ISHC) that we organized (Lisbon, 2022) and that to some extent inspired some parts of this book.

In contrast to the usual random inclusion of the topic in the literature and scientific events, the applications of catalytic reactions focused on a sustainable environment in a diversity of approaches are addressed in this book.

The topic has also contributed to the significance of work that led to recent Nobel Prizes of Chemistry. In 2022, the Nobel Prize was awarded to Barry Sharpless, Morten Meldal, and Carolyn Bertozzi for the development of click chemistry and bioorthogonal chemistry. The set of criteria for a reaction or a process to meet in the context of click chemistry includes, among others, the operation under benign conditions such as those that are environmentally friendly (e.g., preferably under air and in water medium). In 2021, the Nobel Prize was awarded to Benjamin List and David W.C. MacMillan for the development of asymmetric organocatalysis, which relies on environmentally friendly organocatalysts.

The book illustrates the connections of catalysis with a sustainable environment, as well as the richness and potential of modern catalysis and its relationships with other sciences (thus fostering interdisciplinarity) in pursuit of sustainability.

At last, but not least, we should acknowledge the authors of the chapters for their relevant contributions, prepared during a particularly difficult pandemic period, as well as the publisher, John Wiley, for the support, patience, and understanding of the difficulties caused by the adverse circumstances we are experiencing nowadays and that constituted a high activation energy barrier that had to be overcome by all of us... a task that required rather active catalysts.

We hope the readers will enjoy reading its chapters as much as we enjoyed editing this book.

Armando Pombeiro
Manas Sutradhar
Elisabete Alegria

Part V

Organocatalysis

28

Sustainable Drug Substance Processes Enabled by Catalysis

Case Studies from the Roche Pipeline

Kurt Püntener, Stefan Hildbrand, Helmut Stahr, Andreas Schuster, Hans Iding, and Stephan Bachmann

Pharmaceutical Division, Synthetic Molecules Technical Development, Process Chemistry and Catalysis, F. Hoffmann-La Roche Ltd, Basel, Switzerland

28.1 Introduction

As a pioneer in healthcare, Roche has been committed to improving lives since the company was founded 1896 in Basel (CH). Delivering new drugs for innovative healthcare solutions is a tremendous undertaking with a journey, starting from understanding a disease fundamentally through the discovery and development of the therapeutic all the way to the manufacture and administration of the product. Adding to this challenge nowadays is the increasing complexity of the chemical structures of the drug candidates, with most of them bearing multiple chiral centers, complex ring systems, and a manifold of different functional groups that were introduced for optimal drug targeting, drug metabolism and pharmokinetics (DMPK), and toxicological profiles. Furthermore, increasing pressure on the pharmaceutical industry to provide affordable drugs and ultimately cost-efficient therapeutic modalities to enable treatments be accessible to patients worldwide contributes to this challenge. Whereas cost-efficiency in times past was associated with the direct cost-of-goods or manufacturing cost, more recently it has become evident that this term must include additionally the environmental impact of the drug substance (DS) manufacture. At present, it is compellingly self-evident that natural resources are limited and represent a precious commodity for everyone. To this end, the chemical industry has created new metrics on how to measure the "greenness" of their chemical processes and, ultimately, to didactically guide chemists on essential process optimizations. As one of the most prominent metrics, the process mass index (PMI = kg of input materials (solvents and reagents) needed to manufacture 1 kg of DS) has made its way into use in the fine chemical, agricultural, and pharmaceutical industries [1]. This index is used to systematically monitor the sustainability of DS processes and is determined from the very first good laboratory practices (GLP) toxicology supply campaigns through to launch of supplies to highlight the impact that process improvements have had during the course the various drug development phases. Most importantly, the process steps and operations which demonstrate the highest potential for overall resource savings along the way direct as a consequence the process research and development activities.

Catalysis for a Sustainable Environment: Reactions, Processes and Applied Technologies Volume 3, First Edition. Edited by Armando J. L. Pombeiro, Manas Sutradhar, and Elisabete C. B. A. Alegria.
© 2024 John Wiley & Sons Ltd. Published 2024 by John Wiley & Sons Ltd.

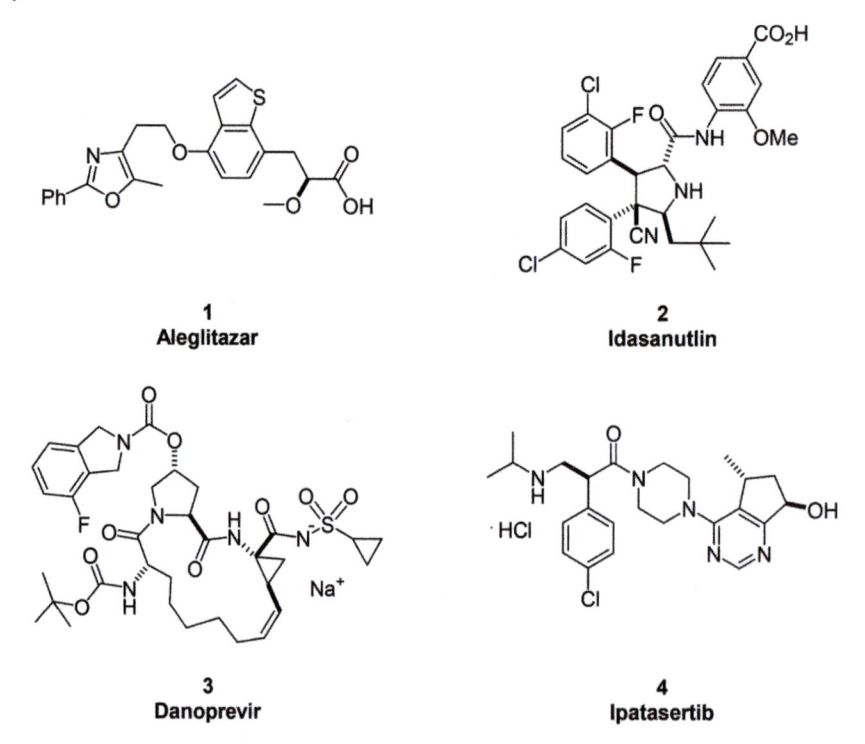

Figure 28.1 The structures of drug candidates **1** (aleglitazar), **2** (idasanutlin), **3** (danoprevir), and **4** (ipatasertib).

To encourage Roche chemists to constantly strive for most sustainable DS processes and to recognize the contributions they have made in this field, the Roche Environmental Awareness in Chemical Technology (REACT) award was introduced. On the basis of the green chemistry principles established by Anastas and Warner [2], the main selection criteria for winning the REACT award are the following: i) shift from avoid/high uncertain to recommended/usable solvents based on the Roche solvent selection guide based on ACS classifications [3]; ii) reduction of PMI factor; iii) introduction of new reaction types, leading to cleaner, more efficient reactions with high atom efficiency; and iv) reduction of energy consumption.

The following four case studies outline Roche's effort to establish most sustainable processes for its investigated drug candidates **1–4** (Figure 28.1). The project teams of danoprevir (**3**), ipatasertib (**4**), and idasanutlin (**2**) were distinguished with REACT awards were achieved back in 2010, 2011 and 2014 [4]. Both cutting-edge process chemistry and catalytic methodologies contributed equally to this success. Although the focus for this casebook chapter is the impact catalysis has had on establishing sustainable DS processes, the contributions from process chemistry have also been fundamental, particularly in delivering efficient syntheses of the substrates for the targeted catalytic transformations and the ensuing downstream chemistry. The reported case studies highlight these contributions and, from a holistic point of view, the attributes a DS process must fulfill to be considered as sustainable.

28.2 Case Studies

28.2.1 Aleglitazar

Aleglitazar (RG1439) (**1**), a potent, balanced dual peroxisome proliferator activated PPARα/γ agonist, showed insulin-sensitizing and glucose-lowering activity and favorable effects on lipid profiles [5]. Clinical studies were run to determine whether **1** might reduce cardiovascular morbidity and mortality among patients with type 2 diabetes mellitus (T2DM) suffering from recent acute coronary syndrome (ACS) when this regimen was added to standard treatment. However, due to safety signals and lack of efficacy, the further development of **1** was halted in 2013.

Aleglitazar (**1**) and edaglitazone (BM131258) (**5**) [6] shared formylbenzothiophene **6** as a common intermediate (Scheme 28.1). When the PPARα/γ program was initiated, **5** was in advanced

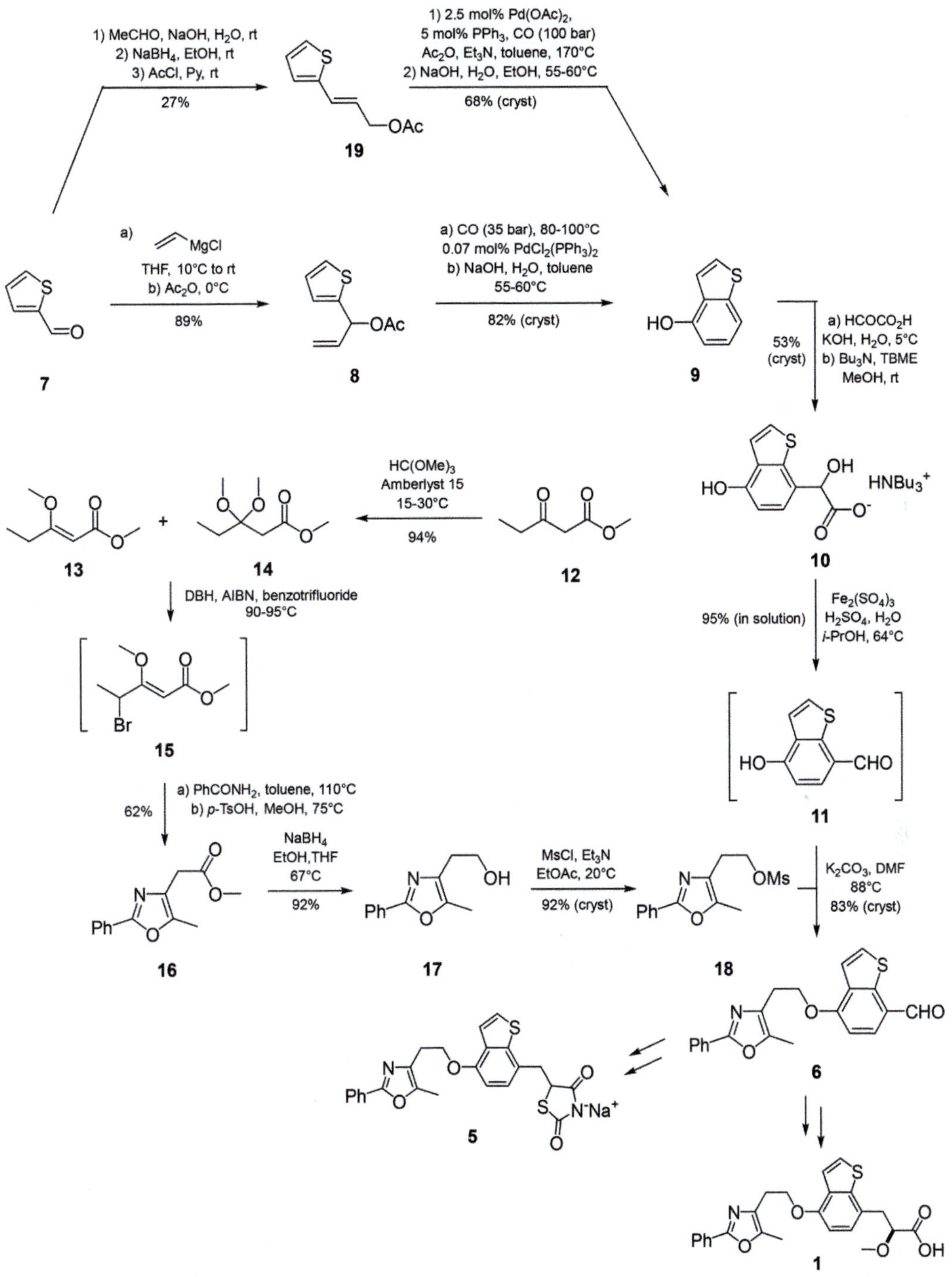

Scheme 28.1 1st and 2nd generation synthesis of **9** for the manufacture of **1** and **5** via formylbenzothiophene **6**.

clinical development and, as a consequence, the synthesis of **6** was already well established and the building block available in multi-kg quantities. Accordingly, both the Discovery Chemistry synthesis of **1** as well as any new envisaged route toward **1** could benefit from readily available **6** or an advanced intermediate thereof.

The synthesis of **6** comprised a total of eight steps (Scheme 28.1) [7, 8]. In the main branch, 2-formylthiophene (**7**) was converted into racemic tetrabutylammonium 2-hydroxy-2-(4-hydroxyphenyl) acetate (**10**), which, after iron assisted reductive decarboxylation and coupling with oxazole mesylate **18**, provided **6** in an overall 32% yield (four steps). Mesylate **18** itself was readily accessible in four steps (49% yield) from 3-oxovalerate **12**.

Key intermediate **9** in the synthesis of **6** was initially accessed following a procedure reported by Hidai et al [9]. In a palladium catalyzed cyclocarbonylation reaction, 3-(2-thienyl)allyl acetate (**19**) (prepared in three steps (27% yield) from **7**) was treated under CO (100 bar) and, in the presence of Ac_2O, Et_3N and a catalytic amount of $Pd(OAc)_2/PPh_3$ in toluene, delivered the product, after the saponification of phenol **9**, in 68% yield (two steps). The harsh reaction conditions employed in the cyclocarbonylation step and the high catalyst loading (2.5 mol%) called for significant process improvements to render the sequence feasible on a technical scale. A breakthrough was achieved when 1-(2-thienyl)allyl acetate (**8**) was employed in place of its regioisomer **19** for the cyclocarbonylation reaction. The acetate isomer **8** was readily accessible (87% yield) by addition of vinyl magnesium bromide to **7** followed by acylation with acetic anhydride and the cyclocarbonylation could be carried out under much milder conditions at significantly lower catalyst loadings (0.07 mol%). Because trace amounts of **19** were detected by high-performance liquid chromatography (HPLC) during the conversion of **8** into **9**, it is very likely that **8** isomerizes prior to cyclization into **19**, indicating that the overall process presumably follows the mechanism as proposed by Hidai. Introducing **8** as a synthesis equivalent for **19** enabled three steps to be cut and furnished **9** in a four-fold higher overall yield (73 vs 18%).

To introduce the aldehyde function into **6**, benzothiophene **9** was treated with glyoxalic acid/KOH followed by the addition of Bu_3N to readily isolate the 7-hydroxyl carboxylate formed as its ammonium salt **10**. In the next step, **10** was subjected to iron assisted reductive decarboxylation to deliver 4-hydroxyl-2-formyl-benzothoiophene (**11**) which, in the final process, was converted without isolation into **6** (83% yield) through coupling with oxazole mesylate **18**. In four steps, **18** was obtained in 48% yield from methyl 3-oxovalerate **12**. The ketoester was treated with trimethyl orthoformate in the presence of amberlyst-15, which delivered a 2:1 mixture of enolester **13** and its ketal **14**. Without separation, subsequent bromination with 1,3-dibromo-5,5-dimethylhydantoin/2,2′-azobis(2-methylpropionitrile) (DBH/AIBN) furnished the intermediate bromo enolester **15**, which was converted (in the presence of benzamide and trace amounts of acid) into the phenyl oxazole **16**. After sodium borohydride reduction, the alcohol **17** formed was converted into **18** via treatment with $MsCl/Et_3N$.

With **6** in hand as an advanced building block, the end game of the Discovery Chemistry synthesis became quite straightforward. This employed the Evans methodology based on boron mediated diastereoselective aldol reactions that promoted the selective introduction of the corresponding chiral centers into key intermediate **22** (Scheme 28.2). (S)-4-Benzyl-2-oxazolidinone (**20**) was deprotonated at −78 °C with butyl lithium and treated with methoxyacetyl chloride, furnishing N-acylated oxazolidinone **21** isolated in 94% yield after chromatographic purification. Subsequently, **21** was condensed with **6** in the presence of dibutylboron triflate at -78 °C, which provided a 91:6:3 mixture of the Evans-*syn*, Evans-*anti*, and non-Evans-*syn* aldol products **22**. Because the non-Evans-*syn* product delivers the undesired enantiomer of **1** after the subsequent transformations, it was chromatographically removed from the crude reaction mixture. The obtained Evans-*syn* and

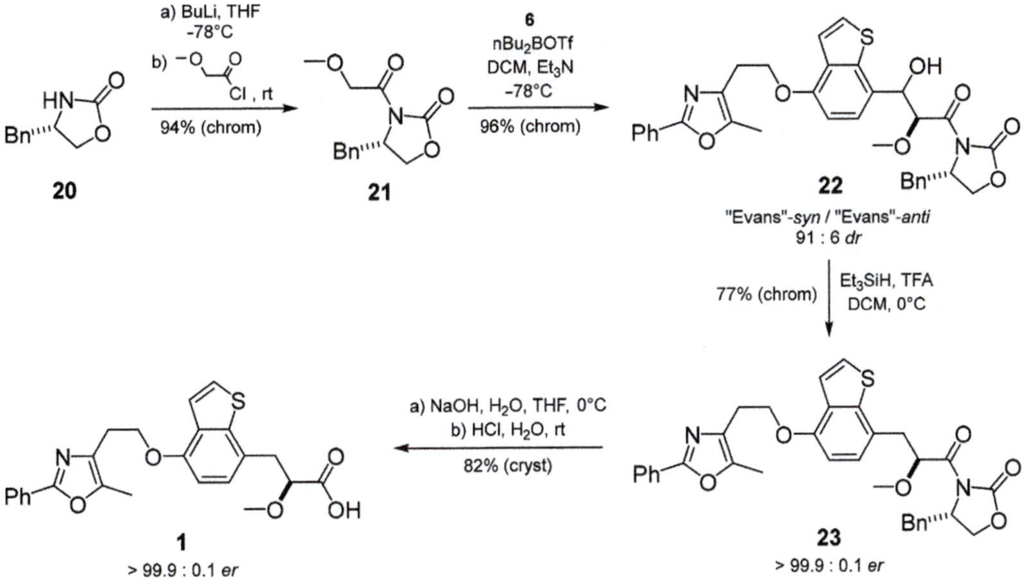

Scheme 28.2 Discovery Chemistry end game synthesis of **1**.

Evans-*anti* aldol product mixture (91:6) was then treated with triethylsilane/trifluoracetic acid (TFA) to promote dehydroxylation and deliver diastereomerically pure **23**. After chromatographic purification and subsequent saponification, **1** with >99.9:0.1 *er* and 99.9 area-% in an overall yield of 56% (from **20**) was obtained.

Although the Discovery Chemistry route was quite short, it suffered from i) the employment of three chromatographic purifications; ii) the need to run two steps under cryogenic conditions; iii) the use of dibutylboron triflate which was expensive and not available in large quantities at that time; and iv) the handling of stoichiometric amounts of a chiral auxiliary that was used to install the stereo-information in **1**.

However, to ensure timely DS supplies for initial toxicology and clinical studies, the Discovery Chemistry route was retained. Work was preliminarily focused to reduce the number of chromatographic purification steps to shorten lead times. Step one product **21** could be purified by crystallization rendering the first chromatography obsolete. The second chromatography, needed to remove the undesired non-Evans-*syn* aldol product and avoid contamination of **1** with its (*R*)-enantiomer, was intentionally skipped as the *er*-value of crude **1** could be enhanced by crystallization downstream. A simple silica gel filtration, nevertheless, was required to achieve a substrate quality suitable for the next steps. Additionally, in the subsequent dehydroxylation step, the amount of SiO_2 could be reduced by replacing the chromatography by a filtration instead delivering **23** in an acceptable *dr* 97:3 quality. After the final saponification, crude **1** was recovered with a corresponding 97:3 *er*. After crystallization from ethyl acetate pure **1** (>99.9 area-%) was isolated with >99.9:0.1 *er* in an overall yield of 45–50% from carbamate **20**. According to this protocol, two batches of **1** (0.5 and 5.8 kg) were produced enabling a timely commencement of GLP tox and clinical phase 1 studies.

Although the three chromatographic purifications could be reduced to two silica gel filtrations leading to a significant decrease in eluent solvent and SiO_2 consumption, it was not possible to omit these waste intense treatments entirely. As a result, process research work was initiated to determine a new route toward **1** that would i) rely on the installment of the chiral center through catalytic methodologies; ii) make use of readily available **6**; iii) be short to remain competitive with the existing route; and iv) deliver **1** in high yield and with a low PMI. Most promising endgame variants were considered to rely upon the asymmetric reduction of a propenoic acid derivative or of a racemic 2-methoxy-3-oxo-propanoic derivative under dynamic kinetic control [10]. The later approach indeed was successful to deliver *syn*-**24** (cf. Scheme 28.3) from its 3-oxo precursor in high yield (99%) as well as selectivity (97:3 *dr*, >99.8:0.2 *er*) [11] and furnish the envisaged target compound **1** after reductive dehydroxylation and saponification. However, introducing such a racemic 2-methoxy-3-oxo-propanoic unit starting from **6** proved cumbersome and this approach was abandoned. Ultimately, prime emphasis was put on approaches that comprised the chiral reduction of 2-methoxypropenoic acid derivatives that would yield either **1** directly or furnish the target product after subsequent saponification of an ester thereof (Scheme 28.3). Such an approach had the highest potential to outperform the Discovery Chemistry route. On the other hand, there was a risk of failure as neither 2-methoxypropenoic acid nor its ester variants had been reported as substrates for asymmetric (transfer) hydrogenations or biocatalysis. Nonetheless, based on Noyori's pioneering work on Ru-BINAP catalyzed asymmetric hydrogenation (AH) of α/β-unsaturated acids [12], as well as a few other reported rhodium and ruthenium based catalyst systems [13], a tailor-made catalyst system to promote hydrogenation of 2-methoxypropenoic acids or a derivative

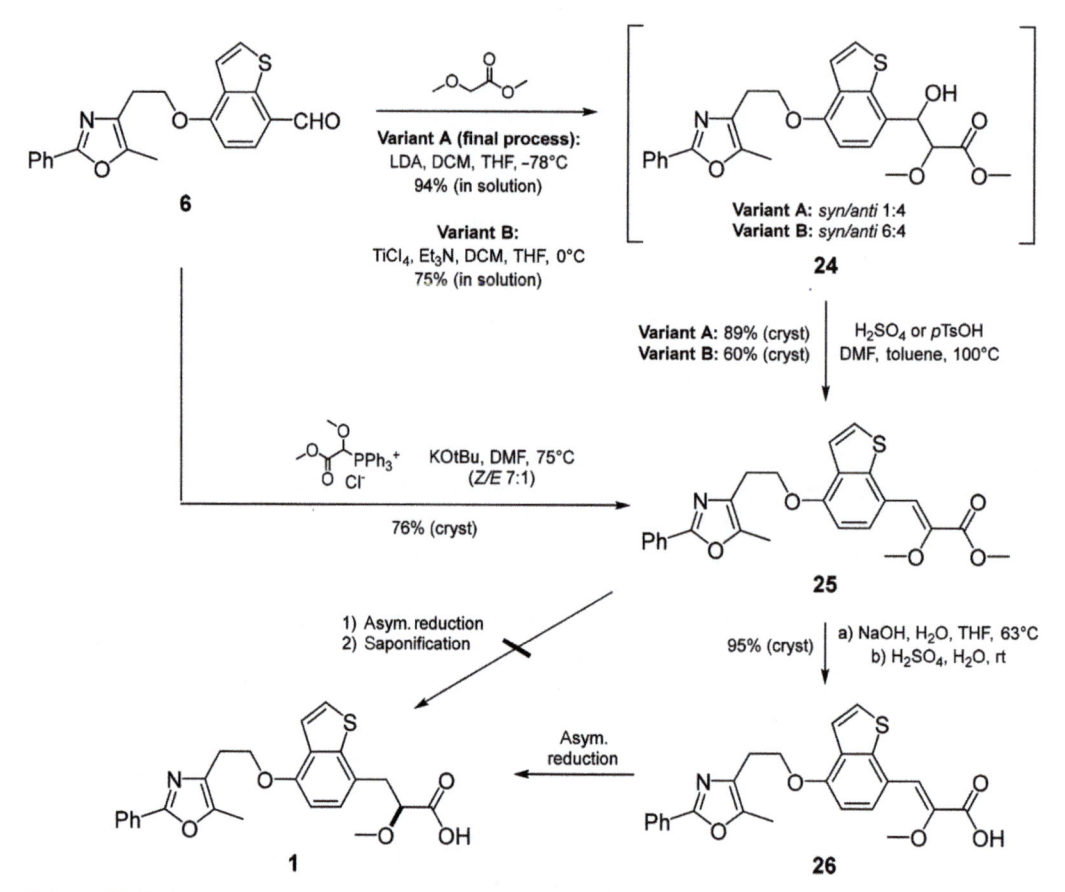

Scheme 28.3 Investigated routes toward propenoic acid derivatives as substrates for asymmetric reductions.

thereof and to install the desired chirality in **1** seemed reasonable. Notably though, the efficiency of the reported systems at the time was greatly substrate dependent, and for most reported catalysts, a high catalyst loading, drastic conditions, or long reaction times were required for satisfactory results. As a saving grace, if the *er* in the asymmetric reduction of 2-methoxypropenoic acid derivatives were not particularly high, crystallization of **1** or a diastereomeric ammonium salt thereof would still have provided a viable option to upgrade the optical purity of the target product.

To acquire a rapid proof-of-concept, propenoic ester **25** was prepared under Wittig conditions from **6** and (1,2-dimethoxy-2-oxoethyl)triphenylphosphonium chloride (DBU, THF, 22 °C, *Z*/*E*-ratio 5:1). After chromatography, pure **25** was converted into **26** via saponification.

Preliminary screening results with either enzymes or chiral metal catalysts under hydrogen pressure for the reduction of **25** or **26** clearly demonstrated that only the hydrogenation approach employing ruthenium or rhodium catalysts and acid **26** as a substrate delivered useful conversions and selectivities (up to 95:5 *er*) [14]. Furthermore, employing pure *Z*-acid led to constantly higher *er*-values than its *E*-isomer or mixtures thereof. Consequently, the selective preparation of pure **26** was needed to achieve the highest *er*. In addition, a high substrate quality was deemed crucial for ensuring low catalyst loadings in the AH step, to save catalyst costs and permit facile removal of residual metal contamination in **1**. Initial work was focused on improving the yield and selectivity in the Wittig reaction. Changing the reaction conditions from DBU/THF to KOtBu/DMF and increasing the reaction temperature to 75 °C brought noticeable improvements. Ester **25** and its *E*-isomer were formed in a 7:1 ratio, whereupon 76% of pure **25** precipitated after cooling the reaction mixture. The mother liquor containing **25** and its *E*-isomer (ratio 1:6) was treated with 2-methyl-5-*t*-butylthiophenol/AIBN at 100 °C to isomerize the *E*- into its *Z*-isomer. An additional 8% of pure **25** (total yield 84%) were thereby obtained. A less waste intensive and technically more feasible route to pure **26** was identified by treating aldehyde **6** with methyl acetoxyacetate in the presence of LDA at -78 °C or alternatively with a $TiCl_4/Et_3N$ mixture at 0 °C yielding the *syn*/*anti* aldol products **24** in ratios of 1:4 (94% yield) and 6:4 (75% yield) respectively, after chromatography. Beneficially, the crude *syn*/*anti*-**24** mixture upon acid catalyzed 1,2-elimination of water furnished exclusively the thermodynamically more stable *Z*-ester **25** independent of the *syn*/*anti* ratio of employed. Finally, the more preferred LDA variant delivered **25** in one step (89% yield) without purification of the aldol intermediate, independent of whether sulfuric acid or *p*TsOH, employed later, was used to promote water elimination. With an efficient process for the selective synthesis of pure **25** and an optimized procedure for the subsequent saponification affording crystalline *Z*-acid **26** (95% yield) now in place, the optimization of the AH process was addressed. For this a wide range of rhodium, iridium and ruthenium catalysts were screened whereby, [Ru(OAc)$_2$((*S*)-TMBTP)] (**27**) [13b] and [Ru(OAc)$_2$((*S*)-1-Naphtyl-MeOBIPHEP)] (**28**) [15] emerged as favorable hits. These two catalysts delivered crude **1** with 95:5 and 85:15 *er* respectively under the screening conditions (Scheme 28.4). Interestingly, none of the iridium catalysts tested provided any conversion. The few rhodium based catalysts that delivered decent *er*-values (up to 90:10 *er*) were finally dropped on account of the associated high metal cost and ultimately because of the superior performance of the significantly less expensive catalyst **27**. In parallel, a process was also developed with catalyst **28** containing (*S*)-1-Naphtyl-MeOBIPHEP, a member of the Roche proprietary atropisomeric MeOBIPEP ligand family [16], to create a third party independent route.

Key for high catalyst performance was the employment of 0.2 eq of NaOMe as base and a 3:2 solvent mixture of MeOH/DCM. Under these conditions, with catalyst **27** and substrate-to-catalyst ratios (S/C) of up to 10,000, a full conversion was achieved in six hours at 30 bar H_2 and 40–60 °C, delivering crude **1** with 97:3 *er*. Following acidification with HCl and two crystallizations, acid

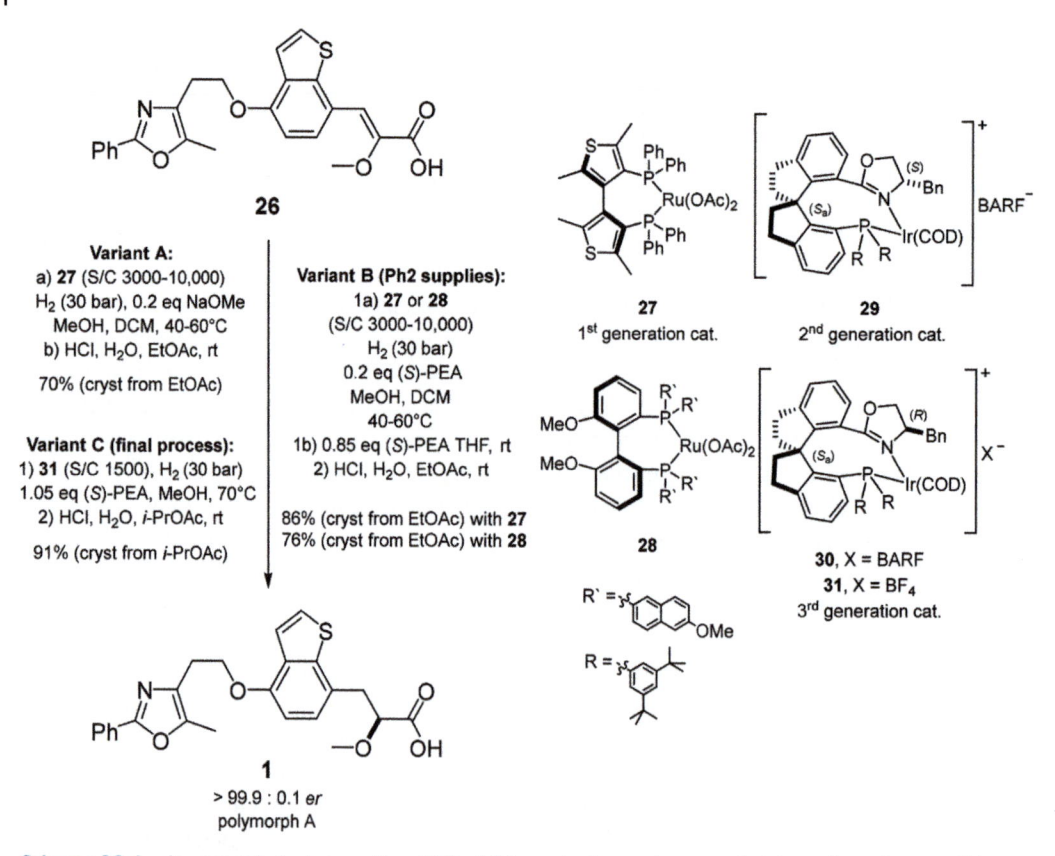

Scheme 28.4 Asymmetric hydrogenation (AH) of **26** employing various catalyst types/reaction conditions and downstream operations.

1 was isolated in 70% yield and >99.9:0.1 *er*. As expected, the *er*-enhancement was rendered more efficient when crude **1** was first crystallized as its (*S*)-phenylethylamine (PEA) salt. After HCl treatment crude **1** with 99.5:0.5 *er* was recovered. A subsequent crystallization created pure **1** with >99.9:0.1 *er* on a lab scale with 20% higher yield (90%) compared to Variant A conditions. To streamline the hydrogenation process even further, (*S*)-PEA was employed in lieu of NaOMe as base without affecting reaction rate or selectivity. After hydrogenation, the remaining 0.8 eq of (*S*)-PEA was added to deliver the proper salt constitution.

In summary, 6.7 kg of **1** were produced in 72% yield (four steps from **6**) for clinical phase 2 supply employing catalyst **27** (S/C 3000) and one (*S*)-PEA salt crystallization for *ee*-upgrading. In a follow-up campaign, 19.6 kg of **1** were obtained in 56% yield, this time benefitting from the more readily accessible Roche catalyst **28** (S/C 3000). Herein, a second (*S*)-PEA salt crystallization step was added to deliver enantiopure **1** (>99.9:0.1 *er*). Finally, both processes produced **1** in comparable purity (>99.9 area-%) and with residual ruthenium levels of <5 ppm.

As **1** was classified as a highly active compound with maximum exposure levels of <0.7 μg/m^3, it needed to be handled in Roche's High Containment Development Plant (HCDP 400-L train). To minimize the occupancy time of the plant and make it available for other high potent development compounds, the time/space yield for **1** had to be shortened significantly. A detailed analysis of the process bottlenecks revealed that an increase in yield and a reduction of the number of crystallization steps required to be conducted in the HCDP plant were key to having improved throughput. The catalyst played here a pivotal role. The higher the *ee*-induction, the simpler the *ee*-enhancement and the higher the yield and throughput would be. In addition, a simple catalyst switch would be the fastest and easiest process modification to be implemented. Consequently, in the ensuing years (2003–2009), the literature was closely monitored to identify catalyst systems suitable for AH of α/β-unsaturated acids that could potentially deliver crude **1** with superior *er* (>95%). For this, we acquired and prepared in-house new catalysts [17] over the course of this work and profiled them periodically on **27**. However, there was little success until 2009, when a breakthrough was realized with Zhou's Ir/SIPHOX catalysts (eg. **29**) [18] received directly from Nankai University for initial trials. These catalyst types delivered exceptional high *er* (up to 99.7:0.3) in AHs of α-alkyl or aryl substituted cinnamic acids and tiglic acids and, to our great satisfaction, also in first runs with **26** [19]. In the set of catalysts provided by Zhou et al., [Ir((S_a,S)-SIPHOX)(cod)]BARF (**29**) performed the best and delivered crude **1** with 99:1 *er* under slightly modified reaction conditions. Interestingly, [Ir((S_a,R)-SIPHOX)(cod)]BARF (**30**) (the reported miss-match combination) outperformed **29**, rendering **1** with outstandingly high *er* up to 99.9:0.1. As no difference in catalyst performance was noticed with BARF or BF_4 as catalyst counter ions, the more accessible [Ir((S_a,S)-SIPHOX)(cod)] BF_4 (**31**) catalyst was selected for further development. Careful process optimizations enabled unsustainable DCM to be eliminated as a co-solvent by employing 1.05 eq of (*S*)-PEA from the outset, which sufficiently enhanced the solubility of **26** and its (*S*)-PEA salt in pure methanol. Finally, with **31** at S/C 1500, 30 bar hydrogen pressure, 70 °C in methanol, and the presence of 1.05 eq (*S*)-PEA, full conversion was achieved within seven hours, yielding crude **1** with 99% selectivity and >99.9:0.1 *er*. Although the *er* of crude **1** would have met the specifications already, the (*S*)-PEA crystallization was retained in the process as was the final crystallization. The first part provided the best control over the impurity profile and for the depletion of residual iridium and the second part obtain the desired polymorph (form A). With the higher quality of crude **1**, the downstream crystallizations proceeded with improved yields and superior efficiency. In the registration campaign, enantiopure **1** was isolated with 62% yield (four steps from **6**), >99.9 area-% purity and a double time-space yield. Consequently, the HCDP occupancy time for the manufacturing of **1** reached very acceptable levels.

Overall, the new AH process enabled the preparation of **1** in a very robust and atom-efficient manner. The use of stoichiometric amounts of a chiral auxiliaries for setting up the chiral center and the three chromatographic purification steps (which created large volumes of solvent and silica gel wastes) could be skipped entirely. The new process benefitted greatly from the identification of **31** as the catalyst that furnished almost perfect *ee*-induction (>99.9:0.1 *er*). Furthermore, it also delivered the hydrogenation substrate **26** in a reaction sequence that, in contrast to the first Wittig approach, generated only very minimal amounts of waste and thereby contributed to a significantly improved PMI of 281 (from **6**).

28.2.2 Idasanutlin

Idasanutlin (RG7388) (**2**) is a potent oral inhibitor of the mouse double minute 2 homolog (MDM2) protein antagonist and is being investigated at Roche as a potential treatment for a variety of solid tumors and hematologic malignancies [20].

As depicted in Scheme 28.5, **2** contains a pyrrolidine carboxamide core with four contiguous stereocenters with an all-*anti* relative configuration (2*R*,3*S*,4*R*,5*S*). The original medicinal chemistry synthesis for **2** employed an azomethine ylide-based [3+2] cycloaddition reaction approach with electron-deficient olefins, a logical synthetic disconnection [21]. Idasanutlin (**2**) was prepared in six steps and 16% overall yield starting from 2-(4-chloro-2-fluorophenyl) acetonitrile (**32**) and

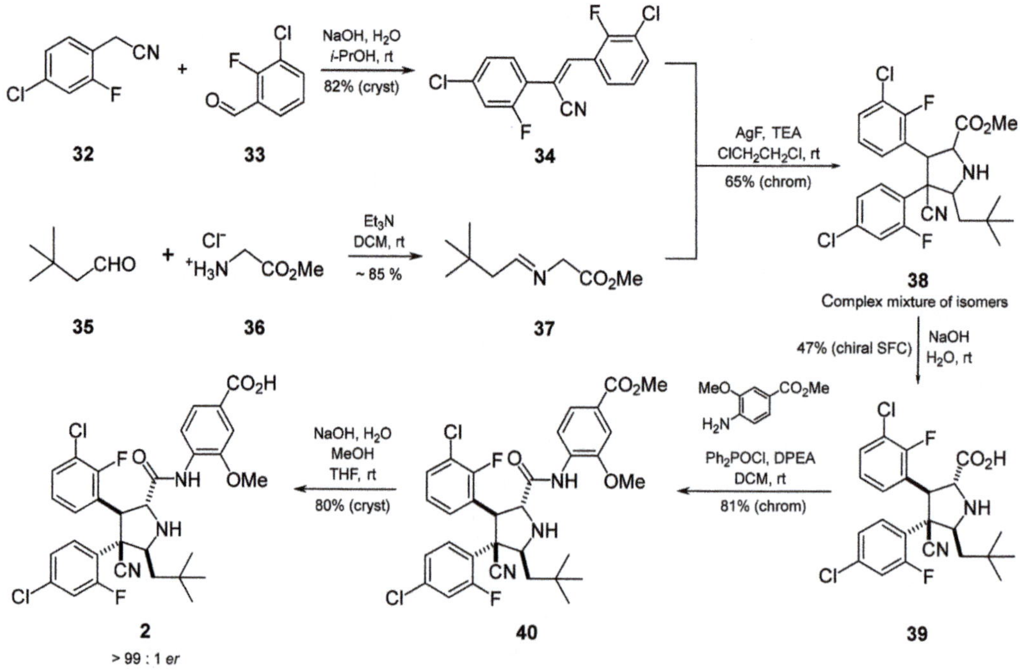

Scheme 28.5 Discovery Chemistry synthesis of **2**. Reproduced with permission from Ref [23a].

employed silver fluoride for the key cycloaddition reaction to construct the racemic pyrrolidine core (Scheme 28.5) [22].

This process was not suitable for larger scale preparation due to several reasons: i) the use of stoichiometric amounts of AgF (1.2 g Ag were used to create 1 g of final product **2**); ii) the use of large quantities of chlorinated solvents (76 g to make 1 g of final product **2**); and iii) the need for two chromatographic purifications and an additional chiral supercritical fluid chromatography (SFC) for enantiomer separation.

Intense process chemistry efforts toward a more efficient and scalable process to support clinical studies resulted in the elaboration of a catalytic asymmetric [3+2]-cycloaddition reaction of stilbene **34** and imine **41** using a catalytic amount AgOAc with (R)-MeOBIPHEP (**42**) as a chiral ligand. In the presence of 1.0 mol% of AgOAc and 1.1 mol% of **42**, the reaction of **34** and **41** in 2-MeTHF (0 °C, 15 hours) afforded a mixture of diastereoisomers **43a-c** (75 area-% by HPLC in total) and the ring-opened Michael addition product **44** (11 area-% by HPLC) (Scheme 28.6) [23].

Upon treatment of this reaction mixture directly with micronized anhydrous LiOH, the pyrrolidine diastereoisomers **43a** and **43b**, as well as the Michael addition product **44**, were converted to the thermodynamically most stable desired ester **43c** (84:16 er), which was isolated by crystallization from n-heptane/2-MeTHF in 97% yield. Hydrolysis of **43c** with LiOH in iPrOH followed by filtration of the insoluble racemate, furnished a filtrate comprising enantioenriched **2** (as its lithium salt). Subsequent acidification of the filtrate with AcOH and isolation by crystallization from iPrOH/water afforded **2** with >99:1 er and 44% overall yield from **34** (36% from **32**).

The Ag-catalyzed process previously outlined was effectively employed to produce more than 100 kg of **2** for early clinical trials. Despite the successful supply campaigns, this process nevertheless was not considered appropriate for a potential commercial manufacturing process for the following reasons: i) efficient and complete epimerization could only be achieved with finely milled

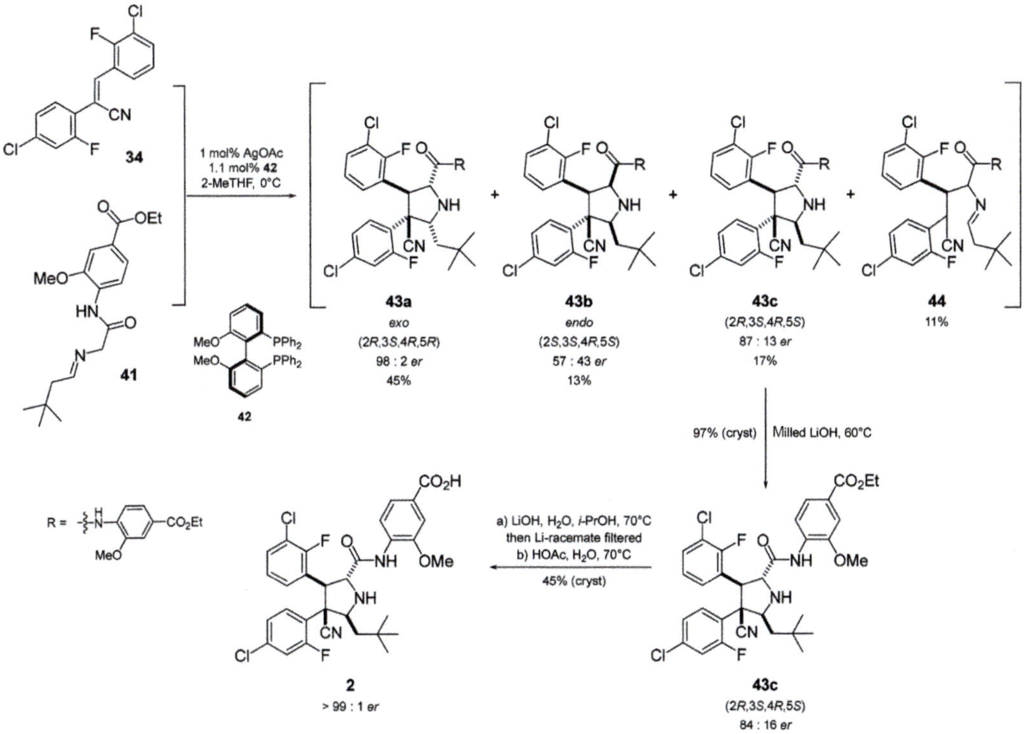

Scheme 28.6 Ag-catalyzed process for the manufacture of **2**. Adapted from Ref [23a] and Ref [23b].

and anhydrous LiOH; ii) silver oxide and other fine particles precipitated during the hydrolysis process and this lead to serious centrifuge clogging during filtration of the racemate; iii) laborious intermediate reactor cleaning was necessary due to the formation of Ag deposits on the reactor walls, and iv) the modest enantioselectivity of the cycloaddition reaction lead to an erosion of the yield due to the necessity for enantioenrichement via racemate removal.

In an effort to address the disadvantages of the Ag process and to improve the selectivity of the [3+2] cycloaddition reaction, the use of Cu catalysts was investigated. After intense screening, CuOAc as the pre-catalyst and readily available (*R*)-BINAP (**46**) as a chiral ligand were found to be the most suitable combination for the asymmetric [3+2] cycloaddition reaction. The commercial manufacturing process ultimately developed for **2** is depicted in Scheme 28.7. The process is a convergent, 4+1 step synthesis starting from readily accessible non-complex starting materials. Imine **41** was prepared by condensation of 4-(2-amino-acetylamino)-3-methoxy-benzoic acid ethyl ester hydrochloride (**45**) with 3,3-dimethylbutyraldehyde (**35**) in the presence of Et₃N in TBME at room temperature. After separation of precipitated Et₃N·HCl by filtration, **41** was obtained by crystallization from *n*-heptane in 94% yield and excellent purity (99.9 area-%). Z-stilbene **34** was prepared by a Knoevenagel condensation of 3-chloro-2-fluoro-benzaldehyde (**33**) and 2-(4-chloro-2-fluoro-phenyl) acetonitrile (**32**) in EtOH/water in the presence of catalytic amounts of NaOMe. The product **34** precipitated from the reaction mixture and was isolated by filtration. The only observed side product detected in the reaction mixture was the corresponding *E*-stilbene (up to 6 area-% by HPLC) which did not precipitate from the reaction mixture and was completely removed into the mother liquor during product filtration. This optimized manufacturing process delivered **34** in 91% yield and 99.9 area-% purity containing <0.05 area-% (by HPLC) of *E*-stilbene.

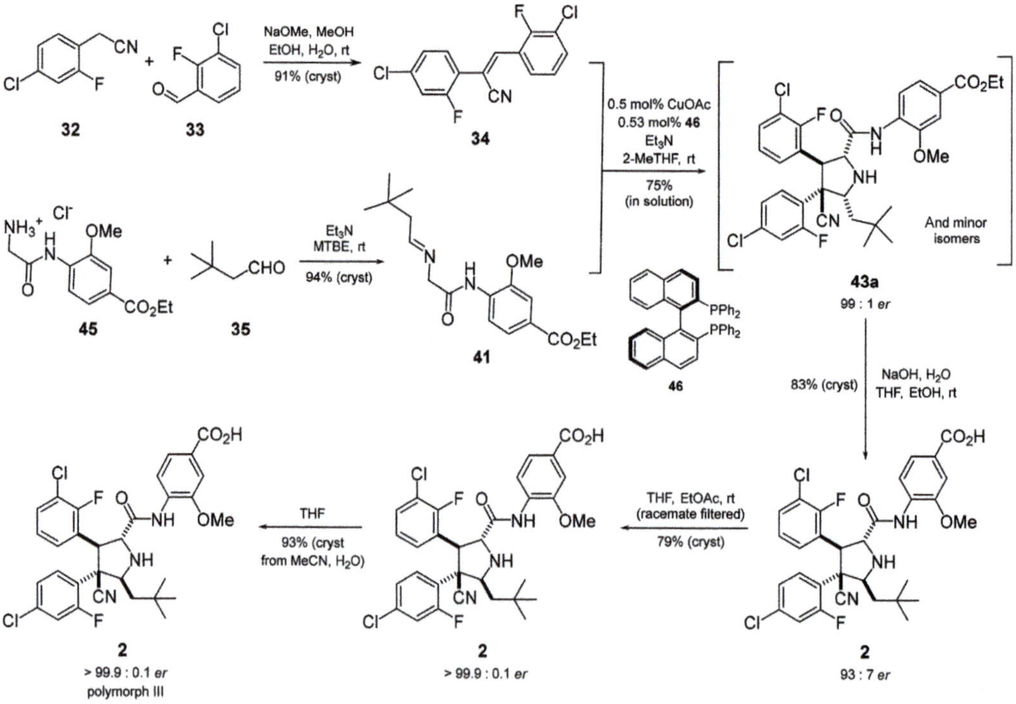

Scheme 28.7 Cu-catalyzed process for the manufacture of **2**. Adapted from Refs [23a] and [23b].

Tight control of *E*-stilbene levels in **34** is crucial to minimize the formation of the (3*R*)-stereoisomers in the next step, as this would be the only stereocenter which cannot be epimerized during downstream chemistry. The subsequent asymmetric [3+2] cycloaddition reaction was conducted using 0.5 mol% CuOAc and 0.53 mol% of ligand **46** in 2-MeTHF at room temperature for five hours. The cycloaddition reaction proceed predominately via an *exo* transition state resulting in an exceptional *er* of 99:1 in the main product **43a** (75 area-%), along with minor isomers **43b** (5 area-%, 40:60 *er*), **43c** (7 area-%, 86:14 *er*) and the ring-opened Michael addition product **44** (3 area-%). The *exo* selective major product **43a** contains all the structural features of **2**, except the configuration at the C-5 carbon in the pyrrolidine core [23].

Detailed experimental as well as theoretical investigations of reaction products from the cycloaddition reaction resulted in the identification of reaction conditions to epimerize the C-5 chiral center with the concomitant hydrolysis of the ester in **43a** to yield the desired acid functionality in **2**. Taking advantage of the fact that **2** is thermodynamically the most stable of all the potential diastereoisomers, the hydrolysis/isomerization method was fine tuned to direct all the minor diastereoisomers toward **2** as well. The process was conducted at room temperature with aqueous NaOH (1.7 eq) in THF/EtOH as the solvent for 15 hours. Telescoping the cycloaddition reaction into the hydrolysis/isomerization step delivered **2** with 93:7 *er* and 83% yield (after neutralization with AcOH and subsequent isolation by crystallization from *i*-PrOH/water).

Enantiopure **2** was then obtained by profiting of the significantly poorer solubility of *rac*-**2** vs its pure enantiomer in THF/EtOAc. Removal of *rac*-**2** by simple filtration of the THF/EtOAc suspension of enantioenriched **2** followed by solvent exchange of the filtrate from THF/EtOAc to MeCN/water delivered **2** in 79% yield and excellent quality (>99.8 w/w-%, >99.9 area-%, >99.9:0.1 *er*).

Finally, a recrystallization from MeCN/water furnished **2** in 55% overall yield (from nitrile **32**) as the desired polymorph (form III) required for the tablet manufacturing process.

The main characteristics of the developed manufacturing process for **2** include: i) the short and high-yielding convergent route using readily accessible starting materials; ii) the highly selective cycloaddition/isomerization/hydrolysis sequence using a cheap chiral Cu(I)/BINAP catalyst system; and iii) the application of the principles of Green Chemistry such as atom economy, minimization of input materials (PMI = 137), minimization of waste and energy consumption, and the avoidance of undesirable solvents. This process has already been successfully applied to produce more than 1.5 metric tons (MT) of **2**.

28.2.3 Danoprevir

Danoprevir (RG7227) (**3**) is an orally administered 15-membered macrocyclic peptidomimetic inhibitor of HCV protease and was originally developed for the treatment of hepatitis C [24]. The compound was discovered by InterMune then licensed to Roche for development and commercialization. In 2013, **3** was licensed to Ascletis Pharma by Roche for co-development in China. Five years further on, **3** was approved for commercialization in China under the trade name Ganovo® for the treatment of hepatitis C.

During clinical development, several production campaigns were conducted to manufacture **3** using different synthetic routes and process variants. The initial, highly telescoped four-step manufacturing route is shown in Scheme 28.8.

The 15-membered macrocyclic compound features five stereogenic centers and one double bond with *syn* configuration giving rise to 64 potential stereoisomers for **3**. The synthetic strategy was to introduce the stereogenic centers by commencing with the enantiopure vinylcyclopropane **47**, Boc-L-hydroxyproline **48** and the nonenoic acid **51**. The tripeptide **52** was formed by three highly selective and efficient peptide coupling steps. Subsequently the 4-fluoroisoindoline building block **53** was introduced to create the tripeptide carbamate **54** which was then cyclized by ring-closing metathesis (RCM) employing Zhan's pentacoordinated ruthenium catalyst **55** [25] to give the macrocycle **56**. The two final stages to conclude the synthesis sequence involved introducing the cyclopropane sulfonamide building block **58** and isolation of **3** as the sodium salt.

Using the synthetic pathway shown in Scheme 28.8, more than 200 kg of **3** were manufactured to supply the first clinical studies. However, several major downsides needed to be overcome for the manufacture of a projected market demand of several metric tons (MT). First, the peptide coupling sequence was conducted using potentially hazardous, environmentally unsustainable, and relatively expensive reagents (EDC, HOBT, and TBTU). Furthermore, substrate **54** used for the RCM was not crystallized but employed as a solution in toluene. Impurities and reaction byproducts originating from the carbamate formation had to be removed by extractive work-up and on production scale. Additionally, **54** was not consistently obtained in the desired quality, which had an impact on the performance of the subsequent key RCM step. In particular, impurity **60**, formed from 4-fluoroisoindoline (**53**) and CDI (Figure 28.2), had to be well controlled as, in the RCM step, it triggered epimerization at the β-position to the ester group of **54**, generating impurities of type **A** and **B** shown in Figure 28.2 and thus affecting the purity and yield of **56** [26].

The most significant limitation for a commercial process, however, was the high dilution under which the RCM had to be conducted. To attain a reasonable yield and purity, the substrate concentration in this reaction was limited to a maximum of 1 w/w-%. At higher concentration, the amount of various dimeric impurities (Figure 28.3), as well as unspecified oligomers, significantly increased, with marked negative impact on yield and product purity. These dimeric impurities had

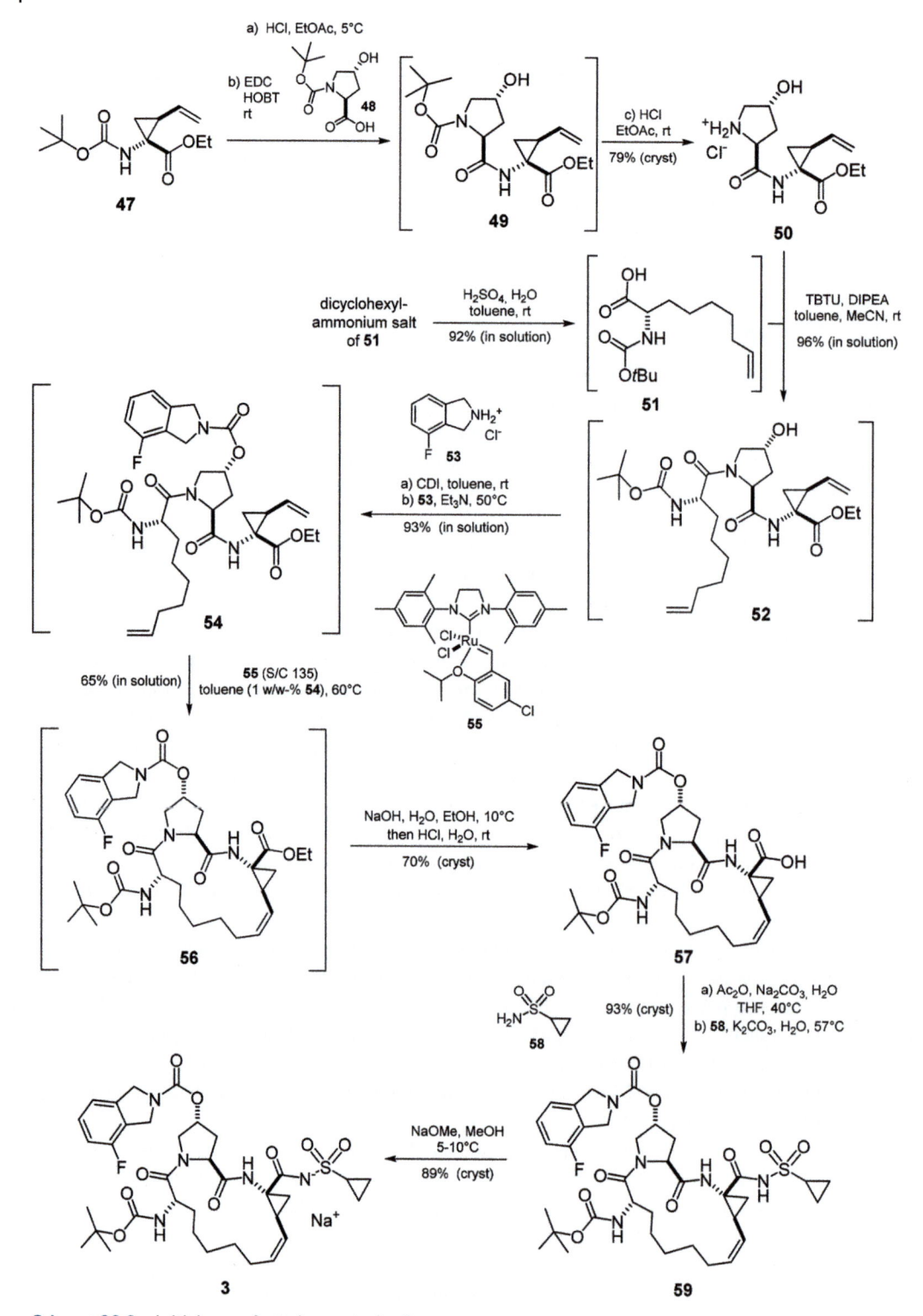

Scheme 28.8 Initial manufacturing route for **3**.

Figure 28.2 Side product **60** (originated from **58** with CDI) furnishing type A and B by-products.

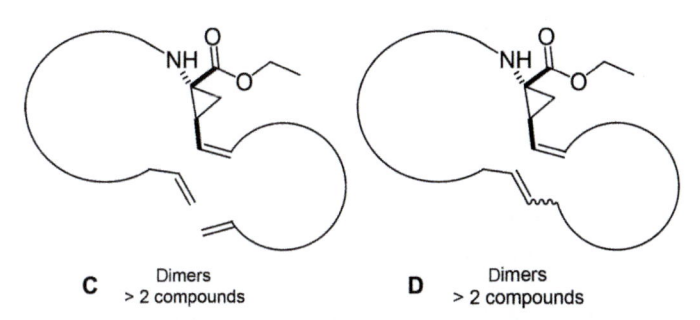

Figure 28.3 Dimers formed in ring-closing metathesis (RCM).

also the potential to be carried over to **3** if not appropriately controlled. Another downside was the rather high catalyst load (S/C 135). A reduction was desirable in order to reduce cost, to facilitate the control of elemental impurities in **3** as well as to reduce transition metal waste.

As a first step to address these flaws to the existing manufacturing process, the peptide coupling sequence was rearranged and further developed (Scheme 28.10). By introducing the 4-fluoroisoindoline (**53**) one synthetic step earlier, the dipeptide carbamate **62** could be obtained as a new isolated intermediate in high quality (98 w/w-%) by crystallization [27]. Furthermore, the undesired coupling reagents EDC/HOBT and TBTU could be eliminated by using pivaloyl chloride to activate **48** and **51** by forming the mixed anhydrides. As such, the same coupling methodology could be applied for the synthesis of both dipeptide **49** and tripeptide carbamate **54** [27]. The latter was still processed further as a solution, but the quality of this substrate no longer contained impurities that induced epimerization in the RCM.

Having an improved quality of the RCM substrate **54** in hand, the next stage was to optimize the RCM reaction. The high catalyst load in this step could be reduced by slowly dosing the dissolved catalyst **55** to the reaction mixture at reaction temperature. In this way, the catalyst loading could be improved from S/C 135 to 750.

The most pressing problem remained the low concentration in the RCM step, which could be overcome by introducing the structurally modified RCM substrate **63** (Scheme 28.10). Scientists at Boehringer Ingelheim Pharmaceuticals had demonstrated that the concentration in similar macrocyclic RCM reactions could be substantially increased by selectively introducing a protecting group at the vinylcyclopropane amide nitrogen of the diene substrate [28]. Based on these findings, a screening of various protecting groups was conducted and the benzoyl group was selected as the protecting group of choice. The benzoyl group could be selectively introduced by deprotonaton of

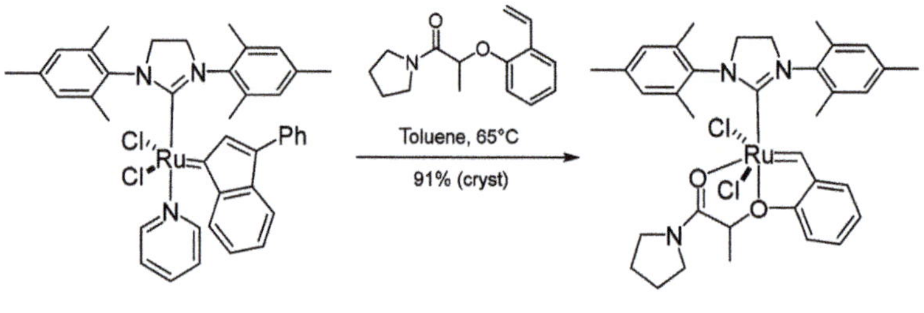

Scheme 28.9 Synthesis of ring-closing metathesis (RCM) catalyst **64**.

tripeptide carbamate **54** with lithium *t*-butoxide in the presence of benzoyl chloride and later easily removed during the saponification of ester **65**. The modified substrate for the RCM allowed this reaction to be conducted at a concentration of 10 vs 1 w/w-%, significantly improving the overall efficiency of the manufacturing process [29]. In addition to these improvements, to improve the IP-position, the Roche proprietary hexacoordinated ruthenium RCM catalyst **64** (Scheme 28.9) was developed and manufactured in kg quantities [30].

These individual improvements were implemented into a telescoped process starting from dipeptide carbamate **62** and furnishing the macrocyclic carboxylic acid **57** in a sequence of six chemical transformations (Scheme 28.10).

The development of the final sequence was straightforward. A telescoped process was developed to avoid the isolation of **59**, thus further improving manufacturing efficiency [31]. In addition, of note was the development of a flow process for the preparation of cyclopropane sulfonamide building block **58** which facilitated the manufacture of 0.5 MT in excellent quality (100 w/w-%) [32].

The economic and ecologic rewards derived from the described improvements were realized already during the development phase. The new process was implemented for the manufacture of 1 MT of **3** destined for late stage clinical development. Compared to the original process (Scheme 28.8) the overall yield was increased from 27–29% to 38–45%. Hazardous and not inherently biodegradable reagents EDC, TBTU, and HOBt were replaced by employing pivaloyl chloride instead and the consumption of a total of 2.4 MT of these reagents was avoided. For the RCM step, the manufacturing productivity was increased by a factor of 10. As a consequence, the consumption of 273 MT of toluene was averted. Overall, process improvements resulted in a decrease of the PMI factor from 645 to 341. The process was later transferred to Ascletis Pharma to enable commercial manufacturing of **3**.

28.2.4 Ipatasertib

Ipatasertib (RG7440) (**4**) is a potent small molecule Akt kinase inhibitor currently being tested in phase III clinical trials for the treatment of metastatic castration-resistant prostate cancer and triple negative metastatic breast cancer [33]. Ipatasertib (**4**) is a complex molecule with three stereocenters which were assembled in a convergent, 7+1 step synthesis from four starting materials: acid **68**, *rac*-**71**, formamidine acetate (**77**), and *N*-Boc-piperazine (**78**) (Scheme 28.11, [34, 35]). The three stereocenters of **4** are constructed by enzyme kinetic resolution using a nitrilase, a diastereoselective ketoreductase (KRED) reduction, and metal catalyzed asymmetric hydrogenation (AH).

The Discovery Chemistry route utilized (*R*)-pulegone (chiral pool approach) to establish the stereocenter in **74** (eleven steps) [36]. To shorten the sequence, a scalable process relying on enzymatic ester hydrolysis was introduced, generating **74** in nine steps [37]. Thereby, the lipase applied at 100 kg

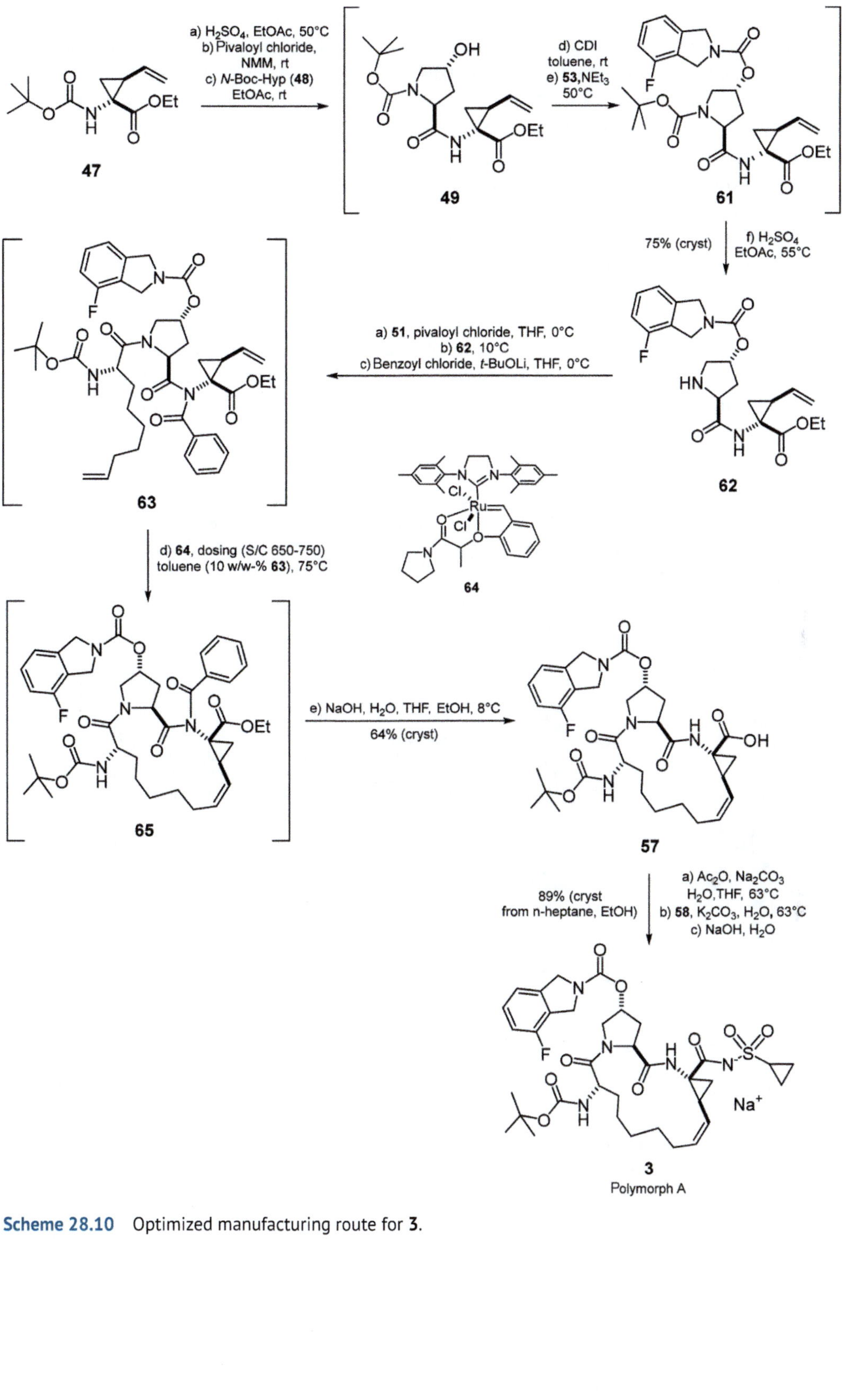

Scheme 28.10 Optimized manufacturing route for **3**.

reaction scale required a high loading of 12 w/w-% enzyme powder (AYS Amano 30G), reflecting the low enzyme activity. The high enzyme loading led to a tedious and time consuming isolation. The final commercial resolution process embarked on a nitrilase at low enzyme loading to deliver (R)-71 from its racemate (Scheme 28.11). The undesired (S)-enantiomer was hydrolyzed to the corresponding carboxylic acid **80** (Scheme 28.12) and depleted in the aqueous extractions upon work-up. This commercial synthesis enabled the production of **74** in only four steps.

On a laboratory scale, two equally efficient processes (Scheme 28.12) were optimized that relied on either lipase CRL III from *Candida rugosa* or a mutant nitrilase from *Acidovorax facilis*. These enzymes were discovered by screening more than 250 hydrolases (lipases, esterases, and proteases), and close to 100 nitrilases, respectively. The lipase process formed a mixture of acids as side products, and required pH control via base addition, yet displayed a very high enzyme selectivity (E > 100) despite only a moderate activity. The nitrilase forms the (S)-acid and ammonia as products and therefore the reaction is pH neutral. The nitrilase displays a good selectivity (E ~50) and a good activity. The remaining (R)-nitrile **71** is the resolution target and as the undesired enantiomer is hydrolyzed. Therewith, the chiral purity of **71** can be controlled precisely by the conversion degree

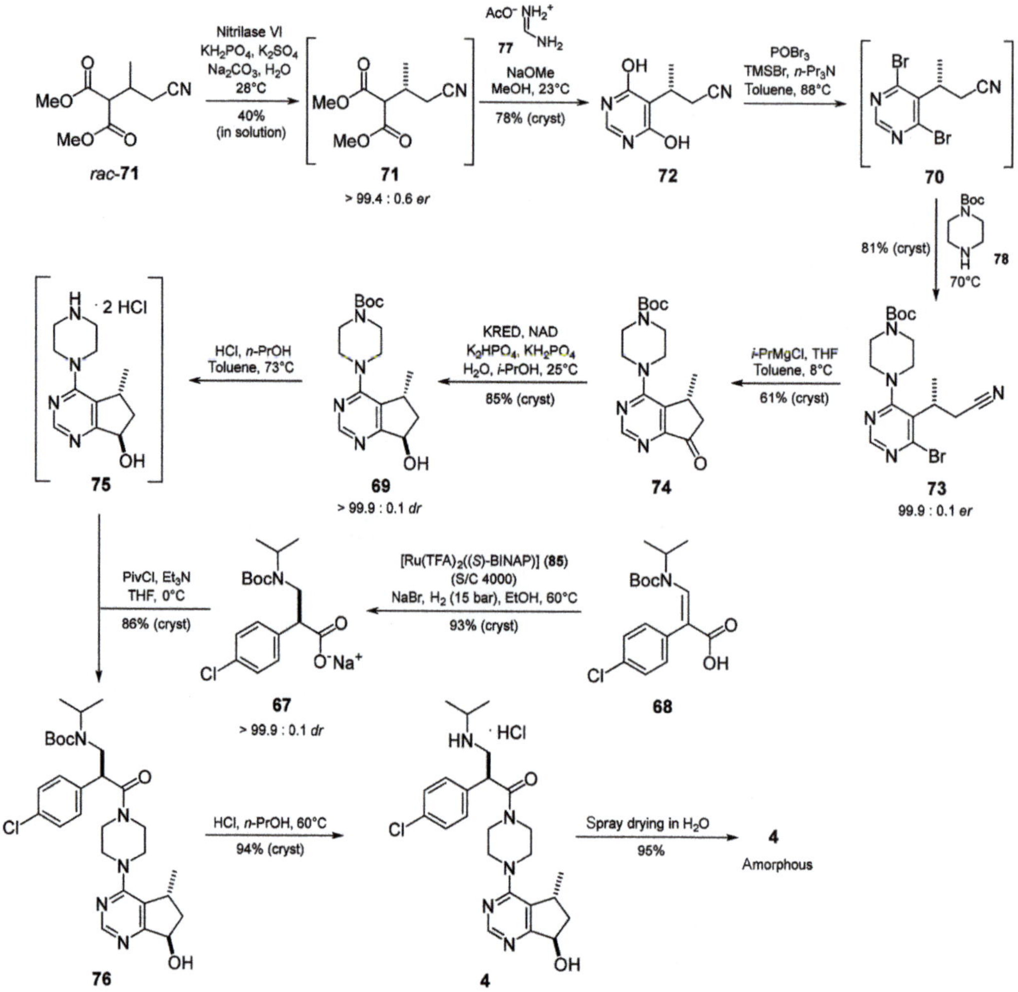

Scheme 28.11 Optimized manufacturing route for **4**. Adapted from Refs [34a] and [35].

to ensure the depletion of the undesired enantiomer to the specified level. Both enzymes successfully resolved racemic **71** at 20 w/w-% substrate loading with yields in the range of 40–47%.

The process selection was finally based on the straightforward heterologous expression of the nitrilase in *E. coli*, whereas the lipase necessitates a yeast, *Pichia*, as the production host. The preferred enzyme production in *E. coli* significantly shortens the fermentation development timelines and facilitates its optimization via enzyme engineering, especially for the applied rational evolution strategy. In addition to the obvious targets (such as increased activity, stability, and selectivity), a stable liquid formulation of the nitrilase was targeted to simplify the process and reduce the sensation risk of handling enzymes as powders. The chosen liquid formulation was based on glycerol which had an inhibitory effect on the activity rate of nitrilase I under the process conditions (see Table 28.1), and added a further challenge to the enzyme engineering.

To select the starting template for enzyme evolution, an existing mutant library of the nitrilase I was screened with regard to initial activity and process stability. Subsequently, the evolvability and positive acceptance of amino acid exchanges was investigated by the saturation mutagenesis of

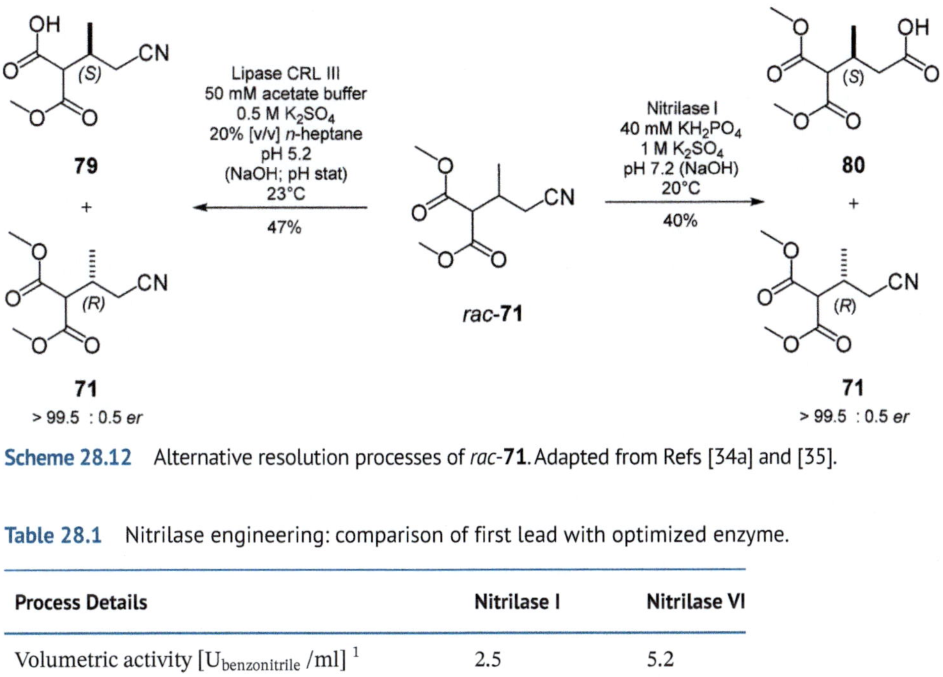

Scheme 28.12 Alternative resolution processes of *rac*-**71**. Adapted from Refs [34a] and [35].

Table 28.1 Nitrilase engineering: comparison of first lead with optimized enzyme.

Process Details	Nitrilase I	Nitrilase VI
Volumetric activity [$U_{benzonitrile}$/ml] [1]	2.5	5.2
FIOP [2] (use-test activity)	1	5
$er_{(71)}$ use-test (43 hours)	99:1	99.4:0.6
$er_{(71)}$ stability test (120 hours)	99.3:0.7	99.6:0.4
E	~60	>100
Inhibition effect of glycerol [3]	−6.20%	−0.60%
Thermostability [$T_{m50\%}$]	55 °C	52 °C
Expression level factor	1	1

1) U: unit = 1 μmol min^{-1}.
2) FIOP: fold improvement over parent.
3) % of reduced activity in the presence of 2.5 v/w-% glycerol final concentration in use-test.

seven amino acid positions which were selected on the basis of substrate docking and modeling. One mutant, nitrilase II, possessed increased process stability and all the activity hits were variants thereof. Thus, nitrilase II was chosen as the template for the following mutagenesis studies. In total, 2,647 variants from five mutant libraries were screened and 54 hit-variants were characterized in detail with respect to their stability under the process conditions (lower enzyme loadings at extended reaction time), thermostability, initial activity with benzonitrile as substrate, process related activity, enzyme selectivity, inhibitionary effect of glycerol, and protein expression.

The final tailor-made nitrilase VI exhibits a higher activity, an increased selectivity, minor glycerol inhibition at 2.5 w/w-% glycerol concentration in the ultimate process, a slightly inferior thermostability, a high process stability, and the processability as a stable liquid formulation (Table 28.1). The enzyme engineering was performed in parallel to process development, which increased significantly the complexity of both tasks.

Additional key achievements on this step were the depletion and control of residual enzyme. For this purpose, the nitrilase resolution, conducted in an aqueous buffer (pH 8.7 to 9.4), and affording crude **71** with 99.4:0.6 er, was acidified with 30% sulfuric acid (pH 1.6 to 2.2) to denature the enzyme. Subsequent TBME addition enabled the precipitated enzyme to be filtered efficiently as protein flakes. Product extraction with TBME and azeotropic water removal (<1000 ppm), induced the precipitation of residual proteins which were removed by a second filtration. The protein content (<1000 ppm) was measured in the downstream product **69** using the Bradford assay. The nitrilase process was validated successfully and enabled the manufacture of intermediate **71** on a commercial scale, producing ~11 metric tons (MT) of material in 40% yield, 97.5 w/w-% assay, 99.5 area-% purity, and 99.4:0.6 er.

In the next step, diester **71** was cyclized with amidine **77** to the corresponding pyrimidine **72**. The reaction was performed in methanol with NaOMe as a base. Quenching with aqueous HCl precipitated the product, which was isolated by filtration. To prepare for further functionalization of the pyrimidine core, **72** was di-brominated to **70**, which was reacted directly with piperazine **78** to yield intermediate **73**. Dibromination required quite harsh reaction conditions, using a mixture of $POBr_3$, TMSBr, and n-Pr_3N. With a stoichiometry of 1.05 to 1.15 eq of **78**, a fast conversion to **73** occurred without noticeable bis-addition. In the subsequent step, the cyclopentyl ring was created by Br-Mg exchange, followed by intramolecular cyclization. The imine intermediate was hydrolyzed during an acidic aqueous work-up to produce the corresponding ketone **74**.

The second stereocenter was introduced by diastereoselective reduction of the keto function in intermediate **74** – again promoted by biocatalysis. For initial clinical DS supplies on up to a 100 kg scale, the commercial KRED (KRED-NADPH-101) was applied in the reduction step. For the in situ cofactor (NADPH) regeneration, glucose dehydrogenase (GDH) was applied to oxidize D-glucose as hydride source to gluconic acid (coupled enzyme approach, Scheme 28.13, left arrow).

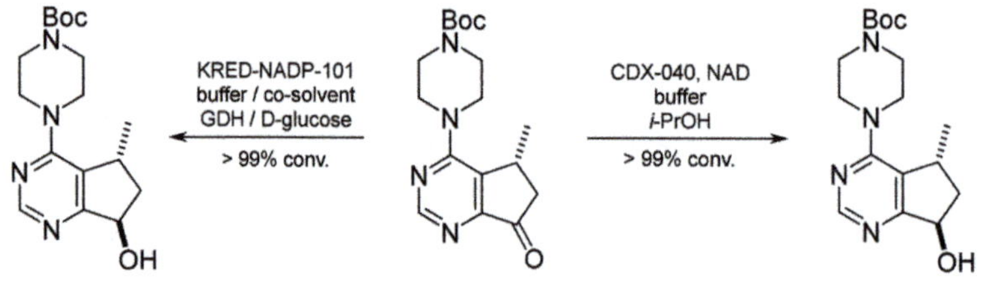

Scheme 28.13 Alternative ketoreductase (KRED) processes. Adapted from Refs [34a] and [35].

The commercially available KRED was replaced by an engineered one (CDX-040) that accepted *i*-PrOH as hydride source (coupled substrate approach for cofactor regeneration; Scheme 28.13, right arrow) for the manufacturing process.

The initial process required a high enzyme and cofactor loading, and delivered colored **69** (>98 area-%), which required a charcoal treatment to obtain colorless **4**. As a result, our internal KRED library (>250) was screened to identify a more efficient enzyme. The best hit with regard to selectivity and activity was clearly still the implemented KRED-NADP-101, a KRED not tolerating *i*-PrOH as hydride source. In a subsequent process optimization, a slurry to slurry biphasic process was introduced, enabling at least a four-fold increased space-time-yield and a three-fold reduced KRED loading, all at an increased substrate concentration of 10 w/w-%. Surprisingly, the GDH and NADP loadings could not be reduced. In addition, the color formation could be completely precluded by changing the cosolvent from DMSO or PEG to TMBE. This significantly improved the homogeneity of the stirred reaction mixture and reduced the solubility of **74**, which was key to avoid the formation of color. The process optimization is summarized in Table 28.2.

In summary, the KRED process was successfully optimized with regard to efficiency, robustness, and coloration of intermediate **69**, but the cofactor recycling aspect remained unsatisfactory.

To enable a coupled enzyme approach, the screening efforts were extended. A commercial panel of engineered KREDs (>350) included one highly active enzyme with good diastereoselectivity (~95:5% dr) that accepted high concentrations of *i*-PrOH. After initial optimization experiments,

Table 28.2 Summary table of ketoreductase (KRED) processes including enzyme engineering.

Ketoreductase	KRED-NADP-101		1st round P1B02	2nd round P1B06	3rd round P1F01	4th round P1F01	5th (Final) round CDX-040
Diastereo-selectivity [*dr*]	>99.9:0.1	>99.9:0.1	98:2	99.6:0.4	>99.9:0.1	>99.9:0.1	**>99.9:0.1**
Substrate loading [w/w-%]	≤5	10	20	20	20	10	**10**
Time	≥2 d	1 d	1 d	1 d	1 d	2.5 d	**<20 h (1 d)**
KRED loading [S/E][1]	33	100	100	100	100	33	**50 (75)**
cofactor	NADP	NADP	NADP	NAD	NAD	NAD	**NAD**
Cofactor recycling approach	Coupled-enzyme (GDH)		**Coupled-substrate (*i*-PrOH)**				
GDH[2] loading[S/E][1]	28	33	n.a.	n.a.	n.a.	n.a.	**n.a.**
cosolvent	DMSO	TBME	*i*-PrOH	*i*-PrOH	*i*-PrOH	*i*-PrOH	***i*-PrOH**
conc.[v/w-%]	20	20	20	40	40	8	**8**
T [°C]	30	25	40	40	40	20	**25**
Appearance	colored	white	colored	colored	colored	white	**white**

Cell border double lines indicate undesirable values for a commercial process.

1) S/E: substrate to enzyme ratio in w/w
2) GDH: glucose dehydrogenase (2nd process enzyme).

the pH-dependent diastereoselectivity could be improved to a ratio of 98:2 along with a substrate-to-NADP ratio of 1000 with a KRED P1B02 loading of S/E 100 at 20 w/w-% substrate concentration. This hit was regarded as a suitable candidate for enzyme engineering.

Incorporating a strategy for statistical analysis of protein sequence activity relationship (ProSAR) [38] enabled, after only two engineering rounds and screening a few hundred variants, the evolution of an absolute diastereoselective variant P1F01. This variant faciliated a process change from the slurry to slurry to a slurry to solution by increasing the i-PrOH concentration to 40 v/w-% and the temperature to 40 °C. The P1F01 variant tolerated even 60 °C and 50 v/w-% i-PrOH. In addition, the cheaper cofactor NAD was tolerated by P1F01. The increased solubility in 40 v/w-% i-PrOH of ketone **74** at elevated temperature, however, led to a colored product **69**, although the purity (>98 area-% by HPLC) otherwise would have been acceptable. Despite the excellent enzyme process performance, the color formation required the reduction of ketone **69** to be performed as a slurry to slurry process. This was realized at 20 °C with 8 v/w-% i-PrOH compromising the process performance and requiring a reduced substrate concentration (10%), an increased enzyme loading (S/E 33), and an extended reaction time (60 hours) to achieve the high specified conversion (<0.5 w/w-% **74**).

To improve the process performance, an additional enzyme engineering and combinatorial round created the final variant CDX-040, permitting a reaction time <20 hours at an enzyme loading of S/E 50 (<1 w/w-% **74**). The process optimization and development is summarized in Table 28.2.

The final process was catalyzed by an engineered KRED in water with 3.5 eq of i-PrOH as a hydride source and NAD as a cofactor. The reaction operated well in a range of 4 to 12 L of water per kg of **74**. The crude product **69** was filtered off, washed with water, dissolved in toluene, and dried by azeotropic distillation (which assisted the removal of the insoluble enzyme by filtration). Finally, intermediate **69** was crystallized by the addition of n-heptane. Residual protein (<1000 ppm) was monitored by a Bradford assay of the isolated intermediate **69**. This process was successfully applied on commercial scale with the production of ~5 MT of **69** with 85% yield, 99.4 w/w-% assay, 99.8 area-% purity, and >99.9:0.1 dr.

The third stereocenter in the side branch of this convergent synthesis was introduced by AH of starting material **68** employing a highly active and selective ruthenium catalyst. The initial four step synthesis of enantiopure **67** required the use of Evans's auxiliary, which on large scale production would have caused unacceptable high waste formations [37]. Introduction of the new AH route shortened the initial synthesis of **67** to three steps. In the first generation hydrogenation process, [RuCl$_2$(S)-BINAP] (**81**) (S/C 1500) was employed as a catalyst in combination with LiBF$_4$ to activate **81** by chloride abstraction. A H$_2$ pressure of 30 bar at 60 °C was applied to reach full conversion overnight (Table 28.3, entry 1) [39]. Even though the reaction suffered from inconsistent er ranging from 99.5:0.5 to 93.0:7.0, the subsequent crystallization depleted the undesired (R)-enantiomer efficiently, leading to amino acid **67** with a reproducible chiral purity of 99.5:0.5 er. Removal of residual Ru required the employment of expensive SiliaMetS as a metal scavenger. Ultimately, its use could be obviated by isolating **67** as the sodium salt, which was achieved via the addition of NaOEt to the reaction mixture directly after hydrogenation. In this way, **67** was isolated as a white powder in >90% yield and with very low residual Ru levels (<5 ppm). For further process improvements, more than 50 chiral diphosphines were tested. However, representatives from the Josiphos, Skewphos, DIOP, and Duphos ligand families [40] provided incomplete conversions (even at high catalyst loadings) and low to moderate er (<90:10). In contrast, catalysts prepared from [Ru(COD)(TFA)$_2$]$_2$ (**82**) and atropisomeric diphosphines such as (S)-BINAP (**83**) or

Table 28.3 Selected catalysts and additives in asymmetric hydrogenation (AH) of **68** [34]. Adapted from Ref [34a].

Entry	Catalyst		Additive [1]	Conv. [%]	Yield [%]	er [2]
1 [3]	[RuCl$_2$((S)-BINAP)]	**81**	LiBF$_4$	99	87	93.0:7.0
2 [4]	[Ru(TFA)$_2$((S)-MeOBIPHEP)]		NaBr	>99	95	98.0:2.0
3 [5]	[Ru(TFA)$_2$((S)-BINAP)]	**85**	NaBr	>99.5	92	98.8:1.2
4 [5]	[Ru(TFA)$_2$((S)-BINAP)]	**85**	-	29	n.d.	90:10

1) Additive/Ru 20:1.
2) Determined by chiral high-performance liquid chromatography (HPLC) analysis of the crude reaction mixture.
3) S/C 1500, EtOH, 30 bar H$_2$, 60 °C.
4) S/C 5000, EtOH, 18 bar H$_2$, 60 °C.
5) S/C 10,000, EtOH, 18 bar H$_2$, 60 °C.

(S)-MeOBIPHEP (**84**) proved to be both highly active and enantioselective (≥98:2 er) (Table 28.3, entries 2 & 3) [34].

Finally, [Ru(TFA)$_2$((S)-BINAP)] (**85**) was selected for further development based on its superior performance, catalyst cost considerations, and the fact that **85** could be applied directly in production campaigns bypassing a feasible but cumbersome in situ preparation. The synthesis of **85** was originally established at Roche in the late 80s [15a]. One batch of **85** from these early days was employed from hit finding to a first process development phase thanks to its remarkable stability when stored under argon at room temperature. The original catalyst synthesis involved undesired solvents such as Et$_2$O and n-pentane and not having a crystallization procedure for **85** in place that would allow for its straightforward isolation and scale-up. Development work provided a process that allowed to run its synthesis in THF in lieu of Et$_2$O, followed by EtOH addition to promote crystallization that delivered catalyst **85** in 85% yield and excellent quality (Scheme 28.14). Ultimately, this streamlined process enabled an expeditious production of **85** at a contract manufacturing organization in multi-kilogram quantities.

Notably, when handling **85** in air, its color changes within 18 hours from yellow-orange to grayish-green, yet no negative impact on its performance was noticed. Nevertheless, after one week of storage under air, the conversion in the hydrogenation reactions decreased from >99.5% to 87% and the yield decreased from 92% to 60%, but the quality of isolated **67** remained

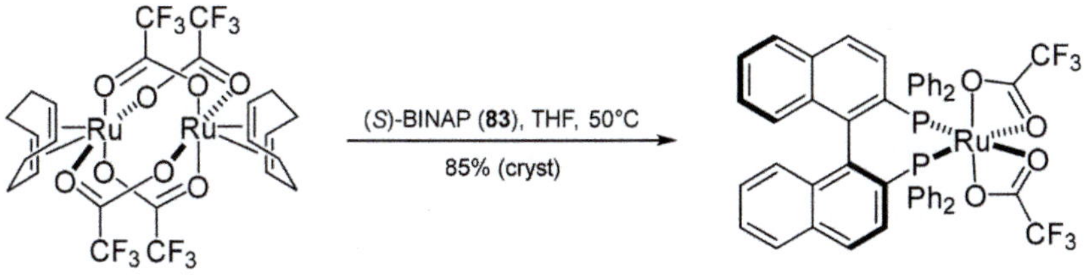

Scheme 28.14 Optimized synthesis of catalyst **85**.

unaffected and high (99.9 area-%, 99.7:0.3 er). Interestingly, **85**, once boosted with NaBr, performed well even at very low catalyst loadings (S/C 10,000) [34, 35]. Other halides and halide sources, such as HBr or HCl/EtOH, increased the catalyst's performance even further, but the NaBr process was the most practicable and avoided the potential formation of mutagenic EtCl or EtBr impurities. Finally, the hydrogenation most reliably ran at a Ru/NaBr ratio of 1:20 and S/C 10,000 (>99% conv. after 12 hours, 60 °C, 18 bar H$_2$), delivering crude **67** with constant high er (Table 28.3, entry 3). Under otherwise identical conditions but in the absence of NaBr, the conversion stalled at 29% conversion (Table 28.3, entry 4). To render the process more robust on technical scales, a somewhat higher catalyst loading of S/C 4,000 at 15 bar H$_2$ was employed. Residual amounts of ruthenium were efficiently depleted in the subsequent crystallization unit operation (<3 ppm). The remaining starting material and the undesired (R)-enantiomer of **67** were also completely purged thereby, affording **67** in high yield (93%) and excellent quality (>99.9 area-%).

The product isolation process was optimized next. In particular, the low bulk density (~0.1 kg/L) of **67** caused a low throughput in the drying step and demanded large storage capacities. An adapted crystallization procedure was thus developed whereby seed crystals were generated in situ by the addition of 0.2 eq of NaOEt (sufficient to induce nucleation), followed by aging of the formed crystals for two hours, and then the addition of 1 eq of NaOEt to complete the crystallization. With this procedure, a total of 5.5 MT of **67** with a 3.5-fold increased bulk density were manufactured in 93% yield, >99.9 w/w-% assay, >99.9 area-% purity, and >99.9:0.1 er [35]. The final three step synthesis of **67** (Scheme 28.15) (~74% overall yield, PMI 100) compares favorably with the first generation hydrogenation variant (PMI 150), and even more noticeably with the initial, chiral. auxiliary based process (PMI 575), leading overall to significantly reduced waste formation and increased throughput [39].

The penultimate step required the coupling of the acid salt **67** with piperazine **69** (Scheme 28.11). Activation of **67** was accomplished with pivaloyl chloride to form a mixed anhydride. In a separate reactor, the Boc-protecting group of intermediate **69** was removed with HCl in n-PrOH. The excess of HCl was neutralized with Et$_3$N and the reaction mixture containing the mixed anhydride of **67** was added. The work-up after reaction completion was carried out by exchanging the solvent to toluene by distillation followed by extractions with aqueous HCl, aqueous NaOH and water to

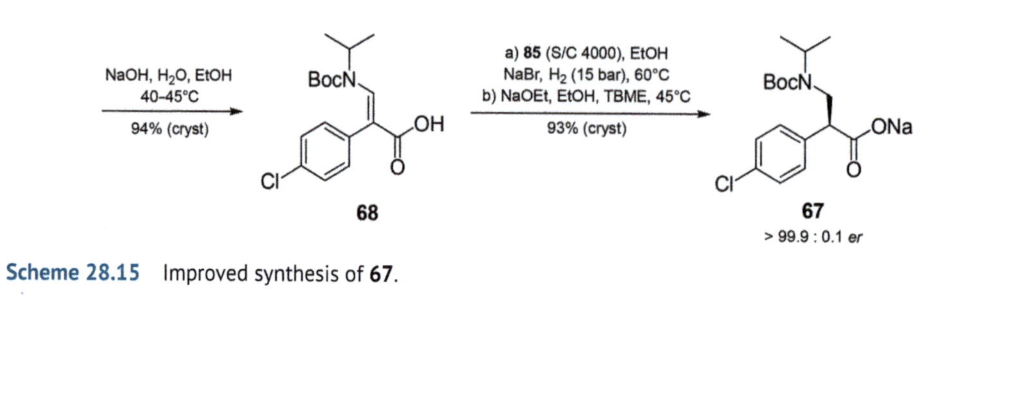

Scheme 28.15 Improved synthesis of **67**.

remove acidic and basic impurities such as unreacted **67** and **75**. Product **76** was crystallized by the addition of *n*-heptane to the toluene solution.

In the final chemical step, the Boc-protecting group of **76** was removed with HCl in *n*-PrOH. The pH was adjusted to 5.5 with aqueous NaOH to acquire the desired mono-HCl salt. After concentration and removal of water by distillation, precipitated NaCl was removed by filtration and the product was isolated by distillative exchange of *n*-PrOH with EtOAc, from which the product precipitates. Lastly, the solid state properties (e.g. particle size distribution, amorphous form) were adjusted by spray drying of **4** as an aqueous solution.

In the late stage development phase, prior to process validation, the overall process was significantly improved regarding robustness, efficiency, and sustainability. In the last three manufacturing campaigns up to validation, the PMI had been reduced by a factor of five (from 1,393 to 269), mainly driven by significant solvent reductions and a greater than three-fold improved overall yield (from 3 to 10%).

28.3 Conclusions

The reported case studies here demonstrate strikingly what impact catalysis can provide in constructing complex DS molecules by enabling new routes or shortcutting existing ones with markedly improved efficiencies. Catalysis has played thereby a key role in connecting building blocks, introducing functional groups and chirality in **1–4**, and their associated intermediates with excellent selectivities, excellent throughputs, and high catalyst turnover frequencies. To achieve this, in-depth catalyst research and process development work was fundamental to determine tailor-made catalyst solutions for each one of the four case study molecules.

Conspicuously, the presented show cases also provided evidence that a most productive catalytic step per se will fail its application in DS manufacturing if or when the substrate and catalyst accessibility, the downstream chemistry, and/or the catalyst depletion is too cumbersome. This is particularly the case when the entire process is inefficiently long or requires tedious (chromatographic) purifications and/or unstainable reagents, solvents, and reaction conditions. Clearly, the key to be able to offer spot-on catalytic solutions is to have ready access to well-diversified enzyme / metal catalyst libraries, state-of-the-art catalyst prediction tools, enzyme engineering, metal catalyst synthesis / enzyme fermentation, optimization and scale-up expertise, and adequate facilities on hand.

Obviously, the delivery of a high performing and most sustainable DS process requires dedicated process research and development work early on and the steady drive to search and implement process optimizations along all development phases. These preliminary efforts result in improved PMI values, reduced energy consumptions, and the elimination of unsustainable solvents and reagents together with DS cost and lead time reductions and a sustainable use of natural resources. Accordingly, patients and likewise the entire society benefits from the outcome. Two cases, idasanutlin **2** [23b] and ipatasertib **4** [35], were recognized by the Swiss Chemical Society, which awarded the associated process development teams with the Sandmeyer Award in 2017 and 2020 for the development of economical, scalable, and sustainable processes. In a larger context made transparent via the Dow Jones Sustainability Indices (DJSI), F. Hoffmann-La Roche Ltd. was ranked for the thirteenth year running as one of the most sustainable pharmaceutical companies in terms of economic, social, and environmental performance worldwide. To maintain this standard, Roche stays fully committed to further fortify the implementation of catalysis, particularly biocatalysis in water and non-noble metal catalysis, especially with the use of recyclable catalysts and other resource saving methodologies and technologies. As further promotion of this, Roche has also created a fund to support research projects with academia and industry targeted to invent products such as new catalysts and catalyst systems with the potential for break-through performance, or to promote so far unprecedented transformations all aimed to enhance the sustainability of our current and future DS processes even further.

References

1 (a) Sheldon, R.A. (2018). *ACS Sustain. Chem. Eng.* 6: 32–34. (b) Jimenez-Gonzalez, C. et al. (2011). *Org. Process Res. Dev.* 15: 912-917.
2 Anastas, P.T. and Warner, J.C. (1998). *Green Chemistry: Theory and Practice*. New York: Oxford University Press.
3 The Roche solvent selection guide (based on https://www.acs.org/greenchemistry/research-innovation/tools-for-green-chemistry/solvent-selection-tool.html from ACS), classifies solvents in four different groups as follows based on a composite score including categories such as safety, health, and environment: recommended, usable, avoid, and high uncertain.
4 Wuitschik, G. et al. (2022). *Curr. Res. Green Sustain. Chem.* 5: 100293.
5 Mohr, P. et al. (2009). *Bioorg. Med. Chem. Lett.* 19: 2468–2473.
6 Fürnsinn, C. et al. (1999). *Br. J. Pharmacol.* 128: 1141–1148.
7 For clarity, herein a chemical transformation or a cascade thereof (a telescoped process) counts as a step from one isolated and characterized intermediate to the next, eventually reaching the final product. Intermediates that are not isolated, either as a crude or in a purified form, but are kept in solution for further processing, are depicted in brackets. Purification methods are reported for intermediates or final products that have been isolated by crystallization (cryst), silica gel chromatography (chrom), distillation (dist), or supercritical fluid chromatography (SFC). Otherwise, the step products were further converted from their crude form.
8 (a) Goehring, W., Hoffmann, U., Scalone, M. et al. (2006). PCT International Publication. Patent application WO2005000844. (b) Junghans, B., Scalone, M. and Zeibig, T.A. (2002). US Patent US6482958.
9 Hidai, M. et al. (1989). *Tetrahedron Lett.* 30: 95–98.
10 As alternatives, the asymmetric reduction of 2-oxo-propanoic acid derivatives to furnish a chiral 2-hydroxy-propanoic acid derivative as well as asymmetric epoxidation approaches were not considered further, as selective etherifications were reported by our Discovery Chemistry colleagues to fail for heavily decorated, benzothiophene type molecules such as **1**.
11 Ratovelomanana-Vidal, V. et al. (2010). *Org. Lett.* 12: 3788–3791.
12 Takaya, H., Noyori, R. et al. (1987). *J. Org. Chem.* 52: 3176–3178.
13 (a) Ohata, T. et al. (1996). *J. Org. Chem.* 61: 5510–5516. (b) Sannicolo, F. et al. (2000). *J. Org. Chem.* 65: 2043–2047. (c) Chan, A.S.C. et al. (2000). *J. Am. Chem. Soc.* 122: 11513–115144. (d) Yamagishi, T. (1996). *Tetrahedron: Asymmetry* 7: 3339–3342.
14 Püntener, K., and Scalone, M. (2005). PCT International Publication WO2005030764.
15 (a) Heiser, B. et al. (1991). *Tetrahedron Asymmetry* 2: 51–62. (b) Chan, A.S.C. et al. (1994). *Inorg. Chim. Acta* 223: 165–167.
16 (a) Schmid, R. et al. (1988). *Helv. Chim. Acta* 71: 897–929. (b) Schmid, R. et al. (1991). *Helv. Chim. Acta* 74: 370–389. (c) Schmid, R. et al. (1996). *Pure Appl. Chem.* 68: 131–138. (d) Schmid, R. and

Scalone M. (2015). (*R*)- and (*S*)- 2,2′-Bis(diphenylphosphino)-6,6′-dimethoxy-1,1′-biphenyl. In: *e-EROS Encyclopedia of Reagents for Organic Synthesis* (eds. L. Paquette, P. Fuchs, D. Crich, and P. Wipf). Hoboken: John Wiley and Sons.

17 (a) Maligres, P.E., Krska, S.W. et al. (2004). *Org. Lett.* 6: 3147–3150. (b) Zhou, Q.-L. et al. (2005). *Angew. Chem. Int. Ed.* 44: 1118–1121. (c) Zhou, Q.-L. et al. (2006). *Adv. Synth. Catal.* 348: 1271–1276. (d) Chan, A.S.C. et al. (2007). *Adv. Synth. Catal.* 349: 517–520. (e) Houpis, I.N. et al. (2005). *Org. Lett.* 7: 1947–1950. (f) Kim, T.-J. et al. (2005). *Organometallics* 24: 4824–4831. (g) Minnaard, A.J. et al. (2005). *Angew. Chem. Int. Ed.* 44: 4209–4212. (h) Chen, W. et al. (2007). *Angew. Chem. Int. Ed.* 46: 4141–4144. (i) Scrivanti A. et al. (2006). *Tetrahedron Lett.* 47: 9261–9265.

18 Zhou, Q.-L. et al. (2008). *J. Am. Chem. Soc.* 130: 8584–8585.

19 Püntener K. and Scalone M. (2010). PCT International Publication WO2010108861.

20 (a) Vu, B. et al. (2013). *ACS Med. Chem. Lett.* 4: 466–469. (b) Ray-Coquard, I. et al. (2012). *Lancet Oncol.* 13: 1133–1140. (c) Ray-Coquard, I. et al. (2015). *Curr. Med. Chem.* 22: 618–626.

21 (a) Wang, C.-J. et al. (2020). *Acc. Chem. Res.* 53: 1084–1100. (b) Adrio, J. and Carretero, J.C. (2019). *Chem. Commun.* 55: 11979–11991. (c) Zhang, J. et al. (2016). *Angew. Chem. Int. Ed.* 55: 6324. (d) Hashimoto, T. and Maruoka, V. (2015). *Chem. Rev.* 115: 5366–5412. (e) Adrio, J. and Carretero, J. C. (2014). *Chem. Commun.* 50: 12434–12446. (f) Garner, P. et al. (2007). *Tetrahedron Lett.* 48: 3867–3870. (g) Pandey, G. et al. (2006). *Chem. Rev.* 106: 4484–4517. (h) Savic V. et al. (2005). *Tetrahedron: Asymmetry* 16(12): 2047–2061.

22 (a) Shu, L. et al. (2016). *Org. Process Res. Dev.* 20: 2050–2056. (b) Fishlock, D., Gu C., Shu, L. et al. (2014). PCT International Publication WO2014128094.

23 (a) Pankaj, R. et al. (2016). *Org. Process Res. Dev.* 20: 2057–2066. (b) Fishlock, D., Diodone, R., Hildbrand, S. et al. (2018). *Chimia* 72: 492–500.

24 Jiang, Y. et al. (2014). *J. Med Chem.* 57: 1753–1769.

25 Zhan, Z.-Y.J. (2007). PCT International Publication Patent WO2007003135.

26 Wei, X., Farina, V. et al. (2006). *J. Org. Chem.* 71: 8864–8875.

27 Gantz, F. and Stahr, H. (2008). PCT International Publication WO2008128921.

28 (a) Shu, C. et al. (2008). *Org. Lett.* 10: 1303. (b) Farina, V., Senanayake, C.H. et al. (2009). *Org. Process Res. Dev.* 13: 250–254.

29 Scalone, M. and Stahr, H. (2010). PCT International Publication WO2010015545.

30 (a) Püntener, K. and Scalone, M. (2009). PCT International Publication WO2009124853. (b) Doppiu, A., Karch, R., Püntener, K. et al. (2010). PCT International Publication WO2010127964.

31 Hildbrand, S., Püntener, K. and Scalone, M. (2009). PCT International Publication WO2009080542.

32 Hildbrand, S. (2009). PCT International Publication WO2009053281.

33 (a) Manning, B.D. et al. (2007). *Cell* 129: 1261–1274. (b) Blake J. F. et al. (2012). *J. Med. Chem.* 55: 8110–8127.

34 (a) Gosselin, F. et al. (2017). *Org. Lett.* 19: 4806–4809. (b) Gosselin, F., Iding, H., Reents, R., Scalone, M. (2015). PCT International Publication W2015073739.

35 Schuster, A. et al. (2021). *Chimia* 75: 605–613.

36 Lane, J.W., Remarchuk, T. et al. (2014). *Org. Process Res. Dev.* 18: 1641–1651.

37 Remarchuk, T., Spencer, K.L. et al. (2014). *Org. Process Res. Dev.* 18: 1652–1666.

38 Huisman, G.W. et al. (2007). *Nat Biotechnol.* 25: 338–344.

39 Remarchuk, T. et al. (2014). *Org. Process Res. Dev.* 18: 135–141.

40 Bachmann, S. et al. (2013). *Org. Process Res. Dev.* 17: 1451–1457. and references cited therein.

29

Supported Chiral Organocatalysts for Accessing Fine Chemicals

Ana C. Amorim[1] and Anthony J. Burke[2,3,4]

[1] *Centro de Química de Coimbra, Institute of Molecular Sciences, Departamento de Química, Faculdade de Ciências e Tecnologia, Universidade de Coimbra, Coimbra, Portugal*
[2] *Faculty of Pharmacy, Universidade de Coimbra, Pólo das Ciências da Saúde, Azinhaga de Santa Comba, Coimbra, Portugal*
[3] *Chemistry and Biochemistry Department, School of Science and Technology, University of Évora, Rua Romão Ramalho 59, Évora, Portugal*
[4] *LAQV-REQUIMTE, University of Évora, Institute for Research and Advanced Studies, Rua Romão Ramalho, 59, Évora, Portugal*

29.1 Introduction

Organocatalysis' position as the third pillar of catalysis, along with metal based catalysis and biocatalysis was cemented with the award of the Nobel Prize in chemistry on the 6th of October 2021 to Banjamin List and David MacMillan for their work in the area of asymmetric organocatalysis [1]. Over the last 20 years since the discovery of the proline based catalysts of List and the imidazolidinone catalyst of Macmillan (as well as a host of other types of organocatalysts), organocatalysts have been exploited with tremendous success in the preparation of a myriad of important, and challenging target compounds such as strychnine (a poison), paroxetine (for treating depression), oseltamivir (for respiratory infections), (−)-tashiromine (a naturally occurring indolizidine alkaloid), and aliskiren (a renin inhibitor). Prior to the seminal work of List and MacMillan in 2000, there was very little activity in the area. However, the pioneering work from two teams (Hajos and Parrish; and Eder, Sauer, and Wiechert), who independently showed that a key triketone substrate could be transformed to the Wieland-Miescher ketone using L-proline (Scheme 29.1) and used to access several natural products as well as steroids, should not be undervalued as this seminal work reported in the early 70s remains a watershed development that eventually led to the discoveries that received the Nobel Prize award in 2021 [2].

Despite the fact that many types of organocatalysts are known at the current time [3], and despite the fact that organocatalysis brings so many benefits to the synthesis lab (such as stability in air, lack of toxicity, and renewability), they have one major downside in that they require greater loading of the catalyst compared to those reactions catalyzed by metals. This limitation becomes all the more inconvenient when the organocatalyst is more expensive or very difficult to make, and recovery via chromatography or other purification techniques becomes difficult, or even when the reaction is run at a large scale. As a solution to this problem, many types of organocatalysts have now been immobilized to various types of support that allows their easy recovery and recycling. Immobilization has also been employed to improve catalyst stability. In this chapter we will look

Catalysis for a Sustainable Environment: Reactions, Processes and Applied Technologies Volume 3, First Edition. Edited by Armando J. L. Pombeiro, Manas Sutradhar, and Elisabete C. B. A. Alegria.
© 2024 John Wiley & Sons Ltd. Published 2024 by John Wiley & Sons Ltd.

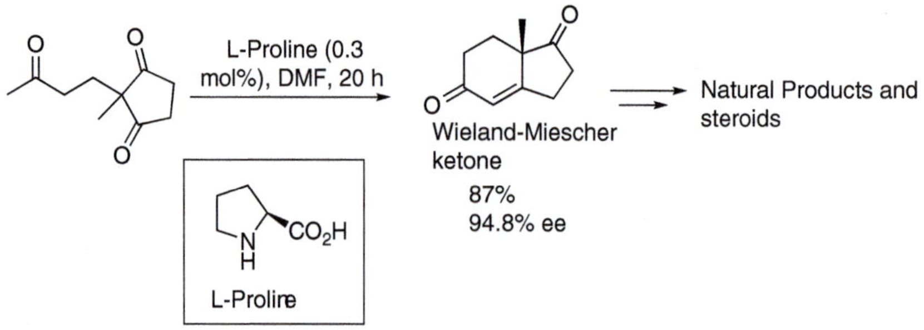

Scheme 29.1 The L-Proline catalyzed synthesis of the Wieland-Miescher ketone as reported by Hajos and Parrish [2] (this was found to be more efficient than with the conditions used by Wiechert et al.).

at some of the organocatalysts that have been successfully immobilized and applied for the synthesis of key targets over the last 10 years. Of course, one of the main driving forces for the immobilization of organocatalysts to solid-supports has been their utilization as part of the catalytic reactor system in flow-chemistry set-ups, which is becoming a standard tool in many pharmaceutical labs.

This review is not exhaustive but will give the reader a general view on what has been achieved during this period. The reader is also encouraged to consult a number of informative reviews on the topic [3, 4]. In the very insightful review by Cozzi in 2006, he stated that only catalysts that can promote multi-reactions, and are highly active, should be considered for immobilization [4–7]. Some of the issues that we will address, through relevant examples, will be (i) the type of support; (ii) the type of linker; (iii) catalyst loading on the support; and, of course, (iv) the reaction conditions.

As a final point, something that was not stated in the interviews and articles that appeared after the awarding of the Nobel Prize in October 2021, and relevant in the context of this review, is the fact that organocatalysis is ideal for immobilization to various supports, and more importantly for application in continuous flow systems.

29.2 Organocatalyst Immobilizations

As mentioned previously, the ability to conduct organocatalytic reactions using immobilized catalysts brings several key advantages, like easy reaction recovery and reusage, greater activity and stability in some cases compared to their homogenous equivalents, diminished product contamination by the catalyst, and easy adaptation to continuous flow systems. To date many types of supports have been used, that range from polymers to silica and aluminosilicate supports to magnetic nanoparticles. In this section we will look at the application and effectiveness of several of the most active organocatalysts that were used in the last 20 years.

29.2.1 Proline Immobilizations

Considering proline's prowess as one of the most privileged organocatalysts available, naturally it was one of the first catalysts to be immobilized to solid supports, and several examples of their use in organocatalytic reactions are known [3–6]. Some examples of their use in the last 10 years are given in this section.

In 2012, Pericàs et al. reported asymmetric aldol reactions using a polystyrene (PS) supported proline catalyst that afforded interesting cyclic aldol products in high yields and with excellent stereoseletivities (Scheme 29.2) [8]. The catalyst was derived from hydroxyproline, and incorporated with a 1,2,3-triazole linker. 10 mol% loading was found to work well. After establishing the best reaction conditions, the team proceeded with a study of the reaction scope. Best results were obtained using the more electron poor aldehydes. These workers also showed that for the prototype reaction with benzaldehyde and cyclohexanone, the catalyst could be recycled up to seven times with little or no drop in the yields, or the stereoselectivities. Due to the excellent results obtained, the prototype reaction was tested with this catalyst (0.54 mmol g^{-1}) in a continuous flow system, using a vertically oriented packed-bed reactor at a flow rate of 25 μL min^{-1} (residence time = approximately 26 minutes). The conversion dropped off after 30 minutes but the diastereoselectivity remained constant up to 45 minutes. Gratifyingly the set-up afforded 4.87 g of the aldol product with a diastereoselectivity of 92% de, and an ee of 97% in a 30 hour run.

An interesting approach was reported by Pothanagandhi and Vijayakrishna, who developed both imidazolium based cross-linked poly(ionic liquids) and polyeletrolytic resins (PILs) bearing proline (as its conjugate base, which was non-covalently attached to the resin) and used them in a variety of reactions that included both Michael additions and Bayliss-Hillman reactions [9]. The imidazolium based achiral PIL resins were prepared by copolymerizing 1-vinyl-3-ethylimidazolium bromide with a cross-linker using 2-ethoxythiocarbonyl-sulfanylpropanoic acid ethyl ester using the reversible-addition-fragmentation-chain-transfer (RAFT) technique. The results were very satisfactory, in the case of the Michael addition between dimethyl malonate and chalcone when the reaction was conducted under neat conditions with 10 mol% catalyst (as poly(ViEIm)(dvb)-pro, where ViEIm = 1-vinyl-3-ethylimidazolium and dvb = divinylbenzene) at room temperature for 24 hours gave the product with a yield of 95% (there was no mention of the enantioselectivity, although both D and L-proline gave the same yield); and in the case of the Baylis Hillman reaction between methyl vinyl ketone and p-nitrobenzaldehyde under neat conditions with 10 mol% of the same catalyst at room temperature gave both entantiopodes with a yield of 94%, but unfortunately with ees of 9% (L-Pr) and 7%, respectively. There was no mention of catalyst reuse.

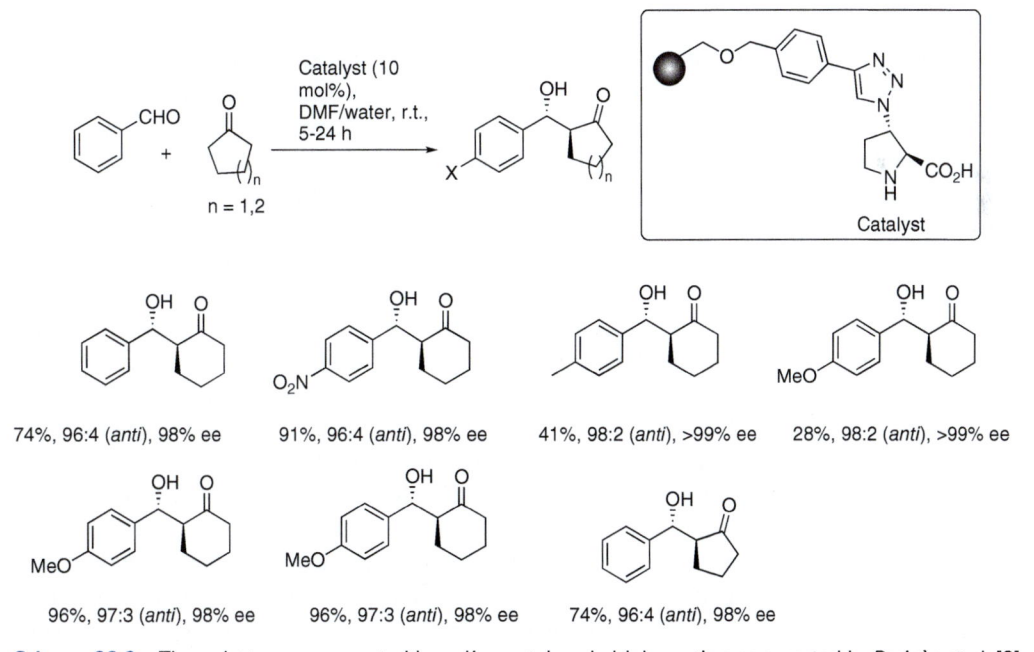

Scheme 29.2 The polystyrene supported L-proline catalyzed aldol reaction as reported by Pericàs et al. [8].

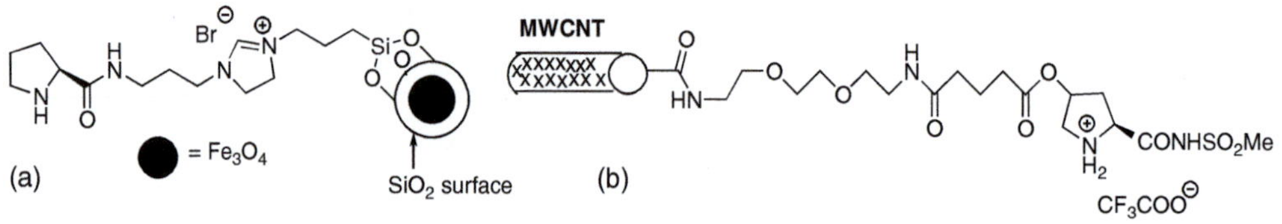

Figure 29.1 (a) The magnetic nanoparticle supported L-proline catalyst developed by Kong et al [11]. (b) The multi-walled carbon nanotube (MWNT) supported proline catalysts developed by Chronopoulos et al. [12].

Proline has also been successfully immobilized to magnetic nanoparticles. Some good examples are included in the 2015 review by Angamuthu and Tai [10]. One interesting example was the report by Kong et al. on the application of an L-proline supported ionic liquid (IL)-modified magnetic nanoparticles for recyclable use in direct aldol reactions in water (Figure 29.1a) [11]. The catalyst consisted of the proline unit grafted to a magnetic ferrite (Fe_3O_4) core via an imidazolium linker. The catalyst was active for the reactions in water without the need for organic solvents. It was suggested that the imidazolium unit facilitated the entry of hydrophobic reactants to active sites in water and stabilized the enamine intermediate. Cyclohexanone, cyclopentanone, and benzaldehydes were used, with yields of up to 96%, diasterometric excesssees of up to 99:1, and ees of up to 89%, with just 10 mol% of catalyst. Furthermore, the catalyst could be recycled four times with marginal drops in the yield and no deceases in either the diastereo- or the enantioselectivity, using cyclohexanone and 2-nitrobenzaldehyde as the test system (12 hours, 30 °C).

In 2015 Chronopoulos et al. described the application of proline tethered units onto multi-walled carbon nanotubes (MWCNTs) and their application in the standard aldol reaction with acetone and a benzaldehyde derivative [12]. The catalyst was prepared using standard amide coupling procedures (Figure 29.1b), and then neutralized with aqueous $NaHCO_3$ before being screened at a loading of 20 mol% in the bench-mark aldol reaction between acetone and p-nitrobenzaldehyde (water at room temperature, 72 hours), to give the aldol in a yield of 94%. It must be noted that the yields were very low in organic solvents. The catalyst was recycled four times and there was a steady drop off in the yield to 64% after the fifth cycle was completed. The enantioselectivity was low (i.e. 16% ee in the first cycle dropping off to 13% ee for the fifth cycle). The authors did not discuss the reasons for these low enantioselectivities, but this may have been due to back-ground catalysis from a non-proline sidechain on the MWCNT surface. Interestingly, the Jørgensen-Hayashi diphenylprolinol unit (see the following section) was also attached to the nano-material, and also gave a high yield (84% after 96 hours), with a low enantioselectivity (14% ee).

In 2013, Pericàs et al. reported a fluorous tagged proline catalyst that was used in fluorous phase catalytic aldol reactions, and as mooted by the authors to possess purported aldolase behaviour [13]. The concept of fluorous phase catalysis was pioneered by Horváth, Cornilis, and Curran in the mid-90s as an alternative approach to recycling expensive catalysts, including organocatalysts [13], but in the last decade there appears to have been a decline in its application. The proline was tagged with a $C_{10}F_{21}$ ponytail using a 1,2,3-triazole linker and initial assessment of the reaction conditions showed that the reaction between acetone and 4-nitrobenzaldehyde in perfluorohexane (C_6H_{14}) at 0 °C at a pressure of 1 bar for 16 hours gave the aldol product with 70% yield and 76% ee (Scheme 29.3). The scope of the reaction on the aldehyde component was then examined and gave highly satisfactory results over a large range of electron poor and rich benzaldehyde derivatives, with very good yields and enantioselectivities in the range 68–96% ee. Gratifyingly the catalyst could be recycled five times with only a noticeable change in the yield on the fifth cycle. The enantioselectivity remained constant at 77% ee.

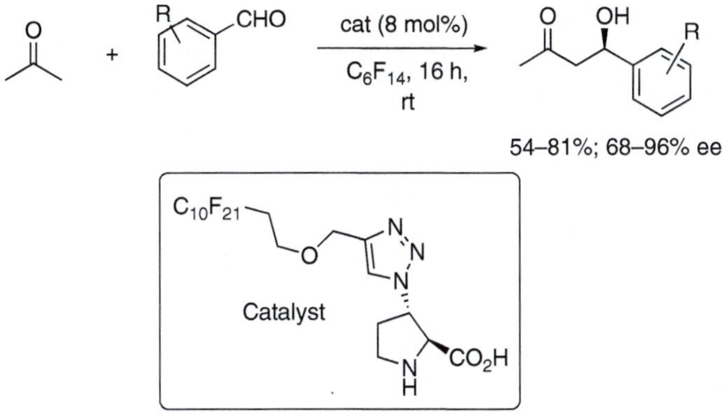

Scheme 29.3 Fluorous phase aldol catalyzed reactions with an L-Proline tagged catalyst as reported by Pericàs et al [13].

29.2.2 Diphenylprolinol Silyl Ether (Jørgensen-Hayashi Organocatalyst) Immobilizations

In 2012, Pericàs et al. reported the application of polystyrene-immobilized diphenylprolinol silyl ether catalysts for the α-amination of aliphatic aldehydes (Scheme 29.4) [14]. A variety of immobilized diphenylprolinol silyl ether catalysts were screened for the α-amination of propanal with dibenzyl azodicarboxylate (DBAD) using 10 mol% catalyst and carboxylic acid co-catalysts and it was found that the *tert*-butyldimethylsilyl (TBS) catalyst gave the best results (92% yield, 93% ee, with 10 mol% AcOH in CH_2Cl_2 (DCM) at room temperature for 45 minutes) for the final aminoalcohol product, obtained via a simple reduction of the aminoaldehyde adduct with $NaBH_4$. The scope of

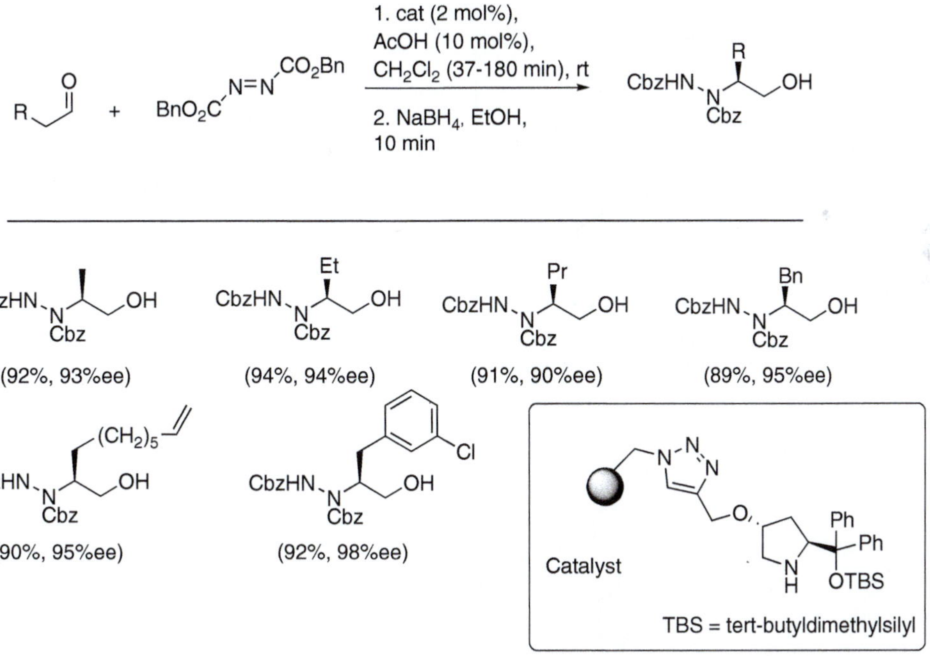

Scheme 29.4 Polystyrene supported prolinol catalysts for the α-amination of aldehydes as reported by Pericàs et al [14].

the reaction was also investigated, and a screening study of a variety of aldehyde products showed that the reaction was tolerant of a wide range of substrates, as both the yields and ees were high. The catalyst could also be recycled up to six times with little alteration in the yield or the enantioselectivity, demonstrating the potential of this catalyst. Finally, the authors performed the reaction under continuous flow conditions, using a packed bed reactor to house the catalyst, and a flow system comprising 2 pumps. One pump was responsible for pumping the aldehyde as a 1.25 M solution in DCM to a T-mixer at a flow rate of 0.075 mLmin^{-1}, and the other the DBAD (0.25 M, at the same flow rate) also to the T-mixer. The mixture was then passed into the catalytic reactor (containing 300 mg of the immobilized catalyst) at a flow rate of 0.15 mL min^{-1} (residence time = 6 min) and the effluent was collected at 0°C, reduced to the aminoalcohol and analysed. The conversion was quantitative for the first six hours, then dropped off to 90% and 87% after the seventh and eighth hours, respectively. Gratifyingly the ee remained constant at approximately 85%. This was interesting work, but it's funny that these workers did not access the useful aminoalcohol products by breaking the N–N bond.

In 2019, Szcześniak et al. reported the application of a key organocatalyzed Michael addition using a prolinol catalyst supported on Wang resin as the key step for the synthesis of the antidepressant active pharmaceutical ingredients (APIs) of (−)-paroxetine (Paxil®, Seroxat®) and the selective serotonin reuptake inhibitor (+)-femoxetine (developed by Ferrosan pharmaceuticals in the 70s) (Scheme 29.5) [15]. After preparing the immobilized catalyst, the reactions with two aldehydes and two nitroolefins were investigated under batch conditions.

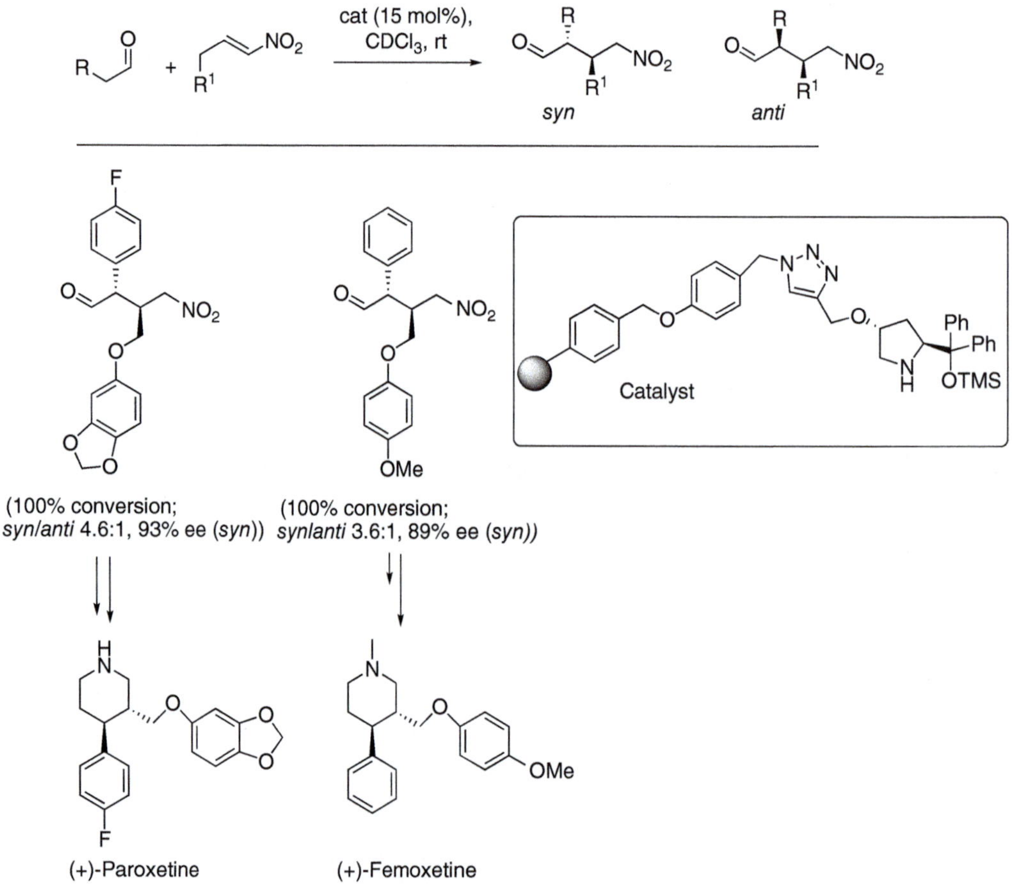

Scheme 29.5 The key asymmetric step in the synthesis of the antidepressant agents (−)-paroxetine and (+)-femoxetine as reported by Szcześniak et al [15].

The reactions afforded both products in very good yields and with high enantioselectivities for the *syn*-isomer (Scheme 29.5). The reaction was also conducted under continuous flow conditions, using a packed-bed reactor to immobilize the catalyst. Full conversion and a diasterometric ratio of 3/1 was achieved when the reaction was run for 11 hours with 20 mol% benzoic acid in $CHCl_3$ at a flow rate of 0.005 mL min^{-1} (operation time = approximately 17 hours). Once the Michael adducts were obtained in sufficient quantities, they were transformed into the anti-depression compounds (−)-paroxetine (35%, 4-steps from the Michael adduct, and 80% ee) and (+)-femoxetine (27%, 5-steps, 75% ee) via a multi-sequential pathway (Scheme 29.5). This also represents a general strategy for the asymmetric synthesis of piperidine scaffolds with *trans*-stereochemistry at C3 and C4.

29.2.3 Organocatalysts Based on Immobilized Pyrrolidines

In 2013 Pericàs et al. (pioneers in the exploitation of continuous flow for organocatalysis, as we see in this chapter) reported the use of a PS supported 1,2,3-triazole linked 3-trifluoromethylsulfonyl-aminopyrrolidine catalyst for asymmetric *anti*-Mannich reactions in continuous flow (Scheme 29.6) [16]. After preliminary optimization studies with isovaleraldehyde and *N*-PMP ethyl glyoxylate,

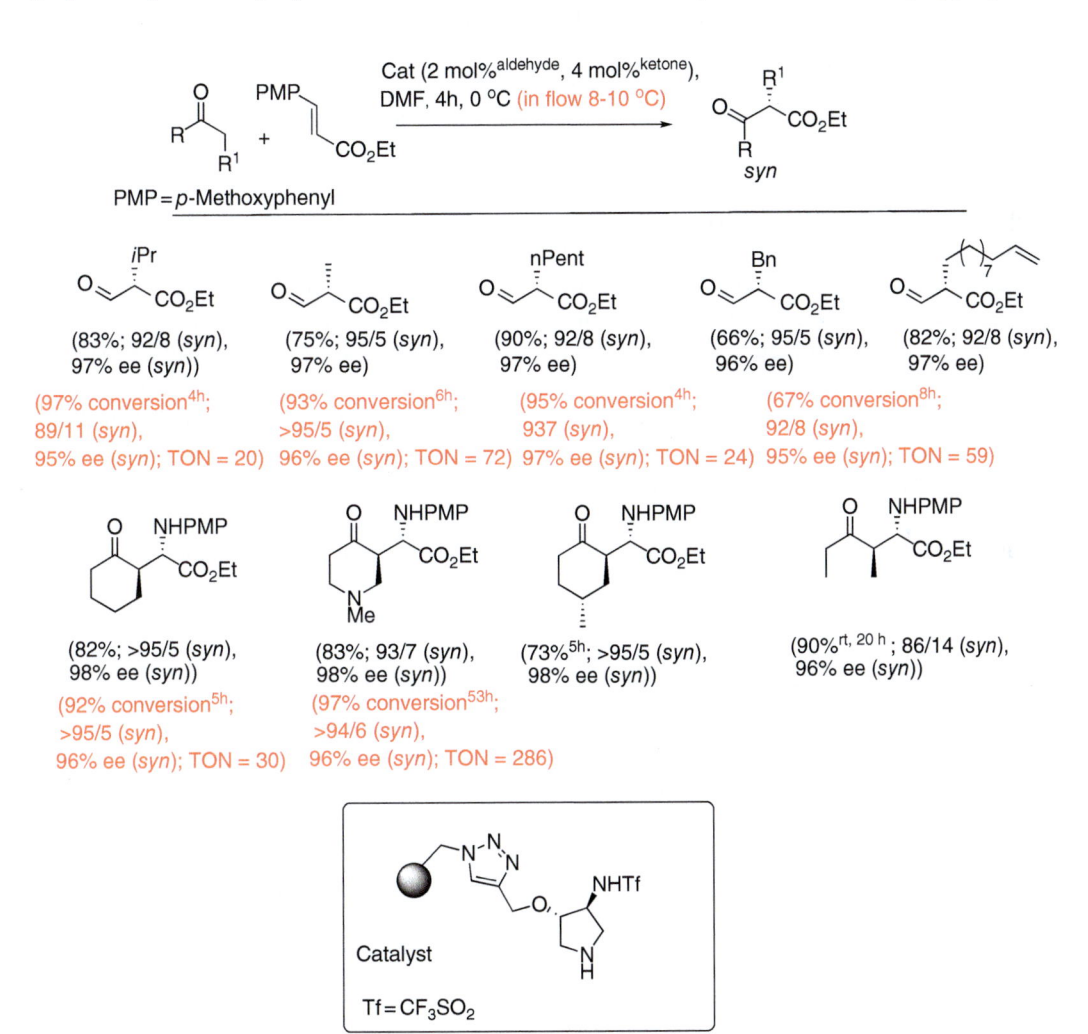

Scheme 29.6 The asymmetric anti-Mannich reaction with a supported 3-aminopyrrolidine catalyst as reported by Pericàs et al [16]. (The results in red pertain to the continuous flow experiments).

the best conditions were 2 mol% of the immobilized catalyst, in THF at 0 °C for four hours (95% conversion, 91/9 [syn], 96% ee). Polymer swelling is an important issue as regards mass transfer effects, and both THF and DMF (93% conversion at 4 mol% catalyst loading, at 0 °C for four hours, giving 92/8 (syn) and 97% ee) showed excellent properties in this regard.

The reaction scope on the keto unit was then evaluated using both aldehyde and ketone substrates and N-PMP ethyl glyoxylate (unfortunately this component was not varied). Moreover, it was observed that the polymer supported catalyst gave better results than the homogenous counterpart. Recycling and reuse of the catalyst was assayed using cyclohexanone (in DMF) and the reaction could be recycled up to five times, with no drop in the conversion nor in the stereoselectivity which also remained constant. In the case of the continuous flow experiments, the catalyst (0.23 mmol on 500 mg resin), was charged to a packed-bed reactor and the carbonyl (0.52 M) and the N-PMP ethyl glyoxylate (0.26 M) reagents introduced to the system at a total flow rate of 0.2 mL min^{-1} (residence time = 12 minutes) (Scheme 29.6). The products were obtained with essentially the same yields and stereoselectivities as the batch conditions. The turn-over-numbers (TONS) were in the range 22 to 72. However, to demonstrate the robustness of this system, some experiments were performed for more than two days (see Scheme 29.6) and these proved to be exceptionally efficient, achieving TONs of >285 (see Scheme 29.6) and yields of products in the order of 68.15 mmol.

The 5-(pyrrolidine-2-yl)tetrazole catalyst was reported independently by Arvidsson, Ley and Yamamoto in 2004, and immediately was found to be more effective than proline in a number of organocatalyzed reactions [17]. It became clear soon afterwards that it was an ideal catalyst for continuous flow reactions as demonstrated by Odedra and Seeberger in 2009, who demonstrated its application in Aldol and Mannich reactions [18]. It was also of interest to immobilize this efficient catalyst to solids so that it could be integrated into catalytic reactors. As a good example, in 2016, Massi et al. integrated this catalyst into a monolithic flow microreactor (in a monolithic reactor the catalyst occupies the space of the reactor in the form of a monolith; which is a structured material possessing a regular or irregular network of channels, which is generally formed within the reactor) by attaching it to a polystyrene resin [19]. (These workers have also previously worked with silica-supported pyrrolidinyl-tetrazole catalysts, which were accessed in both batch and continuous-flow conditions [20]). After attaching the tetrazole to the pyrrolidine unit by standard chemistry to form a styryl-functionalized pyrrolidinyl-tetrazole, it was polymerized in a glass tube to form the polystyrene monolithic catalyst using standard Fréchet-Svec conditions (Scheme 29.7a). The immobilized catalyst was then investigated in a simple bench-mark reaction with cyclohexanone and p-nitrobenzaldehyde, using 10 mol% of the catalyst and a variety of solvents and conditions. When the reaction was performed in water for four hours at room temperature with a catalyst loading of 10 mol% the reaction gave a yield of 95%, a diastereoselectivity of 10/1 (syn) and an ee for the major anti-isomer of 95%. Subsequently, for the key reactions in flow, the polymerization to form the monolithic microreactor was conducted in a stainless-steel column (the Boc deprotection was carried out by sequentially flowing TFA/THF and Et$_3$N/THF solutions through the column). The scope of the reaction was then investigated, and the best conditions were conducting the reaction at 25 °C at a flow rate of 10–20 mL min^{-1} for five hours to give the aldol in quantitative yield, with good diasteroselectivities and ees (Scheme 29.7b).

It should also be pointed out that the reaction could be maintained five days on stream without any degradation of the catalyst, which was probably due to the resistance of the tetrazole unit toward any known side-reactions.

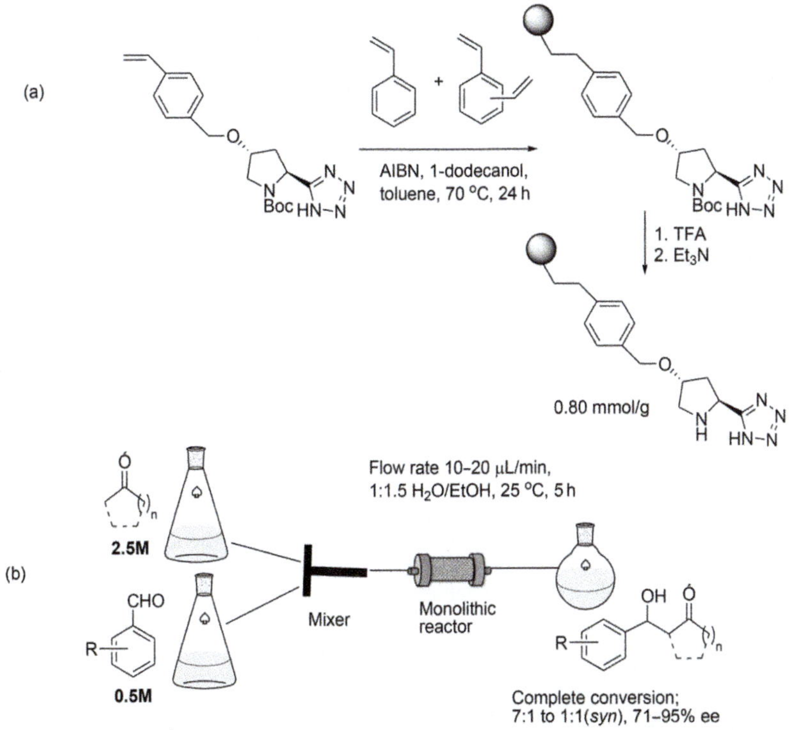

Scheme 29.7 (a) Preparation of the monolithic 5-(pyrrolidine-2-yl)tetrazole catalyst for an asymmetric aldol reaction reported by Massi et al (b) Organization of the flow microreactor used by Massi et al in an aldol reaction [19].

29.2.4 Organocatalysts Based on Immobilized Imidazolidinones

These robust and easy to prepare organocatalysts, also known as MacMillan catalysts, are one of the most effective catalysts for a variety of asymmetric organocatalytic transformations and of course were the grounds for awarding the Nobel prize to David MacMillan [1]. This category of organocatalyst was first synthesized by MacMillan et al. at the beginning of the millennium, been originally employed in the enantioselective Diels-Alder reaction [21]. Its application increased significantly over time and, consequently, modifications and immobilizations to different supports were explored [22]. Some recent examples of immobilized chiral imidazolidinone catalysts using continuous-flow approaches are described.

Back in 2014, Mandoli et al. reported the first example of a chiral imidazolidinone polymer-based monolithic reactor for a cycloaddition reaction leading to isooxazoline compounds which are present in diverse pharmacologically active compounds [23, 24]. After the MacMillan monomer preparation and copolymerization inside a high-performance liquid chromatography (HPLC) column (loading = 0.51 mmolg^{-1}), the catalyst was used in the form of tetraflouroborate salt to synthesize the isoxazoline target compounds, through a 1,3-dipolar addition of N-benzyl-C-phenyl and crotonaldehyde using wet CH_3NO_2, under continuous flow conditions (at a flow rate of 2 µLmin^{-1} at room temperature). This reaction was a success, affording the product in 71% yield with an ee of 90% (*endo*) and a dr of 91:9 (*endo/exo*) (Scheme 29.8a). Additionally, two other reactions were tested exhibiting the reactional scope of this monolithic reactor. A Diels-Alder

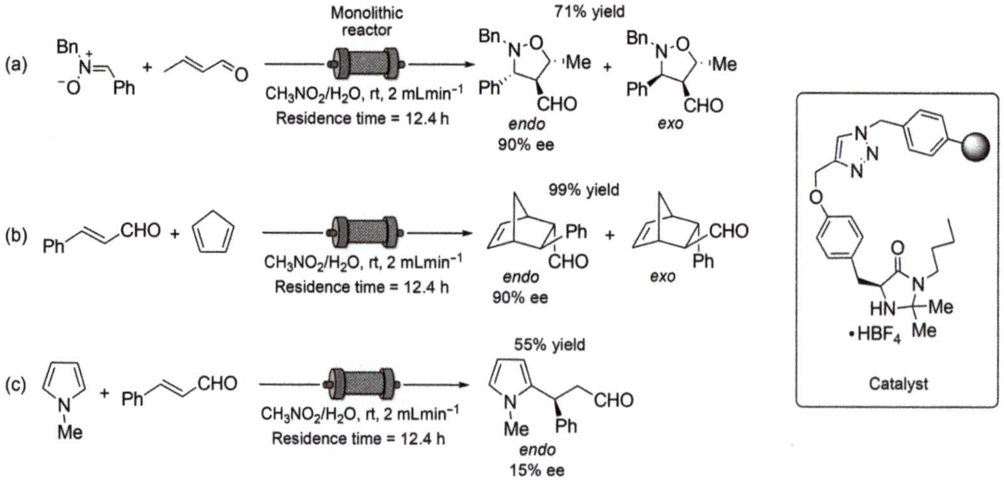

Scheme 29.8 Enantioselective catalytic reactions using a imidazolidinone polystyrene supported monolithic reactor under continuous flow conditions. (a) 1,3-Dipolar cycloaddition of N-benzyl-C-phenyl nitrone and crotonaldehyde; (b) Diels-Alder cycloaddition between cinnamaldehyde and cyclopentadiene; (c) Friedel-Crafts alkylation of N-methyl pyrrole with cinnamaldehyde.

cycloaddition between cyclopentadiene and cinnamaldehyde (Scheme 29.8b) gave an impressive 99% yield, 90% ee (*endo*) and 47/53 dr, and the Friedel-Crafts alkylation between N-methyl pyrrole and cinnamaldehyde (Scheme 29.8c) provided the product in 55% yield and 15% ee (*endo*). This latter reaction gave less satisfactory results suggesting the need for optimization studies. Experiments under batch conditions were also performed indicating a loss in catalytic activity in a reasonably short period of time when compared with this continuous flow set-up. Also, a decrease in the yield could be verified after one cycle.

Three years later, a polymer supported chiral N-picolimidazolidinone catalyst was developed by Benaglia et al. [25] for enantioselective imine reduction reactions. The products of these reactions (chiral amines) are important pharmacophores that are present in many pharmaceutical and agrochemical compounds. The previously prepared catalyst was recovered and reused during six reaction cycles, under batch conditions, affording the desire amine with good yields and ees. After the sixth cycle, a slight decrease in its catalytic activity occurred (76% yield and 87% ee). A considerable number of imines was then tested, giving the corresponding amines in good yields (62–99% yield) and ees (55–97%) (Scheme 29.9a). For the continuous flow approach (Scheme 29.9b), the polystyrene supported organocatalyst (at a loading of 0.57 mmol g^{-1}) was packed into a HPLC column. Then, a mixture of the imine and HSiCl$_3$ in dry DCM was pumped inside the reactor at 0.4 mLh^{-1} at room temperature. After treating the effluent from the reactor with a 10% solution of NaOH the resultant amine product was obtained with excellent yield and enantioselectivity (up to 91% ee).

29.2.5 Other Amino Acid and Peptide Type Catalysts

Beyond the already mentioned catalysts, other successful amino acids and peptide type organocatalysts have been immobilized and tested in asymmetric flow processes. In this section we present some examples, mainly focused on supported primary amino acids and peptide derivatives. A more thorough overview can be found in the literature [26].

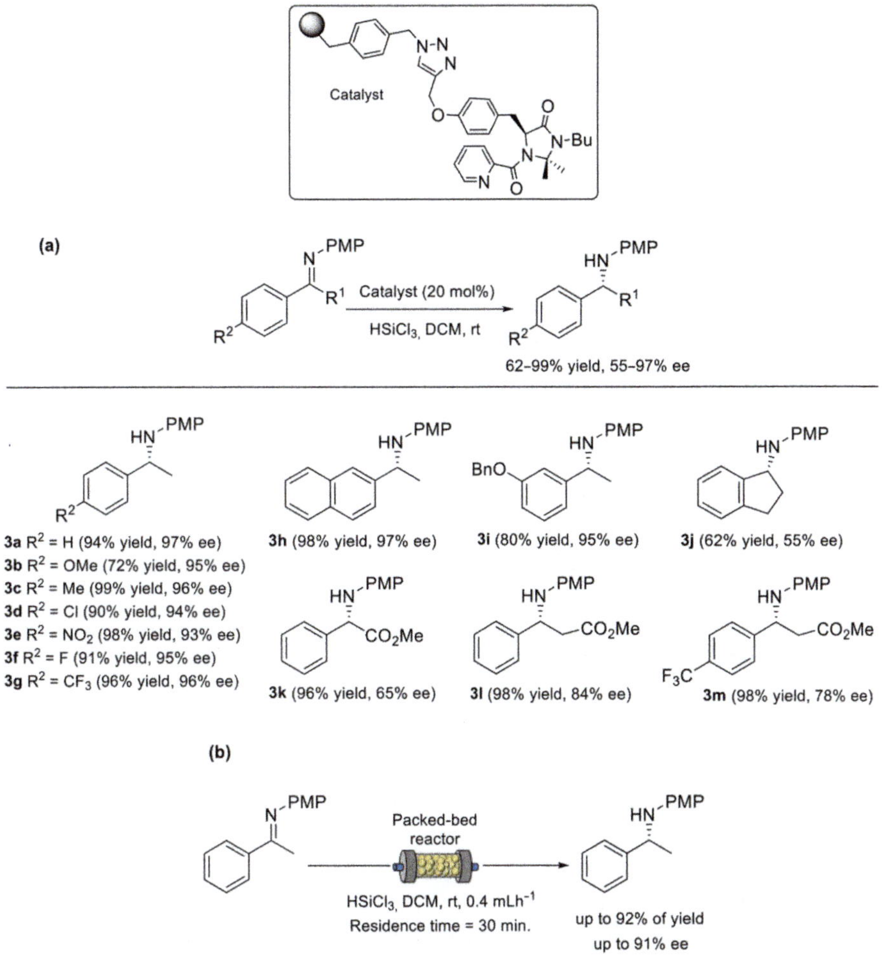

(a)

3a R² = H (94% yield, 97% ee)
3b R² = OMe (72% yield, 95% ee)
3c R² = Me (99% yield, 96% ee)
3d R² = Cl (90% yield, 94% ee)
3e R² = NO₂ (98% yield, 93% ee)
3f R² = F (91% yield, 95% ee)
3g R² = CF₃ (96% yield, 96% ee)

3h (98% yield, 97% ee)
3i (80% yield, 95% ee)
3j (62% yield, 55% ee)
3k (96% yield, 65% ee)
3l (98% yield, 84% ee)
3m (98% yield, 78% ee)

Batch reaction: 62–99% yield, 55–97% ee

(b) Packed-bed reactor, HSiCl₃, DCM, rt, 0.4 mLh⁻¹, Residence time = 30 min. up to 92% of yield, up to 91% ee

Scheme 29.9 Imine reduction with HSiCl₃ using a PS immobilized picolimidazolidinone organocatalyst under (a) batch and (b) continuous flow conditions.

29.2.5.1 Supported-primary Amino Acid Catalysts

Primary amino acids have been a very successful type of organocatalyst for several bench-mark asymmetric transformations [27], catching the attention of many research groups.

In 2014 Pericás et al. reported the application of a PS-supported threonine derivative as an immobilized organocatalyst in three-component *anti*-Mannich reactions [28]. It was found in preliminary tests that a mixture of DMF and CH₂Cl₂ (1:1) was the best solvent system for the bench mark reaction as DMF gave higher stereoselectivity and DCM afforded better results regarding sweeling ability and reaction rate. The reactional scope was explored by using multiple aniline and aldehyde derivatives. The reactions were completed after 3-8h providing the *anti*-adducts, which are advanced synthons for the preparation of biologically active molecules, in moderate to very good yields (62 to 96%), enantio- (up to 95%) and diastereoselectivities (up to 8:92) (Scheme 29.10). The recyclability of this immobilized catalyst was also verified, providing conversions up to 99% and very good enantio- and diastereoselectivities during the first three cycles. However, a decrease in all three parameters was perceptible in the fourth cycle. A small number of reactions were additionally performed under continuous flow system, using a vertical

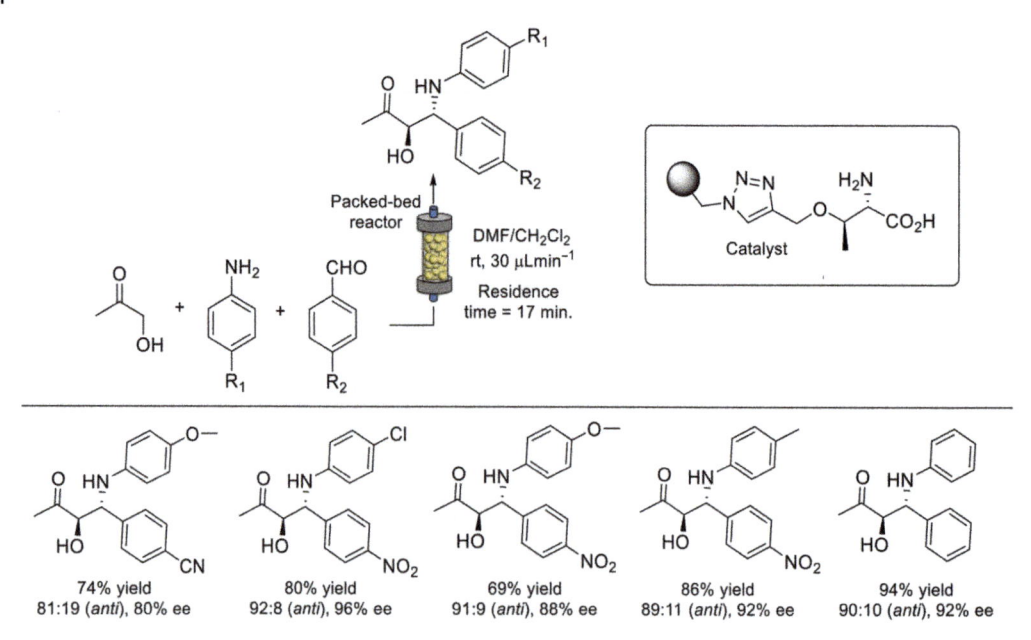

Scheme 29.10 Three-component anti-Mannich reaction catalyzed by a polystyrene-supported threonine derivative under continuous flow conditions as described by Pericàs et al

mounted packed-bed reactor, with a loading of 1.02 mmol g^{-1} of catalyst, and a flow rate of 30 μL min^{-1} at room temperature with a mixture of DMF and CH_2Cl_2 as solvent system. Five target *anti*-Mannich adducts were successfully obtained in good yields and with high diastereo- and enantiomeric purities (Scheme 29.10).

29.2.5.2 Supported-peptide Derivative Catalysts
Due to their modular nature, readily designable structure and inherent chirality, synthetic peptides are of great interest for applications in organocatalytic syntheses. These compounds have a wider structural diversity and are easier to immobilize, as compared to proline.

A curious study using such compounds was made by Fülöp et al., who reported the catalytic activity of *N*-terminal prolyl-peptides in the asymmetric α-amination of aldehydes [29]. Some prolyl-peptides were prepared, but H-Pro-Asp-NH showed the best results in batch reactions, affording a quantitative conversion (100%), a good enantioselectivity of 79% ee and a productivity of 0.5 mmol product×mmol catalyst^{-1} × h^{-1}. With all the optimizations in hand, the researchers looked at the continuous flow approach. The dipeptide was first immobilized in a polystyrene-polyethylene glycol grafted copolymer TentaGel – that is an ideal robust and pressure stable resin for flow-based purposes, with a loading of 0.48 mmol g^{-1}. Then, 200 mg of the catalyst were housed in a packed-bed column and the reaction components consisting of the aldehyde and DBAD in $CHCl_3$ were passed as a continuous stream through the column at a flow rate of 0.1 mL min^{-1} at room temperature. The α-hydrazino aldehyde formed was immediately treated with $NaBH_4$ to form the corresponding alcohol, in order to avoid racemization (very similar to the strategy described by Pericàs in Scheme 29.4). The product, which is a precursor to a variety of biological active compounds, was obtained with a conversion of 86% and a remarkable ee of 90%. The scope of the reaction was studied using different aldehyde derivatives (Scheme 29.11). The reactions proceeded with conversions in the range 86–100%, and ees between 90 and 99%.

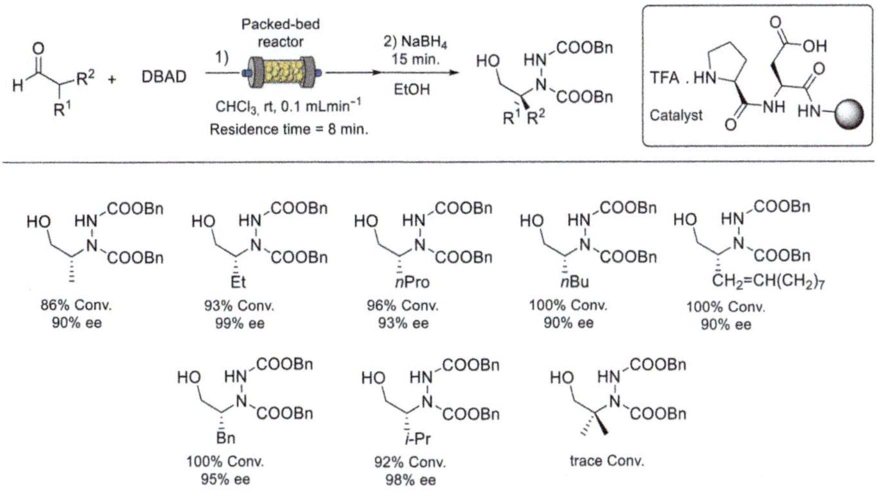

Scheme 29.11 α-Amination of aldehydes using a H-Pro-Asp-NH-TentaGel catalyst in a continuous-flow reactor.

29.2.6 Immobilized Amino-Cinchona Based Organocatalysts

Cinchona alkaloids and their derivatives have a prominent position in organocatalysis, being some of the most active and stereoselective catalysts of the last decade. 9-Amino-9-deoxy-*epi*-cinchona catalysts have been used to catalyze several chemical transformations [30]. Their excellent versatility and enantioselectivity, plus the fact that their synthesis is not straight-forward turn the immobilization of these compounds into a very attractive option from an industrial point of view.

In 2015, Puglisi et al. described an efficient immobilized 9-amino-9-deoxy-*epi*-quinine catalysts for a Michael addition reaction [31]. The general strategy to prepare the catalysts involved the introduction of a linker and then the attachment to a polystyrene support or simply the direct attachment to the solid support with no linker. Preliminary studies using the bench-mark Michael addition reaction of isobutyralde and *trans*-β-nitrostyrene, suggested that the use of the linker and a loading of 0.7 mmol g^{-1} of catalyst gave the best results (>99% yield, 95% ee). Recycling experiments of the catalyst showed that besides the excellent ees up until the sixth cycle, the yield started to decrease after the third cycle. The general scope of these catalysts was demonstrated using distinct α-branched aldehydes or ketones and nitroolefins as reagents, affording the products in moderate to good yields and very high enantioselectivities (Scheme 29.12a). Continuous flow conditions were also applied to these immobilized organocatalysts. For this, a stainless-steel column was packed with the polymer supported catalyst (250 mg) and a mixture of β-nitrostyrene, isobutyraldehyde and benzoic acid in toluene was pumped at room temperature into the reactor with a flow rate of 0.1 mL h^{-1}. The Michael adducts were obtained with very good yields (70–99%) and very high enantioselectivities (85–93% ee) (Scheme 29.12b). The general applicability of this immobilized organocatalyst was demonstrated using two other reactions with different activation pathways.

29.2.6.1 Cinchona Picolinamide Derivatives

Another example of the successful application of cinchona-based organocatalysts, is that of the cinchona picolinamide types, that are often used effectively in hydrosilylation reactions [32, 33].

The cinchona picolinamide catalyst developed by Burke and Benaglia is one example to highlight [33]. A chiral picolinamide derived from 9-amino-*epi*-cinchonine immobilized on a polystyrene

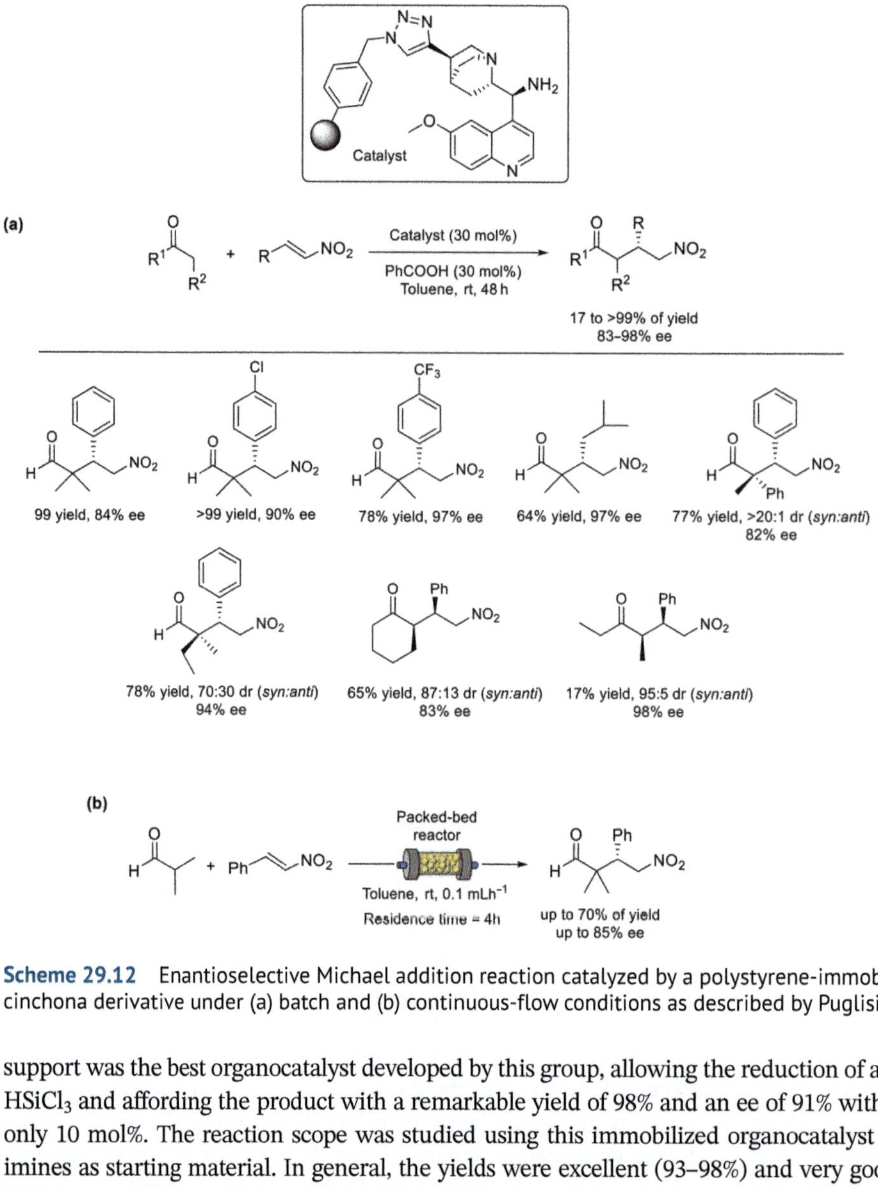

Scheme 29.12 Enantioselective Michael addition reaction catalyzed by a polystyrene-immobilized amino cinchona derivative under (a) batch and (b) continuous-flow conditions as described by Puglisi et al

support was the best organocatalyst developed by this group, allowing the reduction of an imine with $HSiCl_3$ and affording the product with a remarkable yield of 98% and an ee of 91% with a loading of only 10 mol%. The reaction scope was studied using this immobilized organocatalyst with various imines as starting material. In general, the yields were excellent (93–98%) and very good enantioselectivities were achieved (84–87% ee) (Scheme 29.13a). Furthermore, the first continuous flow set-up for the stereoselective synthesis of chiral amines using trichlorosilane was demonstrated, using the immobilized cinchona-based organocatalyst shown previously. Therefore, a packed-bed reactor was prepared with 170 mg of the catalyst. Then, a solution of the imine and $HSiCl_3$ in CH_2Cl_2 was injected into the reactor at a flow rate of 0.02 mL min^{-1} at room temperature. The effluent was treated with an aqueous 10% NaOH solution. The product was first obtained with 98% yield and 47% ee, but unfortunately the ee dropped drastically to 20%, after 150 hours of continuous production (Scheme 29.13b).

29.2.6.2 Cinchona Squaramide Derivatives

These compounds were first studied as organocatalysts by Rawal [34] in 2008, who discovered that they have high catalytic activity for asymmetric Michael reactions. Since then, they have been applied frequently in this type of reaction and, in some cases, they have been evaluated and tested under continuous-flow conditions.

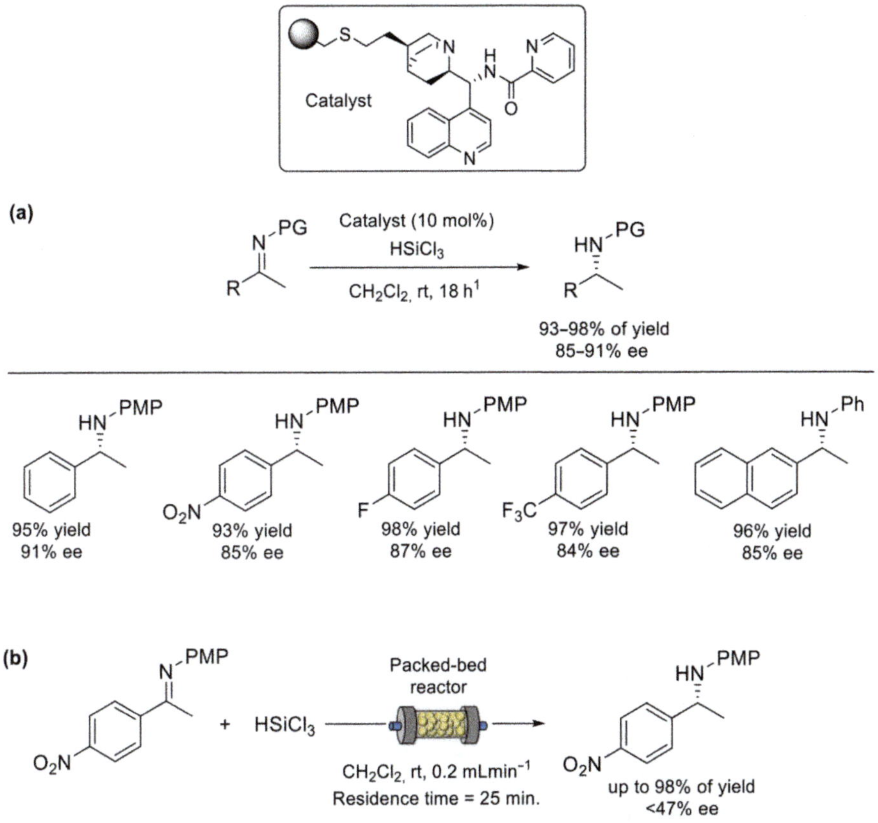

Scheme 29.13 Stereoselective reduction of imine derivatives with HSiCl₃ using solid-supported chiral picolinamides developed by Burke and Benaglia in (a) batch and in (b) continuous flow set-up.

Szekely et al. [35] reported in 2019 the synthesis and application of cinchona squaramide-modified permethyl-β-cyclodextrin catalyst for the asymmetric Michael addition. Cyclodextrins (CD) are perfect platforms for the development of immobilized organocatalysts as they are obtained from renewable resources, are robust and stable. With that in mind, the catalysts were prepared by the attachment of the cinchona amine unit to the CD using the squaramide moiety as a linker. The activity of these catalysts was evaluated using 1,3-diketones and *trans*-β-nitrostyrenes for a Michael addition under batch conditions. As can be seen from Scheme 29.14a, most products were obtained with high yields (up to 95%) and high enantioselectivities (up to 99% ee). For the continuous flow mode, the best reaction from the batch approach was chosen. The continuous flow set-up was performed in a coiled reactor coupled with a membrane separation unit, used to purify the product, and recycle the catalyst (Scheme 29.14b). A fresh solution of the reactants was pumped into the reactor (at room temperature) and then into the membrane (at 50 °C), where it met the retentate stream, which allowed the full recyclability of the catalyst (100%) and 50% of the solvent (2-MeTHF). The product was successfully obtained with 100% conversion, 98% purity, and 99% ee.

29.2.7 Other Organocatalysts

In this final section, we will briefly look at two other families of organocatalysts that have also been used successfully in supported organocatalytic reactions.

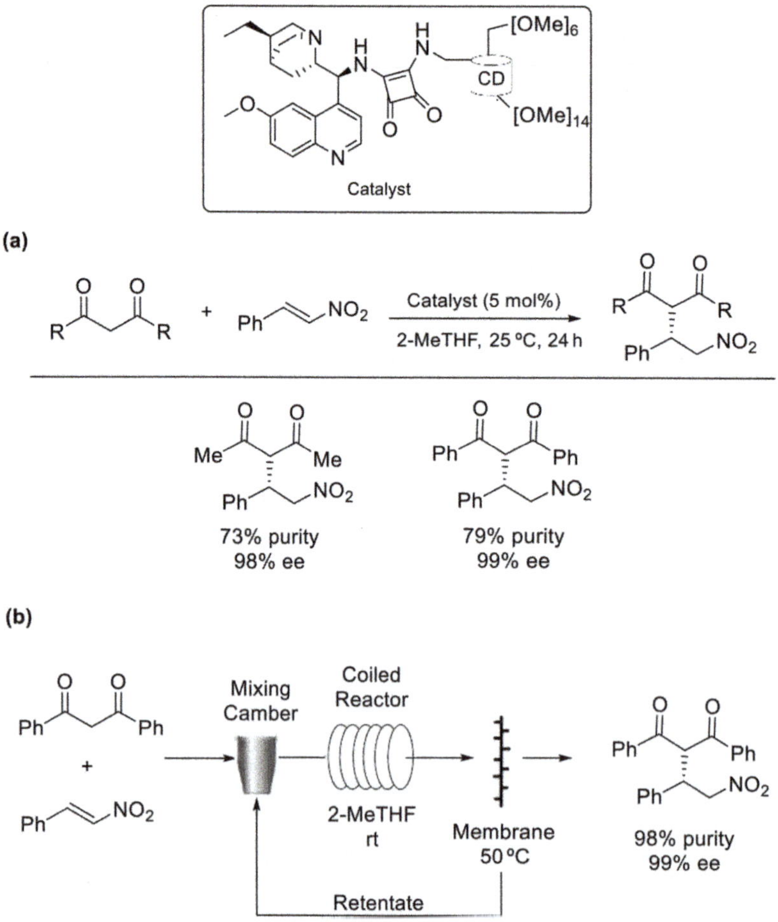

Scheme 29.14 Michael addition of a 1,3-diketone with nitrostyrene catalyzed by cinchona squaramide-modified permethyl-β-cycodextrins under (a) batch and (b) continuous-flow modes.

29.2.7.1 Phosphoric Acid Catalysts

1,1′-Bi-2-naphthol (BINOL)-derived phosphoric acid (PA) organocatalysts have been applied with great success in asymmetric reactions over the last number of years [36]. These catalysts are ideal for bifunctional catalysis, due to the presence of a Brønsted acidic (P–OH) and Lewis basic site (P=O). They are air stable and easily stored. In 2014, Pericàs et al. reported on an enantioselective continuous-flow production of 3-indolylmethanamines catalyzed by a PS-immobilized PA [37]. The catalyst was immobilized at a loading of 0.26–0.37 mmol g^{-1} in four steps starting from (R)-6-hydroxymethyl-2,2′-bis(methoxymethyloxy)-1,1′-binaphthalene. The catalyst was then used in an aza-Friedel–Crafts (loading = 10 mol%) with a tosylimine and indole presenting very good results; the aza-Friedel–Crafts adduct was obtained with a yield of 77% and an ee of 94% (the absolute configuration was assigned using single-crystal x-ray crystallography) after six hours at room temperature (Scheme 29.15). The catalyst could be recycled 14 times and reactivated by a simple acidic wash. The scope was then evaluated, and again good results were observed; except that sterically hindered and electron rich tosyl imines, like those derived from o-methylbenzaldehyde and p-methoxybenzaldehyde gave lower yields. When the cyclohexyl substituted imine was used the results were poor: 44% yield and 14% ee. The scope of the reaction on the indole component was also

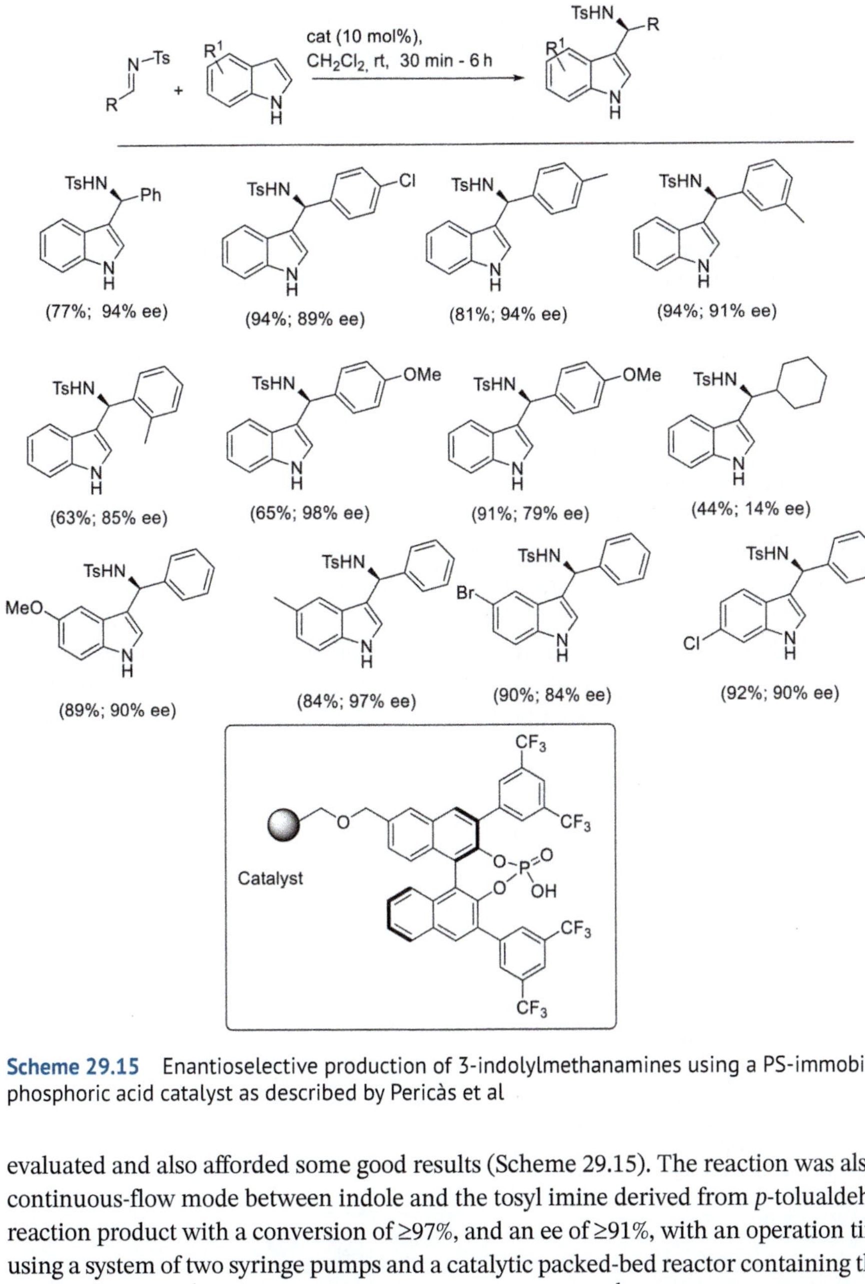

Scheme 29.15 Enantioselective production of 3-indolylmethanamines using a PS-immobilized chiral phosphoric acid catalyst as described by Pericàs et al

evaluated and also afforded some good results (Scheme 29.15). The reaction was also conducted in continuous-flow mode between indole and the tosyl imine derived from p-tolualdehyde to give the reaction product with a conversion of ≥97%, and an ee of ≥91%, with an operation time of six hours using a system of two syringe pumps and a catalytic packed-bed reactor containing the catalyst (360 mg; 0.25 mmol g^{-1}) at a combined flux rate of 0.2 mL min^{-1} to produce the indole product with an isolated yield of 80% (3.6 g) and a calculated TON of 102. Moreover, the catalyst loading for the global process was only 0.8 mol% (at a residence time of 9.3 minutes), which was 12 times lower than the loading for the batch process. This catalytic reactor was also used for a small amino-indole library synthesis that is not discussed here.

29.2.7.2 Isothiourea Catalysts

The final immobilized organocatalyst system that we will consider is the benzotetramisole (BTM) isothiourea class that was pioneered by Birmann [38] and championed over the last decade by

Smith [39]. Once again, Pericàs et al. have been instrumental in their application of this unique efficient and somewhat underrated catalyst system in heterogenous catalytic reactions, when they applied a polystyrene BTM-isothiourea to an interesting [8+2]-cycloaddition reaction that gave rise to cycloheptatriene-fused pyrrolidine derivatives [40]. This is a very useful method synthetically and was first introduced by von Doering and Wiley in 1960, but, unfortunately, there have been few asymmetric approaches reported in the intervening years. For this formal [8+2] annulation an azaheptafulvene was reacted with phenylacetic acid in the presence of a PS-supported-triazole tethered BTM-isothiourea (at 10 mol% loading) with pivaloyl chloride and 1,8-diazabicyclo[5,4,0]undec-7-ene (DBU) at 0 °C in DCM giving the target product with >95% conversion, a dr of 96/4, and an enantioselectivity of 91% ee (Scheme 29.16). The scope of the reaction was then investigated, for both the acetic acid and the azaheptafulvene components. When a broad range of phenylacetic acid derivatives were evaluated the cycloadducts were obtained in yields of 65–80%, with high enantioselectivities in the range of 90–97% ee and excellent diastereoselectivity (>20:1). The absolute configuration of the cycloadduct was assigned through recourse to single crystal x-ray crystallography. When the structure of the azaheptafulvene component was varied, it was obvious that the annulation was insensitive to the electronic nature of the aromatic N-substituent, as the cycloadducts were obtained in yield of 70–85%, diastereoselectivities of >20:1 and ees in the range 90–98%. Recycling experiments using the standard reaction (see previously), demonstrated that there was no significant drop in the stereoselectivity from the first to the seventh cycle, but there was a drop off in the yield becoming more extreme on the sixth and seventh cycles (approximately 50 and 40%, respectively). Finally, the accumulated TON for these recycling experiments was 44.7. A putative mechanism was presented by the authors which was essentially a non-concerted polar type [8+2] cycloaddition process.

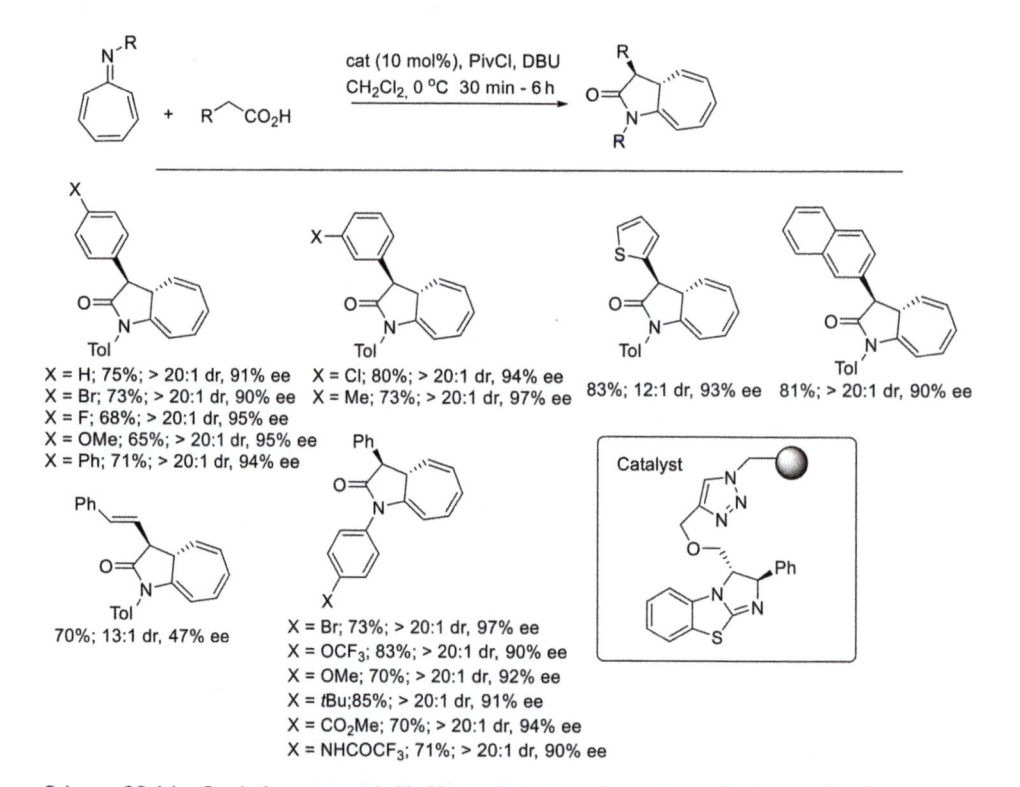

Scheme 29.16 Catalytic asymmetric [8+2] annulation reactions using a PS-immobilized chiral benzotetramisole catalyst as described by Pericàs et al.

29.3 Conclusions

In the autumn of 2022, organocatalysis was instated as the third pillar of catalysis with the award of the Nobel prize in chemistry to Benajamin List and David MacMillan for pioneering, conceptual, and developmental work in this new field. In this chapter, we have considered some of the major advances that have been undertaken in the last 10 years on the immobilization of these catalysts, with the principal goals of activation/stabilization, recycling, and reuse. As we have seen in this monograph, the ability to immobilize organocatalysts not only brings benefits in terms of catalyst recycling and reuse (reducing waste and costs) that is particularly beneficial for industry, but, more interestingly, immobilized organocatalysts have been the main staple of many continuous flow procedures, which has now become a standard industrial tool for accessing important APIs. In the coming years, we expect that immobilized organocatalysts will have greater impact on industrial processes, particularly from a drug development and production point of view.

References

1 (a) The Nobel Prize in Chemistry 2021. https://www.nobelprize.org/prizes/chemistry/2021/summary; (b) Boerner, L.K. (2021). Pioneers of asymmetric organocatalysis win 2021 Nobel Prize in Chemistry. *Chem. Eng. News*. https://cen.acs.org/people/nobel-prize/Asymmetric-organocatalysis-List-MacMillan-Nobel-Prize-Chemistry-organic-synthesis/99/web/2021/10; (c) Rouhi, A.M. (2004.). A renaissance in organocatalysis. *Chem. Eng. News* 6: 41–45; (d) Durrani, J. (2021). How organocatalysis won the Nobel prize. *Chem. World* 18: 22–27.
2 (a) Hajos, Z.G. and Parrish, D.R. (1974). *J. Org. Chem.* 39: 1615–1621; (b) Eder, U., Sauer, G., and Weichert, R. (1971). *Angew. Chem. Int. Ed.* 10: 496–497.
3 Benaglia, M. (2021). *Organocatalysis – Stereoselective Reactions and Applications in Organic Synthesis*. Berlin/Boston: De Gruyter.
4 Cozzi, F. (2006). *Adv. Synth. Catal.* 348: 1367–1390.
5 Kristensen, T.E. and Hansen, T. (2010). *Eur. J. Org. Chem.* 3179–3204.
6 Krištofíková, D., Modrocká, V., Mečiarová, M., and Šebesta, R. (2020). *ChemSusChem.* 13: 2828–2858.
7 (a) Atodiresei, I., Vila, C., and Rueping, M. (2015). *ACS Catal.* 5: 1972–1985; (b) Rodríguez-Escrich, C. and Pericas, M.A. (2019). *Chem. Rec.* 19: 1872–1890.
8 Ayats, C., Henseler, A.H., and Pericàs, M.A. (2012). *ChemSusChem.* 5: 320–325.
9 Pothanagandhi, N. and Vijayakrishna, K. (2017). *Eur. Poly. J.* 95: 785–794.
10 Angamuthu, V. and Tai, D.-F. (2015). *App. Cat. A. Gen.* 506: 254–260.
11 Kong, Y., Tan, R., Zhao, L., and Yin, D. (2013). *Green Chem.* 15: 2422–2433.
12 Chronopoulos, D.D., Kokotos, C.G., Tsakos, M. et al. (2015). *Mat. Lett.* 157: 212–214.
13 Miranda, P.O., Llanes, P., Torkian, L., and Pericàs, M.A. (2013). *Eur. J. Org. Chem.* 6254–6258.
14 Fan, X., Sayalero, S., and Pericas, M.A. (2012). *Adv. Synth. Catal.* 354: 2971–2976.
15 Szcześniak, P., Buda, S., Lefevre, L. et al. (2019). *Eur. J. Org. Chem.* 6973–6982.
16 Martin-Rapún, R., Sayalero, S., and Pericàs, M.A. (2013). *Green Chem.* 15: 3295–3301.
17 (a) Hartikka, A. and Arvidsson, P.I. (2004). *Tetrahedron Asymm.* 15: 1831–1834; (b) Cobb, A.J.A., Shaw, D.M., and Ley, S.V. (2004). *Synlett* 558–560; (c) Torri, M., Nakadai, M., Ishihara, K. et al. (2004). *Angew. Chem. Int. Ed.* 43: 1983–1986.
18 Odedra, A. and Seeberger, P.H. (2009). *Angew. Chem. Int. Ed.* 48: 2699–2702.
19 Greco, R., Caciolli, L., Zaghi, A. et al. (2016). *React. Chem. Eng.* 1: 183–193.
20 Bortolini, O., Caciolli, L., Cavazzini, A. et al. (2012). *Green Chem.* 14: 992–1000.

21 Ahrendt, K., Borths, C., and MacMillan, D. (2000). *J. Am. Soc.* 122: 4243–4244.
22 (a) Raimondi, L., Faverio, C., and Boselli, M. (2021). Chapter 4: Chiral imidazolidinones: a class of privileged organocatalysts in stereoselective organic synthesis. In: *Organocatalysis Stereoselective Reactions and Applications in Organic Synthesis* (ed. M. Benaglia), 177–196. Berlin/Boston: Walter de Gruyter GmbH; (b) Deepa and Singh, S. (2021). *Adv. Synth. Catal.* 363: 629–656.
23 Chiroli, V., Benaglia, M., Puglisi, A. et al. (2014). *Green Chem.* 16: 2798–2806.
24 Kumar, G. and Shankar, R. (2020). *ChemMedChem.* 16: 430–447.
25 Porta, R., Benaglia, M., Annunziata, R. et al. (2017). *Avd. Synth. Catal.* 359: 2375–2382.
26 Triandafillidi, I., Voutyritsa, E., and Kokotos, C. (2021). Chapter 2: Recent advances in reactions promoted by amino acids and oligopeptides. In: *Organocatalysis Stereoselective Reactions and Applications in Organic Synthesis* (ed. M. Benaglia), 29–84. Berlin/Boston: Walter de Gruyter GmbH.
27 Coeffard, V., Greck, C., Moreau, X., and Thomassigny, C. (2015). Chapter 12: Other amino acids as asymmetric organocatalysts. In: *Sustainable Catalysis: Without Metals or Other Endangered Elements*, Part 1 (ed. M. North), 297–308. Cambridge: Royal Society of Chemistry.
28 Ayats, C., Hensele, A., Dibello, E., and Pericàs, M. (2014). *ACS Catal.* 4: 3027–3033.
29 Ötvös, S., Szloszár, A., Mándity, I., and Fülöp, F. (2015). *Adv. Synth. Catal.* 257: 3671–3680.
30 Burke, A. and Hermann, G. (2021). Chapter 3: Amino-cinchona derivatives. In: *Organocatalysis Stereoselective Reactions and Applications in Organic Synthesis* (ed. M. Benaglia), 85–175. Berlin/Boston: Walter de Gruyter GmbH.
31 Porta, R., Benaglia, M., Coccia, F. et al. (2015). *Adv. Synth. Catal.* 357: 377–383.
32 Barrulas, P., Genoni, A., Benaglia, M., and Burke, A.J. (2014). *Eur. J. Org. Chem.* 7339–7342.
33 Fernandes, S., Porta, R., Barrulas, P. et al. (2016). *Molecules* 21: 1182–1190.
34 Malerich, J., Hagihara, K., and Rawal, V. (2008). *J. Am. Chem. Soc.* 130 (44): 14416–14417.
35 Kisszekelyi, P., Alammar, A., Kupai, J. et al. (2019). *J. Catal.* 371: 255–261.
36 Orlandi, M. (2021). *Basic Principles of Substrate Activation through Non-covalent Bond Interactions in Organocatalysis: Stereoselective Reactions and Applications in Organic Synthesis* (ed. M. Benaglia). Berlin/Boston: De Gruyter.
37 Osorio-Planes, L., Rodríguez-Escrich, C., and Pericàs, M.A. (2014). *Chem. Eur. J.* 20: 2367–2372.
38 Birman, V.B., Uffman, E.W., Jiang, H. et al. (2004). *J. Am. Chem. Soc.* 126: 12226–12227.
39 McLaughlin, C. and Smith, A.D. (2021). *Chem. Eur. J.* 27: 1533–1555.
40 Wang, S., Rodríguez-Escrich, C., and Pericàs, M.A. (2017). *Angew. Chem. Int. Ed.* 56: 15068–15072.

30

Synthesis of Bio-based Aliphatic Polyesters from Plant Oils by Efficient Molecular Catalysis

Kotohiro Nomura[1] and Nor Wahida Binti Awang[2]

[1] *Department of Chemistry, Tokyo Metropolitan University, Hachioji, Tokyo, 1920397, Japan*
[2] *Faculty of Applied Sciences, Universiti Teknologi MARA Sarawak Branch, 94300 Kota Samarahan, Sarawak, Malaysia*

30.1 Introduction

Development of sustainable functional polymers from renewable feedstocks has been an attractive subject in the fields of polymer chemistry, synthetic chemistry, and green sustainable chemistry [1–7]. There are many reports of chemicals derived from oxygen-rich molecular biomass (e.g. carboxylic acids, polyols, furans) by using ring-opening polymerization and condensation polymerization [5, 6, 8]. Hydrocarbon-rich molecular biomass (linseed, sunflower, soybean, castor, palm, and olive oils; vegetable oils and fatty acids) have been the most abundant and low-cost molecular biomass, obtained as fatty acids or as fatty acid methyl esters (FAME) obtained from vegetable oils by chemical modifications. Studies of bio-based advanced polyesters (exhibiting tunable mechanical properties and biodegradability) [9], in particular long chain aliphatic polyesters, have been promising semicrystalline materials as alternatives to linear polyethylene. It has been recognized that the precise polymerization technique provides new strategy and methodology for design of the macromolecular architectures [5–20].

Two major routes, condensation polymerization and acyclic diene metathesis (ADMET) polymerization and subsequent hydrogenation (Scheme 30.1), have been considered for this purpose [4, 5, 12, 14–16]. The approach of adopting (living) ring-opening polymerization (ROP) of cyclic monomers has also been considered, but is not introduced in this chapter due to limited monomer scope compared to the two methods previously mentioned. In this chapter, recent results in the synthesis of bio-based aliphatic polyesters by adopting two approaches are summarized in detail. In particular, we wish to introduce important points that must be considered for the goal of adopting these approaches.

Catalysis for a Sustainable Environment: Reactions, Processes and Applied Technologies Volume 3, First Edition. Edited by Armando J. L. Pombeiro, Manas Sutradhar, and Elisabete C. B. A. Alegria.
© 2024 John Wiley & Sons Ltd. Published 2024 by John Wiley & Sons Ltd.

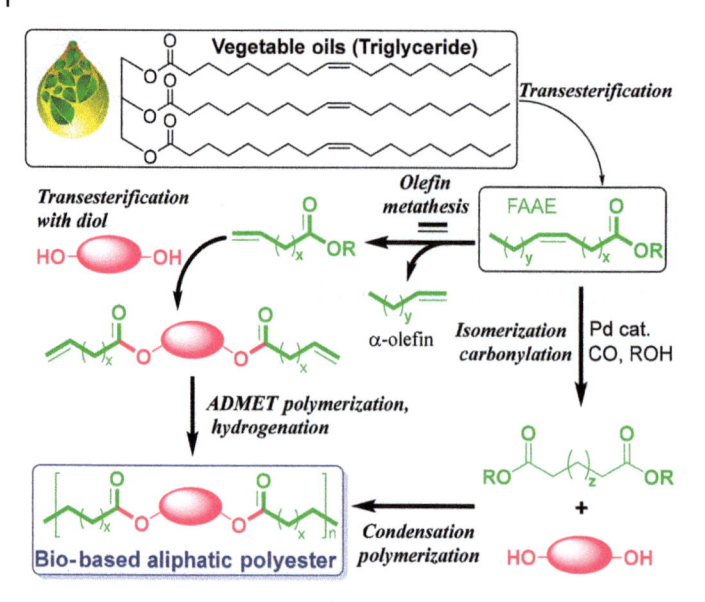

Scheme 30.1 Two approaches for synthesis of bio-based polyesters: (a) isomerization carbonylation and condensation polymerization, (b) olefin metathesis and hydrogenation.

30.2 Synthesis of Bio-Based Aliphatic Polyesters by Condensation Polymerization

Low melting temperatures (T_m values) in entirely aliphatic polyesters with short and middle chain lengths (except for short chain aliphatic polyesters such as polylactic acid and polybutylene succinate) present a problem in thermoplastic processing [15, 21, 22]: undesired softening at elevated ambient temperatures [15, 23]. This problem could be overcome by the introduction of longer, crystallizable methylene segments [15, 24, 25]. It has been known that T_m values in the polyesters in the formula, $[-O-(CH_2)_x-O-C(O)-(CH_2)_y-C(O)-]_n$, are strongly affected by methylene length (y) and whether there is an even or odd number in the diacid monomer unit [15, 23, 26]. This trend is due to the opposite/identical directions of dipoles between ester groups in close layers in the crystalline phase [15, 21, 27]; this also affects the crystalline structures. The extent of the odd–even effect, however, decreases with increasing number of methylene units and the T_m values become dependent on the total number of methylene groups and their distribution within the polymer chain [21]. T_m values higher than 100 °C (eventually close to linear polyethylene) could be achieved by increasing the percentage of long [21, 28], crystallizable methylene segments [15, 21, 25].

Polyesters are prepared by polycondensation through the esterification of dicarboxylic acids with diols or transesterification of the diesters with diols (AA, BB type monomers) or ω-hydroxy acids (AB type monomers), as shown in Scheme 30.2; the use of activated monomers (acid chlorides, anhydrides) is also considered. Synthesis of the polymers by ADMET polymerization [29–32] and subsequent hydrogenation is an alternative route [4, 9, 12, 15, 16, 33–39]. In contrast, synthesis by ring-opening polymerization (ROP), considered as an alternative route, could not be applied to large ring lactones due to their lower availability and low ring strain (considered to be the driving force) [40]. Many examples for catalyzed ROP of various lactones including large membered rings, exemplified by synthesis of high molecular weight poly(pentadecalactone) ($M_n > 150,000$; M_n = number average molecular weight) using Al-salen catalyst [41, 42], have been reported [41–45] and progress will provide possibilities in the future.

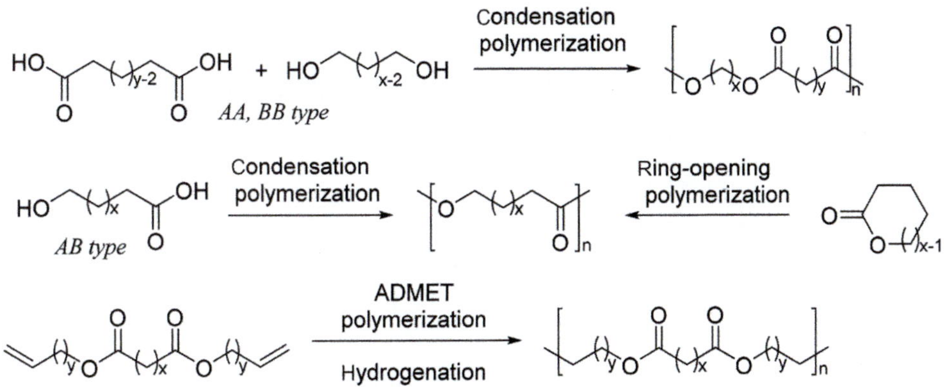

Scheme 30.2 Typical methods for the synthesis of aliphatic polyesters.

It is important to note that the precise control of molar ratios with very high conversions should be required for the synthesis of high molecular weight polymers (or, for certain applications, several 10^4 g mol^{-1}) through condensation polymerization [46, 47]. Therefore, these polymerizations are generally conducted at a high temperature for long reaction times with efficient removals of byproducts (small molecules such as water) to reach a high degree of polymerization (DP$_n$) to obtain high molecular weight polymers. For example, in the synthesis of poly(ethylene terephthalate) (PET), condensation polymerization of purified terephthalic acid with ethylene glycol (excess amount) has been conducted under reduced pressure at elevated temperatures up to 290 °C [47]. Due to the severe difficulty of removing diols owing to their high boiling points (e.g. 1,12-dodecanediol, 189 °C/12 mmHg; 1,16-hexadecane diol 197–199 °C/3 mmHg), this method cannot be considered for the synthesis of long chain aliphatic polyesters. A precise stoichiometric balance (the diol and the diacid or the diacid derivative) should be required to obtain of high molecular weight polymers. Precise control (purity) of monomers with methylene repeat units should be equally important, because we consider the odd-even effect toward the T_m values in the resultant polymers with α,ω-diacid repeat units with various chain lengths as mentioned previously [48–50]. For instance, precise stoichiometry of diols (algae oil) and diesters (C$_{17}$ and C$_{19}$) to obtain polyesters via polycondensation requires high molecular weight ($M_n = 4.0 \times 10^4$) and melting temperatures ($T_m = 99$ °C) [51].

30.2.1 Synthesis of Bio-Based Aliphatic Polyesters by Condensation Polymerization and Dehydrogenative Condensation

Pd-catalyzed isomerization methoxycarbonylation, as depicted in Scheme 30.3, yielded not only dimethyl 1,19-nonadecanedioate exclusively from methyl oleate (MO) [24], but also dimethyl 1,ω-carboxylate from unsaturated fatty acid esters (methyl oleate, methyl erucate, methyl linoleate) [24, 52]. Reduction of the resultant linear terminal C$_{19}$ or C$_{23}$ diacid esters with LiAlH$_4$ yielded corresponding diols. Long-chain linear polyesters prepared by the polycondensation in the presence of Ti(OBu)$_4$ possessed high molecular weights and high T_m values (102, 107 °C) [25]. The approach thus clearly demonstrates a possibility of synthesizing semicrystalline polyesters with high T_m values [24, 25]. However, the stochiometric reduction of diesters by LiAlH$_4$ was employed to obtain diols because the catalytic hydrogenation of the esters generally requires harsh conditions (high temperature, high H$_2$ pressure).

Direct synthesis from linear 1,ω-diols by Ru catalyst containing a pincer ligand (called Milstein catalyst) was demonstrated by Robertson et al. (Scheme 30.3) [53]. The polycondensation could be

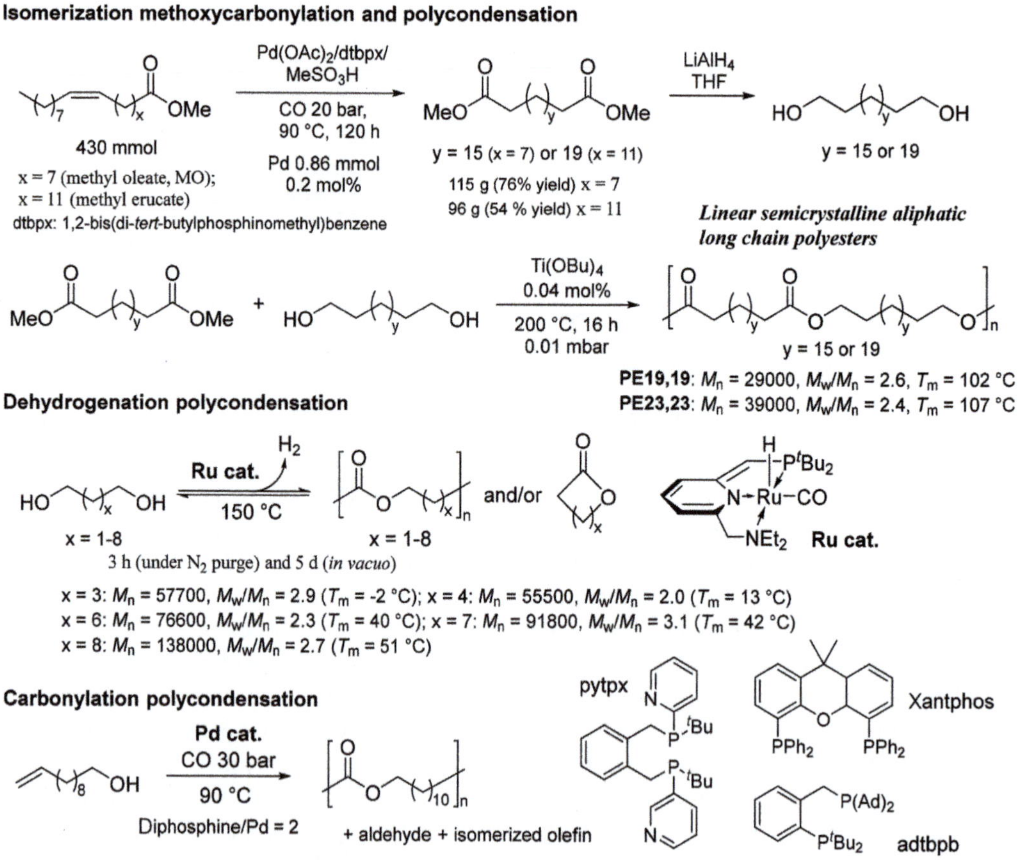

Scheme 30.3 Synthesis of aliphatic polyesters by palladium catalyzed isomerization methoxycarbonylation and polycondensation [25, 51], dehydrogenation polycondensation [53], and carbonylation polycondensation [55].

achieved by dehydrogenation of 1,ω-diols to afford hydroxy monoaldehyde, which was converted to esters by subsequent dehydrogenation from the hemiacetal formed by the addition of alcohol. The ruthenium catalyst has been known to exhibit highly efficient conversion of alcohols to esters [54]. The synthesis, however, requires rather severe conditions (150 °C for five days) for the completion and by-produced cyclic ester. The resultant semicrystalline polyesters from 1,10-decane diol or 1,9-nonane diol possessed high molecular weights (M_n = 138,000, 91,800), but rather low T_m values [53]. The direct synthesis from undecen-1-ol was demonstrated by carbonylation polymerization using a palladium catalyst containing diphosphine ligand (Scheme 30.3) [55]. The reaction in the presence of Pd-xantphos catalyst seemed to be preferred for the synthesis of high molecular weight linear polyesters (M_n = 10,300–17,400) without the formation of aldehyde and/or isomerized olefin byproducts; the phosphine ligands play the key role.

Self-metathesis of erucic acid by **G2** followed by catalytic hydrogenation gave 1,26-hexacosane-dioate, and subsequent treatment with LiAlH$_4$ yielded the diol. Polycondensation of 1,26-hexacosanedioate with the diol in the presence of Ti(OBu)$_4$ gave the polyester (PE-26.26), which possesses a T_m value of 114 °C (Scheme 30.4) [56]. The T_m value is rather high compared to that reported previously for a similar polycondensation by Meier (104 °C) [57], in which PE-26.26, PE-12.26, and PE-4.26 were prepared by polycondensation to explore the effect of a alkoxy methylene spacer on thermal properties [57]. Self metathesis of undecanoic acid followed by hydrogenation by Pd/C to

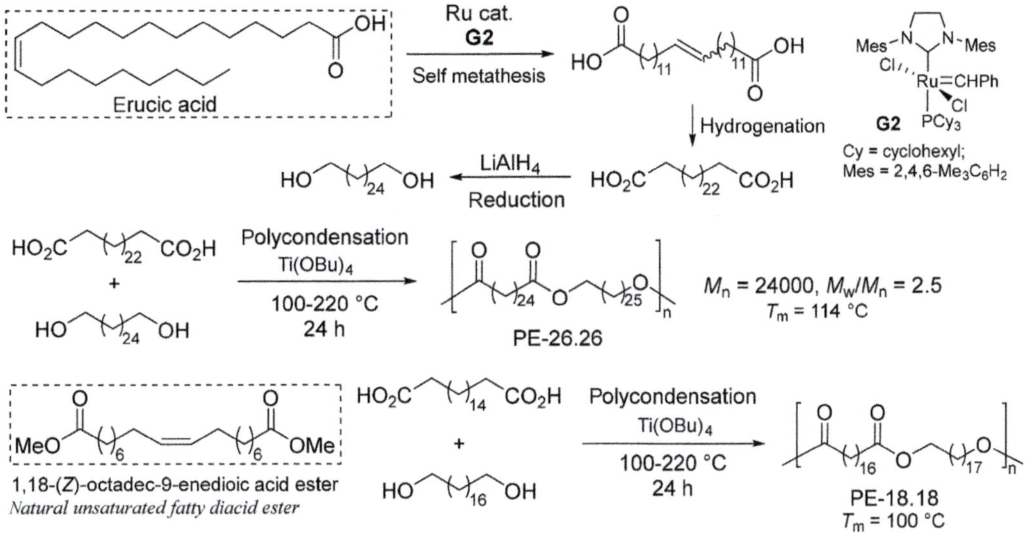

Scheme 30.4 Synthesis of aliphatic long chain polyesters [56, 58].

give 1,20-eicosanedioic acid, and reduction with LiAlH$_4$ yielded eicosane-1,20-diol. The subsequent polycondensation afforded the polyester (PE-20.20) possessing a T_m value of 108 °C [58].

Synthesis of linear polyesters based on 1,18-(Z)-octadec-9-enedioic acid, a natural unsaturated fatty diacid present in its esterified form, was reported by Roumanet et al. (Scheme 30.4) [59]. In particular, poly(1,18-octadecylene 1,18-octadecanedioate) prepared from 1,18-octadecanedioic acid and 1,18-octadecane diol possessed high melting temperature (T_m = 100 °C) and the possibility of using these polyesters as an alternative to petroleum-based synthetic polymers was suggested [59]. A biodegradation process was also highlighted for the resultant polyesters, such as poly(1,18-(Z)-octadec-9-enylene 1,18-(Z)-octadec-9-enedioate) and poly(1,18-octadecylene 1,18-octadecanedioate) [59].

More recently, Mecking et al. demonstrated closed-loop chemical recycling of polyester (Scheme 30.5) [60]. PE18,18 was prepared by the condensation polymerization of 1,18-octadeca dicarboxylic acid (derived from methyl oleate by biorefining) and 1,18-octadecane diol (prepared from the corresponding diacid ester, according to Scheme 30.3) in the presence of T(O^{n}Bu)$_4$. The condensation polymerization of the diol with diethyl carbonate (DEC) in the presence of LiH gave polycarbonate (PC18, M_n = 90,000, M_w/M_n = 2.7). The resultant PC18 was treated with 10 wt% KOH in ethanol (at 120 °C for 24 hours) to yield 1,18-octadecanediol (purity 99% after recrystallization from MeOH) exclusively (yield 98%); the reaction of PC18 with MeOH at 150 °C for 24 hours also recovered the diol (Scheme 30.5) [60]. The recovered diol was treated with DEC to give the high molecular weight PC18 (*recycled* PC18), which possessed similar properties to the virgin polymer (M_n = 70,000, M_w/M_n = 3.4) [60]. Moreover, the resultant reaction mixture from PE18,18 treated in MeOH (150 °C, 12 hours) was polymerized in the presence of Ti(O^{n}Bu)$_4$ to afford high molecular weight PE18,18 (M_n = 79,000, M_w/M_n = 1.9) [60]. These results clearly demonstrate a possibility of closed-loop chemical recycling

30.2.2 Synthesis of BioBasd Aliphatic Polyesters by Acyclic Diene Metathesis (ADMET) Polymerization and Subsequent Hydrogenation

The ADMET polymerization approach [29–32] has also been considered, and three commercially available Ru-carbene catalysts, RuCl$_2$(PCy$_3$)$_2$(CHPh) (**G1**; Cy = cyclohexyl) RuCl$_2$(PCy$_3$)(IMesH$_2$) (CHPh) [**G2**; IMesH$_2$ = 1,3-bis(2,4,6-trimethylphenyl)imidazolin-2-ylidene] and RuCl$_2$(IMesH$_2$)

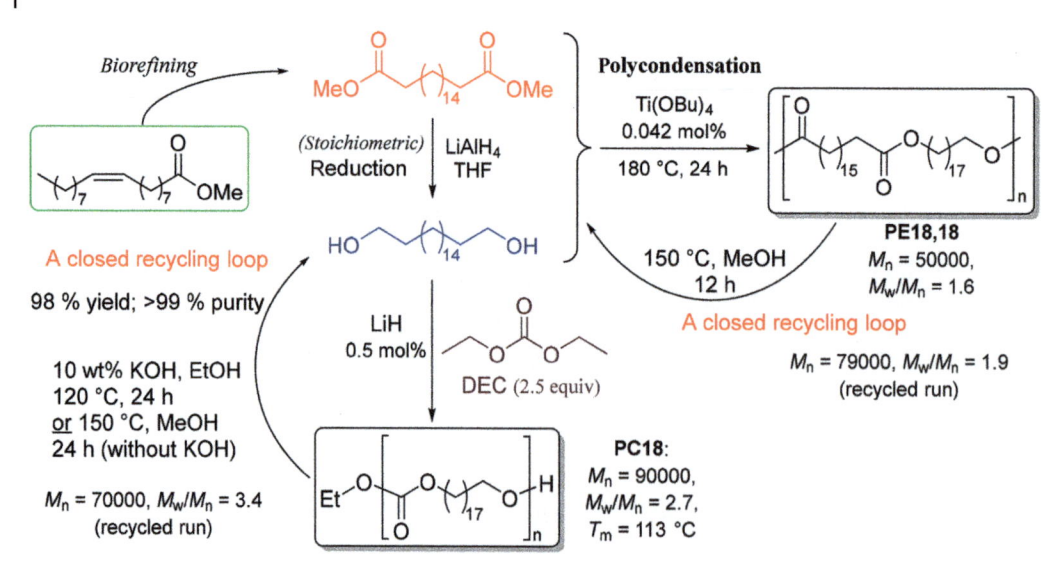

Scheme 30.5 Closed-loop chemical recycling of long chain aliphatic polyesters by polycondensation and depolymerization [60].

(CH-2-O^{i}Pr-C$_6$H$_4$) (**HG2**), shown in Scheme 30.6, have been employed (probably due to their better functional group tolerance and insensitivity to moisture) [61–66].

The polymerization of 10-undecenoic acid and 10-undecenol (**1**, derived from castor oil) in the presence of **G2** or **HG2** gave high molecular weight polymers, especially with **G2** (M_n = 22,000, 26,500). The M_n value was controlled by the addition of a mono-olefin (methyl 10-undecenoate, stearyl acrylate) used as a chain transfer agent for the end modification; polymerization in the presence of acrylate containing an oligomeric ethylene glycol segment was also studied for the synthesis of triblock copolymers [33].

The polymerization of dianhydro-D-glucityl bis(undec-10-enoate) (**2**) derived from castor oil (10-undecenoic acid) and glucose (isosorbide) was also demonstrated in the presence of **G1**, **G2** (Scheme 30.6) [34]. The M_n values in the resultant polymers, poly(**2**), were low, and the value improved by conducting the polymerization under a nitrogen purge for the removal of the ethylene byproduct (run 2 vs. run 3, run 6 vs. run 7, Table 30.1). Due to the decomposition of the catalyst, there was accompanying olefin isomerization (probably by ruthenium-hydride) [67] generated in situ; the degree was measured by gas chromatography (GC) after transesterification of poly(**2**) by treating it with MeOH/ H$_2$SO$_4$ (depolymerization by the ester cleavage). The degree of isomerization was affected by the Ru catalyst rather than the polymerization temperature or nitrogen purging during the reaction [35]. The results also demonstrated the possibility of depolymerization of poly(**2**) by the treatment of acidified methanol. The degree of the isomerization was extensively reduced by the addition of benzoquinone during the polymerization [37]. The synthesis of aliphatic polyamides by adopting the ADMET polymerization of monomers derived from 10-undecenoic acid and 1,ω-diaminoalkane was also demonstrated [35].

Copolymerization of the linear α,ω-diene monomer (**3**) and the *n*-hexyl branched α,ω-diene monomer (**4**), derived from castor oil and vernonia oil, was conducted by **G1** at 85 °C and subsequent hydrogenation in the presence of Pd/C (10%, H$_2$ 50 bar) yielded the corresponding long chain aliphatic polyesters containing a certain percentage of branching (Scheme 30.6) [39]. The resultant polymers were considered as a mimic of linear low-density polyethylene (LLDPE) and very-low-density polyethylene (VLDPE), but the composition was not uniform confirmed by differential scanning calorimetry (DSC) thermograms (multiple T_m values) [39].

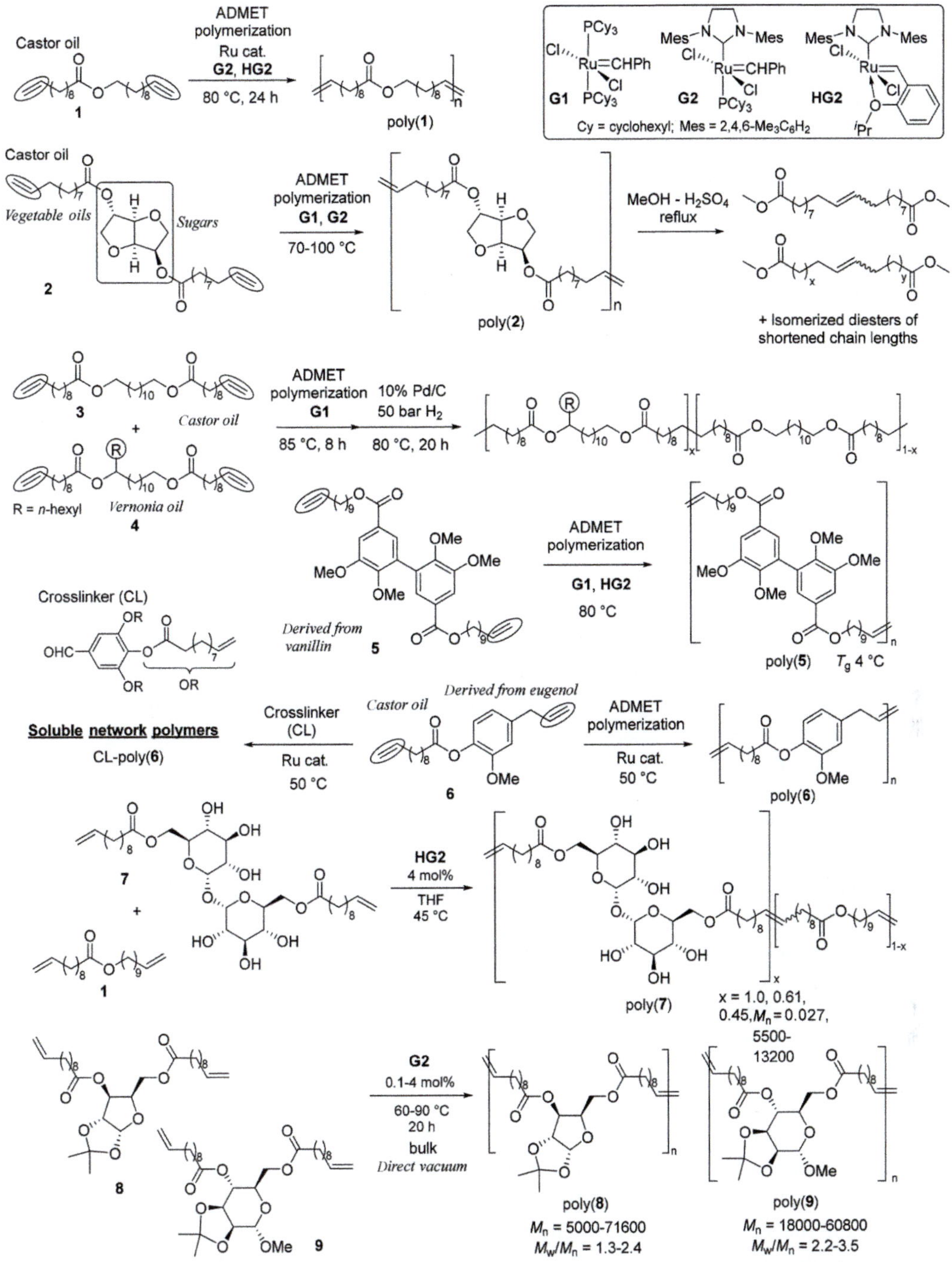

Scheme 30.6 Acyclic diene metathesis (ADMET) polymerization of monomers derived from plant oil (castor oil) and sugar [34, 39, 68, 69, 71, 73, 74].

The polymerization of α,ω-diene monomers containing phenyl units derived from vanillin (**5**) [68] and eugenol (**6**) [69] in the presence of ruthenium catalysts was also reported. The resultant polymer, denoted as poly(**5**), possessed a rather high molecular weight (M_n = 10,000, M_w/M_n = 1.6) and a T_g value of 4 °C due to its C_{12} aliphatic chain [68]. The resultant polymers from 4-allyl-2-methoxyphenyl

Table 30.1 Synthesis of aliphatic polyester, poly(2), by acyclic diene metathesis (ADMET) polymerization using Ru-carbene catalysts[1] [34].

run	cat.[1]	temp./°C	N_2 purge[2]	M_n[3]	M_w/M_n[4]	degree of isomerization[5]/ %
1	G1	70	No	4,400	1.57	3
2	G1	80	No	4,750	1.56	4
3	G1	80	Yes	6,600	1.77	3
4	G1	100	No	5,000	1.61	42[5]
5	G2	70	No	6,000	1.71	49
6	G2	80	No	6,100	1.61	69
7	G2	80	Yes	8,400	1.75	76
8	G2	90	No	6,200	1.65	66

1) Conditions: 1.0 mol% Ru cat, five hours.
2) Continuous nitrogen purging during polymerization.
3) Gel permeation chromatography (GPC) in tetrahydrofuran (THF) vs. polystyrene standards.
4) Percentage of isomerized diesters observed with gas chromatography-mass spectrometry (GC-MS) after transesterification.
5) Unidentified side-products.

10-undecenoate (**6**) by **G2** possessed high molecular weight with unimodal molecular weight distribution [M_n = 10,300–12,700, M_w/M_n = 1.64–1.97] as well as uniform composition as amorphous materials based on DSC thermogram (T_g −9.6 °C) [69]. The reaction conditions (e.g. type of catalyst, amount, time) were optimized to obtain high molecular weight polymers, and conducting the polymerization using **G2** (1.5–2.0 mol%) at 50 °C seemed to be preferred [69]. The polymerization of **6** in the presence of 5-formylbenzene-1,2,3-triyl tris(undec-10-enoate) as a crosslinker gave rather high molecular weight polymers (M_n = 13,600, M_w/M_n = 2.3) and formation of soluble network polymers was suggested from the ^{1}H NMR spectra based on the disappearance of resonances assigned to terminal olefins [69]. These polymerizations were conducted at 50 °C to avoid rapid catalyst decomposition [4, 29, 70–72].

The polymerization of trehalose diundecenoate (**7**), prepared from trehalose by selective enzymatic esterification (the primary hydroxyl groups in the 6 and 6′-positions) by treatment with vinyl undecenoate, by **HG2** (4.0 mol%, 45 °C) gave high molecular weight semicrystalline material (M_n = 13,200) with a high T_m value (T_m = 156 °C, Scheme 30.6) [71, 73]. In the copolymerization with undec-10-en-1-yl undec-10-enoate (**1**), both the M_n and the T_m values decreased upon increasing the percentage of **1** in the resultant copolymer [71, 73]. The ADMET polymerization of 1,ω-dienes (**8,9**), prepared from D-xylose and D-mannose and a castor oil, by **G2** under bulk conditions (no solvent) gave high molecular weight polymers [poly(**8**), Ru 0.1 mol%, M_n = 7.14–7.16 × 10^4, M_w/M_n = 2.2–2.3; poly(**9**), M_n = 3.24 × 10^4, M_w/M_n = 2.4, Scheme 30.6] [74]. The resultant polymers were amorphous materials and the M_n value was affected by the reaction temperature and the monomer/Ru molar ratios; conducting the polymerization at 90 °C with low Ru loading (0.1 mol%) seemed preferable to obtain high molecular weight polymers [74].

The polymerizations of symmetrical 1,ω-dienes (**2,10**) esters with isosorbide and glucarodilactone, respectively, by **G2** (1.0 mol%), were conducted (80 °C, 16 hours) in the presence of methyl-10-undecenoate as the end-capping (chain transfer) reagent [75]. Synthesis of the copolymers, poly(**2**-*co*-**10**)s, with various molar ratios were also explored [75]. Poly(**2**) is thermally stable and exhibits a lower glass transition temperature (T_d = 369 °C, T_g = -10 °C) than poly(**10**) (T_d = 206 °C, T_g = 32 °C) or poly(**2**-*co*-**10**)s. The copolymers featured a rubbery material that possessed a low

Young's modulus with an average elongation at break of 480% and 640%, respectively, and shape memory properties [75]. The polymerization of 2,5-bis(hydroxymethylfuran) undecenoate (**11**) and the copolymerizations with **2** and **10** were also explored (Scheme 30.7) [76]. Replacement of **2** with **11** led to a significant loss of the elasticity and shape memory in the copolymers; these suggest that **2** (isosorbide as the middle) plays a role in mechanical performance [76].

It has been known that T_m value in the aliphatic polyesters [-(CH_2)$_n$-CO-O-]$_m$ is affected by number of the methylene unit (n). Copolymerization of undec-10-en-1-yl undec-10-enoate (**1**) and undeca-1,10-diene (UDD) with various molar ratios in the presence of Ru catalyst (**G1**) gave polyesters (M_n = 7,000–10,300). Subsequent olefin hydrogenation by another Ru catalyst (H_2 40 bar, 110 °C, two days) afforded various long chain aliphatic polyesters (Scheme 30.8) and placed ester groups in different methylene spacing units [38]. As shown in Figure 30.1, a linear relationship between the T_m values and number of ester groups per methylene units differed from those in polyesters with precisely controlled methylene units. The results are due to a random distribution of the ester group in the polyethylene chain; the T_m values decreased linearly with increases in the mole fractions of ester groups [38]. A similar relationship was observed in the hydrogenated copolymers of dianhydro-D-glucityl bis(undec-10-enoate) (**2**) with 1,9-decadiene, prepared by tandem ADMET copolymerization and subsequent hydrogenation (Scheme 30.8) [72].

The polymerizations of di(icos-19-en-1-yl)tricosanedioate (**12a**) and di(tricos-22-en-1-yl)tricosanedioate (**12b**) by **G1** and following hydrogenation using another Ru catalyst gave the corresponding polyesters (PE-38.23, and PE-44.23 (Scheme 30.9) [56]. As shown in Figure 30.2, a linear

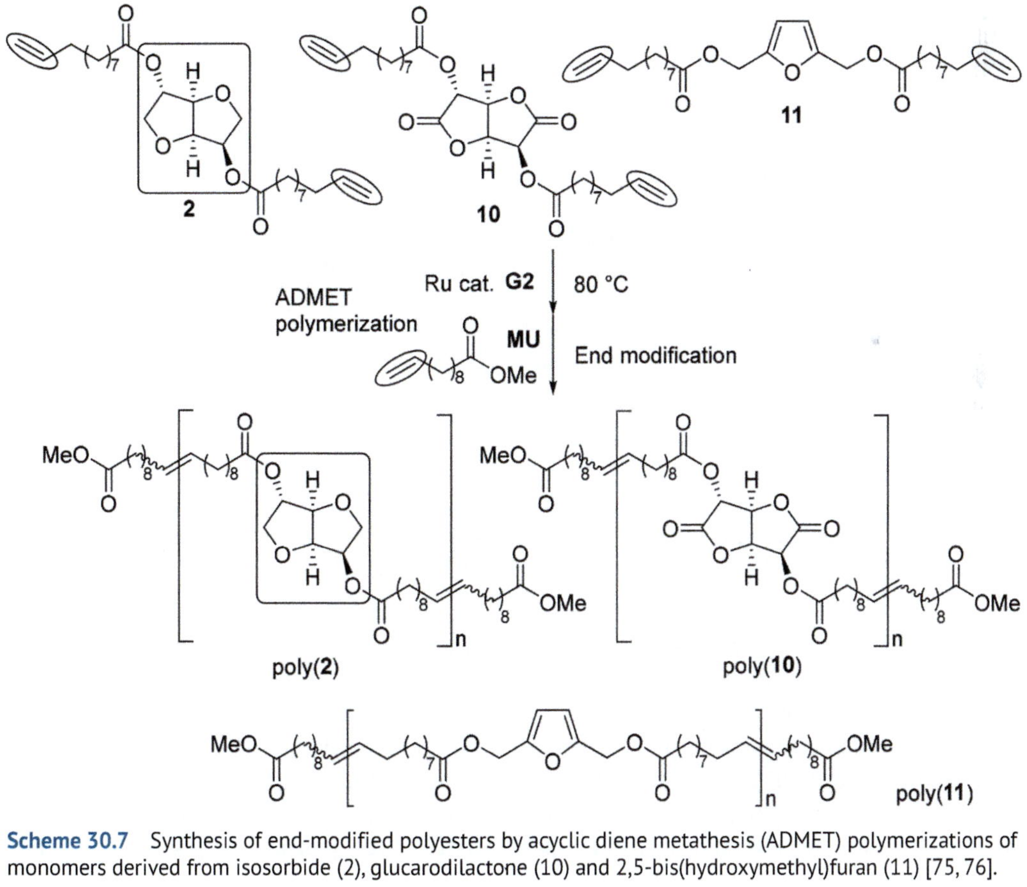

Scheme 30.7 Synthesis of end-modified polyesters by acyclic diene metathesis (ADMET) polymerizations of monomers derived from isosorbide (**2**), glucarodilactone (**10**) and 2,5-bis(hydroxymethyl)furan (**11**) [75, 76].

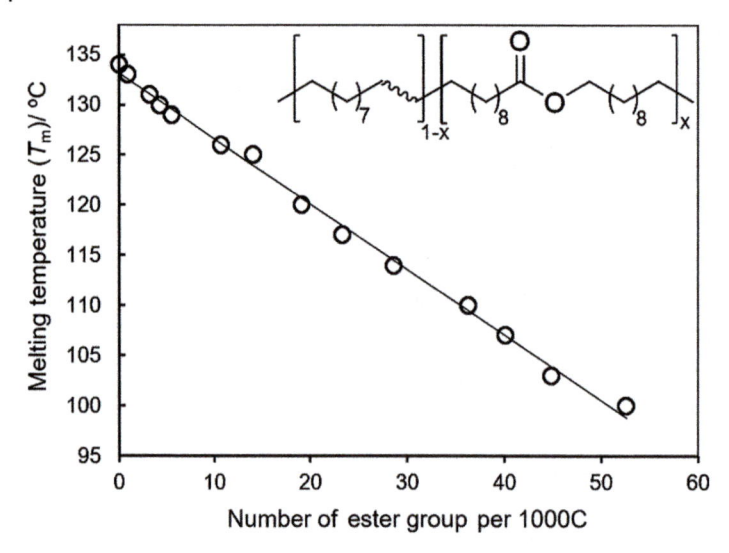

Figure 30.1 Plots of T_m value vs. number of ester groups per methylene units [38].

Scheme 30.8 Synthesis of aliphatic polyesters by acyclic diene metathesis (ADMET) copolymerization undec-10-en-1-yl undec-10-enoate (1) with undeca-1,10-diene (UDD) or dianhydro-D-glucityl bis(undec-10-enoate) (2) with 1,9-decadiene (DCD) and subsequent hydrogenation [38, 72].

relationship between the T_m value and the number of ester groups per methylene unit in [-O-(CH$_2$)$_x$-O-C(O)-(CH$_2$)$_{21}$-C(O)-]$_n$ (PE-23.x), prepared by ADMET polymerization and the subsequent hydrogenation or condensation polymerization, was demonstrated [56, 58, 77]. Moreover, the T_m value in (PE-23.x) reached a constant value upon increasing number of methylene units (x) [56]. In contrast, the resultant polymer (PE-20.20) from undec-10-en-1-yl undec-10-enoate (**9**) by ADMET polymerization and subsequent hydrogenation possessed a T_m value of 103 °C (M_n = 28,000, M_w/M_n = 1.9) [58], which was rather low compared to the polyester (PE-20.20) prepared by polycondensation

30.2 Synthesis of Bio-Based Aliphatic Polyesters by Condensation Polymerization | 669

Scheme 30.9 Synthesis of aliphatic long chain polyester prepared by acyclic diene metathesis (ADMET) polymerization and subsequent hydrogenation [38, 39, 56, 58].

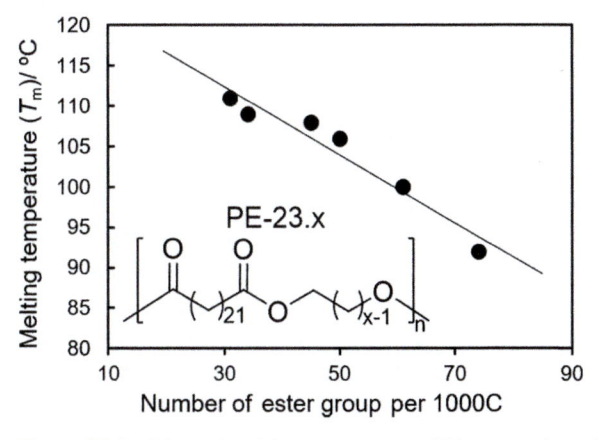

Figure 30.2 Plots of melting temperature (T_m) vs. number of ester groups per methylene unit in [-O-(CH$_2$)$_x$-O-C(O)-(CH$_2$)$_{21}$-C(O)-]$_n$ (PE-23.x) [56].

of 1,20-eicosanedioic acid and eicosane-1,20-diol ($T_m = 108\,°C$). The results thus suggest that, as described [15], precise control of the microstructure plays the role.

As described previously, internal olefinic double bonds in the resultant polymers prepared by ADMET polymerization must be hydrogenated for synthesis of saturated aliphatic bio-based polyesters, possessing higher T_m value (semicrystalline materials) as well as better stability [14–16, 24, 25, 28, 33–39, 68–70, 75, 77–80]. However, as exemplified in Schemes 30.8 and 30.9 [38, 39, 56, 58], the hydrogenation generally requires harsh conditions [81–88]. The catalytic hydrogenations (after isolation of unsaturated polymers) were carried out in the presence of RuCl$_2$(PCy$_3$)$_2$(CHOEt) (4.0 MPa, 110 °C, 48 hours) [38, 56] or 10 wt% of Pd/C (H$_2$ 5.0 MPa, 80 °C, 20–72 hours) [39, 58]. Moreover, the hydrogenations using RhCl(PPh$_3$)$_3$ [84, 85], RuHCl(CO)(PCy$_3$)$_2$ [86], or Pd/C [87] under high pressure (3.1–13.8 MPa, 80–90 °C, two to five days), or Ir(COD)(PCy$_3$)(pyridine) (COD = 1,5-cyclooctadiene, called Crabtree's catalyst) under rather mild conditions (c. Ir 2 mol%, c. H$_2$ 2 MPa at room temperature,

20 hours) [88] were also reported for synthesis of saturated polymers after the ADMET polymerization. p-Toluenesulfonyl hydrazide has been widely used in the laboratory level for this purpose [81–83], even with polymers prepared by ring opening metathesis polymerization [89, 90]. No reports concerning the tandem systems (hydrogenation without isolation of unsaturated polymers especially after ADMET polymerization) were studied for the synthesis of the saturated polyesters until recently.

30.2.3 One Pot Synthesis of Bio-Based Long Chain Aliphatic Polyesters by Tandem ADMET Polymerization and Hydrogenation. Depolymerization by Reaction with Ethylene

Recently, one pot synthesis of rather high molecular weight polymers by ADMET polymerization of monomers (**2, 13–15**) derived from castor oil and sugars using **G2** or **HG2** and the subsequent tandem hydrogenation was demonstrated (Scheme 30.10). The polymerization temperature (50 °C) was lower than those conducted previously (80–100 °C) [70] to prevent the rapid decomposition of the catalyst [91–94]. Instead, the ethylene byproduct produced during the polymerization was removed from the reaction medium [32, 70]. The resultant polymers, poly(**2**), possessed higher M_n values (M_n = 11,900–14,000) [70] than those reported previously (M_n = 4,400–8,400) [34]; the polymerization by **HG2** gave higher molecular weight polymers compared to those prepared by **G2** (Table 30.2). No significant differences in the M_n values were observed irrespective of the reaction scale [poly(**2**): M_n = 15,900 (300 mg scale) vs. 15,800 (1.0 g scale)] [70]. Repetitive removal of ethylene should thus be a prerequisite for obtaining high molecular weight polymers; the removal of ethylene by replacement of solvent was also effective for this purpose (with higher M_n values) [72].

The subsequent olefin hydrogenation was conducted under mild conditions (H_2 1.0 MPa, 50 °C) with the addition of a small amount of Al_2O_3 into the reaction mixture after the polymerization. Addition of Al_2O_3 in a small amount (1.0–1.7 wt %) is necessary for the one pot synthesis, and no significant differences in the M_n and the M_w/M_n values were seen before/after hydrogenation (Table 30.2) [70]. Moreover, the hydrogenation should be checked by DSC thermograms, not by ^{1}H NMR spectra, because confirmation of complete hydrogenation seemed difficult due to insufficient accuracy in the ^{1}H NMR spectra. Only one (rather sharp) T_m value was observed after completion of the hydrogenation, because (as previously described), melting temperatures of these polyesters are sensitive toward microstructure.

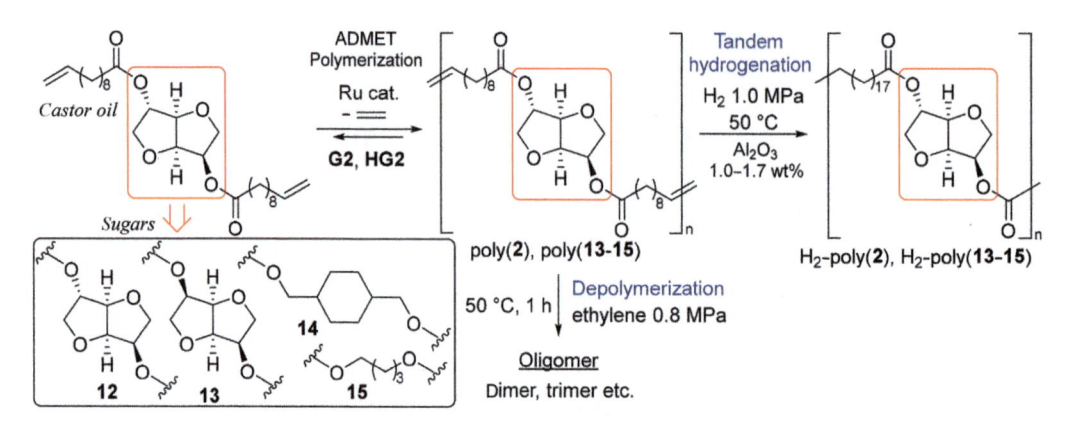

Scheme 30.10 Synthesis of bio-based polyesters by Ru-catalyzed acyclic diene metathesis (ADMET) polymerization and tandem hydrogenation, and reaction with ethylene after the polymerization [70].

Table 30.2 Synthesis of bio-based polyesters by tandem acrylic diene metathesis (ADMET) polymerization and hydrogenation [70].[1]

monomer (mmol)	cat.	Before H_2[1] M_n[3]	M_w/M_n[3]	after H_2[4] M_n[3]	M_w/M_n[5]	""yield[5] / %
2 (0.65)	HG2	14,000	1.42	15900	1.44	87
2 (0.65)	G2	11,900	1.38	13100	1.28	85
2 (2.08)	HG2	13,700	1.48	15800	1.53	93
13 (0.65)	HG2	10,200	1.38	11100	1.39	92
13 (2.08)	HG2	10,900	1.46	11,000	1.53	92
14 (2.09)	HG2	8,600	1.48	9,200	1.51	92
15 (0.71)	HG2	12,800	1.41	13,800	1.45	88
15 (2.36)	HG2	16,400	1.51	16,600	1.48	90

1) Polymerization conditions: Ru cat. 2.0 mol %, monomer 300 mg (in $CHCl_3$ 0.14 mL) or 1000 mg (in $CHCl_3$ 0.34 or 0.38 mL), 50 °C, 24 hours.
2) Sample before hydrogenation.
3) Gel permeation chromatography (GPC) data in tetrahydrofuran (THF) vs. polystyrene standards.
4) Sample after hydrogenation (H_2 1.0 M pa, 50 °C, three hours, Al_2O_3 1.0–1.7 wt %).
5) Isolated polymer yield.

It was revealed that treatment of the reaction solution, after the polymerization, with ethylene (0.8 MPa, 50 °C, one hour) gave oligomers [70]. This means that reaction of unsaturated olefinic double bonds in the polymers with ethylene led to the depolymerization, because the polycondensation proceeds by the removal of ethylene (or propylene) [29–32]. Resonances ascribed to protons/carbons in the terminal olefins were observed (in the ^{1}H and ^{13}C NMR spectra) for the sample after the reaction, whereas the other resonances were retained. The major product after the reaction would be dimers (including trimers and others) on the basis of the integration ratio of internal/terminal olefins [70].

30.3 Concluding Remarks and Outlook

This chapter introduces recent results focused on the synthesis of long chain aliphatic polyesters, focused on condensation polymerization (of dicarboxylic acids with diols) and ADMET polymerization. For monomer synthesis, the catalyst developments for key technologies such as isomerization alkoxycarbonylation, hydrogenation of esters, and others [95] play an essential role for success. Catalyst development for ADMET polymerization, in terms of activity, for synthesis of high molecular weight polymers should also be important. The monomer/polymer design, including end modification for purposes such as making blocks, grafting, and controlled networks, through new methodologies, should be important. Moreover, closed-loop chemical recycling (and upcycling) has been demonstrated recently (Scheme 30.11) [4, 60, 96]; we thus highly that believe extensive progress in these and related projects will be introduced in the near future.

Scheme 30.11 Mechanical recycling and chemical recycling of (bio-based) polyesters [4, 60, 95].

References

1. Gandini, A. (2008). *Macromolecules* 41: 9491–9504.
2. Coates, G.W. and Hillmyer, M.A. (2009). *Macromolecules* 42: 7987–7989.
3. Mülhaupt, R. (2013). *Macromol. Chem. Phys.* 214: 159–174.
4. Nomura, K. and Awang, N.W. (2020). *ACS Sustainable Chem. Eng.* 9: 5486–5505.
5. Gandini, A. and Lacerda, T.M. (2019). Monomers and polymers from chemically modified plant oils and their fatty acids. In: *Polymers from Plant Oils*, 2e. (eds. A. Gandini and T.M. Lacerda), 33–82. Beverly: Scrivener Publishing, LLC.
6. Yao, K. and Tang, C. (2013). *Macromolecules* 46: 1689–1712.
7. Wang, Z., Yuan, L., and Tang, C. (2017). *Acc. Chem. Res.* 50: 1762–1773.
8. Coulembier, O., Degée, P., Hedrick, J.L., and Dubois, P. (2006). *Prog. Polym. Sci.* 31: 723–747.
9. Meier, M.A.R., Metzger, J.O., and Schubert, U.S. (2007). *Chem. Soc. Rev.* 36: 1788–1802.
10. Hillmyer, M.A. and Tolman, W.B. (2014). *Acc. Chem. Res.* 47: 2390–2396.
11. Fortman, D.J., Brutman, J.P., De Hoe, G.X. et al. (2018). *ACS Sustainable Chem. Eng.* 6: 11145–11159.
12. Mutlu, H. and Meier, M.A.R. (2010). *Eur. J. Lipid Sci. Technol.* 112: 10–30.
13. Xia, Y. and Larock, R.C. (2010). *Green Chem.* 12: 1893–1909.
14. Biermann, U., Bornscheuer, U., Meier, M.A.R. et al. (2011). *Angew. Chem. Int. Ed.* 50: 3854–3871.
15. Stempfle, F., Ortmann, P., and Mecking, S. (2016). *Chem. Rev.* 116: 4597–4641.
16. Gandini, A. and Lacerda, T.M. (2019). Metathesis reactions applied to plant oils and polymers derived from the ensuing products. In: *Polymers from Plant Oils*, 2e. (eds. A. Gandini and T.M. Lacerda), 83–108. Beverly: Scrivener Publishing, LLC.
17. Zakzeski, J., Bruijnincx, P.C.A., Jongerius, A.L., and Weckhuysen, B.M. (2010). *Chem. Rev.* 110: 3552–3599.
18. Fenouillot, F., Rousseau, A., Colomines, G. et al. (2010). *Prog. Polym. Sci.* 35: 578–622.
19. Wilbon, P.A., Chu, F., and Tang, C. (2013). *Macromol. Rapid Comm.* 34: 8–37.
20. Thomsett, M.R., Storr, T.E., Monaphan, O.R. et al. (2016). *Green Mater.* 4: 115–134.
21. Korshak, V.V. and Vinogradova, S.V. (1965). *Polyesters*. Oxford, U.K: Pergamon Press.
22. Mandelkern, L. and Alamo, R.G. (2007). Thermodynamic quantities governing melting. In: *Physical Properties of Polymers Handbook* (ed. J.E. Mark), 165–186. New York: Springer.
23. Ishioka, R., Kitakuni, E., and Ichikawa, Y. (2002). Aliphatic polyesters: "Bionolle". In: *Biopolymers*, 4. (eds. A. Steinbüchel and Y. Doi), 275–297. Weinheim, Germany: Wiley-VCH.

24 Quinzler, D. and Mecking, S. (2010). *Angew. Chem. Int. Ed.* 49: 4306–4308.
25 Stempfle, F., Ritter, B.S., Mülhaupt, R., and Mecking, S. (2014). *Green Chem.* 16: 2008–2014.
26 Korshak, W.V. and Vinogradova, S.V. (1953). *Bull. Acad. Sci. USSR, Div. Chem. Sci.* 2: 995–998.
27 Bunn, C.W. (1955). *J. Polym. Sci.* 16: 323–343.
28 Moser, B.R., Vermillion, K.E., Banks, B.N., and Doll, K.M. (2020). *J. Am. Oil Chem. Soc.* 97: 517–530.
29 Selected reviews for acyclic diene metathesis (ADMET) polymerization, see refs. 29-32
Schwendeman, J.E., Church, A.C., and Wagener, K.B. (2002). *Adv. Synth. Catal.* 344: 597–613.
30 Atallah, P., Wagener, K.B., and Schulz, M.D. (2013). *Macromolecules* 46: 4735–4741.
31 Pribyl, J., Wagener, K.B., and Rojas, G. (2021). *Mater. Chem. Front.* 5: 14–43.
32 Chen, Y., Abdellatif, M.M., and Nomura, K. (2018). *Tetrahedron* 74: 619–692.
33 Rybak, A. and Meier, M.A.R. (2008). *ChemSusChem* 1: 542–547.
34 Fokou, P.A. and Meier, M.A.R. (2009). *J. Am. Chem. Soc.* 131: 1664–1665.
35 Mutlu, H. and Meier, M.A.R. (2009). *Macromol. Chem. Phys.* 210: 1019–1025.
36 De Espinosa, L.M., Ronda, J.C., Galià, M. et al. (2009). *J. Polym. Sci.: Part A: Polym. Chem.* 47: 5760–5771.
37 Fokou, P.A. and Meier, M.A.R. (2010). *Macromol. Rapid Commun.* 31: 368–373.
38 Ortmann, P. and Mecking, S. (2013). *Macromolecules* 46: 7213–7218.
39 Lebarbé, T., Neqal, M., Grau, E. et al. (2014). *Green Chem.* 16: 1755–1758.
40 Dubois, P., Coulembier, O., and Raquez, J.-M. (2009). *Handbook of Ring-Opening Polymerization*. Weinheim, Germany: Wiley-VCH.
41 Van der Meulen, I., Gubbels, E., Huijser, S. et al. (2011). *Macromolecules* 44: 4301–4305.
42 Pepels, M.P.F., Koeken, R.A.C., van der Linden, S.J.J. et al. (2015). *Macromolecules* 48: 4779–4792.
43 Pepels, M.P.F., Bouyahyi, M., Heise, A., and Duchateau, R. (2013). *Macromolecules* 46: 4324–4334.
44 Bouyahyi, M. and Duchateau, R. (2014). *Macromolecules* 47: 517–524.
45 Witt, T. and Mecking, S. (2013). *Green Chem.* 15: 2361–2364.
46 Köpnick, H., Schmidt, M., Brügging, W. et al. (2000). Polyesters. In: *Ullmann's Encyclopedia of Industrial Chemistry* (eds. W. Gerhartz and B. Elvers). Weinheim, Germany: Wiley-VCH.
47 Rogers, M.E. and Long, T.E. (2003). *Synthetic Methods in Step-growth Polymers*. Hoboken: Wiley-Interscience.
48 Pepels, M.P.F., Hansen, M.R., Goossens, H., and Duchateau, R. (2013). *Macromolecules* 46: 7668–7677.
49 Le Fevere de Ten Hove, C., Penelle, J., Ivanov, D.A., and Jonas, A.M. (2004). *Nat. Mater.* 3: 33–37.
50 Menges, M.G., Penelle, J., Le Fevere de Ten Hove, C. et al. (2007). *Macromolecules* 40: 8714–8725.
51 Roesle, P., Stempfle, F., Hess, S.K. et al. (2014). *Angew. Chem. Int. Ed.* 53: 6800–6804.
52 Jiménez-Rodriguez, C., Eastham, G.R., and Cole-Hamilton, D.J. (2005). *Inorg. Chem. Commun.* 8: 878–881.
53 Hunsicker, D.M., Dauphinais, B.C., Mc Ilrath, S.P., and Robertson, N.J. (2012). *Macromol. Rapid Commun.* 33: 232–236.
54 Gunanathan, C. and Milstein, D. (2011). *Acc. Chem. Res.* 44: 588–602.
55 Liu, Y. and Mecking, S. (2019). *Angew. Chem. Int. Ed.* 58: 3346–3350.
56 Stempfle, F., Ortmann, P., and Mecking, S. (2013). *Macromol. Rapid Commun.* 34 (1): 47–50.
57 Vilela, C., Silvestre, A.J.D., and Meier, M.A.R. (2012). *Macromol. Chem. Phys.* 213: 2220–2227.
58 Trzaskowski, J., Quinzler, D., Bährle, C., and Mecking, S. (2011). *Macromol. Rapid Commun.* 32: 1352–1356.
59 Roumanet, P.-J., Jarroux, N., Goujard, L. et al. (2020). *ACS Sustainable Chem. Eng.* 8: 16853–16860.
60 Häußler, M., Eck, M., Rothauer, D., and Mecking, S. (2021). *Nature* 590: 423–427.
61 Grubbs, R.H. (ed.) (2003). Books 61–63. In: *Handbook of Metathesis, 1e* Weinheim: Wiley-VCH.

62 Grela, K. (ed.) (2014). *Olefin Metathesis: Theory and Practice*. Hoboken, New Jersey, USA: John Wiley & Sons, Inc.

63 Grubbs, R.H., Wenzel, A.G., O'Leary, D.J., and Khosravi, E. (eds.) (2015). *Handbook of Metathesis*, 2e. Weinheim: Wiley-VCH.

64 Trnka, T.M. and Grubbs, R.H. (2001). *Acc. Chem. Res.* 34: 18–29.

65 Samojzowicz, C., Bieniek, M., and Grela, K. (2009). *Chem. Rev.* 109: 3708–3742.

66 Vougioukalakis, G. and Grubbs, R.H. (2010). *Chem. Rev.* 110: 1746–1787.

67 Schmidt, B. (2004). *Eur. J. Org. Chem.* 2009: 1865–1880.

68 Llevot, A., Grau, E., Carlotti, S. et al. (2015). *Polym. Chem.* 6: 7693–7700.

69 Le, D., Samart, C., Kongparakul, S., and Nomura, K. (2019). *RSC Adv.* 9: 10245–10252.

70 Nomura, K., Chaijaroen, P., and Abdellatif, M.M. (2020). *ACS Omega* 5: 18301–18312.

71 Piccini, M., Leak, D.J., Chuck, C.J., and Buchard, A. (2020). *Polym. Chem.* 11: 2681–2691.

72 Kojima, M., Abdellatif, M.M., and Nomura, K. (2021). *Catalysts* 11: 1098–1106.

73 Piccini, M., Lightfoot, J., Castro, D.B., and Buchard, A. (2021). *ACS Appl. Polym. Mater.* 3: 5870–5881.

74 Hibert, G., Grau, E., Pintori, D. et al. (2017). *Polym. Chem.* 8: 3731–3739.

75 Shearouse, W.C., Lillie, L.M., Reineke, T.M., and Tolman, W.B. (2015). *ACS Macro Lett.* 4: 284–288.

76 Lillie, L.M., Tolman, W.B., and Reineke, T.M. (2017). *Polym. Chem.* 8: 3746–3754.

77 Stempfle, F., Quinzler, D., Heckler, I., and Mecking, S. (2011). *Macromolecules* 44: 4159–4166.

78 Barbara, I., Flourat, A.L., and Allais, F. (2015). *Eur. Polym. J.* 62: 236–243.

79 Dannecker, P., Biermann, U., Sink, A. et al. (2019). *Macromol. Chem. Phys.* 220: 1800400.

80 Barbiroli, G., Lorenzetti, C., Berti, C. et al. (2003). *Eur. Polym. J.* 39: 655–661.

81 Boz, E., Nemeth, A.J., Alamo, R.G., and Wagener, K.B. (2007). *Adv. Synth. Catal.* 349: 137–141.

82 Rojas, G., Inci, B., Wei, Y., and Wagener, K.B. (2009). *J. Am. Chem. Soc.* 131: 17376–17386.

83 Boz, E., Nemeth, A.J., Ghiviriga, I. et al. (2007). *Macromolecules* 40: 6545–6551.

84 Hydrogenation of ADMET polymer using $RhCl(PPh_3)_3$, see refs. 85,86. Inci, B., Lieberwirth, I., Steffen, W. et al. (2012). *Macromolecules* 45: 3367–3376.

85 Baughman, T.W., Chan, C.D., Winey, K.I., and Wagener, K.B. (2007). *Macromolecules* 40: 6564–6571.

86 Sworen, J.C., Smith, J.A., Berg, J.M., and Wagener, K.B. (2004). *J. Am. Chem. Soc.* 126: 11238–11246.

87 Li, H., Rojas, G., and Wagener, K.B. (2015). *ACS Macro Lett.* 4: 1225–1228.

88 Pesko, D.M., Webb, M.A., Jung, Y. et al. (2016). *Macromolecules* 49: 5244–5255.

89 Gibson, V.C. and Okada, T. (2000). *Macromolecules* 33: 655–656.

90 Hou, X. and Nomura, K. (2016). *J. Am. Chem. Soc.* 138: 11840–11849.

91 Lehman, S.E. and Wagener, K.B. (2001). *Macromolecules* 35: 48–53.

92 Courchay, F.C., Sworen, J.C., and Wagener, K.B. (2003). *Macromolecules* 36: 8231–8239.

93 Hong, S.H., Wenzel, A.G., Salguero, T.T. et al. (2007). *J. Am. Chem. Soc.* 129: 7961–7968.

94 Jawiczuk, M., Marczyk, A., and Trzaskowsk, B. (2020). *Catalysts* 10: 887.

95 For example Witt, T., Haußler, M., Kulpa, S., and Mecking, S. (2017). *Angew. Chem. Int. Ed.* 56: 7589–7594.

96 Collias, D.I., James, M., and Layman, J.M. (eds.) (2021). *Circular Economy of Polymers: Topics in Recycling Technologies*. ACS Symposium Series, 1391. Washington, DC: American Chemical Society.

31

Modern Strategies for Electron Injection by Means of Organic Photocatalysts: Beyond Metallic Reagents

Takashi Koike

Department of Applied Chemistry, Faculty of Fundamental Engineering, Nippon Institute of Technology, 4-1 Gakuendai, Miyashiro-machi Saitama, Japan

31.1 Introduction

The transfer of electrons is directly associated with the occurrence of redox reactions. More specifically, compounds that inject electrons into systems are known as reductants, or electron donors. The electron injection process by such compounds is pivotal and fundamental in the activation of inorganic metal salts and the transformation of organic compounds, as exemplified by a tremendous number of organic reactions, such as the Birch reduction, the deprotection of sulfonyl groups, and the pinacol coupling reaction. Classically, metallic reagents (e.g. lithium and samarium iodide) and electrolytic techniques have been employed to inject electrons. In addition, the tetrathiafulvalenes (TTFs) and tetraaminoethylenes (TAEs) have been recognized as strong organic electron donors in the ground state. Recently, the Murphy group has developed exquisitely designed organic electron donors, such as **1** and **2** (Figure 31.1) [1–6]. Indeed, the replacement of highly reactive and hard-to-handle alkali metal reagents and rare metals with organic molecular reagents is a significant topic in synthetic chemistry in the context of operational simplicity, versatility, and sustainability. More recently, organic photoredox catalysts (*OPCs*) exhibiting highly reducing properties have attracted great interest as a game-changing strategy for single-electron injection. In this chapter, recent advancements in organic photocatalytic systems for the reductive transformations of organic substrates are discussed.

31.2 Basic and Advanced Concepts for 1e⁻ Injection by Organic Photoredox Catalysis

In modern organic synthesis, it is essential to control the reaction selectivity. Similarly, controlling the number of electrons to be transferred is a major premise for constructing selective electron injection systems. In this context, metal-free organic photocatalytic systems have received growing attention. More specifically, over the past several years, organic photoredox catalysis has been

Catalysis for a Sustainable Environment: Reactions, Processes and Applied Technologies Volume 3, First Edition. Edited by Armando J. L. Pombeiro, Manas Sutradhar, and Elisabete C. B. A. Alegria.
© 2024 John Wiley & Sons Ltd. Published 2024 by John Wiley & Sons Ltd.

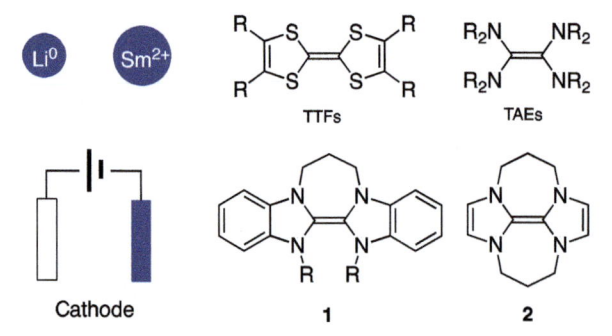

Figure 31.1 Conventional electron-injection strategies for organic synthesis.

regarded as a well-controlled single-electron transfer (SET) system (i.e. a 1e⁻ injection system) [7–10]. In addition, photoredox systems are usually performed under operationally facile reaction conditions, such as under visible-light irradiation below room temperature. Thus, Figure 31.2 outlines two concepts for the 1e injection processes taking place during photoredox catalysis. More specifically, one involves the basic concepts of oxidative quenching and reductive quenching (Figure 31.2a), whereas the other involves advanced concepts based on consecutive photoinduced electron transfer (*conPET*) (Figure 31.2b).

As shown in Figure 31.2, irradiation of the *OPC* (organic photoredox catalyst) with light ($h\nu_1$) produces the excited state of photocatalyst (**OPC*), which can initiate either a 1e⁻ oxidation or a 1e⁻ reduction process. Single-electron transfer (SET) from **OPC* to an electron-accepting (EA) substrate in the first SET event generates a reduced substrate (EA·⁻) together with the oxidized catalyst (*OPC*·⁺), which serves as a highly oxidizing agent. The second SET event involves SET from an electron-donating (ED) substrate to *OPC*·⁺, which regenerates the ground state *OPC*, and produces the oxidized substrate (ED·⁺). This process is known as the oxidative quenching pathway (**OPC*→*OPC*·⁺→*OPC*). In contrast, the process triggered by the reduction of the excited **OPC* species in the first SET event, followed by the second SET event starting from the reduced *OPC*·⁻ is also viable, and is known as the reductive quenching pathway (**OPC*→*OPC*·⁻→*OPC*). The advanced strategy known as *conPET* follows the formation of the reduced species (*OPC*·⁻) generated not only by the reductive quenching pathway (i.e. the blue pathway in Figure 31.2b), but also by means of electrochemical methods (i.e. the purple pathway in Figure 31.2b). Further photoexcitation ($h\nu_2$) generates the excited reduced species (*[*OPC*·⁻]), which serves as a super 1e⁻ reductant. Finally, following these electron transfer processes, *OPC* is regenerated (i.e. the brown pathway in Figure 31.2b). Thus, if the reaction system is designed appropriately, **OPC*, *OPC*·⁻, and *[*OPC*·⁻] can be employed as electron-injection reagents. Although transition metal complexes such as [Ru(bpy)₃]²⁺ and *fac*-[Ir(ppy)₃] (bpy: 2,2′-bipyridine, ppy: 2-pyridylphenyl) are still commonly used as photoredox catalysts [11], the development of new *OPC*s is necessary from the perspective of synthetic electron-injection strategies for sustainable development goals (SDGs).

In the following sections, we describe the recent advances in organic photoredox catalysis. In particular, the molecular design of highly reducing *OPC*s and conceptual designs of *conPET* using *OPC*s are discussed. More recently, an electron-injection system combining organic photoredox catalysis and electrolytic techniques has emerged as a more sustainable strategy, and so this will be also presented.

(a) Basic concepts for 1e⁻ injection by organic photoredox catalysis

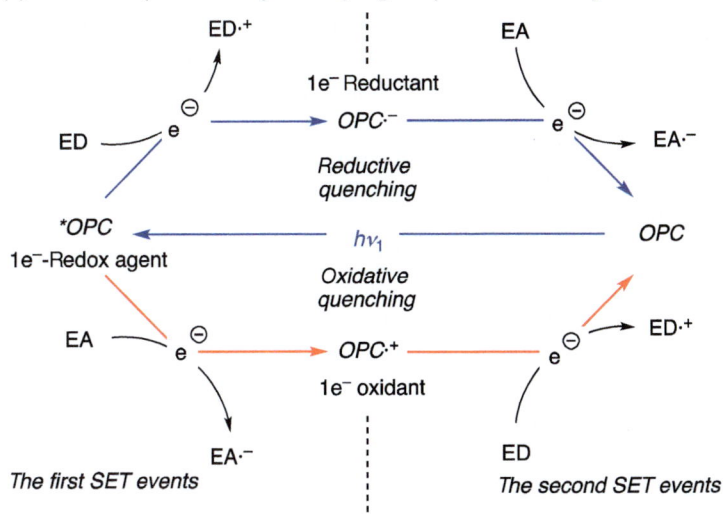

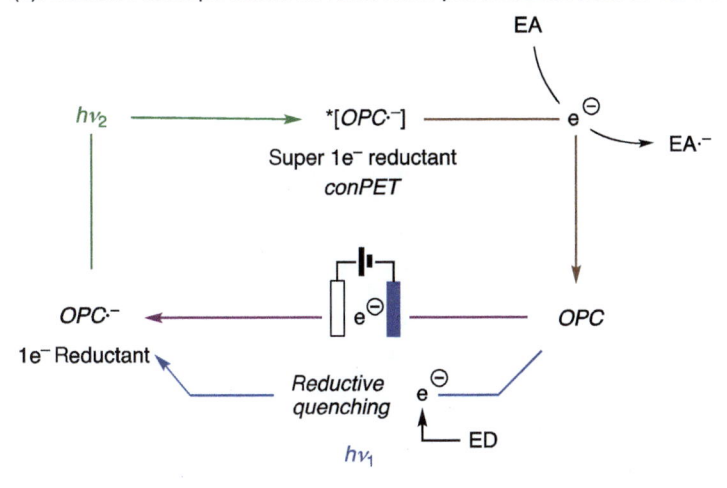

Figure 31.2 (a) Basic and (b) advanced concepts for 1e⁻ injection by organic photoredox catalysis.

31.3 Triarylamine-based Highly Reducing Organic Photocatalysts

As mentioned in the previous section, the oxidative quenching pathway is a basic electron-injection strategy that uses *OPCs*. In particular, by appropriately designing *OPCs*, electron transfer from excited species (*OPC*) to less electrophilic substrates is viable. In other words, *OPCs* exhibit a sufficiently high reducing power even in the absence of sacrificial EDs. In recent years, an increasing number of molecular scaffolds have been reported for organic photocatalysis. Thus, in this section, we discuss several representative examples of recently reported triarylamine-based *OPCs* with high reducing powers. The electron-rich triarylamine unit is a substructure that is frequently found in organic dyes, and many triarylamines exhibit reversible redox behaviors. These facts have therefore encouraged researchers to develop triarylamine-based *OPCs*. It should be noted that in the following systems, the reported overall photocatalytic performance is not always consistent with the reducing power of the photoexcited OPC ($*E_{ox}(OPC)$) because the performance is dependent on the balance of the electron injection and back-electron transfer processes (Figure 31.2). Furthermore, the durability of the catalyst, especially in the case of OPCs, is often more important than the reduction power ($*E_{ox}[OPC]$).

In 2015, the Hawker and Alaniz group reported that 10-phenylphenothiazine (PTH) serves as a highly reducing OPC ($*E_{ox} = -2.1$ V vs. SCE) for the dehalogenative hydrogen transfer reactions of carbon–halogen bonds such as in the case of alkyl and aryl halides (i.e. **3** and **4**) (Figure 31.3) [12]. Aryl bromides, aryl chlorides, and aryl iodides have been employed in the reaction, which is assumed to be triggered by the highly-reducing excited *PTH species, leading to the corresponding aliphatic and aromatic products **5** and **6**, respectively. The Hawker and Fors group has also demonstrated the application of PTH in the atom transfer radical polymerization (ATRP) of alkyl methacrylate **7** in the presence of a traditional ATRP initiator, α-bromocarbonyl **3b** (Figure 31.4) [13].

More recently, Hawker et al. and Miyake et al. extensively studied the catalytic structure–activity relationships for the ATRP process. In particular, the Miyake group conducted diverse modifications of phenothiazine (PTH), phenoxazine (POX), dihydroacridine (ACR), and dihydrophenazine (PAZ) (Figure 31.5) [14–19]. These scaffolds generally exhibit high reducing powers in their excited states (i.e. $*E_{ox} \approx -2.3$ V), and it was determined that their photochemical properties are derived from the connections between the individual scaffolds, the auxiliaries present on the aromatic rings, and the

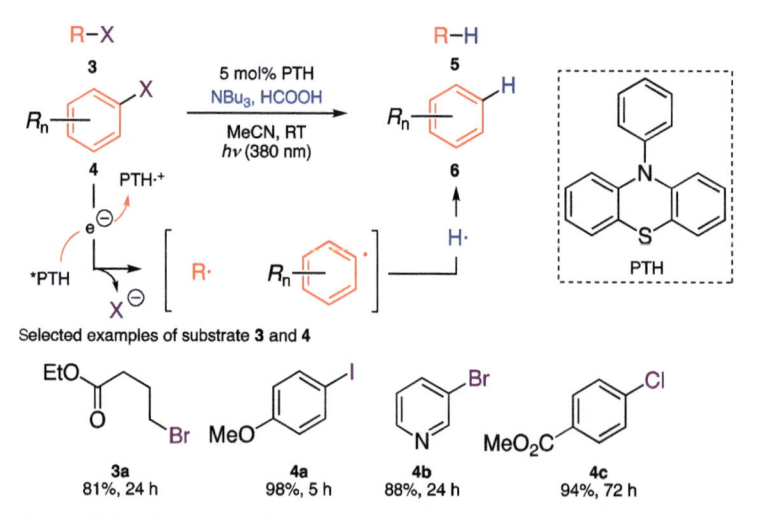

Figure 31.3 Dehalogenative hydrogen transfer of organic halides by 10-phenylphenothiazine (PTH) catalysis.

Figure 31.4 Atom transfer radical polymerization (ATRP) by means of phenylphenothiazine (PTH) catalysis.

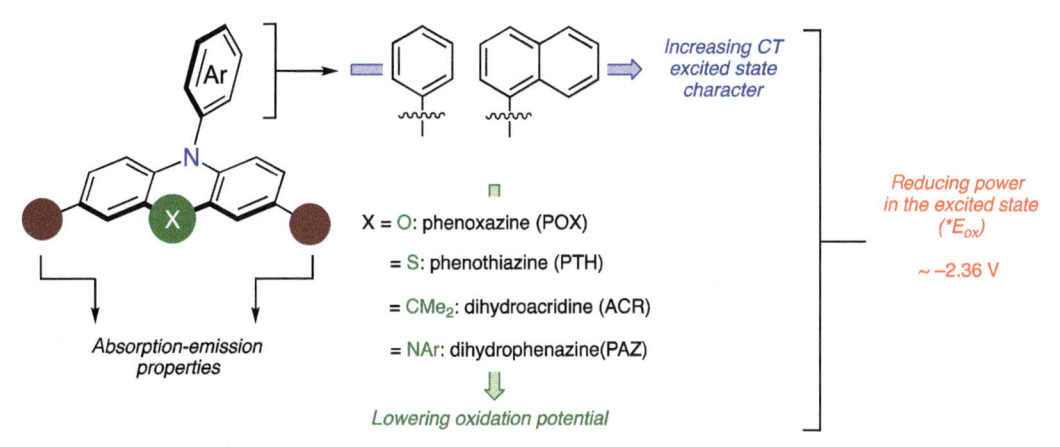

Figure 31.5 Phenothiazine derivatives for use as organic photoredox catalysts (OPCs).

incorporation of heteroatoms. More specifically, a change from a phenyl to a 1-naphthyl moiety promotes a charge transfer (CT)-type excitation. In addition, the heterocyclic components are strongly associated with determining the oxidation potential, absorption properties, and emission properties of these compounds, which can be readily tuned by variation in their substituents.

Furthermore, in 2016 and 2017, Miyake et al. reported that naph-PAZ (*E_{ox} = −2.12 V) served as a good OPC for the ATRP reaction involving α-bromocarbonyl **3b** and methyl methacrylate **7**, in addition to action in the trifluoromethylation of electron-rich heteroarene **8** with CF$_3$I **9**, leading to synthesis of CF$_3$-heteroarene **10** [14, 16]. It was deduced that the key process in these reactions involves generation of the corresponding alkyl radicals via 1e$^-$ injection from the excited OPC (*OPC) to the alkyl halides (Figure 31.6).

Over the past several decades, the catalytic fluoromethylation reaction has developed rapidly due to the importance of the fluoromethyl group as a useful structural motif, especially in the area of medicinal chemistry [20–23]. Although the radical fluoromethylation process has become one of the reliable methods for the preparation of organofluorine compounds, the radical monofluoromethylation process has yet to be reported in detail. As some limited examples, in 2019 and 2021, Koike et al. reported that 1,4-bis(diarylamino)naphthalenes (BDNs) serve as highly reducing OPCs (*$E_{ox} \approx$ −2.0 V) for the radical monofluoromethylation process (Figure 31.7) [24–26]. In particular, BDN itself is effective in the generation of monofluoromethyl radicals via 1e$^-$ injection into the much less electrophilic neutral CH$_2$F-reagent N-tosyl-S-monofluoromethyl-S-phenylsulfoximine **11**. Furthermore, the hydroxy-monofluoromethylation of aromatic alkenes **12** has also been developed. This is a synthetic method for γ-fluoroalcohols **13** from simple alkenes **12**. In general, the efficiency of photochemical reaction depends on the longevity of the excited state and the amount of photoactivated species. In the case of BDNs, despite the short lifetime of the excited state species (τ = 8–10 ns), high fluorescence quantum yields (ϕ = 0.94–0.99) were observed. The efficient formation of the excited state may therefore compensate for the short lifetime of the excited species. In addition, after the first SET event from the excited species (*BDNs) to the electron-deficient substrate (**11**), the formation of a 1e$^-$-oxidized species (i.e. BDNs$^{\cdot+}$) stabilized by the partial delocalization of a hole over the two diarylamino units is a feature of these catalytic systems. These results may therefore act to guide future catalyst design.

Although the OPCs highlighted in the present section exhibit a high reducing power (*E_{ox}) via the oxidative quenching pathway, they do not reach those obtained by alkali metals. However, the use of OPCs provides reaction systems with a high tolerance toward various functional groups, as well as air and water. Thus, the development of super-electron-injecting OPC systems remains an ongoing aim.

Figure 31.6 (a) Atom transfer radical polymerization (ATRP) and (b) the radical trifluoromethylation of electron-rich aromatics by means of naph- dihydrophenazine (PAZ) catalysis.

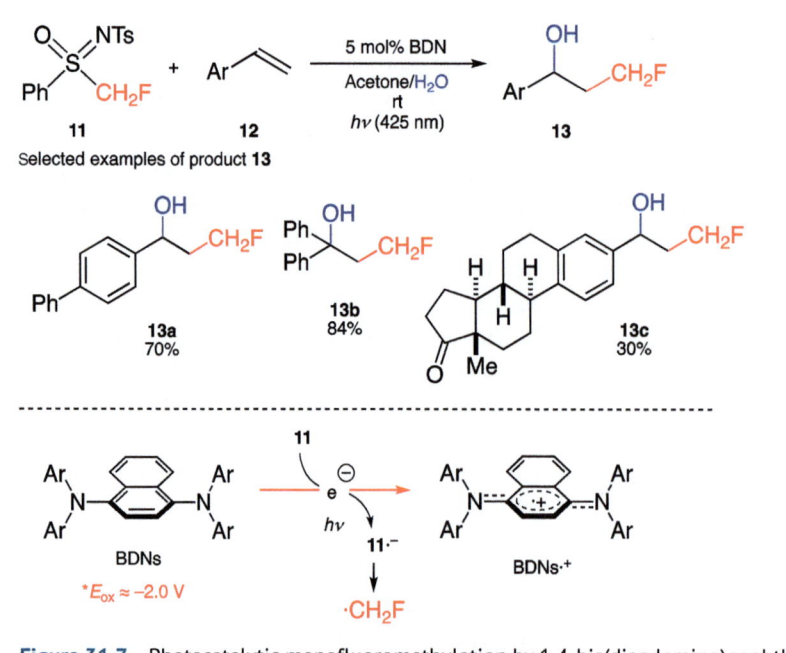

Figure 31.7 Photocatalytic monofluoromethylation by 1,4-bis(diarylamino)naphthalene (BDN) (Ar = Ph).

31.4 Consecutive Photoinduced Electron Transfer (conPET) by Organic Photoredox Catalysis

As shown in Figure 31.2b, advanced concepts, such as *conPET*, provide stronger reducing catalysts than more basic concepts [27]. As an example of a representative *conPET* system, in 2010, König and Ghosh et al. [28] investigated perylene diimides, a class of fluorescent dyes that have been employed as pigments, colorants, photoreceptors, and electronic materials. They found that N,N-bis(2,6-diisopropylphenyl)perylene-3,4,9,10-bis(dicarboximide) (PDI) was effective for the dehalogenative hydrogen transfer of aryl halides and the homolytic aromatic substitution (HAS)-type aryl–aryl cross-coupling between electron-deficient aryl halides and electron-rich heteroaromatics (Figure 31.8).

During this process, photoexcitation ($h\nu_1$) results in the excited *PDI being reductively quenched by Et$_3$N to give PDI$^{·-}$ and the radical cation of triethylamine (Et$_3$N$^{·+}$). The subsequent excitation of a 1e$^-$-reduced species (PDI$^{·-}$) is a key process here, and the ground state of PDI$^{·-}$, which possesses a relatively long lifetime, can be excited again ($h\nu_2$) to generate *[PDI$^{·-}$]. This excited species then undergoes SET to aryl halide **4**, giving the corresponding aryl radical ($^·$Ar) and regenerating ground state PDI. In addition, it was found that the second excited species (*[PDI$^{·-}$]) exhibited a higher reducing power than the 1e$^-$-reduced species (PDI$^{·-}$). It should be noted that when the aryl radical abstracts a hydrogen atom from either Et$_3$N$^{·+}$ or the solvent (S–H), the dehalogenative hydrogen transfer product (i.e. compound **6**) is obtained. In contrast, when the aryl radical reacts with the heteroaromatic moiety **14** and subsequent deprotonation takes place, coupled biaryl product **15** is obtained *via* its corresponding radical and cationic intermediates (i.e. **16** and **17**).

As shown in Figure 31.9, it has also been reported that rhodamine-6G (Rh-6G) [29], 9,10-dicyanoanthrathene (DCA), and 1,8-dihydroxyanthraquinone (Aq-OH) can act as *OPCs* for the *conPET* process [30, 31]. In these systems, the *OPCs* play the roles of 1e$^-$ oxidants and 1e$^-$ reductants, and so they are not always, effective despite their high reducing powers.

Although the short lifetime of the *OPC species in its singlet excited state may be considered a detrimental point, it is not necessary for the photoexcited species in these *conPET* systems to be particularly long-lived species if a sacrificial electron donor is present in the system at a concentration sufficient to quench the short-lived *OPC.

Although the above-mentioned systems are useful for the reaction of electron-deficient aryl halides, they are not appropriate for systems involving electron-rich aryl halides. In this context, in 2020, Nicewicz et al. overturned the concept of the established photocatalytic system based on 9-mesityl acridinium salts (Mes-Acr$^+$), which are well-known to serve as highly oxidizing catalysts (oxidizing power of *[Mes-Acr$^+$]: *$E_{red} \approx +2.2$ V) [32, 33]. They found that the photoexcited *[Mes-Acr$^·$] species generated by means of the *conPET* process could serve as a super-reducing catalyst (*$E_{ox} = -3.36$ V) to reduce tosyl amides **18** and electron-rich aryl halides **20** [34]. Importantly, the reducing power of this species was determined to be comparable to that of lithium (i.e. -3.29 V). As shown in Figure 31.10, this system is useful for the de-tosylation into N–H amines **19**, which is usually performed using alkali metals (Figure 31.10a), and the dehalogenative hydrogen transfer of **20** (Figure 31.10b). More specifically, in the case of the latter reaction, Mes-Acr$^+$ is initially excited by photoirradiation ($h\nu_1$) prior to its reductive quenching by Et$_3$N to give Mes-Acr$^·$ along with the trialkylamine radical cation (R$_3$N$^{·+}$). Subsequently, Mes-Acr$^·$ is excited again ($h\nu_2$) to generate *[Mes-Acr$^·$], which serves as a super-electron injector. As a result, even electron-rich aryl halide **20** can be reduced by *[Mes-Acr$^·$] to give $^·$Ar and regenerate the ground state Mes-Acr$^+$. Finally, the aryl radical abstracts a hydrogen atom from R$_3$N$^{·+}$ to produce the dehalogenative hydrogen transfer product **21**.

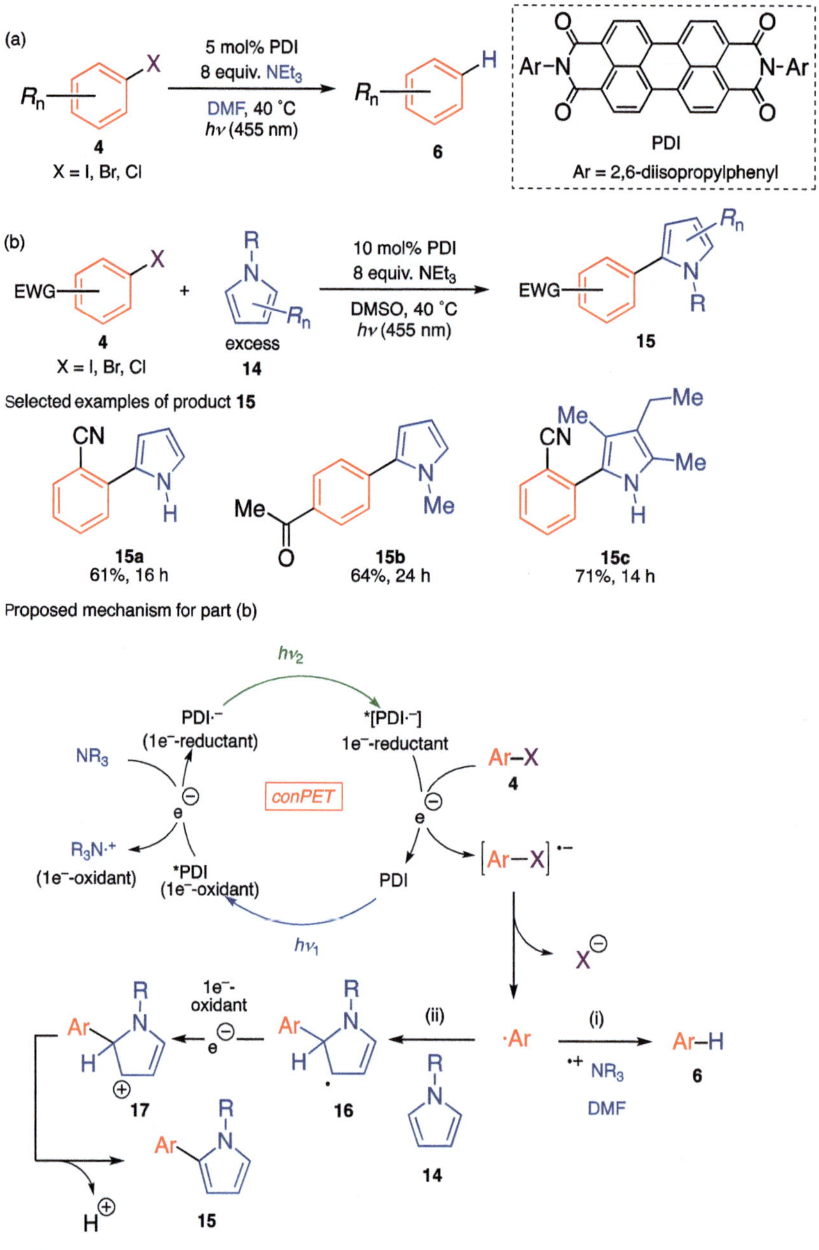

Figure 31.8 (a) Dehalogenative hydrogen transfer and (b) Homolytic aromatic substitution (HAS)-type aryl–aryl cross-coupling by consecutive photoinduced electron transfer (*conPET*) of N,N-bis(2,6-diisopropylphenyl)perylene-3,4,9,10-bis(dicarboximide) (PDI).

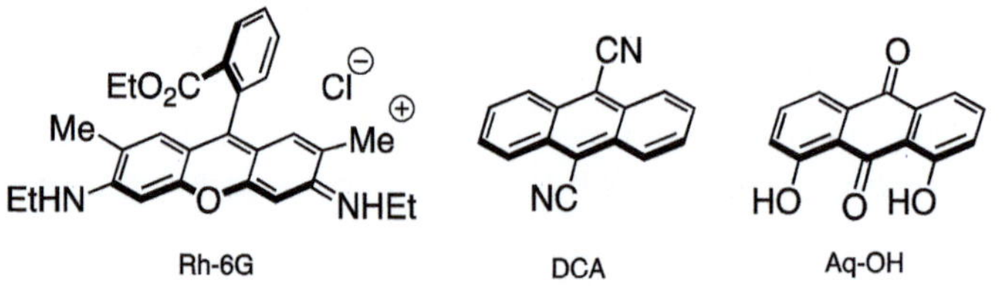

Figure 31.9 Examples of organic photoredox catalysts (*OPCs*) for use in the consecutive photoinduced electron transfer (*conPET*) process.

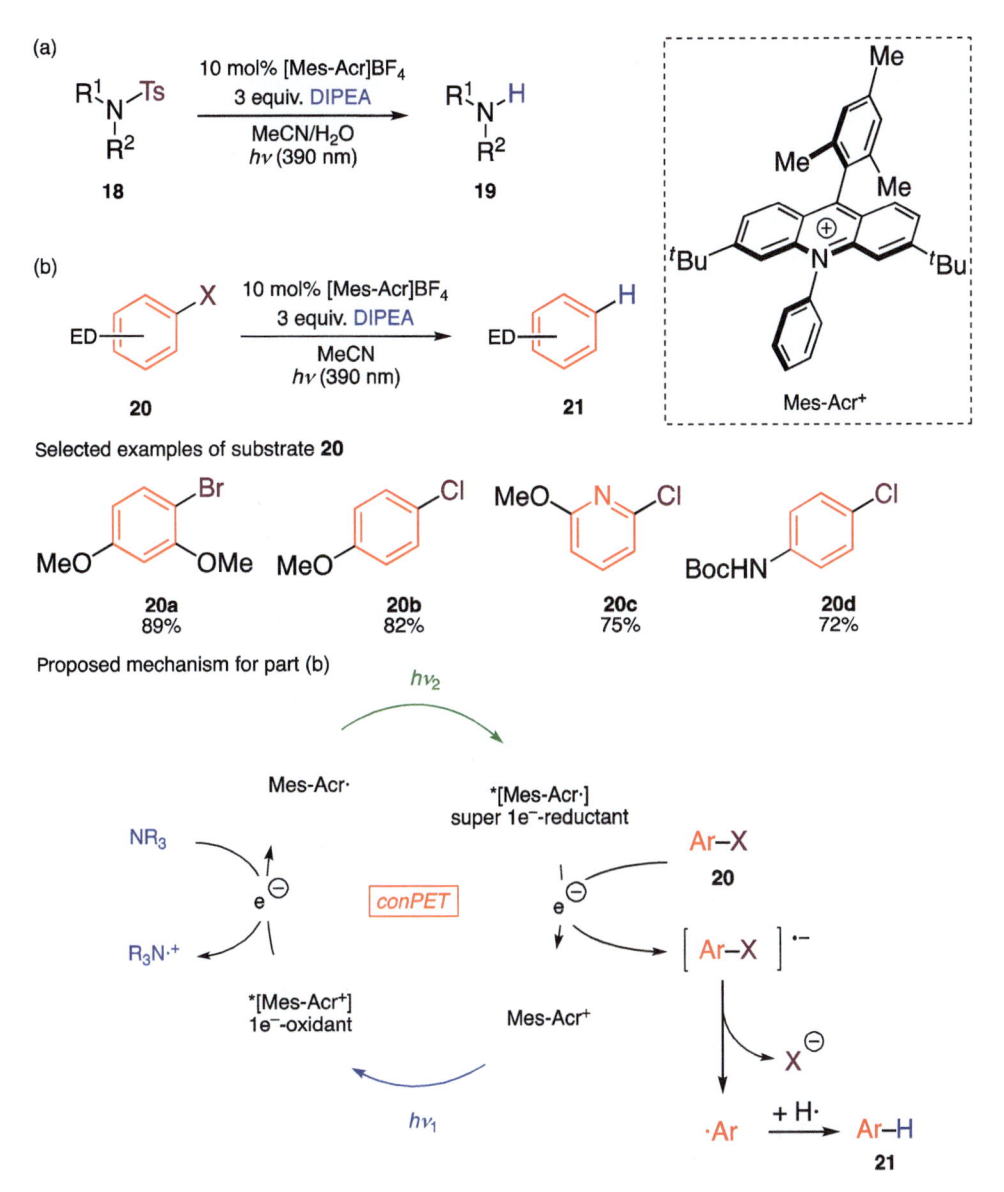

Figure 31.10 (a) Reductive de-tosylation and (b) dehalogenative hydrogen transfer by the consecutive photoinduced electron transfer (*conPET*) of Mes-Acr⁺.

31.5 Consecutive Photoinduced Electron Transfer (*conPET*) by the Combination of Organic Photocatalysis and Electrolysis

As highlighted above (Section 31.4), Mes-Acr⁺ serves as a super-electron donor, where in the presence of sacrificial electron donors (NR_3) results in its reducing power being comparable to those obtained by alkali metal reagents. However, the use of sacrificial electron donors often limits the scope of the reaction because the oxidized species ($R_3N^{·+}$) serves either as an oxidant or as a hydrogen donor. Thus, the development of super-electron-injecting catalytic systems that do not require sacrificial terminal reductants remains of key importance in the context of further sustainability. To address this issue, cathodic reduction has been regarded as the most appropriate solution. Indeed, the recent merging of photocatalysis with electrochemistry has opened up unexplored areas of redox chemistry, with several groups reported the use of oxidative electrophotocatalysis [35–38]. However, in the present section, we describe some recent studies into the reductive organic electrophotocatalysis process [39–42].

More specifically, in 2020, the groups of Lin, Lambert, and Wickens obtained super-electron-injecting systems by merging cathodic reduction with organic photoredox catalysis to reach a potential of approx. −3.3 V [43, 44]. This concept is illustrated in Figure 31.11a, wherein OPCs, DCA and naphthalene monoimide (NpMI), are the main components. In this system, the ground state OPC initially undergoes SET from the cathode to form a 1e$^-$-reduced species ($OPC^{\cdot-}$); this process is an alternative to 1e$^-$ transfer from a sacrificial tertiary alkyl amine via a reductive quenching pathway. Subsequently, the obtained $OPC^{\cdot-}$ is excited by visible light ($h\nu$) to form a super-reducing excited species (*[$OPC^{\cdot-}$]), which promotes the 1e$^-$ reduction of aryl halides (**20**), including aryl chlorides. It should be noted that this system was found to be applicable to various

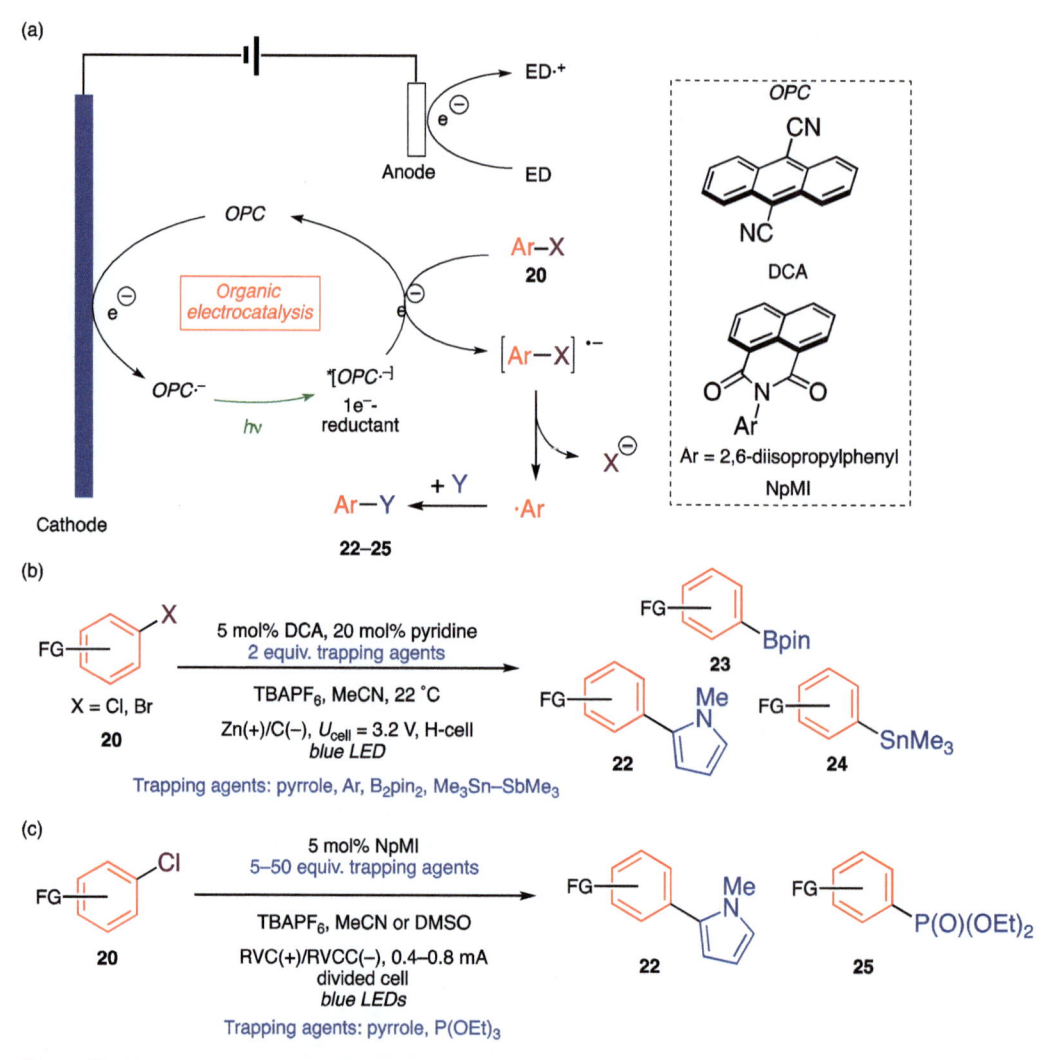

Figure 31.11 A consecutive photoinduced electron transfer (*conPET*)-based arylation process by the combination of organic photocatalysis and electrolysis.

arylation reactions, including a HAS-type aryl–aryl coupling (**22**), and borylation (**23**), stannylation (**24**), and phosphorylation (**25**) reactions. More diverse functionalization of the aryl structure was also reported, presumably because the terminal reductant was changed from an alkyl amine to provide clean and sustainable electricity.

31.6 Summary and Outlook

Alkali metals have been traditionally employed as electron sources for organic synthesis. However, they are highly reactive and dangerous, which is undesirable for a number of reasons. In particular, the water sensitivities of such species and the resulting deterioration in their functionalities are major issues in the contexts of functional group tolerance and sustainability. Thus, an ongoing challenge in the area of synthetic organic chemistry is the development of operationally more facile strategies to produce electron sources. In this context, noble metal-free organic super-reducing photocatalysts have been investigated as a potential alternative. More specifically, the appropriate design of organic photoredox catalyst systems has the potential to yield super-electron injectors that act through single-electron transfer. Although this review provided only a brief overview of such systems, we wish to emphasize the present $1e^-$ injection systems can be applied to a range of synthetically valuable reactions, which are often challenging under alkali metal conditions due to damage to the reactant functional groups. A rational approach to catalyst design therefore involves the evolution of electron-injection strategies in terms of the operational simplicity and green chemistry. Moreover, the studies discussed herein demonstrate that the combination of organic photocatalysis with electrolysis has taken both the photo- and electrochemical synthetic methods to new levels. These modern electron-injection strategies are therefore expected to lead to further important innovations in area of organic synthesis.

References

1 Murphy, J.A. (2014). *J. Org. Chem.* 79: 3731–3746.
2 Doni, E. and Murphy, J.A. (2014). *Chem. Commun.* 50: 6073–6087.
3 Broggi, J., Terme, T., and Vanelle, P. (2014). *Angew. Chem. Int. Ed. Engl.* 53: 384–413.
4 Murphy, J.A., Khan, T.A., Zhou, S.Z. et al. (2005). *Angew. Chem. Int. Ed. Engl.* 44: 1356–1360.
5 Murphy, J.A., Zhou, S.Z., Thomson, D.W. et al. (2007). *Angew. Chem. Int. Ed. Engl.* 46: 5178–5183.
6 Koike, T. and Akita, M. (2021). *Trends Chem.* 3: 416–427.
7 Romero, N.A. and Nicewicz, D.A. (2016). *Chem. Rev.* 116: 10075–10166.
8 Silvi, M. and Melchiorre, P. (2018). *Nature* 554: 41–49.
9 Lee, Y. and Kwon, M.S. (2020). *Eur. J. Org. Chem.* 2020: 6028–6043.
10 Vega-Penaloza, A., Mateos, J., Companyo, X. et al. (2021). *Angew. Chem. Int. Ed. Engl.* 60: 1082–1097.
11 Marzo, L., Pagire, S.K., Reiser, O., and König, B. (2018). *Angew. Chem. Int. Ed. Engl.* 57: 10034–10072.
12 Discekici, E.H., Treat, N.J., Poelma, S.O. et al. (2015). *Chem. Commun.* 51: 11705–11708.
13 Treat, N.J., Sprafke, H., Kramer, J.W. et al. (2014). *J. Am. Chem. Soc.* 136: 16096–16101.
14 Theriot, J.C., Lim, C.-H., Yang, H. et al. (2016). *Science* 352: 1082–1086.
15 Pearson, R.M., Lim, C.H., McCarthy, B.G. et al. (2016). *J. Am. Chem. Soc.* 138: 11399–11407.
16 Du, Y., Pearson, R.M., Lim, C.H. et al. (2017). *Chem. Eur. J.* 23: 10962–10968.

17 McCarthy, B.G., Pearson, R.M., Lim, C.H. et al. (2018). *J. Am. Chem. Soc.* 140: 5088–5101.
18 Sartor, S.M., McCarthy, B.G., Pearson, R.M. et al. (2018). *J. Am. Chem. Soc.* 140: 4778–4781.
19 Buss, B.L., Lim, C.H., and Miyake, G.M. (2020). *Angew. Chem. Int. Ed. Engl.* 59: 3209–3217.
20 Hagmann, W.K. (2008). *J. Med. Chem.* 51: 4359–4369.
21 Zhou, Y., Wang, J., Gu, Z. et al. (2016). *Chem. Rev.* 116: 422–518.
22 Bezencon, O., Heidmann, B., Siegrist, R. et al. (2017). *J. Med. Chem.* 60: 9769–9789.
23 Meanwell, N.A. (2011). *J. Med. Chem.* 54: 2529–2591.
24 Noto, N., Koike, T., and Akita, M. (2019). *ACS Catal.* 9: 4382–4387.
25 Taniguchi, R., Noto, N., Tanaka, S. et al. (2021). *Chem. Commun.* 57: 2609–2612.
26 Noto, N., Takahashi, K., Goryo, S. et al. (2020). *J. Org. Chem.* 85: 13220–13227.
27 Glaser, F., Kerzig, C., and Wenger, O.S. (2020). *Angew. Chem. Int. Ed. Engl.* 59: 10266–10284.
28 Indrajit Ghosh, T.G., Bardagi, J.I., and König, B. (2014). *Science* 346: 725–728.
29 Ghosh, I. and Konig, B. (2016). *Angew. Chem. Int. Ed. Engl.* 55: 7676–7679.
30 Neumeier, M., Sampedro, D., Majek, M. et al. (2018). *Chem. Eur. J.* 24: 105–108.
31 Bardagi, J.I., Ghosh, I., Schmalzbauer, M. et al. (2018). *Eur. J. Org. Chem.* 2018: 34–40.
32 Fukuzumi, S., Ohkubo, K., Ogo, S. et al. (2004). *J. Am. Chem. Soc.* 126: 1600–1601.
33 Fukuzumi, S. and Ohkubo, K. (2013). *Chem. Sci.* 4: 561–574.
34 MacKenzie, I.A., Wang, L., Onuska, N.P.R. et al. (2020). *Nature* 580: 76–80.
35 Huang, H., Strater, Z.M., Rauch, M. et al. (2019). *Angew. Chem. Int. Ed. Engl.* 58: 13318–13322.
36 Zhang, W., Carpenter, K.L., and Lin, S. (2020). *Angew. Chem. Int. Ed. Engl.* 59: 409–417.
37 Capaldo, L., Quadri, L.L., and Ravelli, D. (2019). *Angew. Chem. Int. Ed. Engl.* 58: 17508–17510.
38 Hong Yan, Z.-W.H. and Xu, H.-C (2019). *Angew. Chem. Int. Ed. Engl.* 58: 4592–4595.
39 Liu, J., Lu, L., Wood, D., and Lin, S. (2020). *ACS Cent. Sci.* 6: 1317–1340.
40 Barham, J.P. and Konig, B. (2020). *Angew. Chem. Int. Ed. Engl.* 59: 11732–11747.
41 Lai, X.L., Shu, X.M., Song, J., and Xu, H.C. (2020). *Angew. Chem. Int. Ed. Engl.* 59: 10626–10632.
42 Hossain, M.J., Ono, T., Yano, Y., and Hisaeda, Y. (2019). *ChemElectroChem* 6: 4199–4203.
43 Kim, H., Kim, H., Lambert, T.H., and Lin, S. (2020). *J. Am. Chem. Soc.* 142: 2087–2092.
44 Cowper, N.G.W., Chernowsky, C.P., Williams, O.P., and Wickens, Z.K. (2020). *J. Am. Chem. Soc.* 142: 2093–2099.

32

Visible Light as an Alternative Energy Source in Enantioselective Catalysis

Ana Maria Faisca Phillips[1] and Armando J.L. Pombeiro[1,2],*

[1] *Centro de Química Estrutural, Institute of Molecular Sciences, Instituto Superior Técnico, Universidade de Lisboa, Av. Rovisco Pais 1, Lisboa, Portugal*
[2] *Research Institute of Chemistry, Peoples's Friendship University of Russia (RUDN University), 6 Miklukho-Maklaya Street, Moscow, Russia*
* *Corresponding author*

32.1 Introduction

Present day concerns with our environment and with the preservation of natural resources have led chemists to develop more sustainable methods of synthesis (the so-called "green chemistry") [1]. Energy savings are also important to consider, and, amongst the new developments, photoredox catalysis stands out. Contrary to the classical photochemical reactions, such as the [2+2] cycloaddition of olefins, or the well-known Paterno-Buchi reaction for the synthesis of oxetanes from aldehydes or ketones and olefins which, as a result of molecular orbital symmetry considerations, cannot proceed with thermal activation, but does so in the presence of high energy UV light (100–380 nm) [2], photoredox catalysis works with mild visible light (400–800 nm). In many cases, the reactions proceed at room temperature, which is an important advantage [3]. The lower energy used avoids decomposition and undesirable side reactions [4]. However, many organic molecules do not absorb visible light, and another solution was needed to implement a visible light activation strategy. The way that nature uses various visible light absorbing chromophores/photocatalysts to convert solar energy to chemical energy, as in photosynthesis, undoubtedly inspired many research groups, and hence the photocatalysts were discovered. Photocatalysts have been now applied not only in organic synthesis, including in proton coupled electron transfer, but also in water splitting, solar energy storage, and photovoltaics [5–8].

Photoredox enantioselective catalysis is a fairly new branch of organic synthesis, and most developments have taken place since 2008. Seminal works in this field which sparkled a significant interest and research in this area were early reports by Bach [9], MacMillan [10], Yoon [11], and Stephenson [12], describing highly valuable single-electron transfer processes in synthesis. In 2005, Bach et al. showed for the first time that an organocatalyst could have the dual role of chirality transfer agent and photocatalyst. A chiral xanthone was utilized to induce chirality in a cyclization reaction leading to pyrrolidine ring-formation, while at the same time it transferred light energy to the substrate (by triplet energy transfer [ET]) to make the reaction possible. This process required, however, radiation from a UV source (366 nm) [9]. MacMillan et al. combined a chiral imidazolidinone with a ruthenium photocatalyst, for a highly enantioselective alkylation of enals

Catalysis for a Sustainable Environment: Reactions, Processes and Applied Technologies Volume 3, First Edition. Edited by Armando J. L. Pombeiro, Manas Sutradhar, and Elisabete C. B. A. Alegria.
© 2024 John Wiley & Sons Ltd. Published 2024 by John Wiley & Sons Ltd.

in the presence of visible light, introducing the term metallaphotoredox catalysis [10]. These first two processes are further discussed later in this chapter. The other two reports, although not enantioselective, described new approaches with the use of visible light. Yoon et al. showed for the first time that a [2+2] cycloaddition (of enones) could be made to proceed with visible light irradiation, by using Lewis acid activation [11]. Stephenson et al. described a photoinduced tin-free reductive dehalogenation reaction, based on photoredox catalysis by Ru(bpy)$_3$Cl$_2$ and an amine as hydrogen source [12]. Photodehalogenation is particularly significant for industrial applications, since polyhalogenated compounds are known persistent organic pollutants, which may cause health hazards, being banned by the Stockholm convention [13]. Practical dehalogenation techniques are useful, particularly since the tin hydrides which are otherwise frequently used, have associated toxicity and purification problems.

The power of the new methodologies was soon recognized, and nowadays enantioselective photoredox catalysis is a very active area of research, as several recent reviews on the subject show [14–24]. The photocatalysts are very versatile and allow different new forms of reactivity, because besides having the capacity to convert visible light into chemical energy, they can act both as oxidants and as reductants simultaneously in their excited states [15].

To bring about a photoredox reaction, a photocatalyst has to absorb visible light. This promotes an electron from its highest-occupied molecular orbital (HOMO) to its lowest unoccupied molecular orbital (LUMO). The singlet excited-state catalytic species produced undergoes intersystem crossing (ISC) to a long-lived triplet excited-state. This excited species can lose its energy in a number of ways: i) by single-electron transfer (SET); ii) by ET; iii) by atom or group transfer; and iv) via accelerated elementary organometallic steps [23]. This reactivity of the excited state brings up many possibilities for the development of new transformations with advantages over the conventional ground-state catalysis. In addition, the photocatalyst may also accept the excitation energy from a previously sensitized substrate.

When a substrate accepts the excitation energy of a photocatalyst, a chemical reaction can start. Figure 32.1 shows the four reaction pathways possible, which depend on the redox potentials of photocatalyst and those of the substrates: The excited state of the photocatalyst may be quenched either oxidatively (a) or reductively (b) [3]. The excited state has a very short lifetime, and hence $\Delta G \leq 0$ for the process.

It may also happen that the photocatalyst is quenched by a sacrificial electron donor, typically an amine or ascorbic acid (c) and the photocatalyst will then oxidize a substrate in a ground-state

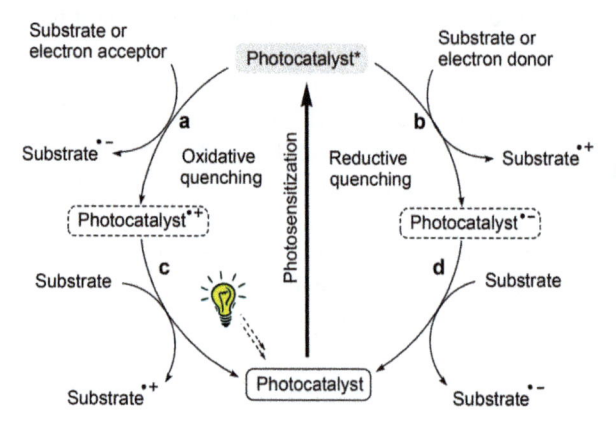

Figure 32.1 Mechanistic pathways of photoredox catalysis. In each case, the activated substrate will engage in further chemical reactions [3, 14].

reaction. Finally there is also the possibility of catalyst quenching by an electron acceptor, such as oxygen, (from air or peroxodisulfate) and the catalyst can then reduce a substrate in a ground state reaction (d). When there is subsequently a fast and irreversible reaction step, endergonic electron transfers are possible with $\Delta G>0$ (up to 500 mV). The majority of photoredox catalysis reactions follow the first path (i), starting with single-electron transfer between the excited photocatalyst (photosensitizer) and the organic substrate or reagent. A transient redox-active intermediate is formed, as well as an ion-radical species that reacts further to give products. The intermediate can return to the ground state by another SET event.

The overall photoredox process can be considered as being net-oxidative, net-reductive, or redox-neutral system, depending on the number of incoming and outgoing electrons. In the redox-neutral reactions, both oxidants and reductants to the sensitizer are incorporated into the product and there is no waste. The system is also insensitive to the sequence of redox events, since both electron-donors and acceptors co-exist in the same reaction vessel. These characteristics make these processes very versatile.

In photoredox chemistry, radicals are generated from nonradical species in a much more controlled manner than in other radical reactions. Because radicals are very reactive and tend to react indiscriminately, providing low selectivity between multiple possible stereo- and regioisomers, high ees can be difficult to achieve, but this is not the case in photoredox catalysis. Low selectivities can also result from the low energy barriers that reactions of odd electron species have [3, 23, 25].

Often the real challenge in these reactions is to overcome the high rate of racemic background reactions [14]. The concentration of reactive radical intermediates remains low and the rate of the unwanted diffusion-controlled radical/radical recombination process is also slow.

The photocatalysts may be inorganic complexes or organic molecules. A selection of the common ones is shown in Figure 32.2. The Ru(II)polypyridine complexes, of which the most commonly employed is the commercially available $Ru(bpy)_3Cl_2$, are quite useful, due to their ease of synthesis, stability at room temperature, and excellent photoredox properties.

To render the photoredox processes enantioselective, a chiral catalyst is used. In some cases the chiral catalyst also plays the role of photocatalyst, but more frequently a dual-catalyst approach is used (a synergistic approach), and the chiral catalyst can be an organocatalyst or a transition metal complex.

Since 2008, there have been a vast number of achievements in this field. The aim of this chapter is to highlight the different types of catalytic activation that are possible in combination with photoredox catalysis to bring about an enantioselective transformation. This is of relevance for the development of future reactions. Besides the description of some historical cases that demonstrate advances in the field, we concentrate on examples that highlight the advantages of using visible light as an energy source in relation (as an alternative) to other methods of synthesis.

The reactions covered are activated by visible light only (wavelength 400–700 nm) and those driven by UV light are excluded (wavelength <400 nm). Examples involving heterogeneous catalysis are not covered, neither are those that include catalysis by enzymes or macromolecular catalysts. Also, due to lack of space, applications in natural product synthesis are not included. After the introduction, the chapter is divided according to the type of catalysis used in the photoredox process: dual chiral organocatalysis and photoredox catalysis first, then dual metal and photoredox catalysis. The latter is subdivided into methods involving transition metal catalysis and those in which the metal acts merely as a Lewis acid, which is followed by a discussion on chiral photocatalysts that is divided into chiral-at-metal catalysts that also act as photocatalysts and organocatalysts that act both as chirality inducers and as photocatalysts. The chapter ends with conclusions.

a) Inorganic photocatalysts

[Ru(bpy)₃]Cl₂
E^0_{red}[Ru(II)*/Ru(III)] ~ −0.83 V
λ_{max} = 452 nm
E^0_{ox}[Ru(II)*/Ru(I)] ~ +0.77 V
λ_{max} = 452 nm

fac-[Ir(ppy)₃]
E^0_{red}[Ir(III)*/Ir(IV)] ~ −1.73 V
λ_{max} = 375 nm
E^0_{ox}[Ir(III)*/Ir(II)] ~ +0.31 V
λ_{max} = 375 nm

[Cu(dap)₂]Cl
E^0_{red}[PC*/PC•⁺] ~ −1.43 V
λ_{max} = 530 nm

a) Organic photocatalysts

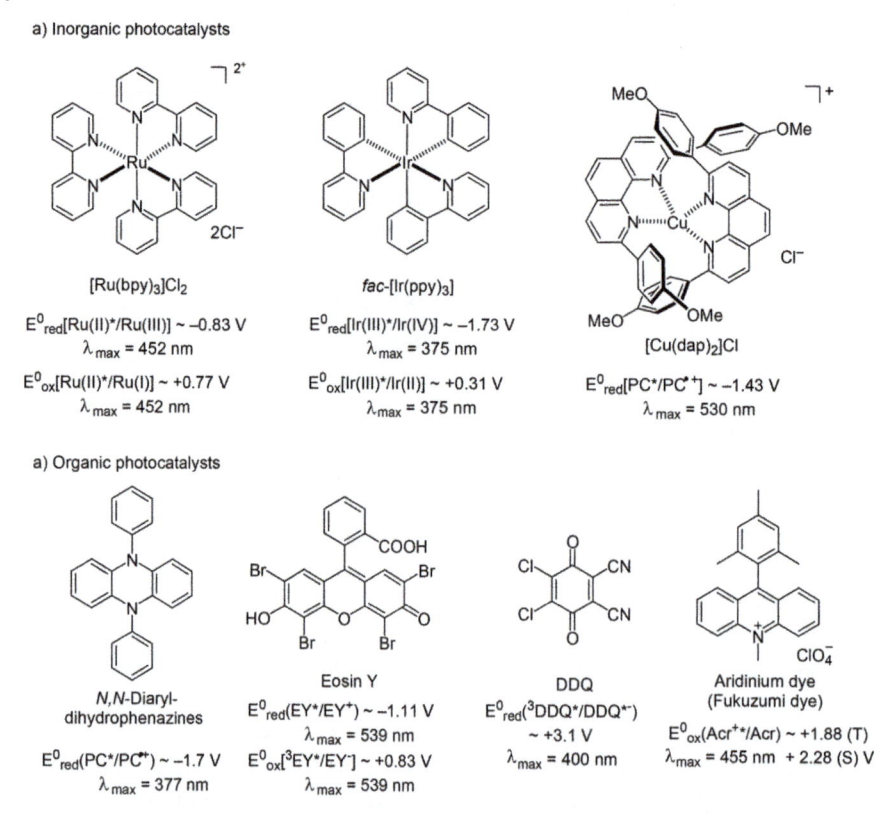

N,N-Diaryl-dihydrophenazines
E^0_{red}(PC/PC•⁺) ~ −1.7 V
λ_{max} = 377 nm

Eosin Y
E^0_{red}(EY*/EY•⁺) ~ −1.11 V
λ_{max} = 539 nm
E^0_{ox}[³EY*/EY⁻] ~ +0.83 V
λ_{max} = 539 nm

DDQ
E^0_{red}(³DDQ*/DDQ•⁻) ~ +3.1 V
λ_{max} = 400 nm

Aridinium dye (Fukuzumi dye)
E^0_{ox}(Acr⁺*/Acr•) ~ +1.88 (T)
λ_{max} = 455 nm + 2.28 (S) V

Figure 32.2 Examples of photoredox catalysts, their redox properties and their maximum wavelength of absorption [3]. Bpy = 2,2′-bipyridine; ppy = 2-phenylpyridine; dap = 6-amidine-2-(4-amidino-phenyl)indole; DDQ = 2,3-dichloro-5,6-dicyano-1,4-benzoquinone; acr = acridinium.

32.2 Dual Chiral Organocatalysis and Photoredox Catalysis

The combination of chiral organocatalysis with photoredox chemistry has not only allowed the development of several novel enantioselective reactions, but it has also made possible milder and more selective versions of known transformations. This section highlights the different modes of organocatalytic activation available that may be combined with PC and important developments in this field.

32.2.1 Chiral Amines as Catalysts

Chiral primary and secondary amines can activate carbonyl substrates in the following ways: i) via formation of nucleophilic enamines, from enolisable aldehydes and ketones with loss of a water molecule (HOMO activation); ii) via the formation of electrophilic iminium ions upon reaction with unsaturated carbonyl compounds (LUMO activation); and iii) via α-iminyl radical cation intermediates, which are generated when enamines undergo single-electron oxidation in the presence of a chemical oxidant (singly-occupied molecular orbital [SOMO] activation). Tertiary amines, of which the cinchona alkaloids catalysts are well-known examples, have quite a different mode of action and are discussed separately. They are covered under phase-transfer catalysts.

A landmark in the field of photoredox catalysis was the development of the enantioselective α-alkylation of aldehydes **1** with alkyl halides **2** by MacMillan et al. in 2008 (Figure 32.3a) [10]. Although being a reaction of great synthetic importance, the direct functionalization of carbonyl substrates, and in particular of aldehydes, had proven difficult until then, even with enamine-based chemistry because, as a result of the low reactivity of alkyl halides, N-alkylation of the Lewis basic amine catalysts and self-aldol condensation tended to predominate [26–28].

Realizing that the main problem resided with the ionic nature of the reaction and its S_N2 pathway, MacMillan et al. used the alkyl bromides as radical precursors rather than as electrophiles, relying on the fact that electron-deficient radicals react rapidly with π-rich olefins, thus allowing the formation of difficult-to-make carbon-carbon bonds. $Ru(bpy)_3Cl_2$ (bpy = 2,2′-bipyridine), capable of generating open-shell species from α-bromo carbonyl compounds, was used as photocatalyst in combination with organocatalyst **3** (Figure 32.1). This photocatalyst was already then well-known as a SET catalyst for inorganic applications, but its use in synthetic chemistry had

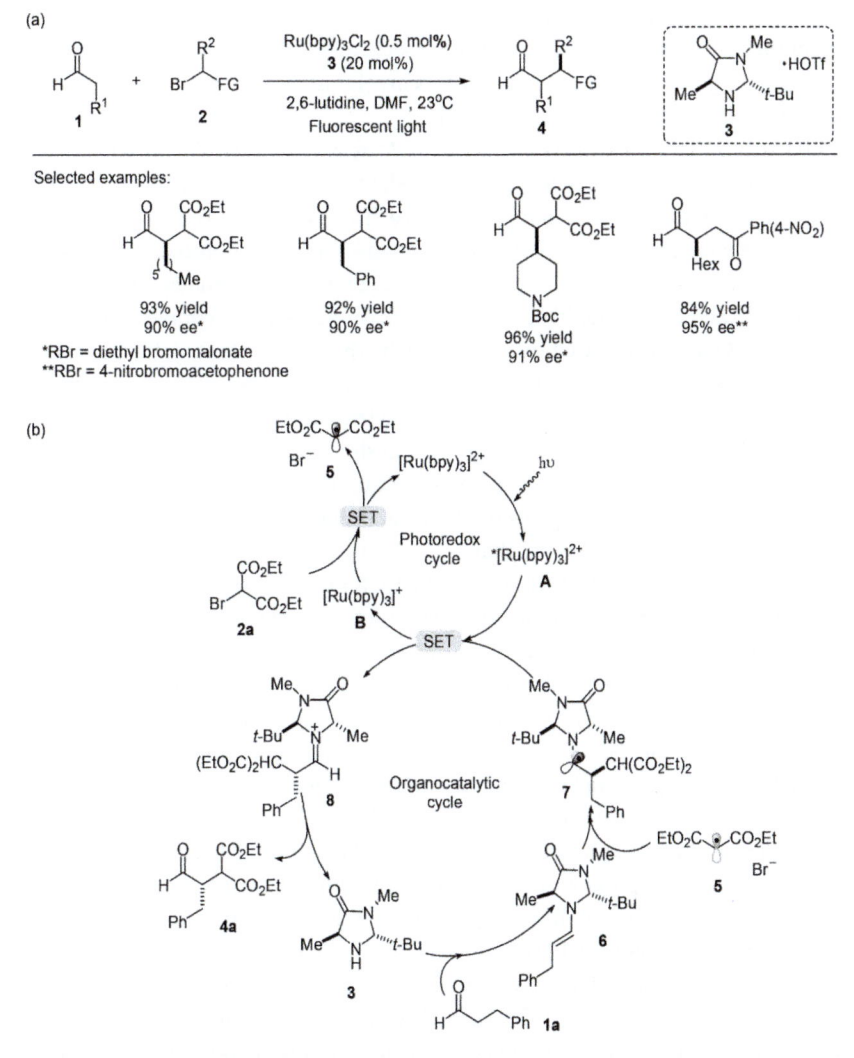

Figure 32.3 (a) Enantioselective alkylation of aldehydes by photocatalysis (PC) [10]. (b) The reaction mechanism proposed for the alkylation reaction [10].

been so far very limited [29, 30]. The alkylated aldehydes **4** were obtained in very high yields and ees (Figure 32.3a).

The reaction mechanism for aldehyde alkylation (Figure 32.3b) involves two catalytic cycles: the first is a photoredox cycle, in which the ruthenium catalyst accepts a photon from a light source (a W fluorescent light bulb) to populate the *Ru(bpy)$_3^{2+}$ metal-to-ligand charge transfer (MLCT) excited state forming **A** [10]. In the first catalytic cycle this high energy intermediate removes a single electron from a sacrificial quantity of enamine, forming the electron-rich Ru(bpy)$_3^+$ (**B**). Ru(bpy)$_3^+$ is known to be a potent reductant [−1.33 V versus saturated calomel electrode (SCE) in CH$_3$CN] [31]. It transfers a single electron to bromocarbonyl substrate **2a** (SET), leading to the formation of the electrophilic alkyl radical **5**. Ru(bpy)$_3^{2+}$ (**1**) is formed concurrently, ready for a new catalytic cycle ($E_{1/2}$ for phenacyl bromide = −0.49 V versus SCE in CH$_3$CN, $E_{1/2}$ = half reduction potential) [32, 33]. The second catalytic cycle, is an organocatalytic cycle, in which the enamine **6** generated from the reaction between a chiral amine and an aldehyde traps the radical **5** generating the electron rich species **7** bearing a new chiral center. Species **7** is finally oxidized by the excited state of the Ru(II) photocatalyst (*Ru(bpy)$_3^{2+}$, **A**), in a SET event, affording the iminium **8**. Upon hydrolysis of the iminium the alkylated substrate **4a** is released, as well as the organocatalyst, which can enter a new cycle. Alternative mechanisms involving radical chain propagation have also been proposed since, for related reactions [34, 35].

The scope of the α-alkylation reaction shown in Figure 32.3a was subsequently expanded to trifluoromethylation [36], cyanoalkylation [37], benzylation [38], and related transformations [39, 40]. Other organocatalysts and photocatalyst combinations have also been tried [3].

The combination of cross-dehydrogenative coupling (CDC) and asymmetric photoredox catalysis is also possible, but it has proven to be rather challenging. In 2017, Pericàs et al. used enamine catalysis with Jørgensen's catalyst to achieve the highly enantio- and diastereoselective coupling of aldehydes with xanthenes (Figure 32.4a) [41]. BrCCl$_3$ was found to be the best oxidant. A base is also needed to scavenge HBr, which is a stoichiometric product of the reaction, and Na$_3$PO$_4$ gave the best combination of yield/er. A range of products was obtained in high yields and ees, but non-conformationally restricted, electron-rich diarylmethanes did not give the desired products, probably due to a lower stability of the corresponding radical. The Ru(bpy)$_3$ complex used to achieve this transformation has potentials that match the corresponding electrochemical properties of the reagents/intermediates. In the photoredox cycle, bromotrichloromethane ($E_{1/2}$ = −0.18 V vs SCE) is reduced by the excited state of the Ru(II) complex ($E_{1/2}$ = −0.81 V vs SCE) to the CCl$_3$ radical (oxidative quenching), which then abstracts a hydrogen atom from a xanthene to generate the xanthyl radical **A**. The xanthyl radical is oxidized to a cation **B** ($E_{1/2}$ = 0.12 V vs SCE) by the Ru(III) complex ($E_{1/2}$ = 1.29 V vs SCE), thus allowing the attack on the enamine **12**, which leads to C–C bond formation yielding **C**. The final product is obtained after enamine hydrolysis and in-situ aldehyde reduction. The reaction mechanism was proposed by the authors based on kinetic experiments and discrete Fourier transform (DFT) calculations. The possibility of a radical chain mechanism where radical **A** ($E_{1/2}$ = 0.12 V vs SCE) would reduce directly BrCCl$_3$ ($E_{1/2}$ = −0.18 V vs SCE) to a CCl$_3$ radical was also considered possible, in which case the photocatalyst would only be required in the first cycle to start a chain.

By merging three catalytic processes (photoredox, enamine and hydrogen-atom transfer (HAT) catalysis), MacMillan et al. developed an enantioselective α-aldehyde alkylation reaction with simple olefins **13** as the coupling partners in 2017 (Figure 32.4b) [42]. The method was suitable for the synthesis of α-functionalized linear and for α-cyclic aldehydes **16**, the latter being obtained by the intramolecular reaction of olefin-tethered aldehydes. 5-*Exo*, 6-*exo* and 7-*endo*-cyclizations were possible. The yields ranged from good to very high, whereas the ees remained very high irrespective of the nature of substituents present in the reactants. The diastereoselectivities (where applicable) were up to >20:1.

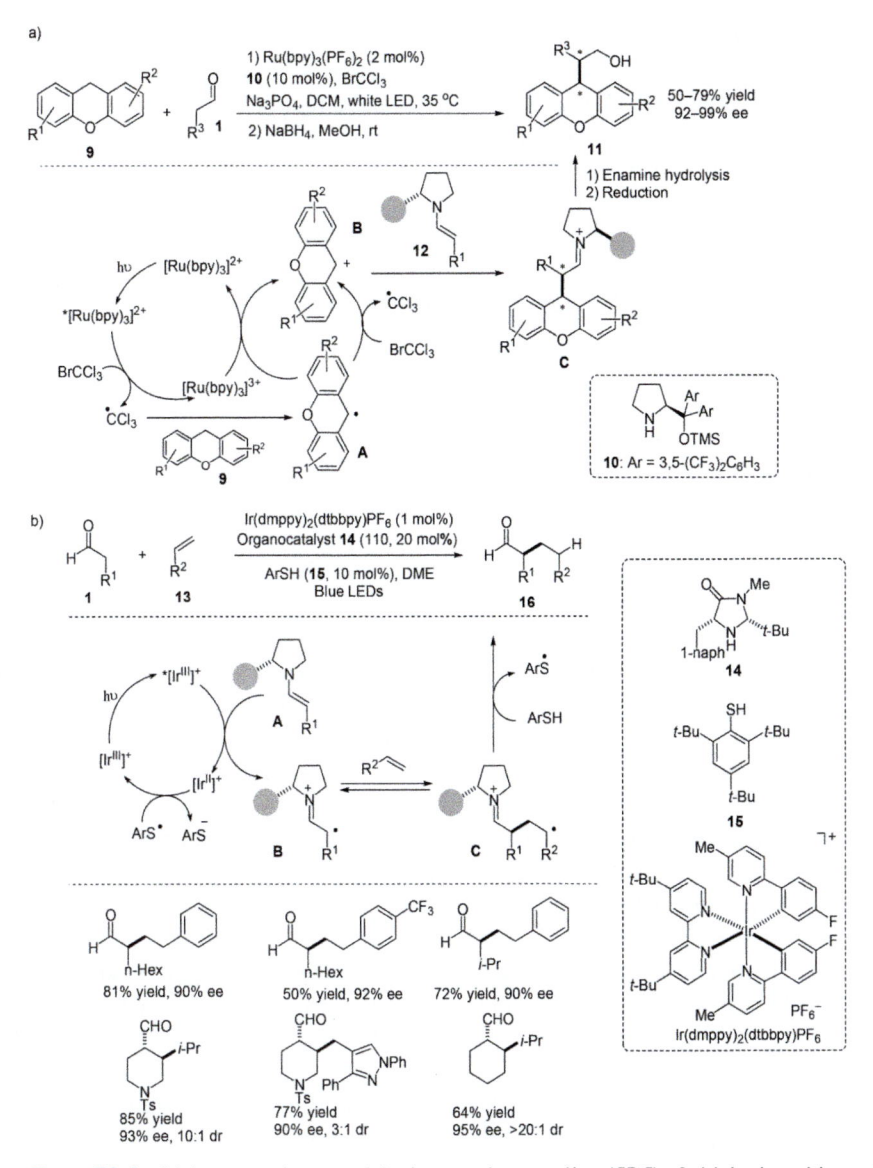

Figure 32.4 (a) Asymmetric cross-dehydrogenative coupling (CDC) of aldehydes with xanthenes [41]. (b) Enantioselective inter- and intramolecular α-alkylation of aldehydes using simple olefins [42]. DME = dimethoxyethane; Dmppy = 2,4-bis(3,5-dimethylphenyl)pyridine; Dtbbpy = (4,4′-di-tert-butyl-2,2′-bipyridine).

According to the reaction mechanism proposed, an enamine intermediate (**A**) is involved and the hydrogen transfer step involves the transfer of a hydrogen atom from **B** to the alkyl radical species formed after olefin addition (**C**). This step is by no means trivial, because the HAT catalyst has to be able to discriminate between trapping of an initial electrophilic 3πe – enaminyl radical intermediate (**B**) or the alkyl radical species generated following olefin coupling. In addition, since the olefin addition is a reversible reaction, the retention or loss of the enantiocontrol gained in the SOMO-addition step is dependent on the kinetic efficiency of the HAT catalyst.

Iridium photocatalyst Ir(Fmppy)$_2$(dtbbpy)PF$_6$ (Fmppy = 2-(4-fluorophenyl)-4-(methylpyridine), dtbbpy = 4,4′-di-*tert*-butyl-2,2′-bipyridine) was found efficient to produce an excited-state complex upon irradiation with visible light ($E_{1/2}$ red [*Ir^{III}/Ir^{II}] = +0.77 V vs SCE). This complex can mediate single-electron transfer from the electron-rich enamine **A**, resulting from the reaction between organocatalyst **14** and the aldehyde, which generates a reduced Ir complex and the critical 3πe–enaminyl radical species **B**. In the olefin addition step (C–C bond formation), which leads to secondary alkyl radical **C**, the stereogenic center is created. Thiophenol **15** (BDE$_{S-H}$ =78 kcal.mol^{-1} [43]) works well as HAT catalyst, and after hydrolysis of the iminium ion, the alkylated aldehyde is released and the organocatalyst is also set free to enter a new catalytic cycle (Figure 32.4b).

There are many known examples of iminium ion activation, particularly for the conjugate addition of soft nucleophiles to the β-carbon atom of unsaturated carbonyl compounds. However, to trap nucleophilic radicals in a stereoselective manner has been more problematic. The reason for this lies in the fact when a radical adds to a cationic iminium ion a reactive unstable α-iminyl radical cation is generated, with a high tendency to undergo β-scission and to regenerate the more stable iminium ion. Melchiorre et al. were interested in developing this type of methodology, for the creation of a chiral quaternary carbon center since, contrary to what happens in ionic reactions, the reactivity of radicals is only marginally affected by steric factors [44]. Nevertheless, examples of this type of approach for the construction of quaternary chiral centers were still very limited [16]. The radical conjugate addition (RCA) between cyclic enones **17** and a variety of benzodioxole-derived radicals or alkyl aryl amines **18** afforded products in high yields and very high ees via the cooperative catalysis of photoredox Ir catalyst Ir[dF(CF$_3$)ppy]$_2$(dtbbpy)PF$_6$ and diamine **19**, an organocatalyst bearing a redox-active carbazole moiety (Figure 32.5a) [45].

In this reaction the organocatalyst operates via the formation of iminium ions **A** and is capable of trapping the photochemically generated carbon-centered radicals stereoselectively by means of an electron-relay mechanism. The carbazole moiety can undergo a rapid intramolecular SET reduction of the unstable α-iminyl radical cation **B** generated initially yielding **C**, which prevents it from breaking down. The nascent enamine intermediate tautomerizes to the more stable imine, thus avoiding a possible competitive back-electron transfer. The carbazole radical cation then undergoes single-electron reduction from the reduced photoredox catalyst, regaining its neutrality, while yielding the quaternary product. Hence the photocatalyst not only creates the nucleophilic radical but it also promotes the final redox process, which is in fact the turnover-limiting step of the overall reaction.

The decarboxylation of aliphatic acids or activated esters is a means of generating highly reactive alkyl radicals for the formation of C–C bonds. Recently the unprecedented enantioselective α-alkynylation of β-ketocarbonyl compounds **21** with propiolic acids **22** was achieved by photocatalytic decarboxylation with visible light and a chiral primary amine catalyst (Figure 32.5b). This transformation, developed by Luo et al. had, as distinctive feature, the intermediacy of an α-imino radical, which could be generated with a photoredox catalytic system involving Ru(bpy)$_3$Cl$_2$ and hypervalent iodine reagent (BI-OH) [46]. The oxidative reagent had been previously used for photochemical decarboxylation [47, 48] but not in enantioselective reactions of this type. In the absence of this dual reagent system no alkynylation took place. DFT calculations suggested that this N–H bond has a pK$_a$ in the case of radical cation **26**, which suggests that the transfer of a proton to generate the α-imino radical is a facile process. A wide range of phenylpropiolic acids bearing either electron-donating or electron-withdrawing groups at the *meta* or *para* position of the arene moiety yielded alkynylation products bearing all-carbon quaternary centers in moderate yield and excellent ees, together with the corresponding alkylation products, which were obtained in low yields and moderate to good ees. The β-ketocarbonyl compounds could be cyclic or acyclic.

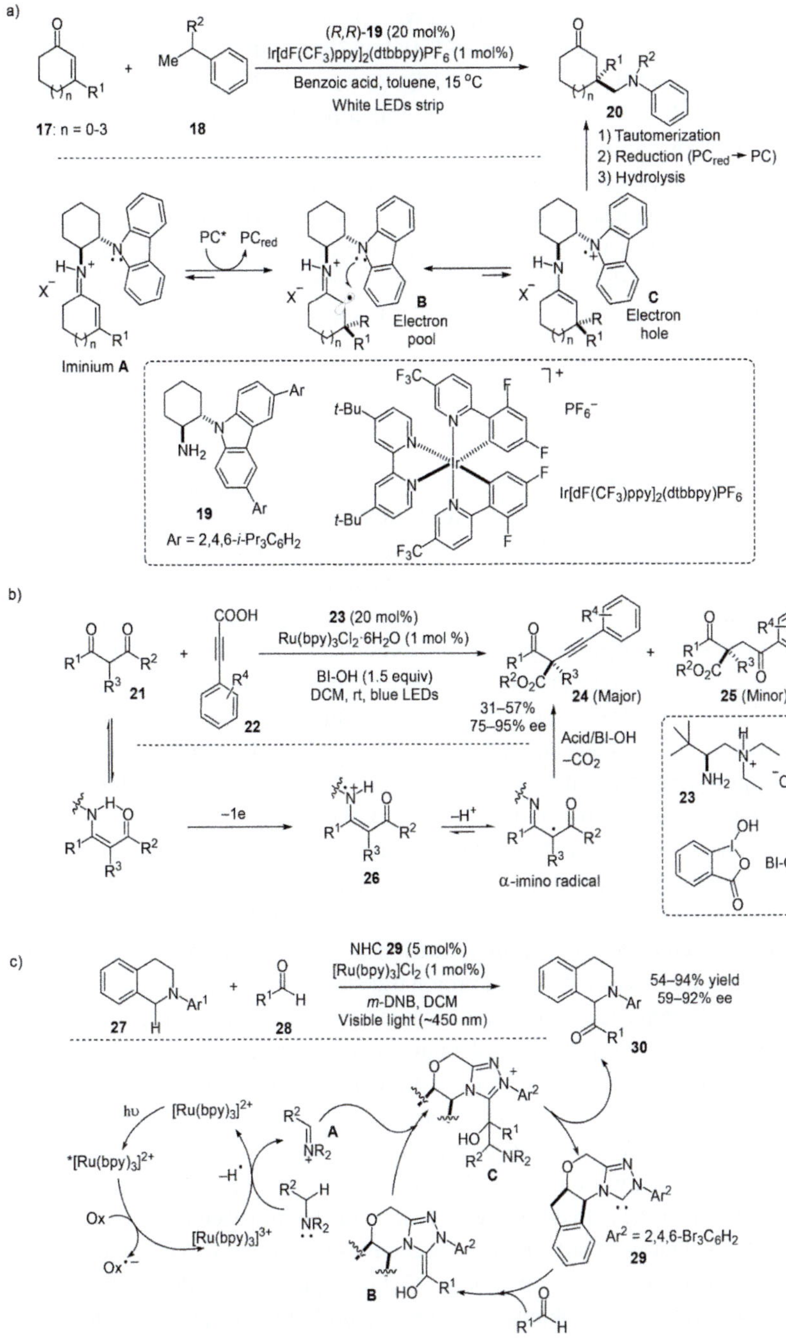

Figure 32.5 (a) Enantioselective 1,4-addition of α-amino radicals to enones [45]. (b) Enantioselective decarboxylative α-alkynylation of β-ketocarbonyl compounds [46]. (c) Asymmetric α-acylation of tertiary amines [49]. DNB = 1,3-dinitrobenzene; dF(CF$_3$)ppy = 2-(2,4-difluorophenyl)-5-(trifluoromethyl)- pyridine.

32.2.2 N-Heterocyclic Carbenes (NHCs) as Catalysts

The catalytic activation modes discussed in the previous section were all examples of covalent catalysis. Another example of this type of catalysis is that obtained when chiral NHCs are used as catalysts. The utilization of these catalysts in combination with photoredox catalysis has allowed the enantioselective acylation of tetrahydroisoquinolines (THIQs) **27** with aldehydes **28** as described by DiRocco and Rovis (Figure 32.5c) [49]. The acylated THIQs **30** were obtained in high yields and ees. The role of the photocatalyst In this case is not to generate a reactive radical, but a reactive iminium ion **A** through the photooxidation of the THIQ with the powerful photochemically generated oxidant $[Ru(bpy)_3]^{3+}$ (1.29 V vs SCE). In this way, single electron abstraction takes place followed by hydrogen abstraction in the presence of *m*-dinitrobenzene (*m*-DNB), acting as oxidative quencher. The imine is then trapped by the nucleophilic Breslow intermediate **B** formed by the reaction between the NHC catalyst and the aldehyde to yield **C**. Elimination of the NHC yields the product. This method is compatible with the presence of other functional groups on the aldehyde, namely thioethers, esters, and protected amines, without compromising the yields or ees.

32.2.3 Chiral Phosphoric Acids as Catalysts

Chiral phosphoric acids (CPAs) have emerged in the last two decades as highly efficient catalysts for many enantioselective transformations. A significant number of reports involve photoredox catalysis. It may be difficult to distinguish strictly between hydrogen bonding catalysis and Brønsted acid/base catalysis when considering reactions in which these catalysts are involved. If an exchangeable proton is required for the final transformation, even if hydrogen bonding interactions are involved, the latter type of catalysis may be considered to be in operation [21, 50].

The first enantioselective example of cooperative hydrogen bonding/photoredox catalysis was described by Knowles et al. in 2013 [51]. It was an intramolecular enantioselective aza-pinacol cyclization of ketone tethered hydrazones **31**, made possible by the action of 1,1′-bi-2-naphthol (BINOL)-derived CPA (CPA) **32**, $Ir(ppy)_2(dtbpy)PF_6$ and Hantzsch ester (HE) **33**, used as a stoichiometric reductant (HE is a series of dialkyl 1,4-dihydro-2,6-dialkyl-3,5-pyridinedicarboxylate) (Figure 32.6a). Although ketones are not reduced by PC due to their high negative reduction potential, the method devised involved concerted proton-coupled electron transfer (PCET), which surpassed this problem through the generation of lower energy neutral ketyl radical intermediates (e.g. **35**, which underwent C–C bond formation producing **36** followed by reduction to yield the desired products **34**) (Figure 32.6a). In PCET, a single electron transfer and a proton transfer occur in a concerted manner. Excellent levels of enantio- and diastereoselectivity were obtained for a range of products.

Azaarenes are important structural components of many natural products, pharmaceuticals and agrochemicals. These substances often bear substituents incorporating chiral centers. Minisci reactions are often used in these fields for the formation of carbon-carbon bonds, by joining carbon radicals to carbon centers adjacent to nitrogen in pyridine rings. However, enantioselective approaches had not been described until a 2018 report by Phipps et al. [52]. In this procedure, prochiral radicals, generated from amino acid derivatives **37**, are added to nitrogen heterocycles **38**, namely pyridines and quinolines (Figure 32.6b). Photoredox catalysis with $Ir(dF(CF_3)ppy)_2(dtbbpy)PF_6$ and a CPA, (*R*)-TRIP (3,3′-bis(2,4,6-triisopropylphenyl)-2,2′-binaphtholate) (**39**) or (*R*)-TCYP ((11b*R*)-4-hydroxy-2,6-bis(2,4,6-tricyclohexylphenyl)-4-oxide-dinaphtho[2,1-d:1′,2′-f][1,3,2]dioxaphosphepin) (**40**) allowed the construction of α-tertiary stereocenters in a highly stereoselective manner producing compounds **41**. The method was applied to the regio- and stereoselective late-stage functionalization of two pharmaceuticals, namely, metyrapone, used for the diagnosis of adrenal insufficiency, and etofibrate, a fibrate [52].

One year later, Jiang et al. described a procedure for the enantioselective addition of alkyl or ketyl radicals to 2- and 4-vinylpyridines **42** also with a combination of Brønsted acid catalysis and photoredox catalysis, although in this case an organic photocatalyst, 5,6-bis(5-methoxythiophen-2-yl)

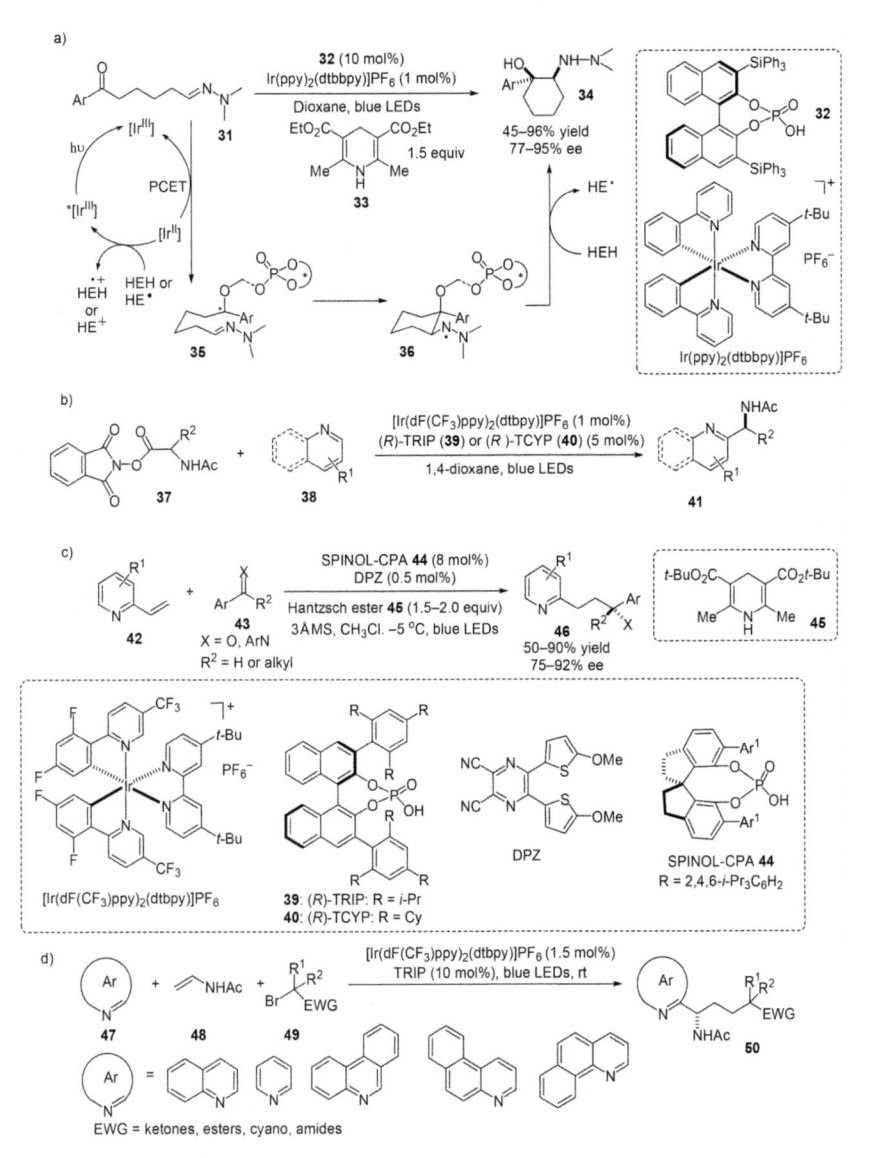

Figure 32.6 (a) Intramolecular enantioselective aza-pinacol cyclization of ketone-tethered hydrazones [51]. (b) Enantioselective Minisci-type addition to heteroarenes [52]. (c) Enantioselective reductive coupling of vinylpyridines with aldehydes, ketones, and imines [53, 54]. (d) Asymmetric synthesis of heterocyclic γ-amino-acid and diamine derivatives by three-component radical cascade reactions [55]. Dtbpy = 4,4'-di-tert-butyl bipyridine.

pyrazine-2,3-dicarbonitrile (DPZ) was used (Figure 32.6c) [53]. Chiral γ-phenyl and aminopropylpyridines **46** were obtained with very high yields and ees, in spite of the much longer distance between the α-amino radical and the nitrogen-coordinated chiral catalyst. In this reaction, the organocatalyst is axially chiral 1,1'-spirobiindane-7,7'-diol (SPINOL)-CPA **44**, which activates the pyridine and the C=X bond (SPINOL refers to a series of axially chiral 1,1'-spirobiindane-7,7'-diols). The α-aminoalkyl and ketyl radicals are generated in situ from the corresponding ketones, aldehydes, and imines, undergoing subsequently conjugate addition to vinylpyridine. 2-Vinyl pyridines with very electron-withdrawing ring substituents, did not react. A reductive coupling strategy was used in this synthesis, with HE **45** playing the role of stoichiometric reductant [53, 54].

A further development on Minisci-type reactions was the three-component reaction between azaarenes **47**, enamides **48** and α-bromo carbonyl compounds **49**, to yield valuable chiral heterocyclic γ-amino acid derivatives **50** in high yields and enantioselectivities (Figure 32.6d) [55]. This Ir-CPA-catalyzed cascade reaction described by Zheng and Studer has very broad substrate scope, being compatible with quinolines, pyridines, and several quinolines with an additional fused benzene ring.

Photoredox catalysis in combination with organocatalysis has also been used to bring about radical cross-coupling reactions. This is not a trivial matter when enantioselective reactions are involved, due to high reactivity of radicals and the possibility of strong racemic background reactions [50]. Chemoselectivity could be a problem, with homocoupling competing [56, 57]. However, this approach has been successfully used a number of times. Jiang et al. obtained β-amino alcohols in which the amino group is primary and the alcohol tertiary, by coupling N-aryl glycines **51** with acyclic and cyclic activated ketones (1,2-diketones (**53**) and N-Boc isatins (**56**)) (Figure 32.7a) [58]. SPINOL-CPAs **53** and **55**, in combination with DPZ, performed well and the products were obtained in high yields and ees. The reactions were compatible with the formation of heteroquaternary stereocenters. This example represents the first catalytic asymmetric method ever reported for the synthesis of enantioenriched 1,2-amino tertiary alcohols. The rational for the reaction design was based on the fact that product formation could probably be achieved via radical coupling (Figure 32.7b). Ketyls **A** can be generated directly from 1,2-diketones with DPZ e.g. $E_{1/2}^{red}$ (benzyl) = −1.169 V and −1.251 V vs. a SCE in CH_3CN and $E_{1/2}^{red}$ (DPZ) = −1.45 V; vs. SCE in CH_2Cl_2. In addition N-aryl glycines can be precursors of α-amino radicals that can generate radical species **B** via single electron oxidative decarboxylation by the visible-light-activated DPZ (N-phenyl glycine, carboxylate ion: Ep = +0.52 [for the N moiety] and +1.09 V [for the COO⁻ moiety] vs. Ag/AgCl in CH_3CN). A similar radical approach was used by Jiang et al. to couple N-aryl amino acids and α-halo carbonyl compounds [59] and to couple 3-chlorooxindoles with N-aryl glycines [60].

A similar SPINOL-CPA **58**/DPZ system could be used to couple N-aryl glycines to a range of α-branched 2-vinylpyridines **59** yielding, by decarboxylation, conjugate addition and enantioselective protonation, γ-chiral aryl amines **60**, in high yields and ees (Figure 32.7c) [61]. The procedure could also be applied with 2-vinylquinolines with SPINOL-CPA **61** instead of **58**.

The versatility of the DPZ/SPINOL-CPA system was further demonstrated when it was utilized for dehalogenation-enantioselective D_2O deuteration of racemic α-halogenated azaarenes (Figure 32.7d) [62]. A large range of α-chiral-α-deuterated quinolones **62** and related heterocycles (benzothiazole, phenanthidrine, pyridine) could be obtained with 80–99% ee and with up to 95% incorporation of deuterium, using this procedure, although with the pyridine-based system the ee was lower (53%). In this process an HE (**64**) acts as the terminal reductant. The reduction-deuteration of azaarene-substituted ketones was also feasible.

2,2-Disubstituted indolin-3-ones, structural entities found present in many alkaloids, may also be obtained via cooperative photoredox catalysis and chiral CPA catalysis. In 2014, Xiao et al. [63] reported the first example of an enantioselective approach to these substances via a cascade aerobic oxidation and semipinacol rearrangement of 2-aryl-3-alkyl-substituted indoles **66** (Figure 32.7e). $Ru(bpy)_3Cl_2 \cdot 6H_2O$ and 3,3′-anthracyl-substituted CPA **67** provided the activation needed for this transformation in the presence of dioxygen and white light emitting diodes (LEDs). In these transformations, mechanistically, the superoxide anion radical adds to a radical cation **A** produced from the indole via single-electron oxidation, to yield species **B**. Proton transfer and O–O cleavage give rise to a tertiary alcohol **C** which undergoes the semi-pinacol rearrangement to yield the final products **68**. This chemistry was further explored by Jiang [64]. The mild reaction

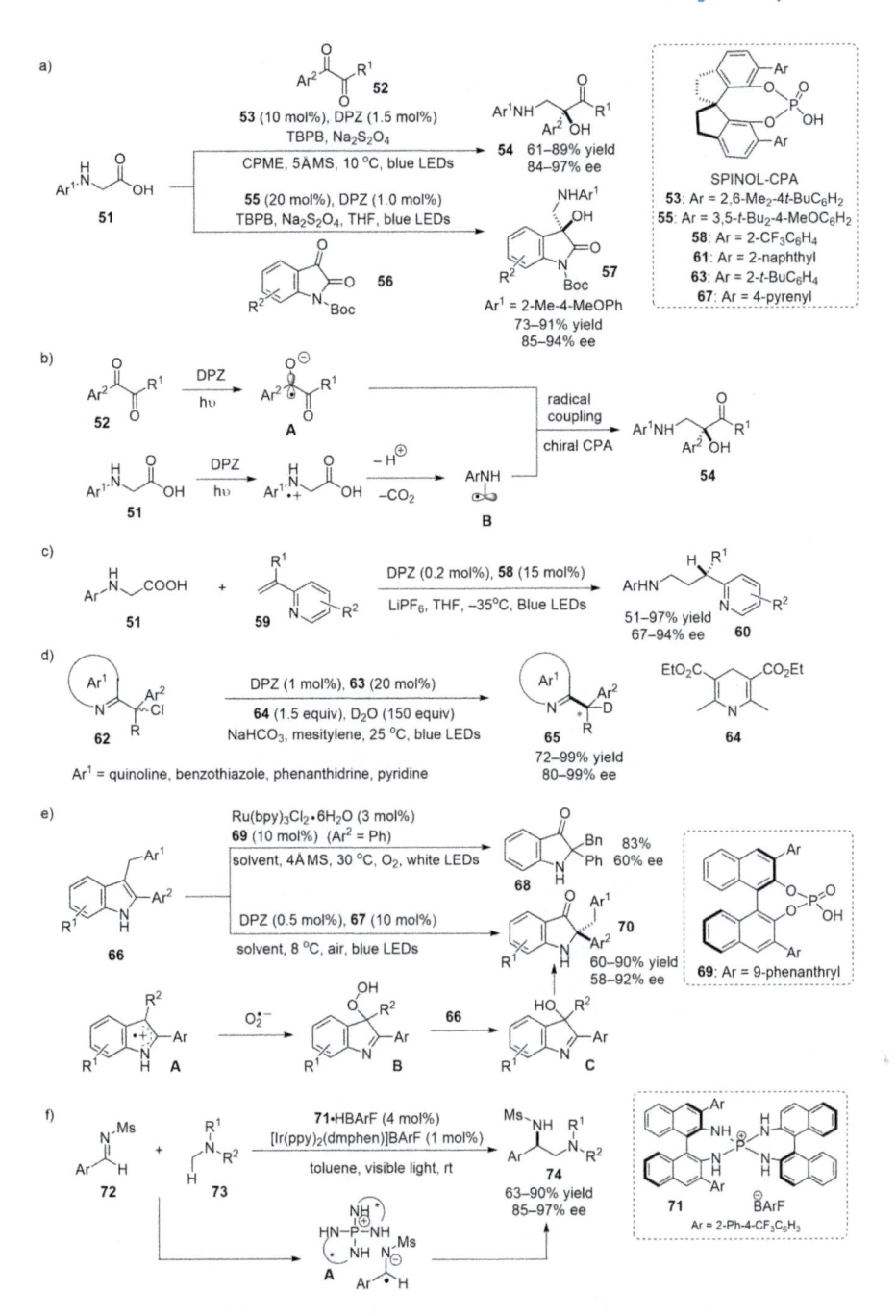

Figure 32.7 (a) Enantioselective radical coupling of activated ketones with N-aryl glycines [58] / Royal Society of Chemistry / CC BY 3.0. (b) The mechanism proposed for the radical coupling [58] / Royal Society of Chemistry / CC BY 3.0. (c) Enantioselective conjugate addition–enantioselective protonation of N-aryl glycines to α-branched 2-vinylazaarenes [61]. (d) Enantioselective α-deuteration of azaarenes with D_2O [62]. (e) Enantioselective synthesis of 2,2-bisubstituted indolin-3-ones by dual photoredox/chiral phosphoric acid (CPA) catalysis [63, 64]. (f) Asymmetric α-coupling of N-arylaminomethanes with aldimines [65]. BArF = $[3,5\text{-}(CF_3)_2C_6H_3]_4B$; CPME = cyclopentyl methyl ether; dmphen = 2,9-dimethyl-1,10-phenanthroline; DIPEA = diisopropyl ethyl amine; dmphen = 2,9-dimethyl-1,10-phenanthroline; TBAPNP = tetrabutylammonium 4-nitrophenoxide; tbpb = tetra-n-butylphosphonium bromide; TIPS-EBX = 1,2-benziodoxol-3(1H)-one.

conditions and green oxidizing agent (dioxygen) make this a valuable method to obtain these substances, previously only attained by a stepwise route with harsher oxidizing agents and reaction conditions. Jiang et al. found that the reactions could be conveniently performed under normal atmospheric conditions (in air) if DPZ was used as catalyst with CPA **69** and obtained a large range of 2,2-disubstituted indolin-3-ones **70** in high yields and ees.

Córdova et al. had previously reported an early example of the utilization of dioxygen as a reagent, in an enantioselective α-oxygenation of aldehydes co-catalyzed by a chiral secondary amine and tetraphenylporphyrin as photocatalyst [66]. An intermediate peroxide was obtained which could be reduced in situ with NaBH$_4$ to afford the corresponding enantioenriched diols.

CPAs have not been the only Brønsted acids explored for cooperative catalysis under photoredox conditions [50]. Ooi et al. developed a *P*-spiro chiral arylaminophosphonium barfate **71** that made possible a highly enantioselective α-coupling between aldimines **72** and *N*-arylaminomethanes **73** (Figure 32.7f) [65]. The resulting chiral 1,2-diamines **74** were obtained in high yields. A reaction mechanism involving chiral cation-directed catalysis was proposed, with ion-pairing between the chiral tetraaminophosphonium ion and an α-amino anion-radical **A** generated photochemically by the action of iridium catalyst [Ir(ppy)$_2$(bpy)]BarF being responsible for the enantiocontrol obtained. This fact was confirmed when a reaction was performed with tetrabutylammonium salt (Bu$_4$N·BArF), which lacks the H-bond-donor ability, in place of the chiral catalyst **71**·HBArF under identical photoredox conditions in the presence of a photocatalyst, which yielded mainly the corresponding *meso* product. The utilization of nonionic Brønsted acids such as the phosphoric acid diester, benzoic acid, and the 3,3′-Ph$_2$-BINOL as H-bond donors was also ineffective and the *meso* product was obtained as the major product along with other unidentified side products, reinforcing the idea that the ionic H-bond donor character is a prerequisite for catalysis in this case.

32.2.4 Miscellaneous

CPAs have been utilized in catalysis both in their acid forms and as Brønsted bases. The latter approach was utilized by Knowles et al. to obtain from tryptamines **75** the three-ring-fused system of pyrroloindolines **76** via the formation of indole radical cation intermediates (Figure 32.8a) [67]. Catalysis in this case takes place via PCET, with the neutral free radical intermediates existing as H-bonded adducts with a chiral Brønsted base. The 2,2,6,6,-tetramethylpiperidine 1-oxyl radical (TEMPO) worked well as the radical source. Several *N*′-Cbz-protected tryptamines were converted into pyrroloindolines in a highly enantioselective manner when H8-TRIP phosphate (**77**) was used as catalyst, by interception of a TEMPO radical in a sequential single-electron oxidation to **A**, radical coupling to **B** and addition processes. TIPS-EBX (1,2-benziodoxol-3(1*H*)-one) is used as the sacrificial oxidant and proton acceptor. Although TIPS-EBX is a poor one-electron oxidant ($E_{1/2} = -1.12$ V vs SCE), and it cannot react directly with the indole by electron transfer reactions, it can do so easily with the Ir(ppy)$_3$ photocatalyst in its excited state (*$E_{1/2} = -1.73$ V vs SCE). The carboxylate base produced in this process is the stoichiometric acceptor for the proton liberated in the catalytic cycle, a step which prevents the formation of TEMPO-H. Interestingly, the products of these reactions, the TEMPO-substituted pyrroloindolines **76**, could be reacted further forming carbocation intermediates via a catalytic single-electron oxidation/mesolytic cleavage under the influence of the iridium catalyst [Ir(dCF$_3$Me-ppy)$_2$(dtbbpy)]-PF$_6$ and blue LEDS and undergo a substitution reaction with a variety of nucleophiles. A range of alkaloid natural products, including (−)-calycanthidine, (−)-chimonanthine and (−)-psychotriasine were obtained via this route in four steps from tryptamine **75**.

Brønsted base (phosphate base) catalysis in combination with photoredox catalysis and HAT also made possible the deracemization of a racemic urea (±**78**) (Figure 32.8b) [68]. Upon photooxidation to **A** and proton transfer, an achiral radical intermediate (**B**) is generated, which undergoes

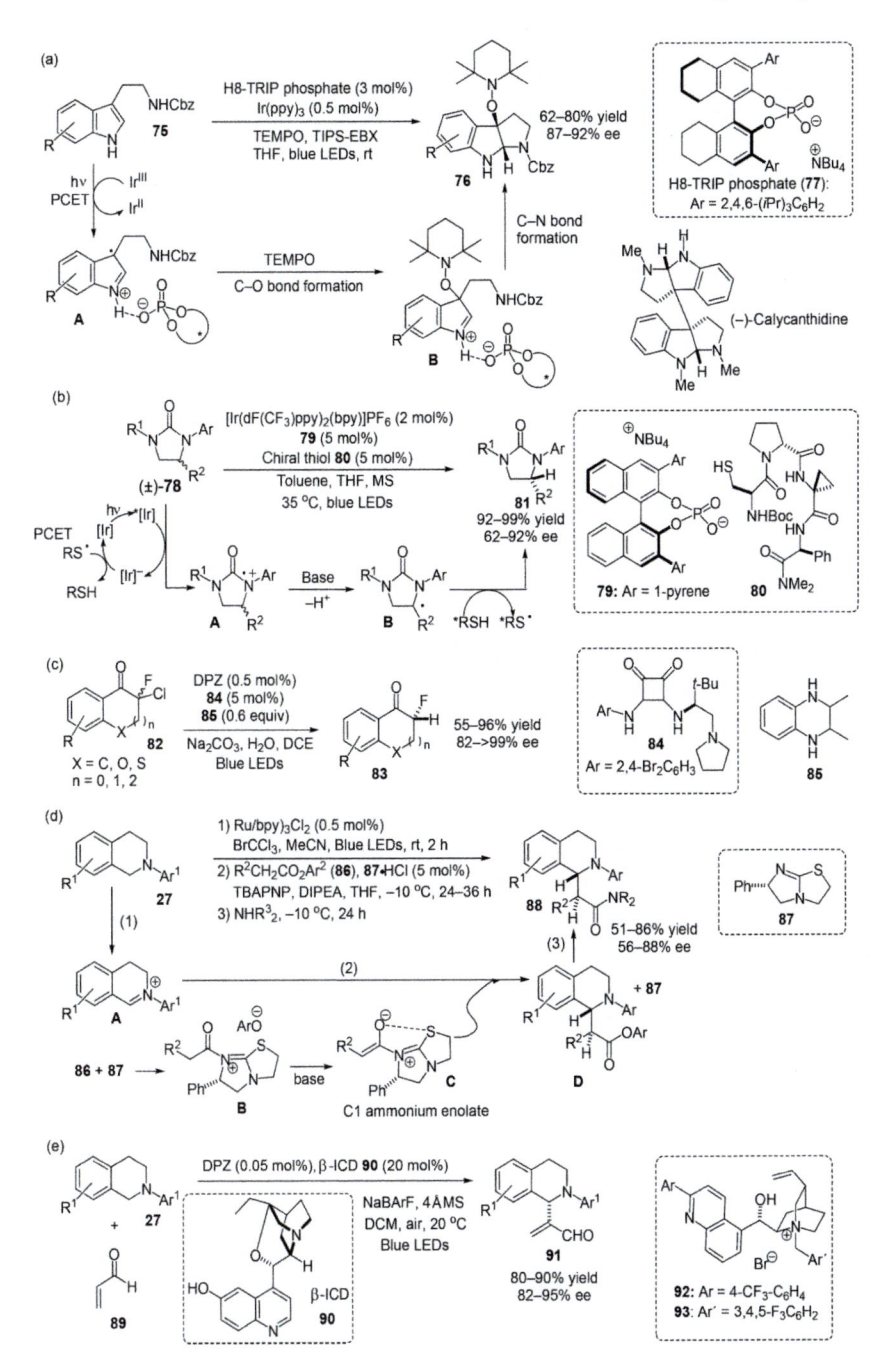

Figure 32.8 (a) Enantioselective synthesis of pyrroloindolines [67]. (b) Light-driven deracemization of ureas [68]. (c) Enantioselective photoredox dehalogenative protonation [72] / Royal Society of Chemistry / CC BY 3.0. (d) Isothiourea-catalyzed enantioselective addition of 4-nitrophenyl esters to iminium ions [73]. (e) Enantioselective aerobic oxidative C(sp^3)–H olefination of amines [75]. NaBArF = sodium tetrakis[(3,5-trifluoromethyl)phenyl]borate; TBADT = tetrabutylammonium decatungstate.

enantioselective HAT when a chiral thiol catalyst is used. Ees up to 88% were obtained for compounds **81** when the method was applied to several ureas [68].

Chiral tertiary α-haloketones are important building blocks for numerous biologically active substances [69]. Direct halogenation can be used, although enantioselective methods for α-chlorination and α-fluorination of aliphatic ketones have been developed relying on enamine catalysis, they are not compatible with aromatic variants, because of the slow enamine formation and the formation of approximately equimolar enamine rotational isomers [70, 71]. In 2019, an alternative was developed, based on enantioselective photoredox dehalogenative protonation (Figure 32.8c) [72]. This method allowed a range of cyclic and acyclic ketones with labile chiral secondary C–F, C–Cl and C–Br bonds at the α-position to be obtained in high yields and good to excellent ees **83** from dihalogenated compounds **82**. DPZ was used as photosensitizer in combination with a chiral H-bonding catalyst, a L-*tert*-leucine-based squaramide amine **84** under visible light. 2,3-Dimethyl-1,2,3,4-tetrahydroquinoxaline (**85**) acted as the terminal reductant in this case.

The synergistic combination of chiral isothiourea/photoredox catalysis was utilized by Smith et al. to couple THIQ (**27**) with aryl esters (**86**) via the generation of C1-ammonium enolates (Figure 32.8d) [73, 74]. The esters obtained are converted in situ to amides by reaction with amines and the amide-functionalized THIQs **88** are obtained in high yields and excellent ers. According to the mechanism proposed, the THIQs are initially oxidized to iminium salts **A** under photocatalytic conditions. The Lewis basic isothiourea is acylated by the active aryl ester to **B** with release of a nucleophilic aryloxide. Enolization of this intermediate results in the formation of an ammonium enolate **C**, conformationally locked due to an n_O to σ^*_{C-S} interaction between the enolate O and the catalyst **87**. This interaction eventually determines the enantioselectivity, since addition takes place preferentially *anti*- to the phenyl stereodirecting group. Nucleophilic attack of the aryloxide on the acyl ammonium, gives the β-amino ester product **D**, releasing at the same time the catalyst for a new cycle. Tetrabutylammonium 4-nitrophenoxide (TBANP) increases the catalytic activity of the isothiourea Lewis base catalyst and increases the polarity of the reaction. The main challenge in this approach was to select an electrophilic reaction partner (in this case the tetrahydroisoquinolinium) to react with the catalytically-generated C(1)-ammonium enolate, which was at the same time compatible with the nucleophilic tertiary amine catalyst as well as with the aryloxide.

Lewis base catalysis was also combined with PC to promote a CDC between THIQs and enals **89** (Figure 32.8e). DPZ was used as photosensitizer, cinchona alkaloid derivative β-isocupreidine (β-ICD) (**90**) as chiral catalyst and air (dioxygen) worked well as oxidant. High yields and ees of α-alkenylated products **91** were obtained for a large range of substrates [75, 76].

The enantioselective aerobic oxidative C(sp^3)–H olefination of tetrahydro-β-carbolines was accomplished under similar conditions [75].

Another interesting example of enantioselective oxidative C–H functionalization of THIQs was their reaction with silyl ketene acetals. β-enamino esters were produced in high yields and ees when photoredox catalysis with [Ru(bpy)$_3$Cl$_2$] was combined with asymmetric anion-binding catalysis obtained with an amide-based thiourea [77].

Cinchona alkaloids were also utilized by Melchiorre et al. as phase transfer catalysts for the asymmetric perfluoroalkylation of β-ketoesters. Quaternary cinchonine ammonium salts **92** and **93** allowed the reaction to be performed under heterogeneous conditions [78].

32.3 Metal Catalyzed Processes

32.3.1 Dual Transition Metal/Photoredox Catalysis

The dual approach utilizing a metal and a photoredox catalyst has been named metallaphotoredox catalysis by MacMillan. Nickel, copper, chromium and more rarely palladium, are the main metals utilized in a synergistic approach with photoredox catalysts [15, 18]. The chirality is provided by chiral ligands, and hence efficient ligand design plays an important role in this area. Besides the ligand, the oxidation state, as well as catalyst excitation, can be used to modulate the reactivity of the metal. The reactions highlighted in this section are distinct from those in which the metal plays simply the role of a Lewis acid, which are discussed later [79].

Metallaphotoredox catalysis is particularly useful for C(sp^3) fragment couplings, traditionally challenging to achieve with transition metals alone [18]. Nickel catalysis works well in these cross couplings, because of the ability of nickel catalysts to undergo facile oxidative addition with alkyl electrophiles at room temperature, their favorable single-electron redox potentials, and the lower susceptibility to undergo β-hydride elimination steps [18]. Traditionally cross-coupling reactions have been achieved with two-electron transmetallation reactions, whereas the dual nickel/photocatalytic approach allows cross coupling to take place via single electron transmetallation. Pioneers in this area were Molander et al. [80] and Doyle, MacMillan et al. [81], who published works in this field in 2014. In the first example, potassium alkoxyalkyl- and benzyltrifluoroborates were cross-coupled with a range of aryl bromides at rt, in the absence of a strong base and under visible light, to yield arylated products in good yields. Only one example of an enantioselective coupling was reported, of a secondary benzyltrifluoroborate (**94**) with **95** and with the chirality provided by chiral ligand **96**, but the ee of cross-coupled product **97** was only moderate (Figure 32.9a). Doyle et al. and MacMillan et al. described a similar cross-coupling reaction proceeding with Ni/photoredox co-catalysis, but employed carboxylic acids instead of organoboron reagents [81]. This procedure was extended to a highly enantioselective decarboxylative C_{sp}^2–C_{sp}^3 cross-coupling with *N*-Boc-protected amino acids **98** and compounds **99** through the use of chiral cyano-bisoxazoline ligand **100** (Figure 32.9b) [82]. This reaction has good functional group tolerance, allowing the presence of carbamate, ether, ester, alkyl chloride, carbonate, indole, and thiophene substituents in a large range of amino acids and aryl halides (or halo-substituted pyridines), with the α-arylated α-amino acids **101** being obtained with ees as high as 96%. This procedure replaces the organometallic reagents that are traditionally used as coupling partners with amino acids. According to the reaction mechanism proposed, photocatalyst-mediated oxidation and decarboxylation of a α-amino acid produces a prochiral α-amino radical **A**. The aryl halide is activated by oxidative addition, which results in the formation of a Ni(II)–aryl complex **B**, that reacts with the newly generated α-amino radical. The resulting diorganonickel(III) adduct **C** undergoes reductive elimination producing the desired coupled product.

The enantioselective nickel/photoredox catalyzed asymmetric reductive cross-coupling of racemic α-chloro esters **102** with aryl iodides **103** was reported for the first time by Mao, Walsh et al. in 2020 (Figure 32.9c) [83]. The best photoredox catalyst for this transformation was 1,2,3,5-tetrakis(carbazol-9-yl)-4,6-dicyanobenzene (4CzIPN) (**104**) and an organic reductant (HE **106**) was used, whereas most reductive cross-coupling reactions use stoichiometric metals. α-Aryl esters **107** were obtained in this case in good yields and high ees.

The highly enantioselective cross-coupling of alkyl benzenes **108** and aryl bromides **99** by photoredox/nickel dual catalysis was made possible with chiral sterically hindered biimidazoline ligands **109** by Lu et al. (Figure 32.9d) [84]. In this C(sp^3)–H bond arylation functional group compatibility was achieved under mild conditions and external redox reagents were not required. A reaction mechanism was proposed based on mechanistic studies and previous findings (Figure 32.9e). An initial oxidative addition of the aryl bromide to in-situ generated Ni(II) complex **A** produces **B** which undergoes single-electron oxidation induced the photoexcited Ir catalyst to give **C** and an active bromine atom. Ir is reduced to Ir(II) simultaneously. The occurrence of an electron transfer process in the initiation of the bromine-free radical was also considered a possibility. The benzylic radical was formed, by an HAT process, from the bromine free radical (BDE$_{HBR}$ = 366 kJ/mol, BDE$_{EtPh}$ = 357 kJ/mol, BDE$_{dioxane}$ = 406 kJ/mol) with 4,4′-dimethoxybenzophenone (DMBP)

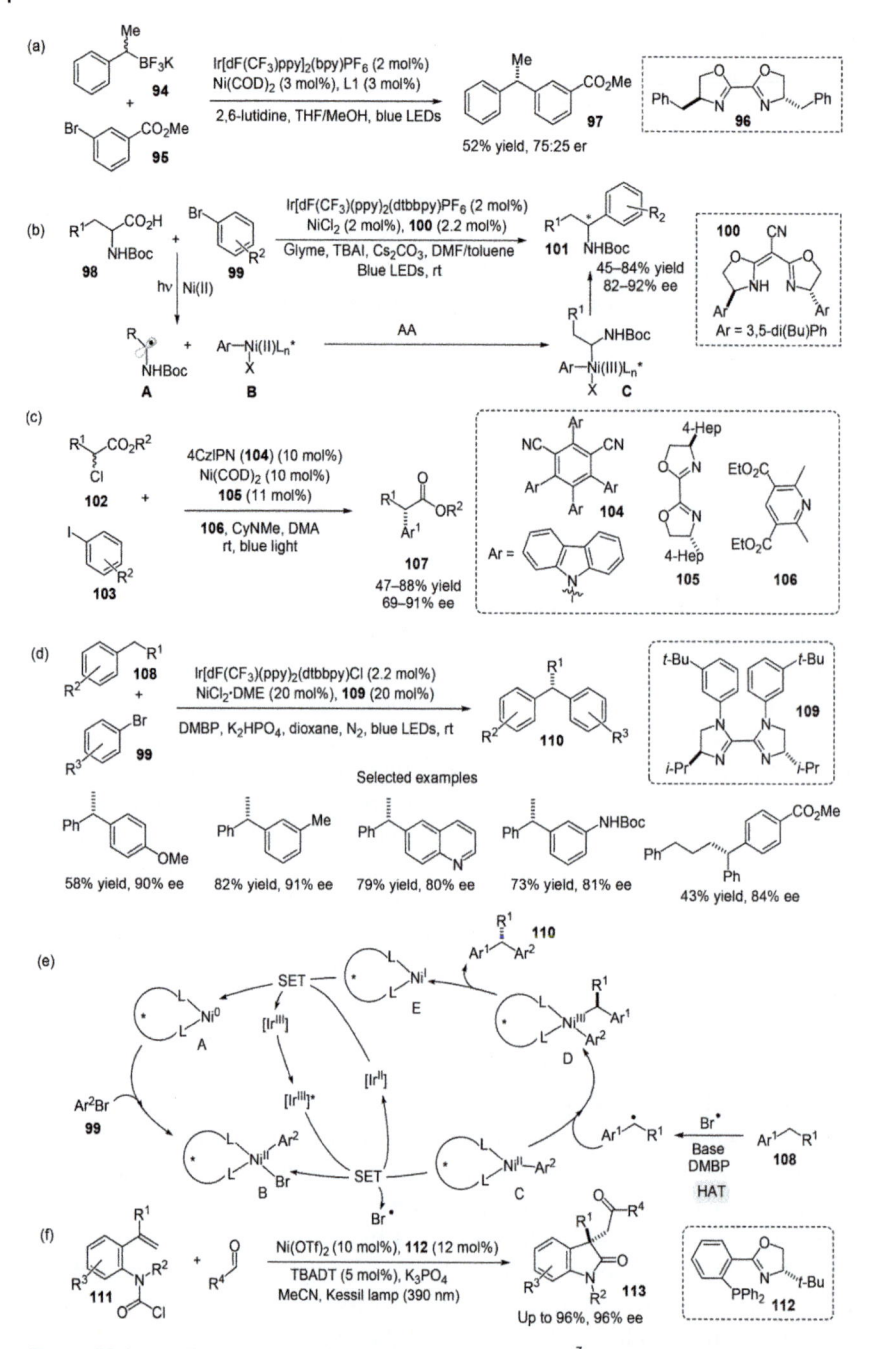

Figure 32.9 (a) Enantioselective coupling of α-carboxyl sp^3-carbons with aryl halides [81]. (b) Enantioselective decarboxylative arylation of α-amino acids [82]. (c) Asymmetric reductive cross-coupling of racemic α-chloro esters with aryl iodides [83]. (d) Enantioselective benzylic CH arylation [84] / Springer Nature / CC BY 4.0. (e) The mechanism proposed for the arylation reaction shown in (d) [84] / Springer Nature / CC BY 4.0. (f) Synthesis of 2,2′disubstituted indoles by enantioselective nickel-photoredox catalysis [85]. COD = 1,5-cyclooctadiene; TBAI = tetrabutyl ammonium iodide; CyNMe = *N,N*-dimethylcyclohexylamine; 4CzlPN =1,2,3,5-Tetrakis(carbazol-9-yl)-4,6-dicyanobenzene.

as a co-catalyst, being trapped by a aryl Ni(II) species **C** to afford Ni(III) complex **D** (BDE = bond dissociation energy). Reductive elimination gives rise to the chiral 1,1-diaryl alkanes and a Ni(I) complex **E**, which can undergo single-electron reduction to regenerate Ni(0) species **A** and the photocatalyst.

Wang et al. developed a procedure for the synthesis of 3,3′-difunctionalized 2-oxindoles [85] based on an asymmetric difunctionalization of alkenes, merging nickel and PC. In this process aldehydes were used as an acyl source via tetrabutylammonium decatungstate (TBADT) catalyzed acyl C–H activation, which was combined with nickel catalysis in the presence of a monophosphine-oxazoline chiral ligand **112** (Figure 32.9f). The desired products **113** were obtained in high ees [85]. This was the first example of dual HAT photochemistry and asymmetric transition metal catalysis in alkene difunctionalization. High ees were obtained for a range of oxindoles produced containing a quaternary stereogenic center. This process, however, required light with a wavelength just under the visible region, 390 nm.

The combination of photocatalysts with copper catalysts has been used to incorporate functional groups into organic molecules, but C–C bond forming reactions have also been particularly successful, and more specifically cyanation reactions. A reactive chiral copper(II) cyanide can be used to trap benzylic radicals enantioselectively, and in this way optically pure benzyl nitriles can be synthesized efficiently. In 2016, this was demonstrated by Liu et al. via single electron oxidation achieved using an electrophilic F$^+$ reagent (NFSI = F–N(SO$_2$Ph)$_2$) as oxidant and as chiral catalyst a copper-bisoxazoline complex [86]. More recently Lin, Liu et al. found that *N*-hydroxyphthalimide (NHP) esters **114** could be used as sources of benzylic radicals in enantioselective decarboxylation reactions without employing electrophilic oxidants, if the reaction was performed under a visible light source in the presence of a photocatalyst (Figure 32.10a) [87]. The reactions ran under mild conditions (rt), with trimethylsilyl chloride (TMSCl) as cyanide source, had broad substrate scope, and high yields and high ees were obtained with a Cu-**115** complex as a catalyst for products **116**. The reaction could also be conducted on a 270 mmol scale, and it was applied to the synthesis of key intermediates for the synthesis of the chiral antidepressant (*R*)-Phenibut and weight-losing drug (*R*)-Lorcaserin. In the mechanistic proposal (Figure 32.10b), the excited photocatalyst transfers an electron to NHP, initiating the decarboxylation to **A** and the formation of L*Cu(II)CN. Radical rebound of the benzylic radical with L*Cu(II)(CN)$_2$ and reductive elimination, yields the product **116** and regenerates the catalyst.

The enantioselective radical cyanoalkylation of styrenes **117** has also been described [88]. In this copper-catalyzed reaction the alkyl group radicals are produced from NHP esters as in the previous methods, with dual copper/photoredox catalysis and TMSCN as a cyanating agent (Figure 32.10c) [88].The method, which represents the first example of photoredox and copper-catalyzed radical difunctionalization of alkenes, is compatible with primary, secondary, and tertiary alkyl substituted NHP esters. The products **119** were obtained with high to excellent ees. A further development was the reaction of *N*-alkoxypyridinium salts with readily available silyl reagents (TMSN$_3$, TMSCN, TMSNCS), which led to δ-azido, δ-cyano, and δ-thiocyanato alcohols in high yields [89]. The products were obtained via a domino process involving alkoxy radical generation, 1,5-hydrogen atom transfer (1,5-HAT) and copper-catalyzed functionalization of the resulting C-centered radical. It was shown that an enantioselective approach was also possible, and δ-C(sp^3)–H cyanation was achieved with TMSCN and a mixture of copper acetate, chiral ligand *ent*-**115** and *fac*-Ir(ppy)$_3$ under irradiation by blue LEDs. 4-Phenyl-2-cyanobutanol was obtained in this way in 74% yield and 86% ee from a *N*-alkoxypyridinium salt of 4-phenylbutanol.

The Nozaki-Hiyama-Kishi (NHK) reaction is the chromium catalyzed allylation of aldehydes. NiCl$_2$ is used as a co-catalyst. The reaction has excellent chemoselectivity for aldehydes over ketones or esters and the *anti* diastereomer is obtained regardless the double bond geometry of the original allylic C–C double bond. One drawback is the stoichiometric amounts generation of toxic chromium

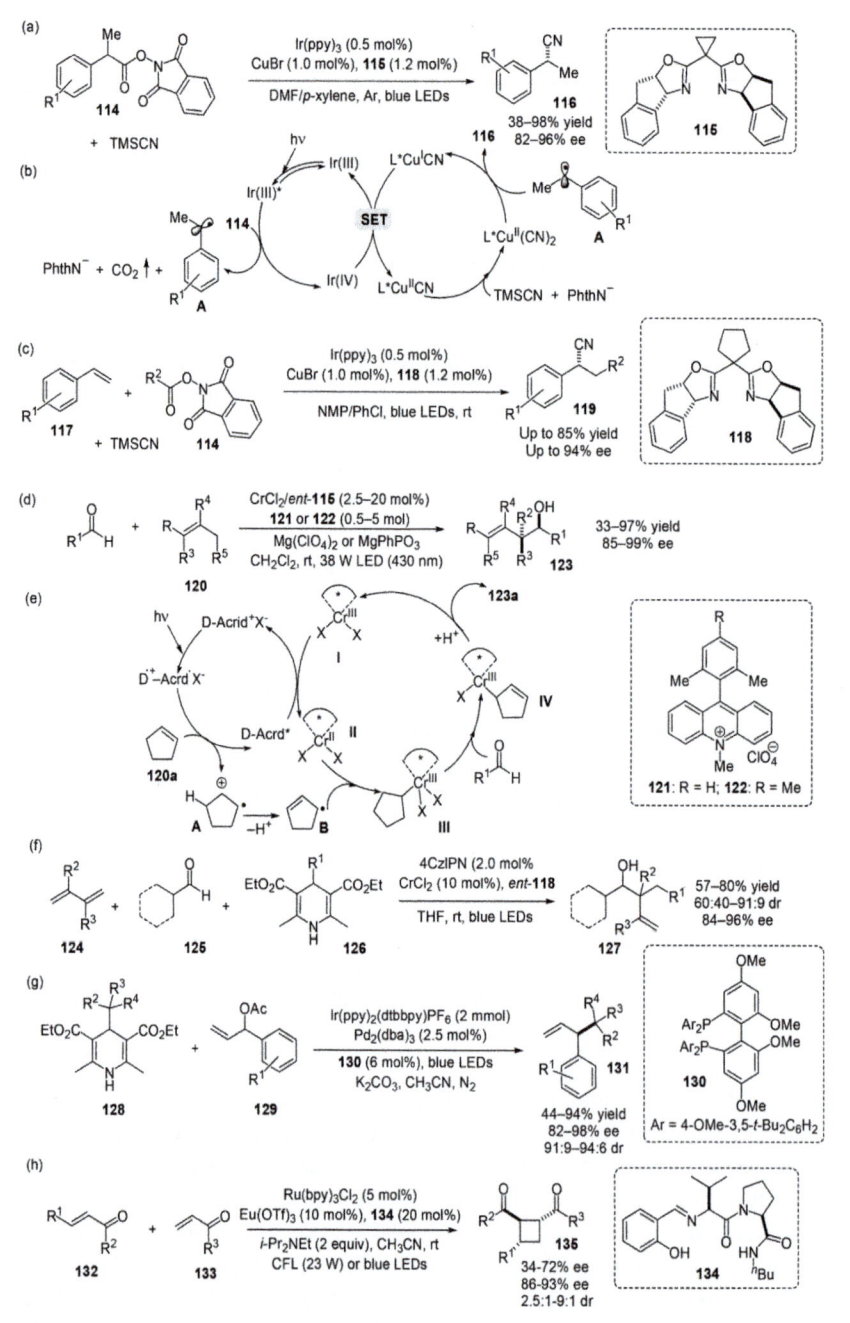

Figure 32.10 (a) Enantioselective decarboxylative cyanation [87]. (b) The mechanism proposed for the decarboxylative cyanation [87]. (c) Enantioselective radical cyanoalkylation of styrenes [87]. (d) Cyanoalkylation of unactivated alkenes [92] / Royal Society of Chemistry / CC BY 3.0. (e) The mechanism proposed for the cyanoalkylation of alkenes [92] / Royal Society of Chemistry / CC BY 3.0. (f) Dialkylation of 1,3-dienes by dual photoredox and chromium catalysis [93]. (g) Enantioselective allylic alkylation with 4-alkyl-1,4-DHPs [94]. (h) Enantioselective [2+2] photocycloadditions of enones using visible light [95]. CFL = compact fluorescent light; Dba = dibenzylideneacetone; NMP = N-methyl-2-pyrrolidinone.

waste. Related reactions employ excess of manganese as a terminal reductant [90]. In 2018, Glorius et al. developed a redox neutral diastereoselective procedure using dual chromium/photoredox catalysis, which expands the scope of the NHK reaction. One enantioselective example was tried but the ee was only moderate [91]. Masamune et al. subsequently developed an enantioselective version of the chromium-photoredox process for alkenes **120** using a chiral bisoxazoline ligand (*ent*-**115**) and an acridinium (acr) photocatalyst (**121** or **122**) (Figure 32.10d) [92]. An interesting aspect of this work is that it reports the unprecedented use of unactivated hydrocarbon alkenes (e.g. **120a**) as precursors to chiral allylchromium nucleophiles for the asymmetric allylation of aldehydes. The homoallylic alcohols produced were obtained with a diastereomeric ratio >20/1 and up to 99% ee.

According to the mechanism proposed (Figure 32.10e), a cationic allyl radical **A** is produced from the alkene via electron-transfer oxidation of the π-bond by a photoexcited electron-donor substituted acr photocatalyst. After deprotonation to **B** it is intercepted by the reduced chiral chromium(II) catalyst **II** to give chiral allyl chromium(III) complex **III**. The new complex reacts with an aldehyde via a six-membered chair transition state to produce an enantiomerically-enriched chromium alkoxide **IV** in a *syn*-selective manner. After protonation, the desired product is obtained, as well as an oxidized Cr(III) complex, regenerated by the reduced form of the photocatalyst.

Subsequently Glorius et al. reported that Cr(III) catalysis with chiral bisoxazoline *ent*-**118** and the same photocatalysts could be used for the regio- and diastereoselective three-component dialkylation of 1,3-dienes **124**, with aldehydes **125** and alkyl HEs **126** for the synthesis of homoallylic alcohols **127** [93]. The ees obtained in this case were also very high (Figure 32.10f).

Although palladium is proven to be one of the most useful metals for catalytic applications, so far there have been very few reports of enantioselective reactions in which Pd is involved in dual catalysis with a photoredox catalyst. Zhang et al. developed a dual catalytic process, in which alkyl radicals generated from 4-alkyl-1,4-dihydropyridines (DHPs) **128** act as the coupling partners to π-allyl palladium complexes bearing chiral phosphine ligands (e.g. **130**) (Figure 32.10g) [94]. A variety of allyl esters **129** are compatible with this procedure, that expands the scope of the traditional Pd-catalyzed asymmetric allylic alkylation reaction and is at the same time an alternative and potential complement to the same. The allylic alkylated products **131** were obtained with high ees, as the chiral phosphine ligand controlled the facial selectivity of π-allylpalladium complex during alkyl radical addition.

32.3.2 Dual Chiral Lewis Acid/Photoredox Catalysis

The utilization of Lewis acids in photoredox catalysis is in great part linked to the cycloaddition reaction [2+2] [18]. Cycloaddition reactions have been known for a long time, with ultraviolet radiation providing the required activation. Enantioselective variants with substoichiometric amounts of catalysts proved to be very difficult to develop, mainly because uncatalyzed background photochemical processes (of catalyst-unbound substrate) were difficult to control. Free substrates produced racemic products or afforded low ees, unless the racemic reaction could be slowed down. Bach et al. devised a solution to this problem, designing reactions in which the catalyst-substrate complex absorbed light at longer wavelengths than the free substrate. A chiral hydrogen-bonding xanthone-based photosensitizer [96, 97] was utilized or a chiral Lewis acid catalyst [98], capable of inducing a bathochromic shift in the bound substrate to 366 nm and in this way high ees were obtained. The methods had a limitation that irradiation with a monochromatic light source that selectively excites the catalyst-substrate complex at a wavelength where absorption by the free substrate is minimized was required for stereocontrol. In addition, with the chiral Lewis acids high catalyst loadings had to be used to overcome the background reactions (typically approx. 50 mol%).

A general strategy to eliminate completely the uncatalyzed background photochemistry had not been found until recently. The works of Back inspired Yoon et al., who found a solution in 2014 using dual chiral Lewis acid/photoredox catalysis [95]. A [2+2] cycloaddition reaction of α,β-unsaturated ketones **132** (2′-hydroxychalcones) with enones **133** could be performed under visible light irradiation, with 10 mol% of Lewis acid europium(III) triflate as catalyst, a chiral Schiff base ligand (**134**) and Ru(ppy)$_3$Cl$_2$ as photocatalyst, affording cyclobutanes **135** with high ees (Figure 32.10h). Lewis acid coordination lowers dramatically the triplet energy of the chalcone substrate and the background absorption in this case was not a problem, since the uncatalyzed reaction could not proceed under these conditions. There was also no reaction in the absence of light.

Subsequently, Yoon et al. applied similar methodology to the enantioselective [2+2] cycloaddition of 2′-hydroxychalcones and dienes, using scandium triflate as catalyst and a chiral bisoxazoline ligand, and Ru(bpy)$_3$(PF$_6$)$_2$ as photocatalyst and visible light [99]. The products were obtained in high yields (66–84%) and ees (up to 98%) and dr 2–4:1. Later it was shown by Yoon et al. that styrenes could also be reacted with 2′-hydroxychalcones with the same set of catalysts to afford products in high yields and ees [100]. The synthesis of unsymmetrical cyclobutanes by heterodimerization of olefins is still a significant challenge, particularly in an asymmetric fashion, and these contributions represent significant advances. In 2019, the enantioselective [2+2] cycloaddition was extended to the reactions of regular cinnamate esters with styrenes [101].

Formal [3+2] cycloaddition of cyclopropyl ketones **136** with styrenes using a gadolinium Lewis acid catalyst and chiral ligand **137** was also achieved with white light activation (Figure 32.11a) [102]. Products **138** were obtained with high ees. This method expands the available technologies for [3+2] cycloadditions, which previously relied on highly activated "donor–acceptor" cyclopropanes (e.g. cyclopropane diesters [103], vinylcyclopropanes [104]), giving access to densely substituted cyclopentanes in an enantioselective manner.

Yoon et al. reported later the conjugate addition reaction of a α-silylamine pronucleophile **139** to a β-substituted Michael acceptor **140** with a related dual catalysis strategy (Figure 32.11b) [105]. The reaction was postulated to proceed via an α-amino radical intermediate. The products **142** were obtained in high yields and ees. One limitation of this method was that for successful Michael addition, the presence of one N-aryl substituent was required. Aliphatic amines underwent protodesilylation, but subsequently there was no addition to the Michael acceptor. Prior to this example there had been only one report of stereocontrol in reactions of α-amino radical intermediates, which was obtained via an intramolecular conjugate addition reaction by Bach et al. using a chiral hydrogen-bonding photosensitizer [9], with pyrrole tethered quinolones, but it was achieved under UV radiation.

Xiao et al. utilized a chiral octahedral complex **144** formed in situ, to catalyze the enantioselective RCA reaction of alkyl and acyl radicals derived from DHPs **126** to enones **143** (the Photo-Giese reaction) under visible light activation to obtain products **145** (Figure 32.11c). With alkyl DHPs, Fukuzumi's acr photocatalyst Mes-Acr$^+$ was required as well. However, with acyl DHPs, no additional external photocatalyst was required [106].

Other metals are compatible with PC and have been utilized as chiral Lewis acids, not only transition metals like scandium [105] and copper [107–109], but also aluminum [110]. A more detailed coverage is not possible due to space limitations, but the potential for future developments seems evident.

32.4 Chiral Photocatalysts

In this section are considered reactions promoted by light and a single catalyst. That is, the chiral catalyst is one that possesses photoredox properties too. They may be either metal complexes chiral at the metal or chiral organocatalysts [111].

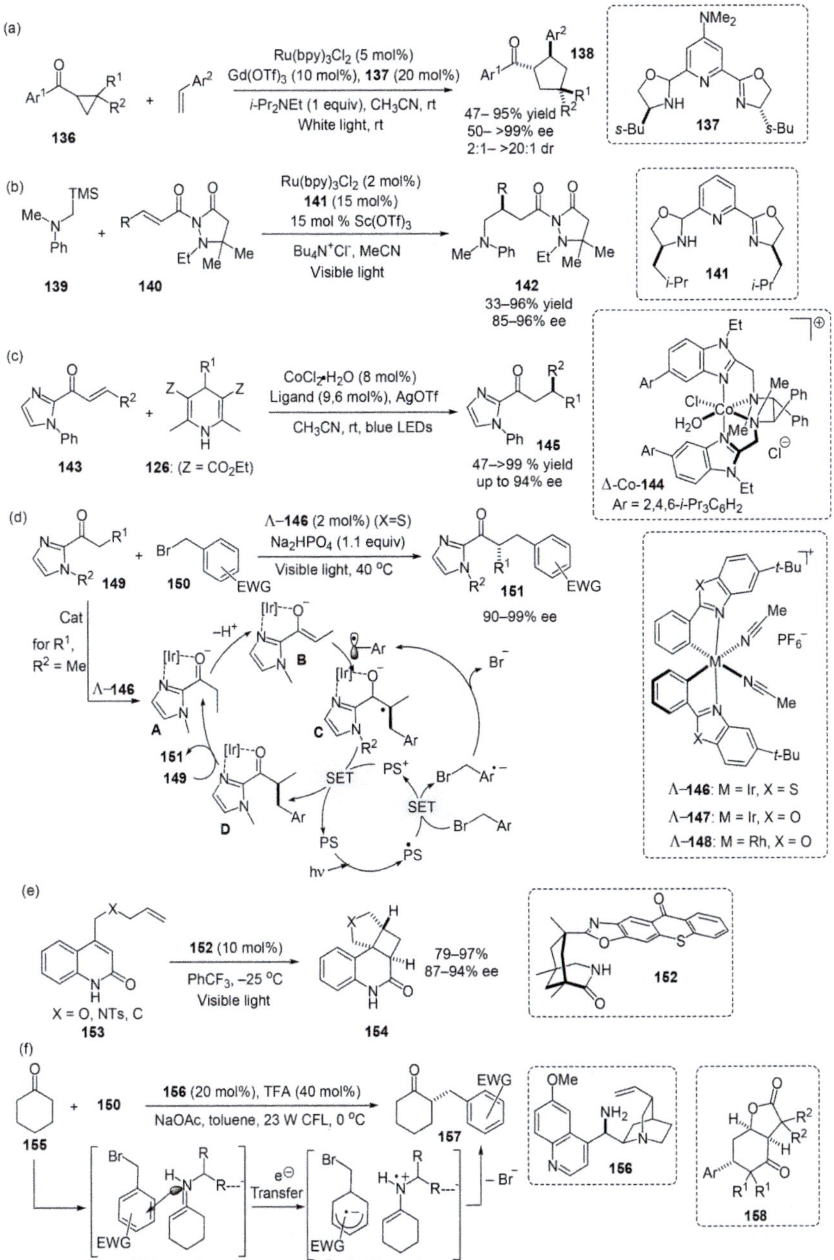

Figure 32.11 (a) Formal [3+2] cycloaddition of cyclopropyl ketones **136** with styrenes [102]. (b) Enantioselective conjugate additions of α-amino radicals derived from α-silyl amines [105]. (c) Enantioselective conjugate additions of α-amino radicals derived from DHPs to enones [106]. (d) The enantioselective alkylation of 2-acyl imidazoles with benzyl bromides [113]. (e) Visible light-mediated enantioselective [2+2] photocycloaddition of 4-substituted quinolones bearing an olefin in the tether [121]. (f) Intermolecular enantioselective α-alkylation of cyclic ketones with alkyl bromides [126].

32.4.1 Chiral-at-Metal Photocatalysts

The first catalysts that performed a dual role were described by Meggers et al. in 2014 [112]. The catalysts, bis-cyclometalated iridium(III) and rhodium(III) complexes, contain only achiral ligands. They are Λ- and Δ-enantiomers (left- and right-handed propellers, respectively), with two cyclometalating 5-*tert*-butyl-2-phenylbenzothiazoles and two acetonitrile ligands, as well as a hexafluorophosphate counterion (e.g. **146–148**) (Figure 32.11d; shows the Λ-enantiomers). The propeller shape chiral geometry is provided by the two cyclometalated ligands, and the acetonitriles are labile and can be replaced by substrate molecules, so that they function as Lewis acids. The iridium catalyst and its enantiopode, independently, were capable of catalyzing the enantioselective alkylation of 2-acyl imidazoles **149** with benzyl bromides **150** and phenacyl bromides, by C_{sp3}–H/C_{sp3}–H coupling, affording products **151** in 90–99% ee in the presence of visible light [113]. The 2-acyl-*N*-methylimidazole substrates tolerate steric, electron donating and electron accepting substituents in the phenyl moiety, but in the alkyl bromide, electron withdrawing substituents were required. In the dark there was no reaction (<5% of the product was observed), neither there was in the absence of catalyst.

According to the mechanism proposed, the iridium catalyst coordinates to the 2-acyl imidazoles in a bidentate fashion as in **A**, after which a nucleophilic iridium(III) enolate complex **B** is formed by deprotonation. A photo-reductively generated electrophilic radical then adds to the double bond generating an iridium-coordinated ketyl radical **C**. This ketyl intermediate undergoes oxidation to a ketone by single electron transfer, regenerating the Ir photosensitizer and complex **D**, which releases the product upon exchange with unreacted starting material, to start a new catalytic cycle. Enolate **B** provides the asymmetric induction and it also serves as the in situ-generated active achiral photosensitizer.

Subsequently, Meggers et al. developed several enantioselective processes to functionalize 2-acylimidazoles with many functional groups. For example, using chiral catalyst Λ-**147**, a highly enantioselective reaction of *N,N*-diaryl-*N*-(trimethylsilyl)methylamines and 2-acyl-1-phenylimidazoles was developed, which afforded the oxidative coupling products in good yields and excellent ees (90–98%) under visible light irradiation with a household lamp. The catalyst facilitates the desilylative oxidation of the α-silyl amines, producing electron rich α-amino radicals which are oxidized by air to iminium cations that undergo a Mannich reaction with the iridium enolate (C_{sp3}–H/C_{sp3}–H coupling) [114]. Chiral catalyst Λ-**146** was also successfully used in a highly enantioselective perfluoroalkylation of 2-acyl imidazoles with perfluoroalkyl iodides (CF_3I, C_3F_7I, C_4F_9I, $C_6F_{13}I$, $C_8F_{17}I$, and $C_{10}F_{21}I$) and perfluorobenzyl iodide, at the α-position of the carbonyl group by C_{sp3}–H/C_{sp3}–H coupling. This redox neutral process also required visible light activation [115]. Meggers, Gong et al. later applied a similar chiral rhodium(III) photoredox catalyst (Λ-**148**) in the aerobic enantioselective dehydrogenative cross-coupling of 2-acyl imidazoles with arylamine, a C_{sp3}–H/C_{sp3}–H coupling leading to α-amino alkylation, that also required visible light activation (blue LEDs) [116]. HEs were the alkyl radical sources. The catalysts could also be applied in their dual capacity in the enantioselective synthesis of 1,2-amino alcohols from tertiary amines and trifluoromethyl ketones via the formation of nucleophilic α-amino radicals and radical-radical recombination [117], in the asymmetric α-amination of 2-acyl imidazoles with arylazido and α-diazo carboxylic esters (by Csp^3–N bond formation) [118], and in the enantioselective radical amination of 2-acyl imidazoles using 2,4-dinitrophenylsulfonyloxy carbamates [119], amongst others. Recently Baik and Yoon described chiral-at iridium complexes as triplet ET photocatalysts in the enantioselective excited-state [2+2] photocycloaddition of 3-alkoxyquinolones.

Although this area of research is very new, a few more examples have been described recently. The interested reader may find further information in Yoon's recent review [111].

32.4.2 Organic Photocatalysts

There are organic molecules that behave not only as chirality transfer agents, but can also absorb light and convey the absorbed energy to reactant molecules to promote chemical reactions [111]. The first examples were described by Bach et al., the chiral xanthones referred to in the introduction and in section 3.2, which absorb radiation in the UV region of the spectrum [9, 97]. These organocatalysts were the subject of several studies by Bach et al., and eventually it was found that if the thioxanthones, e.g. **152** were used as catalysts instead, it was possible to promote some reactions with visible light [120]. In 2014, the enantioselective synthesis of cyclobutane rings by intramolecular [2+2] photocycloaddition of 4-substituted quinolones **153**, bearing an olefin in the tether, was found to lead to tetracyclic compounds **154** with high yields and ees under these conditions (Figure 32.11e) [121]. The same catalyst was used later for intermolecular [2+2] photocycloadditions of 2(1H)-quinolones to electron-deficient olefins, such as acrylates, also produced in high yields and ees [122]. According to the studies performed and reported by Bach et al., these reactions were possible due efficient triplet ET and high enantioface differentiation provided by the two-point hydrogen bonding between the catalyst and the substrates [123].

More recently Melchiorre et al. described another way to perform enantioselective reactions using one organocatalyst and visible light activation [124]. It relies on the formation of an electron donor acceptor (EDA) complex, when an electron acceptor substrate A and a donor molecule D (a Lewis acid and a Lewis base) aggregate in the ground state. Even if the neither of the two molecules absorbs visible radiation, the resulting EDA complex does, and upon excitation an intramolecular SET event takes place generating radicals, which can engage in further chemical reactions. These intermediate charge transfer complexes were first described by Melchiorre et al. [124, 125]. They also found that in reactions proceeding via the formation of transient enamines, which are commonly applied with the ground state organic reactions, the same enamines can absorb light via an intermolecular charge-transfer state (EDA route) and be used to initiate photoactive radical formation in the absence of external photosensitizers. This strategy was successfully employed in enantioselective γ- and α-alkylations of enals and aldehydes, respectively, with bromomalonates, using the Jørgensen–Hayashi aminocatalyst under compact fluorescent light (CFL) irradiation conditions.

Another development was the use of a quinidine-derived primary amine and trifluoroacetic acid (TFA), with photoactivation provided by CFL irradiation (23 W), to promote the intermolecular enantioselective α-alkylation of cyclic ketones with alkyl bromides, which provided the α-alkylated products with high drs and ees [126]. This was the first report of an asymmetric catalytic alkylation of unmodified ketones, e.g. cyclohexanone (**155**), with alkyl halides (**150**) in the presence of chiral amine **156**, and it was possible via EDA complex formation (Figure 32.11f).

More recently, Melchiorre et al. published a more detailed mechanism for the photochemical enantioselective α-alkylation of aldehydes with electron-deficient alkyl halides. This reaction could be performed via the EDA complex, as well as through direct photoactivation of the chiral enamine, which generates an electronically excited state that acts as a photoinitiator in the subsequent radical reaction.

They showed that carbon-centered radical trapping by the enamine, to form the carbon–carbon bond, is the rate determining step.

In 2017, Melchiorre et al. utilized a chiral fluorinated secondary amine catalyst in combination with an achiral Brønsted acid (TFA) to generate chiral iminium ions with α,β-unsaturated aldehydes [127]. The chiral iminium ions could be excited by visible light to enable a photocatalytic enantioselective β-alkylation with alkyl silanes that cannot be achieved via thermal activation.

A recent application of photo-organocatalytic reactions described by Melchiorre et al. is an enantioselective radical cascade reaction brought about by single-electron transfer activation of allenic acids. When reacted with enals, in the presence of a chiral pyrrolidine, $Zn(OTf)_2$ and irradiation at 420 nm by a single highpower (HP) LED afforded complex, polysubstituted bicyclic lactones of type **158** in high yields, drs and ees, via a photoactive chiral iminium ion [128].

Although the field of enantioselective photoredox chemistry with visible light activation is less than two decades old, the variety of transformations achieved so far, some of which could not be included due to length restrictions, show the enormous potential of this technique. Of great importance are the new opportunities to develop greener processes, not only of academic interest but also for industry, since the reactions take place under mild conditions and may be compatible with aqueous environments, they have broad functional group tolerance, site-specific selectivity and are also operationally simple [15, 129]. There have been several advances related to medicinal chemistry with racemic versions: in peptide functionalization, protein bioconjugation, Csp^3–Csp^2 cross-coupling, isotopic labeling, DNA-encoded library technology, microenvironment mapping (μMap), and late-stage functionalization [130]. Flow chemistry and photoenzymatic enantioselective catalysis are other research areas which shows some growth and with obvious potentials for development [131]. Further explorations on photoredox catalysis will not doubt allow the incorporation of chiral technologies into some of the methodologies just described, important for drug discovery.

32.5 Conclusions

The capability of photocatalysts to convert visible light into chemical energy has enabled the development, in the last decade and a half, of several new methods of enantioselective synthesis, some of which have no parallel with other methodologies. Of importance is also the versatility of photoredox chemistry in being combined with other modes of reactivity, such as HAT and PCET. Several chiral catalysts used to render the processes enantioselective have now been found to be compatible with photoredox chemistry, leaving at the disposal of the research chemist a vast arsenal of tools to develop future methodologies. The greenness of the reactions, which take place under mild conditions, are operationally simple to perform, and suitable for late stage functionalization of special molecules, are also a strong appeal, and make enantioselective photoredox catalysis an area of research with great promise for future developments.

Acknowledgements

Support by Fundação para a Ciência e a Tecnologia (FCT), Portugal, in the form of projects UIDB/00100/2020 and UIDP/00100/2020 of Centro de Química Estrutural and project LA/P/0056/2020 of the Institute of Molecular Sciences, is gratefully acknowledged. AMFP also thanks FCT for finance in the form of project EXPL/QUI-QOR/1079/2021. Furthermore, this publication has been supported by the RUDN University Strategic Academic Leadership Program (recipient A.J.L.P, preparation).

References

1 Anastas, P. and Eghbali, N. (2010). *Chem. Soc. Rev.* 39: 301–312.
2 Flemian, I. (1978). *Frontier Orbitals and Organic Chemical Reactions.* London: Wiley.
3 Marzo, L., Pagire, S.K., Reiser, O., and König, B. (2018). *Angew. Chem. Int. Ed.* 57: 10034–10072.

4 Offenloch, J.T., Gernhardt, M., Blinco, J.P. et al. (2019). *Chem. Eur. J.* 9: 3700–3709.
5 Kalyanasundaram, K. (1982). *Coord. Chem. Rev.* 46: 159–244.
6 Juris, A., Balzani, V., Barigelletti, F. et al. (1988). *Coord. Chem. Rev.* 84: 85–277.
7 Balzani, V., Bergamini, G., Marchioni, F., and Ceroni, P. (2006). *Coord. Chem. Rev.* 250: 1254–1266.
8 Sala, X., Romero, I., Rodriguez, M. et al. (2009). *Angew. Chem., Int. Ed.* 48: 2842–2852.
9 Bauer, A., Westkamper, F., Grimme, S., and Bach, T. (2005). *Nature* 436: 1139–1140.
10 Nicewicz, D. and MacMillan, D.W.C. (2008). *Science* 322: 77–80.
11 Ischay, M.A., Anzovino, M.E., Du, J., and Yoon, T.P. (2008). *J. Am. Chem. Soc.* 130: 12886–12887.
12 Narayanam, J.M.R., Tucker, J.W., and Stephenson, C.R.J. (2009). *J. Am. Chem. Soc.* 131: 8756–8757.
13 Martin, R., Higginson, B., Sanjosé-Orduna, J., and Gu, Y. (2021). *Synlett.* 32: (16): 1457–2399.
14 Wang, C. and Lu, Z. (2015). *Org. Chem. Front.* 2: 179–190.
15 Shaw, M.H., Twilton, J., and MacMillan, D.W.C. (2016). *J. Org. Chem.* 81: 6898–6926.
16 Silvi, M. and Melchiorre, P. (2018). *Nature* 554: 4–49.
17 Jiang, C., Chen, W., Zheng, W.-H., and Lu, H. (2019). *Org. Biomol. Chem.* 17: 8673–8689.
18 Zhang, H.-H., Chen, H., Zhu, C., and Yu, S. (2020). *Sci. China: Chem.* 63: 637–647.
19 Saha, D. (2020). *Chem. Asian J.* 15: 2129–2152.
20 Hong, B.-C. (2020). *Org. Biomol. Chem.* 18: 4298–4353.
21 Rigotti, T. and Alemán, J. (2020). *Chem. Commun.* 56: 11169–11190.
22 Prentice, C., Morrisson, J., Smith, A.D., and Zysman-Colman, E. (2020). *Beilstein J. Org. Chem.* 16: 2363–2441.
23 Yao, W., Bazan-Bergamino, E.A., and Ngai, M.-Y. (2022). *ChemCatChem* 14: e202101292.
24 Chan, A.Y., Perry, I.B., Bissonnette, N.B. et al. (2022). *Chem. Rev.* 122: 1485–1542.
25 Brimioulle, R., Lenhart, D., Maturi, M.M., and Bach, T. (2015). *Ang Chem Int. Ed.* 54: 3872–3890.
26 Stork, G., Brizzolara, A., Landesman, H. et al. (1963). *J. Am. Chem. Soc.* 85: 207–222.
27 List, B., Lerner, R.A., and Barbas, III C.F. (2000). *J. Am. Chem. Soc.* 122: 2395–2396.
28 List, B., Čorić, I., Grygorenko, O.O. et al. (2014). *Angew. Chem. Int. Ed.* 53: 282–285.
29 Juris, A., Barigelletti, S., Campagna, S. et al. (1988). *Coord. Chem. Rev.* 84: 185–277.
30 van Bergen, T.J., Hedstrand, D.M., Kruizinga, W.H., and Kellogg, R.M. (1979). *J. Org. Chem.* 44: 4953–4962.
31 Bock, C.R. et al. (1979). *J. Am. Chem. Soc.* 101: 4815–4824.
32 Value was corrected from the $Ag/Ag+ClO_4$ – electrode.
33 Tanner, D.D. and Singh, H.K. (1986). *J. Org. Chem.* 51: 5182–5186.
34 Cismesia, M.A. and Yoon, T.P. (2015). *Chem. Sci.* 6: 5426–5434.
35 Gualandi, A., Marchini, M., Mengozzi, L. et al. (2015). *ACS Catal.* 5: 5927–5931.
36 Nagi, D.A., Scott. M.E., and MacMillan, D.W.C. (2009). *J. Am. Chem. Soc.* 131: 10875–10877.
37 Welin, E.R., Warkentin, A.A., Conrad, J.C., and MacMillan, D.W.C. (2015). *Angew. Chem. Int. Ed.* 54: 9668–9672.
38 Shih, H.-W., Vander Wal, M.N., Grange, R.L., and MacMillan, D.W.C. (2010). *J. Am. Chem. Soc.* 132: 13600–13603.
39 Hong, B.-C., Lin, C.-W., Liao, W.-K., and Lee, G.-H. (2013). *Org. Lett.* 15: 6258–6261.
40 Cecere, G., König, C.M., Alleva, J.L., and MacMillan, D.W.C. (2003). *J. Am. Chem. Soc.* 135: 11521–11524.
41 Larionov, E., Mastandrea, M.M., and Pericàs, M.A. (2017). *ACS Catal.* 7: 7008–7013.
42 Capacci, A.G., Malinowski, J.T., McAlpine, N.J. et al. (2017). *Nat. Chem.* 9: 1073–1077.
43 Venimadhavan, S., Amarnath, K., Harvey, N.G. et al. (1992). *J. Am. Chem. Soc.* 114: 221–229.

44 Fischer, H. and Radom, L. (2001). *Angew. Chem. Int. Ed.* 40: 1340–1371.
45 Murphy, J.J., Bastida, D., Paria, S. et al. (2016). *Nature* 532: 218–222.
46 Wang, D., Zhang, L., and Luo, S. (2017). *Org. Lett.* 19: 4924–4927.
47 Huang, H., Zhang, G., and Chen, Y. (2015). *Angew. Chem., Int. Ed.* 54: 7872–7876.
48 Huang, H., Jia, K., and Chen, Y. (2015). *Angew. Chem., Int. Ed.* 54: 1881–1884.
49 DiRocco, D.A. and Rovis, T. (2012). *J. Am. Chem. Soc.* 134: 8094–8097.
50 Yin, Y., Zhao, X., Qiao, B., and Jiang, Z. (2020). *Org. Chem. Front.* 7: 1283–1296.
51 Rono, L.J., Yayla, H.G., Wang, D.Y. et al. (2013). *J. Am. Chem. Soc.* 135: 17735–17738.
52 Proctor, R.S.J., Davis, H.J., and Phipps, R.J. (2018). *Science* 360: 419–422.
53 Cao, K., Tan, S.M., Lee, R. et al. (2019). *J. Am. Chem. Soc.* 141: 5437–5443.
54 Lee, K.N., Lei, Z., and Ngai, M.-Y. (2017). *J. Am. Chem. Soc.* 139: 5003–5006.
55 Zheng, D. and Studer, A. (2019). *Angew. Chem. Int. Ed.* 58: 15803–15807.
56 Yi, H., Zhuang, G., Wang, H. et al. (2017). *Chem. Rev.* 117: 9016–9085.
57 Xie, J., Jin, H., and Hashmi, A.S.K. (2017). *Chem. Soc. Rev.* 46: 5193–5203.
58 Liu, Y., Liu, X., Li, J. et al. (2018). *Chem. Sci.* 9: 8094–8098.
59 Li, J., Kong, M., Qiao, B. et al. (2018). *Nat. Commun.* 9: 2445.
60 Zeng, G., Li, Y., Qiao, B. et al. (2019). *Chem. Commun.* 55: 11362–11365.
61 Yin, Y., Dai, Y., Jia, H. et al. (2018). *J. Am. Chem. Soc.* 140: 6083–6087.
62 Shao, T., Li, Y., Ma, N. et al. (2019). *iScience* 16: 410–419.
63 Ding, W., Zhou, -Q.-Q., Xuan, J. et al. (2014). *Tetrahedron Lett.* 55: 4648–4652.
64 Bu, L., Li, J., Yin, Y., Qiao, B., Chai, G., Zhao, X., Jiang, Z. (2018). Organocatalytic asymmetric cascade aerobic oxidation and semipinacol rearrangement reaction: a visible light-induced approach to access chiral 2,2-disubstituted indolin-3-ones. *Chem.-Asian J.* 13: 2382–2387.
65 Uraguchi, D., Kinoshita, N., Kizu, T., Ooi, T. (2015). Synergistic catalysis of ionic Brønsted acid and photosensitizer for a redox neutral asymmetric α-coupling of N-arylaminomethanes with aldimines *J. Am. Chem. Soc.* 137: 13768–13771.
66 Córdova, A., Sundén, H., Engqvist, M., Ibrahem, I., Casas, J. (2004). The direct amino acid-catalyzed asymmetric incorporation of molecular oxygen to organic compounds. *J. Am. Chem. Soc.* 126: 8914–8915.
67 Gentry, E.C., Rono, L.J., Hale, M.E. et al. (2018). *J. Am. Chem. Soc.* 140: 3394–3402.
68 Shin, N.Y., Ryss, J.M., Zhang, X. et al. (2019). *Science* 366: 364–369.
69 Thomas, G. (2000). *Medicinal Chemistry: An Introduction*. Hoboken: Wiley.
70 Marigo, M., Bachmann, S., Halland, N. et al. (2004). *Angew. Chem., Int. Ed.* 43: 5507–5510.
71 Kwiatkowski, P., Beeson, T.D., Conrad, J.C., and MacMillan, D.W.C. (2011). *J. Am. Chem. Soc.* 133: 1738–1741.
72 Hou, M., Lin, L., Chai, X. et al. (2019). *Chem. Sci.* 10: 6629–6634.
73 Arokianathar, J.N., Frost, A.B., Slawin, A.M.Z. et al. (2018). *ACS Catal.* 8: 1153–1160.
74 McLaughlin, C. and Smith, A.D. (2021). *Chem. Eur. J.* 27: 1533–1555.
75 Wei, G., Zhang, C., Bures, F. et al. (2016). *ACS Catal.* 6: 3708–3712.
76 Faisca Phillips, A.M., Silva, M.F.G., and Pombeiro, A.J.L. (2020). *Catalysts* 10: 529.
77 Bergonzini, G., Schindler, C.S., Wallentin, C.-J. et al. (2014). *Chem. Sci.* 5: 112–116.
78 Woźniak, Ł., Murphy, J.J., and Melchiorre, P. (2015). *J. Am. Chem. Soc.* 137: 5678–5681.
79 Yoon, T.P. (2016). *Acc. Chem. Res.* 49: 2307–2315.
80 Tellis, J.C., Primer, D.N., and Molander, G.A. (2014). *Science* 345: 433–436.
81 Zuo, Z., Ahneman, D.T., Chu, L. et al. (2014). *Science* 345: 437–440.
82 Zuo, Z., Cong, H., Li, W. et al. (2016). *J. Am. Chem. Soc.* 138: 1832–1835.
83 Guan, H., Zhang, Q., Walsh, P.J., and Mao, J. (2020). *Angew. Chem. Int. Ed.* 259: 5172–5177.
84 Cheng, X., Lu, H., and Lu, Z. (2019). *Nat. Commun.* 10: 3549.

85 Fan, P., Lan, Y., Zhang, C., and Wang, C. (2020). *J. Am. Chem. Soc.* 142: 2180–2186.
86 Zhang, W., Wang, F., McCann, S.D. et al. (2016). *Science* 353: 1014.
87 Wang, D., Zhu, N., Chen, P. et al. (2017). *J. Am. Chem. Soc.* 139: 15632–15635.
88 Sha, W., Deng, L., Ni, S. et al. (2018). *ACS Catal.* 8: 7489–7494.
89 Bao, X., Wang, Q., and Zhu, J. (2019). *Angew. Chem. Int. Ed.* 58: 2139–2143.
90 Thomé, I., Nijs, A., and Bolm, C. (2012). *Chem. Soc. Rev.* 41: 979–987.
91 Schwarz, J.L., Schäfers, F., Tlahuext-Aca, A. et al. (2018). *J. Am. Chem. Soc.* 140: 12705–12709.
92 Mitsunuma, H., Tanabe, S., Fuse, H. et al. (2019). *Chem. Sci.* 10: 3459–3465.
93 Schwarz, J.L., Huang, H.-M., Paulisch, T.O., and Glorius, F. (2020). *ACS Catal.* 10: 1621–1627.
94 Zhang, H.H., Zhao, J.J., and Yu, S. (2018). *J. Am. Chem. Soc.* 140: 16914–16919.
95 Blum, T.R., Miller, Z.D., Bates, D.M. et al. (2016). *Science* 354: 1391–1395.
96 Müller, C., Bauer, A., and Bach, T. (2009). *Angew. Chem. Int. Ed.* 48: 6640–6642.
97 Müller, C., Bauer, A., Maturi, M.M. et al. (2011). *J. Am. Chem. Soc.* 133: 16689–16697.
98 Guo, H., Herdtweck, E., and Bach, T. (2010). *Angew. Chem. Int. Ed.* 49: 7782–7785.
99 Du, J., Skubi, K.L., Schultz, D.M., and Yoon, T.P. (2014). *Science* 344: 392–396.
100 Miller, Z.D., Lee, B.J., and Yoon, T.P. (2017). *Angew. Chem. Int. Ed.* 56: 11891–11895.
101 Daub, M.E., Jung, H., Lee, B.J. et al. (2019). *J. Am. Chem. Soc.* 141: 9543–9547.
102 Amador, A.G., Sherbrook, E.M., and Yoon, T.P. (2016). *J. Am. Chem. Soc.* 138: 4722–4725.
103 Parsons, A.T. and Johnson, J.S. (2009). *J. Am. Chem. Soc.* 131: 3122–3123.
104 Trost, B.M. and Morris, P.J. (2011). *Angew. Chem., Int. Ed.* 50: 6167–6170.
105 Ruiz Espelt, L., McPherson, I.S., Wiensch, E.M., and Yoon, T.P. (2015). *J. Am. Chem. Soc.* 137: 2452–2455.
106 Zhang, K., Lu, L.-Q., Jia, Y. et al. (2019). *Angew. Chem., Int. Ed.* 58: 13375–13379.
107 Pagire, S.K., Kumagai, N., and Shibasaki, M. (2020). *Chem. Sci.* 11: 5168–5174.
108 Li, Y., Zhou, K., Wen, Z. et al. (2018). *J. Am. Chem. Soc.* 140: 15850–15858.
109 Han, B., Li, Y., Yu, Y., and Gong, L. (2019). *Nat. Commun.* 10: 3804.
110 Stegbauer, S., Jandl, C., and Bach, T. (2018). *Angew. Chem., Int. Ed.* 57: 14593–14596.
111 Genzink, M.J., Kidd, J.B., Swords, W.B., and Yoon, T.P. (2022). *Chem. Rev.* 122: 1654–1716.
112 Zhang, L. and Meggers, E. (2017). *Chem.–Asian J.* 12: 2335–2342.
113 Huo, H., Shen, X., Wang, C. et al. (2014). *Nature* 515: 100–103.
114 Wang, C., Zheng, Y., Huo, H. et al. (2015). *Chem. – Eur. J.* 21: 7355–7359.
115 Huo, H., Huang, X., Shen, X. et al. (2016). *Synlett.* 27: 749–753.
116 Tan, Y., Yuan, W., Gong, L., and Meggers, E. (2015). *Angew. Chem., Int. Ed.* 54: 13045–13048.
117 Wang, C., Qin, J., Shen, X. et al. (2016). *Angew. Chem., Int. Ed.* 55: 685–688.
118 Huang, X., Webster, R.D., Harms, K., and Meggers, E. (2016). *J. Am. Chem. Soc.* 138: 12636–12642.
119 Shen, X., Harms, K., Marsch, M., and Meggers, E. (2016). *Chem. – Eur. J.* 22: 9102–9105.
120 Großkopf, J., Kratz, T., Rigotti, T., and Bach, T. (2022). *Chem. Rev.* 122: 1626–1653.
121 Alonso, R. and Bach, T.A. (2014). *Angew. Chem. Int. Ed.* 53: 4368–4371.
122 Tröster, A., Alonso, R., Bauer, A., and Bach, T. (2016). *J. Am. Chem. Soc.* 138: 7808–7811.
123 Zou, Y.-Q., Hörmann, F.M., and Bach, T. (2018). *Chem. Soc. Rev.* 47: 278–290.
124 Arceo, E., Jurberg, I.D., Álvarez-Fernández, A., and Melchiorre, P. (2013). *Nat. Chem.* 5: 750–756.
125 Arceo, E., Bahamonde, A., Bergonzini, G., and Melchiorre, P. (2014). *Chem. Sci.* 5: 2438.
126 Spinnato, D., Schweitzer-Chaput, B., Goti, G. et al. (2020). *Angew. Chem. Int. Ed.* 59: 9485–9490.
127 Silvi, M., Verrier, C., Rey, Y.P. et al. (2017). *Nat. Chem.* 9: 868–873.
128 Perego, L.A., Bonilla, P., and Melchiorre, P. (2019). *Adv. Synth. Catal.* 362: 302–307.
129 Crisenza, G.E.M. and Melchiorre, P. (2020). *Nat. Comunn.* 11: 803.
130 Li, P., Terrett, J.A., and Zbieg, J.R. (2020). *ACS Med. Chem. Lett.* 11: 2120–2130.
131 Mondal, S., Dumur, F., Gigmes, D. et al. (2022). *Chem. Rev.* 122: 5842–5976.

Part VI

Catalysis for the Purification of Water and Liquid Fuels

33

Heterogeneous Photocatalysis for Wastewater Treatment: A Major Step Towards Environmental Sustainability

Shima Rahim Pouran[1],* and Aziz Habibi-Yangjeh[2],*

[1] *Department of Environmental and Occupational Health, Social Determinants of Health Research Centre, Ardabil University of Medical Sciences, Ardabil, Iran*
[2] *Department of Chemistry, Faculty of Science, University of Mohaghegh Ardabili, Ardabil, Iran*
* *Corresponding authors*

33.1 Introduction

Although barely 1% of the total water on the earth is utilizable by humans, water resources are severely affected by anthropogenic activities [1]. If the current trend in water-use and resource management continues, the freshwater crisis will become an explosive issue globally. A major cause of concern is unsustainable tracks in policy planning and implementations in all sectors including agriculture, industry, and domestic, which can greatly affect humans and the environment over the next few decades [2]. In addition to being affected by excessive demands, water bodies around the world receive more than 80% of untreated effluents, which results in water pollution and subsequent health issues [3]. Hence, it is crucial to take serious actions to reverse this worrying path and make practical and effective use of the available freshwater in a sustainable manner. From this perspective, in the recent World Water Development report released by the United Nations (2021), valuing water in all facets of development, has been considered as the most essential factor to achieve the United Nations sustainable development goals (SDGs) [4].

Being a universal solvent, water is susceptible to pollution by a large number of chemicals associated with poorly managed wastes from several point and non-point sources. This impairs water quality and makes it unfit for drinking by exposing humans and other living organisms to various diseases and health risks [5]. As per a study by the Global Burden of Disease (GBD), water pollution-associated diseases were responsible for the death of approximately two million people all over the world in 2015 [6]. To tackle the problems accompanying polluted water bodies, management authorities have integrated resource recovery plans with waste minimization and elimination strategies [7]. A strong emphasis has been placed on persistent and dangerous substances that are difficult to remove through conventional approaches.

Since the construction of Cloaca Maxima in about 600 BCE as one of the ancient sewage systems [8], numerous systems and approaches have been implemented for water decontamination purposes. Alongside the emergence of recalcitrant pollutants into the aquatic environment,

Catalysis for a Sustainable Environment: Reactions, Processes and Applied Technologies Volume 3, First Edition. Edited by Armando J. L. Pombeiro, Manas Sutradhar, and Elisabete C. B. A. Alegria.
© 2024 John Wiley & Sons Ltd. Published 2024 by John Wiley & Sons Ltd.

wastewater treatment systems have also been improved and advanced. Despite all of the advancements and investments in wastewater treatment technologies, the majority lack purifying efficiency combined with cost and energy effectiveness in an environmentally sustainable manner. The conventional approaches for wastewater treatment are the schemes made up of various physical, chemical, and biological methods, which remove a wide variety of pollutants, including organic and inorganic compounds and microbial contamination, through several levels [5]. Nonetheless, these processes are limited on account of being expensive, complex, and time-consuming in addition to involving phase–transformation of pollutants and the generation of second-pollution such as sludge while being ineffective in the case of most recalcitrant pollutants [9]. On this basis, advanced oxidation processes (AOPs) have been established and extensively studied in recent decades. These technologies are acknowledged as powerful wastewater treatment processes, which are capable of removing almost all types of pollutants via the generation of reactive species [10]. However, not all AOPs comply with sustainable wastewater treatment principles. Two main objectives of a sustainable treatment method can be described as follows: first, the system should have no harmful effects on the society and environment and should not lead to some other type of pollution; secondly, the system should be cost- and energy-efficient, and built based on nature-based and environmentally-friendly elements. From this perspective, heterogeneous photocatalysis has stepped forward as one of the most effective AOPs for the decomposition of non-biodegradable compounds, which is blessed with simple operation conditions and can be modified to meet sustainability objectives [11]. The key element, which has the central role in this system, is the photocatalyst. Hence, designing and fabricating highly active photocatalysts for sustainable photocatalytic reactions are continuing concerns within energy and environmental studies.

This chapter is aimed at providing data on the latest scientific achievements in innovative nanostructured photocatalysts with advanced performance for wastewater treatment through the photocatalysis approach. Special emphasis was put on sustainable nanomaterials, which are abundant, naturally available, and can be manufactured, modified, recycled, and recovered with minimal effects on the environment. Significant attention was paid to photocatalytic routes that require bottom-level energy input and can be activated under sunlight and adapted for large-scale treatment units.

33.2 Heterogeneous Photocatalysis

All heterogeneous photocatalytic systems based on semiconductors pivot on the following steps. The photocatalyst is sensitized under light illumination with photons of equivalent or larger energy than the bandgap of the material. Upon this activation, the electron-hole (e^-/h^+) pairs are separated and transferred to the catalytic sites on the surface of the photocatalyst. The optically excited electrons transfer from the valence band (VB) to the conduction band (CB), where they can get trapped in a set of reduction reactions with acceptor ions/molecules such as dissolved oxygen. This process not only avoids the recombination of the photogenerated e^-/h^+ pairs, but also gives rise to reactive species (e.g. superoxide anion radical) that can enter the oxidation cycle. On the other hand, the unpaired holes either oxidize the pollutant molecules immediately or produce some other reactive species, mainly the ˙OH radical, when they come into contact with water molecules or hydroxide anions [12]. Ultimately, the wastewater treatment is accomplished by the radicals generated in this process, which yield simple degradation products (i.e. water and carbon dioxide). It is important to note that the CB and VB energies of the desired photocatalyst should be enough for reduction of oxygen molecules and oxidation of water molecules/hydroxide anions, respectively [13]. Scheme 33.1 represents the basic processes in a photocatalytic reaction.

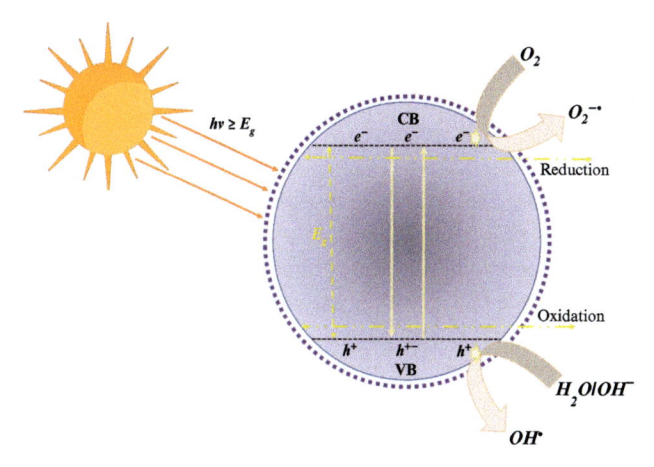

Scheme 33.1 Key parameters of a photocatalytic system.

Some instances that are frequently studied for photocatalytic wastewater treatment are TiO_2 [14], ZnO [15], CeO_2 [16], g-C_3N_4 [17], and α-Fe_2O_3 [18]. There are several fundamental features that are taken into account in a photocatalytic system using any semiconductor. These are (i) a wide absorption profile; (ii) a lengthy excitation period/prolonged separation intervals; and (iii) being structurally robust against photo-corrosion or, in other words, being reusable [19]. On the other hand, because the photocatalytic reactions occur at the interface of the semiconductor surface and the fluid medium, holding broad surface area and facile transfer (adsorption and desorption) of the reactants and products from the photocatalyst surface are also significant factors.

33.3 Sustainable Photocatalysts

From the sustainability perspective, a number of factors are considered significant in heterogeneous semiconductor photocatalysis. One objective is to operate using naturally-available materials or materials that can be synthesized from simple and nature-based ingredients through facile and green approaches. Moreover, the utilized materials need to be abundant, economically viable, able to be easily modified and tuned to hold desired properties, and characterized by robust stability. Additionally, these materials should be nontoxic and not pose any health risk to living organisms or impact on the environment. Most of the semiconductor metal oxides, magnetic photocatalysts, and carbonaceous materials are included in this list.

The efficiency of energy utilization is another factor of concern. Within this realm, solar-driven photocatalysts are increasingly considered the best energy-efficient candidates to run a photocatalytic system. Keeping this in mind, efforts on designing sustainable photocatalysts have been continually accompanied with an energy-efficiency goal via the activation of photocatalyst by solar energy as a clean, renewable, inexhaustible, and freely available solution for energy shortage. Moreover, the integration of photocatalytic systems with photothermal materials could able simultaneous degradation of organic pollutants, while water desalination via solar energy utilization takes place [20, 21]. Under this multifunctional system, the heat generated from the solar-evaporated water can reduce the energy required for the activation of a photocatalyst and magnify the Arrhenius pre-exponential constant, thus increasing the rates of the degradation reactions in the photocatalytic systems [22]. Additionally, the generated heat at the interface region can also boost the rate of the surface reactions and concurrently remove a wide variety of pollutants from microbes to oil contaminants, meeting energy-water-environment nexus principles [23].

33.3.1 Metal Oxide-based Photocatalysts

In the history of photocatalysis, metal oxides were the initially investigated photocatalysts [24]. This group of semiconductors soon emerged as powerful alternatives to execute photocatalytic remediation of contaminated aquatic environments owing to their robust and modifiable properties along with their availability in nature and the fact that they are recyclable, low in toxicity, superior in optical transparency, chemical/photo stable, and, more importantly, energy-efficient [25, 26]. In comparison to bulk materials, the nanosized metal oxides offer a higher surface area to volume ratio and deliver better adsorption efficiencies [27].

The chemical, optical, and catalytic properties of metal oxides are strongly affected by synthesis parameters and growth conditions, which chiefly control the morphology (size and shape), crystalline structure, defect density, and electronic structure [28, 29]. In particular, metal oxides with a strong absorption edge at the visible region and large adsorption capacity are of paramount importance in a heterogeneous photocatalysis process. Hence, the research on designing novel metal oxide-based photocatalysts is undertaken in such a way that satisfies these two criteria through bandgap modulation, doping, surface engineering, defect formation, functionalization, and other approaches [27, 30, 31]. A realistic look at various metal oxides employed so far for the photocatalytic treatment of various wastewater finds that not all follow the sustainability principles. Indeed, the metal oxides (which are expensive, unstable, toxic to humans and the environment, and require impractical approaches for their fabrication) are excluded from the list.

Despite the high stability and external quantum efficiency of the commonly used metal oxide semiconductors for photocatalytic degradation of pollutants (e.g. TiO_2, ZnO, and CeO_2), a big part of this group has wide bandgaps, which in almost all cases are only active under UV irradiation. Several strategies have been developed to tackle this challenge, such as surface modulation [27], heterojunction creation [32], integration with transition metal/s [33], or 2D materials [34]. Therefore, in most cases, instead of pure metal oxide, a composition of the wide-bandgap metal oxide semiconductor with some other semiconductor/s, transition metal/s, or 2D material/s is generally used as a powerful photocatalyst, which can be activated under visible-light illumination. There are tremendous studies on the application of a diverse range of metal oxides for photodegradation of organic contaminants. Details on the physical-chemical properties, merits, and demerits of the well-known metal oxide semiconductors such as CeO_2 [35], Ag-based [36], Bi-based [37, 38], WO_3 [39], SnO_2 [40], Cu_2O [41, 42], and so more are reviewed and reported in numerous studies in the literature. Herein, we give an account of the most studied metal oxide semiconductors (TiO_2 and ZnO) and put forward criteria for their use in sustainable photocatalytic applications.

Titanium dioxide (TiO_2) is recorded as the first semiconductor studied in a photocatalytic system [43]. This n-type metal oxide semiconductor is physically stable, inexpensive, and almost environmentally benign [44]. Based on the phase structure of TiO_2, the bandgap varies between 3.0 eV (pure rutile phase) and 3.2 eV (pure anatase phase), with the photoreactions beginning under UV-A light with $\lambda < 400$ nm [45]. Moreover, the photonic efficiency (ζ) of TiO_2 is limited because of the rapid recombination of the charge carriers, which fades its photocatalytic efficiency in the course of treatment [46]. Given this, adapted strategies should not only solve these optical limitations, but also obey the sustainability rules.

Investigation on expanding of the light absorption profile of TiO_2 towards the visible region is intensive and ongoing. Surface modification by metals [47], organometallic [48], and organic molecules [49], fabrication of bioinorganic composites of TiO_2 [49], and doping by 2D materials [50] are some examples of strategies to improve the efficiency of TiO_2 towards solar energy utilization. The simplest way to enhance the photocatalytic activity of TiO_2 is to make changes in the synthesis parameters without the introduction of any external elements into the phases. For instance, the concurrent enhancement in the crystalline and particle sizes of TiO_2 was observed by varying the calcination temperature (from 300 to 1,100 °C); the active surface was significantly reduced from 101 to 3.25 m^2/g [51]. Reportedly, the increase in the particle size of most semiconductors leads to a reduction in bandgap energy and a red shift towards the visible region [52]. Accordingly, the bandgap energy was reduced upon calcination temperature elevation. Moreover, not only the crystallinity was enhanced at higher calcination temperatures but also the phase composition of TiO_2 was changed from combined brookite/rutile (300 °C) to brookite-to-anatase/rutile (500 °C), then to anatase/rutile (700 °C), anatase-to-rutile (900 °C), and finally to pure rutile (1,100 °C).

One approach to enhance the lifespan of the photoinduced charge carriers is to deposit a noble metal/s onto TiO_2 particles. Because the Fermi level of the most noble metals is below the TiO_2 CB, a Schottky barrier is created at the junction point of TiO_2 and noble metal and the photo-induced electron on the CB_{TiO2} moves to the Fermi level of noble metal, while the valance band holes remain discrete [53]. The photo–deposition or deposition–precipitation methods have been conventionally employed for noble metal-semiconductor synthesis. Despite the facile and fast deposition of metals through these methods, the morphology and size of the metals cannot be controlled, thus affecting the photocatalytic performance of the nanocomposite. Recently, the ligand-exchange approach has been used that involves introduction to the system of an intermediate ligand with a volatile nature (e.g. 3-mercaptopropionic acid) [54] that can take the place of the original ligands on the noble metal nanoparticles and can be simply detached via evaporation at the final step. Under the ligand-absent condition and relatively low temperatures, the metals can be uniformly deposited on the surface of the semiconductor where close Schottky contact is created between the metal and semiconductor. This leads to a shift towards visible wavenumbers and higher catalytic activities of the fabricated nanocomposite under solar irradiation [54].

In an interesting study by Ji et al. [50], TiO_2 nanotubes (TNTs) were deposited on graphitic carbon nitride sheets (g-C_3N_4). The intra-phase transformation of TiO_2 to P25-type with anatase:rutile proportion of 80:20 resulted in hot spots at the rutile/anatase and anatase/titanate interfaces at nanoscales. These heterojunctions facilitated charge transfer whereas the enhanced active sites of g-C_3N_4 sheets improved the redox reactions in the interface. Moreover, the sunlight absorption efficiency of TiO_2 was improved owing to the heterojunction with g-C_3N_4. Figure 33.1a illustrates the role of the created heterojunctions on the photogeneration, segregation, and transfer of the charge carriers in the TNTs/g-C_3N_4 photocatalyst. In this photocatalyst, the e^-/h^+ pairs are generated following the absorption of solar photons with the rutile phase ($E_g = 2.76$ eV). The electrons are transferred through the rutile/anatase and anatase/titanate interfaces towards the surface. Afterwards, the photoinduced electrons are trapped by surface Ti^{4+} cations and give rise to Ti^{3+} species. Upon electron clearance, the unpaired holes can get trapped with surface hydroxyl anions or the water molecules in the interface and generate ·OH radical species, which enter the degradation process (Figure 33.1b).

Wurtzite ZnO is another well-known n-type semiconductor and bears a wide bandgap of 3.37 eV [15]. The nontoxicity, cost-effectiveness, and favourable electro-optical properties put this semiconductor forward as a versatile and sustainable photocatalyst for immense applications. However, surface and structural modifications are required to tune the band structure of ZnO and extend the

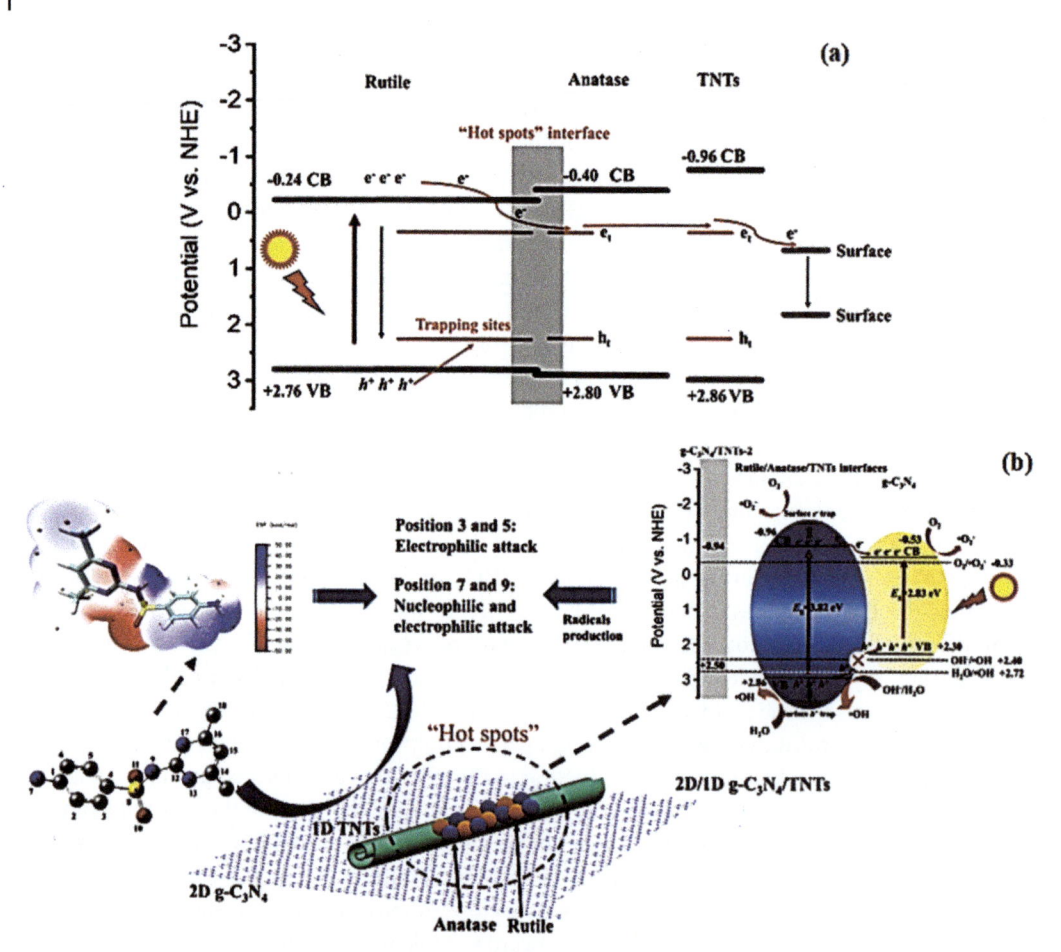

Figure 33.1 (a) Schematic representation of (a) photocatalysis mechanism using TNTs/g-C$_3$N$_4$ and (b) photocatalytic hot spots at the rutile@anatase@titanate heterojunctions. Ref [50] / Elsevier.

separation time of the charge carriers [15]. Additionally, it is essential to empower ZnO against photo-corrosion for effective photocatalytic and practical applications [55].

As discussed, the integration of a noble metal and semiconductor can greatly influence the transfer of photoinduced charge carriers. The integration of ZnO with another semiconductor with an appropriate band position has been extensively studied as a facile and sustainable route to enhance its activity under solar irradiation. ZnO/CeO$_2$ can be simply synthesized through a co-precipitation method and give rise to a solar active photocatalyst using optimized wt.% of CeO$_2$. As per the report by Wolski et al. [56], the CeO$_2$/ZnO sample with 8 wt.% of CeO$_2$ showed the highest adsorption and photodegradation efficiencies of ciprofloxacin compared to those with lower and/or higher studied wt.%. Figure 33.2 represents a plausible mechanism of disinfection over Ag/SnO$_2$/ZnO nanocomposite [57]. In this system, the created heterojunction between SnO$_2$ and ZnO plays a vital role in the segregation and transfer of the photogenerated charge carriers. Upon photocatalyst solar-excitation, the electrons move from the CB$_{ZnO}$ to CB$_{SnO2}$ and holes transfer from VB$_{SnO2}$ to VB$_{ZnO}$ due to the differences in their work functions until the corresponding Fermi levels are aligned and thermal equilibrium is established. Moreover, the photoinduced electrons

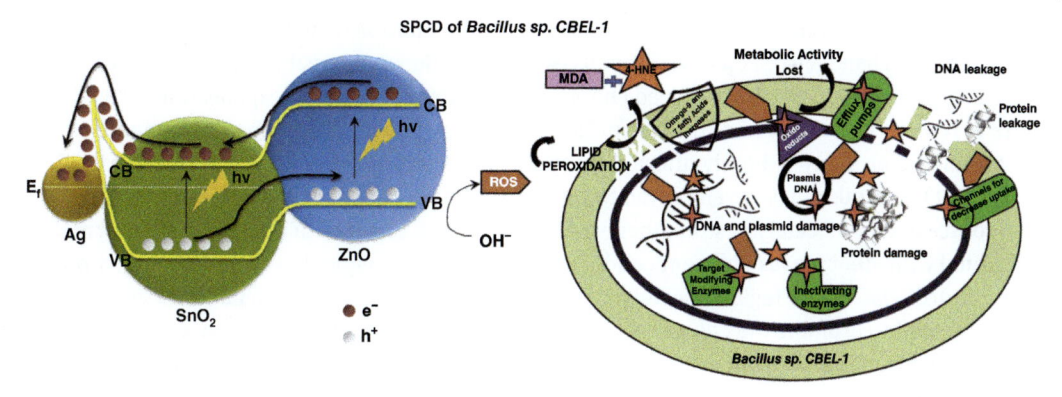

Figure 33.2 Schematic view of solar-photocatalytic decomposition of *Bacillus* sp. CBEL-1 in presence of Ag/SnO$_2$/ZnO [57] / Elsevier.

on CB$_{SnO2}$ can move towards Ag nanoparticles through the Ag/SnO$_2$ interface. Under this condition, an electrostatic field is created at the SnO$_2$/ZnO and Ag/SnO$_2$ interfaces, which effectively separates e/h and enhances the holes accessibility for production of higher reactive oxygen species (ROS).

33.3.1.1 Magnetic Metal Oxide Semiconductors

Heterogeneous semiconductor materials are recognized as highly active photocatalysts at nanoranges, in contrast with the corresponding bulk samples, due to their large active surface and better adsorption features. However, conventional separation methods are generally inefficient in the complete separation of nanoparticles from the treated effluents, which may engender unpleasant health effects or unpredictable environmental consequences. On the other hand, the reusability potential of a photocatalyst is emphasized for sustainable applications. Given this, the facile and effective recovery of photocatalyst nanoparticles can be considered as a key factor from both ecological and economical viewpoints.

Magnetic collection has rapidly become one of the most effective separation approaches in this context. The new generation of photocatalysts was accordingly introduced to fulfill this purpose. The application of magnetic compounds as support [58] or dopant [59], and/or induction of defects [60] and impurities [61], have been among the widely employed strategies to induce magnetism to a photocatalyst. However, the need for additional modifications was erased when magnetic semiconductors entered the scene. In contrast to the photocatalyst nanoparticles, which grow feeble upon the immobilization over inert surfaces, supporting nanoparticles on magnetic compounds not only makes the separation step an easy and affordable process, but also produces a significant enhancement in the photocatalytic properties in most instances once magnetic semiconductor/s are utilized. The fascinating semiconducting properties and magnetism of magnetite (Fe$_3$O$_4$) and hematite (Fe$_2$O$_3$) brought about extensive studies on their photocatalytic applications.

Magnetite or ferrite is a mixed iron [Fe(II), Fe(III)] oxide with an inverse spinel structure of tetrahedral and octahedral coordination, which occurs in a wide variety of geological environments [62]. Being abundantly available, biocompatible, and ferrimagnetic ($\approx$ 90 emu/g), has made Fe$_3$O$_4$ a great nominee to supply magnetism to a photocatalyst [63]. On the other hand, the integration of Fe$_3$O$_4$ nanoparticles with some other supporting substrates and/or metal oxide nanoparticles can diminish the agglomeration of nanoparticles, which originates from the intense dipole-dipole magnetic interactions among the nanoparticles [26, 64]. In addition to its effects on magnetic properties, the small bandgap of Fe_3O_4 results in its easier activation at the visible region [59]. However, the fast recombination of photoinduced charges hinders its photocatalytic applications. The literature is replete with the studies on the surface and energy gap modifications of magnetite as its photocatalytic property is highly affected by synthesis factors and the final size of the particles (Table 33.1). Several routes are employed to get the maximum benefit from magnetite as a magnetism source and photocatalysis strengthener. As per the small bandgap of magnetite, it is normally integrated with other metal oxide semiconductors of wide band energies to shift the light absorption edge towards the visible region. However, to inhibit the centralization of charge carriers in the magnetite and to promote the charge separation and movement, a top-cover such as silica is used to circle the magnetite while preserving its magnetic property and improving adsorption sites [65]. To impede the recombination step, one or more transition metal/s are incorporated into the lattice structure of magnetite wherein the corresponding redox pairs induce oxygen vacancies to capture the photogenerated electrons [66]. In addition, some electron acceptor compounds (e.g. persulphate ions), have been also introduced to the photocatalytic system to scavenge the photoinduced electrons and free up the holes [67].

Another approach is to utilize a support material, normally carbon derivatives such as graphene, to attract the photogenerated electrons of magnetite and transfer them onto the surface. The application of support materials not only expands the lifespan of holes, but also improves the adsorption capacity and thermal/mechanical stability of the magnetic nanocomposite by impeding Fe leaching [26, 32]. For instance, the deposition of cerium-substituted magnetite onto graphene oxide sheets gave rise to a highly visible-light-active magnetic nanocomposite [32]. In this study, the synergistic action of the Fe and Ce redox pairs along with the graphene nanosheets brought about a significant red shift in the light absorption profile and effective separation and movement of the charge carriers (Figure 33.3). In this framework, carbonaceous support materials can be produced from biomass and/or organic waste materials in order to make the procedure in line with SDGs [68].

Hematite (α-Fe_2O_3) is an n-type semiconductor metal oxide that has received a great deal of interest for photocatalytic applications among the iron oxide species. Considering the medium bandgap of 1.9–2.2 eV (which includes most visible light portions), non-toxicity, and thermodynamical stability of α-Fe_2O_3, it has high potential to be employed for sustainable wastewater treatment through photocatalysis. Although α-Fe_2O_3 has lower saturation magnetization than Fe_3O_4, the corresponding nanocomposites with optimal amounts of hematite generally possess enough magnetism to be attracted by an external magnetic field [69]. The right combination of semiconductor materials with α-Fe_2O_3 as per the energy band positions will give rise to highly active magnetic visible-light-triggered photocatalysts. Figure 33.4 depicts the facile transfer of the photogenerated charge carriers in the Fe_2TiO_5/α-Fe_2O_3/TiO_2 nanocomposite [70]. As per the model I band alignment, the short bandgap of α-Fe_2O_3 (2.0 eV) allows the activation under visible light wherein the photoinduced electrons transfer from the $CB_{\alpha\text{-}Fe2O3}$ to $CB_{Fe2TiO5}$ and then to CB_{TiO2} while the holes are moved in an inverse direction between the TiO_2/Fe_2TiO_5 and α-Fe_2O_3/Fe_2TiO_5 heterojunctions. In model II, the photoinduced electrons on the $CB_{\alpha\text{-}Fe2O3}$ and $CB_{Fe2TiO5}$ move to CB_{TiO2}, and holes transfer in opposite directions between the TiO_2/Fe_2TiO_5 and α-Fe_2O_3/TiO_2 heterojunctions.

Table 33.1 summarizes data on novel modified metal photocatalysts based on TiO_2, ZnO, Fe_3O_4, and α-Fe_2O_3 for degradation of organic pollutants. In the summarized studies, efforts were directed to simultaneously reduce the bandgaps, expand visible-light absorption regions, and prolong the separation period.

Table 33.1 Sustainable photocatalysts based on metal oxides.

Photocatalyst	Modification type	Modification effect	Pollutant	Remarks	Ref.
TiO_2	Surface modification of TiO_2 by organic solvents and arranged on a polymeric support	Extension of absorption towards visible region	Crystal violet	Holes on the organic fraction of the surface complex are scavenged by ethanol to generate ROSs	[49]
	Anatase TiO_2/WO_3 nanocomposite	Visible-light activity: WO_3 effects on the preservation of the structure and a full anatase phase	1,4-Dioxane	Total removal within 240 min; Anatase content-dependent efficiency up to 8 mol. % of WO_3 calcined at 800 °C	[71]
	Fe–Pr co-doped TiO_2	Bandgap reduction as per TiO_2 (2.7 vs. 3.2) solar-light active and prolonged separation	Acid orange7 (AO7), Phenol	87% of AO7 and 80% of TOC removal within 60 min; Pr^{3+}/Fe^{3+} decreased TiO_2 bandgap and generated oxygen vacancies where electrons were trapped and extended holes lifetime.	[47]
	g-C_3N_4/TiO_2 nanotubes	Solar-light active, enhanced e^-/h^+ separation transfer. Nanoscale hot spots creation at Anatase/rutile/titanate interfaces	Sulfamethazine	Anatase to rutile proportion: ~80:20;	[50]
	Fe_3O_4/TiO_2 mol.% (1:4)	Enhanced separation and transfer of e^-/h^+ pair	Norfloxacin	Complete removal within 80 min; Safe to environment at applied concentration.	[72]
	$BaTiO_3/TiO_2$ (Wt. ratio: $BaTiO_3$:TiO_2 = 1.2:1)	Continuous charge separation and transport; Accelerated catalytic reactions over the $BaTiO_3$/TiO_2	Rhodamine B	Complete RhB removal at $\Delta T = 12$ °C via induction of successive intermittent changes in the temperature into the $BaTiO_3$-TiO_2 nanocomposite where dynamic internal field engineering attributed to the $BaTiO_3$ pyroelectric effect	[73]
	TiO_2 calcined at 700 °C (mixed anatase/rutile)	High crystallinity and optimal mixed phase with desirable band energy	Methylene blue (MB)	Complete photodegradation of MB (20 ppm) within 180 min at pH 7. Strong inactivation of MS2 bacteriophage, influenza virus, and murine norovirus	[51]

(*Continued*)

Table 33.1 (Continued)

Photocatalyst	Modification type	Modification effect	Pollutant	Remarks	Ref.
ZnO	ZnO/carbon dots	The induced structural defects enhanced the charge transfer and ROS generation due to e^-/h^+ recombination inhibition	Ciprofloxacin (CIP)	98% solar degradation of CIP pH 6.3 ± 0.1 using 0.6 g/L of photocatalyst with 110 min at the rate of 0.30 min^{-1}	[74]
	Core-shell Ag/SnO$_2$/ZnO	Enhanced charge separation and transfer, and e^-/h^+ recombination inhibition	multidrug resistant bacterium, *Bacillus* sp. CBEL-1	Total disinfection using 500 mg/L of photocatalyst under solar irradiation	[57]
	ZnO/CNF	0.3 M Zn(Ac)$_2$ and 1.5 wt% CNF	MB	Total MB removal within 10 min under sunlight	[75]
	Phytogenic ZnO/Au (3 wt.%)	Au-ZnO Schottky contact; Surface plasmon resonance; Increased light absorption profile with the aid of Au, effective separation and transfer of e^-/h^+	MB	88% MB removal within 180 min, 3.2-fold higher activity than pure ZnO under visible-light illumination	[76]
	ZnO/CeO$_2$ 40 wt.%	Creation of a Z-scheme heterojunction, Facilitated separation and transfer of e^-/h^+	Ciprofloxacin (CIP)	Major role of holes in photodegradation	[56]
Fe$_3$O$_4$	SMAO/Fe$_3$O$_4$/ED/rGO	Concurrent transformation of Fe$_2$O$_3$ into Fe$_3$O$_4$ NP by generated electrons through the reaction of rGO/ED/HPV.	Ciprofloxacin (CIP) and ibuprofen (IBF)	88.97% CF removal within 90 min and 83.12% IBF removal within 120 min under visible light irradiation.	[64]
	TiO$_2$/Fe$_3$O$_4$/activated carbon	Ozone on the TiO$_2$ surface acts as electron acceptor to avoid e^-/h^+ recombination. Visible-light absorption enhancement in the presence of FeC; excellent magnetic property,	CECs/ DOC	< 55% CECs and < 45% DOC removals through solar photocatalytic system in the absence of ozone; complete CECs and < 15% DOC removals through solar photocatalytic Ozonation	[77]
	Fe$_3$O$_4$ coated SiO$_2$/ P25/ topcoat: TiO$_2$	Topcoat diminished P25 loss and improved the robustness of the photocatalyst; Higher adsorption capacity.	2-methyl isoborneol (MIB)	No changes in the TiO$_2$ topcoated sample; 2-MIB removal rate constant was 0.042 ± 0.002 min^{-1}	[78]
	C-TiO$_2$/Fe$_3$O$_4$/AC	Magnetic property, reduced bandgap (2.535 eV) due to C doping thus visible light activity. Higher adsorption capacity and facilitated separation of charge carries by AC.	Congo red (CR)	92.9% CR removal within 30 min of simulated sunlight exposure, reaction rate constant of 0.1776 min^{-1}	[79]

Catalyst	Features	Pollutant	Performance	Ref.
ZnFe$_2$O$_4$/TiO$_2$/Cu	Magnetic separation; enhanced visible-light activity (2.62 eV) due to Cu plasmon absorption, e$^-$/h$^+$ separation, and transfer.	Naproxen (NPX)	80.73% NPX (10 mg/L) removal under sunlight within 120 min at pH 4.	[80]
Z-scheme Ag$_3$PO$_4$/Fe$_3$O$_4$-BAB	e$^-$/h$^+$ recombination inhibition via Ag$_3$PO$_4$-Fe$_3$O$_4$ heterojunction; charge transfer through BAB	Bisphenol A (BPA); Sulfamethoxazole (SMX); 2,4-dichlorophenol (2,4-DCP); rhodamine B; Phenol; Naproxen.	95.6% BPA, 97% SMX, 94% 2,4-DCP, 85% RhB, 50% phenol, and 29% NPX removals using 1 g/L of photocatalyst and 0.5 mm PDS within 60 min at pH 6.5 under visible light with reaction constant of 0.006 min^{-1}	[81]
BiOI/Fe$_3$O$_4$ (molar ratio of 2:1)	Magnetic separation, enhanced stability, effective separation of e$^-$/h$^+$ through BiOI/Fe$_3$O$_4$ heterojunction.	Bisphenol A	Total BPA (10 mg/L) removal within 30 using 1 g/L of photocatalyst at pH 7 under simulated sunlight	[59]
Graphene-TiO$_2$/Fe$_3$O$_4$	Magnetic separation; negligible Fe leaching, enhanced adsorption capacity, and effective separation and transfer of e$^-$/h.	Cotinine (CTN)	CTN (10 mg/L) complete removal and 80% TOC removal within 75 and 120 min using 0.5 g/L of photocatalyst and 10 mg/L of ozone at natural pH under simulated sunlight	[82]
Fe$_{3-x}$Ce$_x$O$_4$/GO	Magnetism, effect of Fe$_3$O$_4$ on enhanced light absorption at visible range, effect of GO on the improved separation and transfer of e/h	Oxytetracycline (OTC)	88% OTC (30 mg/L) removal within 120 min using 0.8 g/L of photocatalyst at OTC natural pH under visible light	[32]
Fe$_2$O$_3$ K$_4$Nb$_6$O$_{17}$/Fe$_3$N/α-Fe$_2$O$_3$/C$_3$N$_4$	Magnetic separation, visible-light activity; smooth photoelectron transport; effect of oxygen vacancies to hamper e$^-$/h$^+$ recombination.	Acetamiprid (AP)	76% AP (50 mg/L) removal within 180 min of visible-light illumination at the rate constant of 0.01 min^{-1}	[83]
Fe$_2$TiO$_5$/α-Fe$_2$O$_3$/TiO$_2$	Magnetic separation, visible-light absorption, enhanced e$^-$/h$^+$ separation and transfer	MB	>%98 MB (10 mg/L) removal using 1 g/L of ternary photocatalyst within 120 min under simulated sunlight, photocatalytic degradation efficiency order: α-Fe$_2$O$_3$ > Fe$_2$TiO$_5$ > TiO$_2$.	[70]
α-Fe$_2$O$_3$/CeO$_2$ NTs	enhanced solar activity and e$^-$/h$^+$ separation and transfer via z-scheme α-Fe$_2$O$_3$/CeO$_2$ heterojunction	Tetracycline (TC)	88.6% TC removal	[84]

Reduced graphene oxide (rGO), ethylene diamine (ED); HPV (H$_4$[PVW$_{11}$O$_{40}$]·32H$_2$O); single metal atom oxide (SMAO); contaminants of emerging concern (CECs); Effluents contain Organic Matter (EfOM); granular activated carbon (AC); bamboo-derived activated biochar (BAB); peroxydisulfate (PDS).

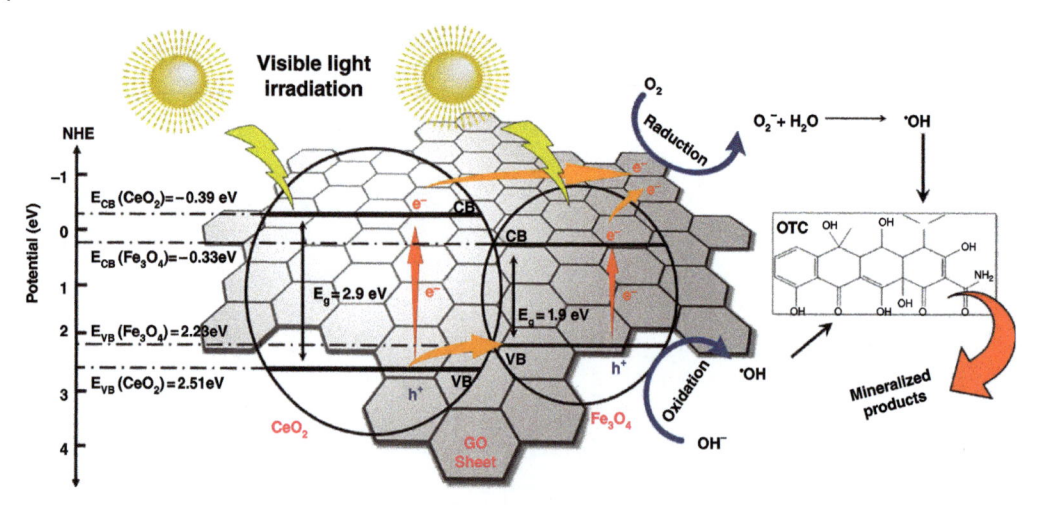

Figure 33.3 The synergistic photocatalytic action of $Fe_{3-x}Ce_xO_4$ and GO for oxytetracycline removal under visible-light irradiation. Ref [32] / Elsevier.

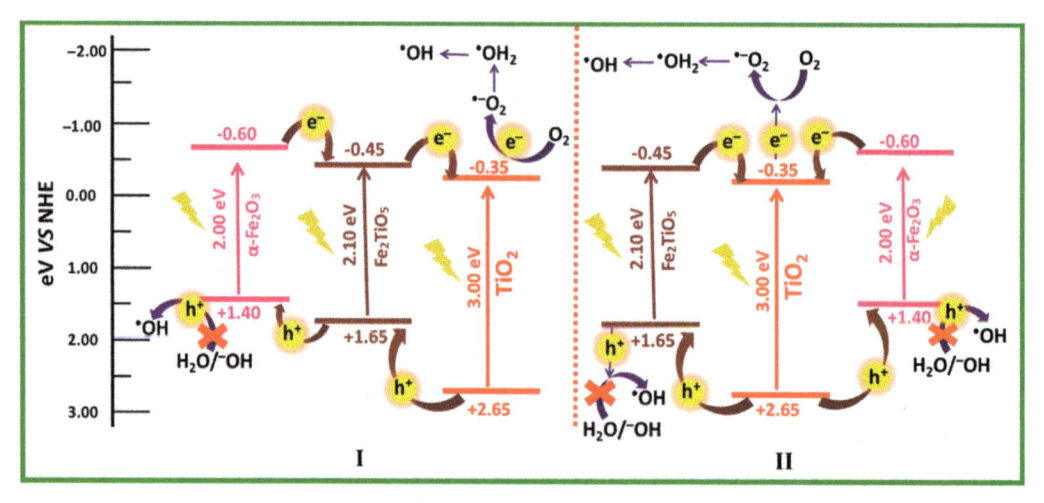

Figure 33.4 Plausible charge transfer route based on the CB potentials of the constituent semiconductors in the ternary $Fe_2TiO_5/\alpha\text{-}Fe_2O_3/TiO_2$ nanocomposite. Ref [70] / Elsevier.

33.3.1.2 Green Synthesis Routes

A sustainable and cost-effective synthesis route that has engaged enormous attention in recent years is the application of a whole plant or its derivatives as natural sources for reducing, stabilizing, and/or capping purposes [85]. These secondary metabolites have unique properties and are ecofriendly and rich sources of polyphenolic compounds that become strongly involved in the reducing of metals and preparation of metal oxides while confining size and stabilizing the structure [86]. A plasmonic photocatalyst was prepared based on Au and ZnO using the aqueous extract of cumin seeds in the absence of surfactant or any chemical agent [76]. The ions of Au^{4+} was reduced to zerovalent Au over the ZnO nanoparticles with the aid of phytomolecules of extract. The Schottky contact created at the interface of Au and ZnO surface resulted in higher photoactivity at the visible region coupled with robust stability of the nanocomposite (Figure 33.5).

Figure 33.5 Suggested mechanism on the reducing and stabilizing effects of polyphenolic compounds [76] / American Chemical Society.

Application of bioingredients instead of toxic chemicals not only endow a mild and safe environment for driving surface modifications but also is a sustainable way, which reverses the undesired consequences. The fruit extract of black grape was used as the natural source of polyphenols to reduce graphene oxide sheets prior to TiO_2 deposition onto the planes (Figure 33.6a) [87]. In this process, the sonically exfoliated GO in ethanol was mixed with grape extract and refluxed for several hours at 80 °C until the brownish colour changed to black, indicating the GO reduction. The resultant powder undergoes a hydrothermal reaction with tetrabutyltitanate solution to give rise to TiO_2/rGO nanocomposite. The mechanism through which the GO is reduced by the polyphenolic content of grape fruit extract is represented in Figure 33.6b. The 85.4% decrease in the oxygen content of rGO compared to that of GO confirmed the successful reduction process. Under visible-light illumination, about 80% of the TOC of bromophenol blue solution was removed over the resultant TiO_2 nanocomposite, which was 30% more than that of TiO_2 alone. The employed nanocomposite particles were separated by centrifuge, regenerated using a dilute acid solution, dried, and reused for fresh dye solution for six successive photocatalytic experiments. The results revealed the high stability and potential reusability of the photocatalyst.

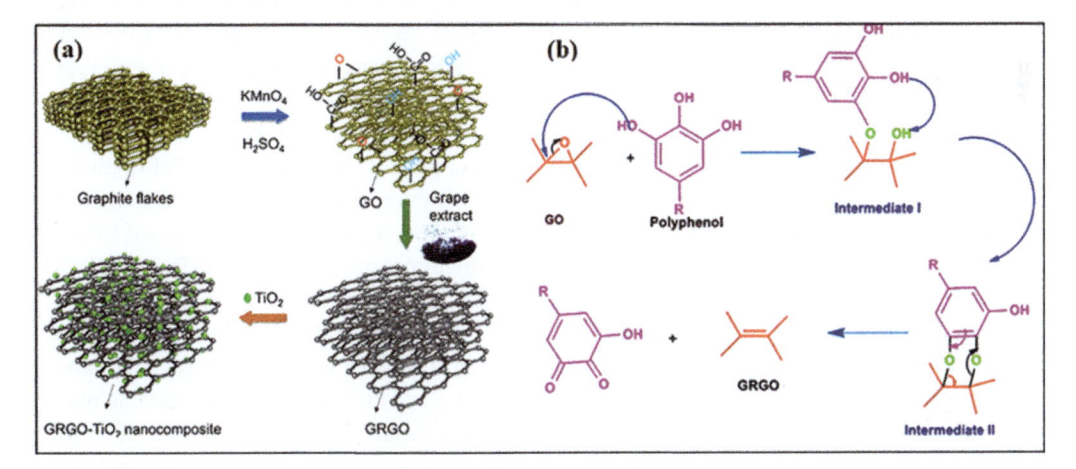

Figure 33.6 (a) Schematic representation of the synthesis process of TiO_2 deposited-grape extract mediated graphene oxide (GRGO); (b) The mechanism of GO reduction via polyphenolic content of grape fruit extract. Reproduced with permission from Ref [87].

33.3.2 Carbonaceous Photocatalysts

Even though metal oxide-based photocatalysts have great efficiencies for the photooxidation of organic pollutants under sunlight exposure, their application has been restricted due to being expensive in some measure and their surface alteration under some gaseous media [88]. The introduction of carbonaceous materials as nature-based, abundant, economical, and metal-free catalysts opened up a new line towards sustainable photocatalysis. Carbonaceous materials offer a wide array of configurations such as carbon nanotubes (CNTs), carbon quantum dots (GQDs), fullerene, activated carbon, biochar, graphene oxide, and graphitic carbon nitride (g-C_3N_4) [89]. Over the past decade, there has been a tremendous increase in the studies on the photocatalytic utilization of carbon-based materials. This stems from the fact that carbon-based materials are low-cost, environmentally-benign, well-disposed, and stable candidates for the fabrication of photocatalysts with versatile features. Moreover, they can be prepared from natural precursors for the synthesis of active photocatalytic composites, which is one of the main criteria for sustainable photocatalysis. In this connection, there is a growing body of literature on the synthesis of carbon-based photocatalysts from natural materials. The preparation of CQDs from pear juice through a one-pot hydrothermal procedure [90], CQDs from corncob biomass [91], activated carbon from date seeds [92, 93], activated carbon from basil leaves (*Ocimum basilicum*) [94], *Eriobotrya japonica* leaf for biochar preparation [95], and liquefied larch sawdust for fabrication of multi-walled CNTs [96] are just a few from many of the studies on biomass-derived carbon-based materials [97–99].

The development of heterojunction/s in the carbon-based nanocomposites along with the broad visible-light absorption delivers excellent performance in photocatalytic systems. Literature is overflowing with studies on carbon-based nanomaterials with various dimensionality and accordingly unique and specific properties beneficial for photocatalytic applications. Herein, a brief statistic on a number of distinguished carbonaceous materials including carbon dots (zero-dimensional), CNTs (one-dimensional), and graphitic carbon nitride (two-dimensional), which have significant potencies for sustainable photocatalytic decontamination of wastewater is provided as follows.

The fluorescent carbon nanoparticles that were noticed during the synthesis and purification of single-walled CNTs, were named CQDs (carbon quantum dots, as mentioned previously). This class of quantum dots was soon become the center of attention for a wide range of applications owing to being environmentally sound, safe, cost-effective, and easy to prepare. Though CQDs were initially synthesized by accident [100], several cost-effective and simple routes have been thereafter developed and modified. In the top-down route, the CQDs are synthesized through structural disintegration of the larger carbonaceous materials such as carbon nanotubes, graphite, or activated carbon [101]. The fabrication of CQDs using carbohydrates or some other molecular precursors via various synthesis methods is called a bottom-up approach [102]. Data on the carbon sources and synthesis methods of CQDs reported in the literature is given in Table 33.2. In general, the CQDs and their modified derivatives are fabricated via microwave heating, ultrasonication, arc-discharge, laser ablation, hydrothermal, solvothermal, and oxygen plasma treatment procedures [103]. The existence of hydroxyl and carboxylic functional groups on the surface of CQDs increases its solubility in aqueous systems and facilities the surface modifications and doping with some other chemical/biological compounds (Figure 33.7a) [104]. In addition to its high solubility in water, the powerful solar-light harvesting properties and effective charge transfer of CQDs have listed this group of carbonaceous materials among the potent candidates for the sustainable photocatalytic treatment of recalcitrant wastewater. The unique electro-optical behavior results from the integration of the quantum confinement effect of small-sized nanoparticles ($\leqslant$ 10 nm) and intrinsic electrical features of carbon materials [105]. Several properties of CQDs, including

Table 33.2 Carbon-based materials for photocatalytic removal of pollutants.

Nanocomposite	Carbon source	Synthesis method	Properties	Pollutant	Operational condition	Outcome	Ref.
ZnO/CNTs	Purchased	CVD[1]/sol-gel	d_{CNT}: 8–15 nm; L_{CNT}: 30–50 μm;	MB	$[MB]_0$: 50 mg/L; Cat: 1 g/L; T: 60 min dark–240 min sunlight	98% MB removal within 150 min over ZnO/CNTs (10 wt.%)	[127]
Au/CNTs/TiO$_2$	Purchased	Sonication-assisted sol-gel	d_{CNT}: 18 nm; Au and TiO$_2$ deposition on CNTs	MB	lamp (λ_{max} = 254 nm): 66 W/1.37 mWcm^{-2}; MB: 5×10^{-5} M; Cat: 0.3 g/L; T: 150 min.	78.9% MB removal using 0.1%Au-2%CNT/TiO$_2$ sample within 150 min.	[128]
Ti-NTs/CNTs	Toluene/ethanol	CVD	$d_{CNT} \approx 100$ nm; Deposition of CNTs into Ti-NTs	Rhodamine B (RhB) λ_{max} = 554 nm	Hg-Xe lamp (1000 W); Cat: 1 g/L; $[RhB]_0$: 1.6 mg/L; T: 60 min dark–30 min light.	RhB photodegradation rate constants of 3.3×10^{-3} and 3.3×10^{-2} min^{-1} over Ti-NTs and the NC.	[111]
TiO$_2$ NRs/CNTs	-	CVD	$d_{CNT} \approx 10$–21 nm; MB	MB	$[MB]_0$: 25 mg/L; Cat: 0.02 g T: 300 min sunlight.	97.3% MB removal within 300 min	[129]
CQDs/TiO$_2$	L-glutamic acid & m-phenylenediamine	Hydrothermal	$d_{CQDs} \approx 10$ nm	Methyl orange (MO)	$[MO]_0$: 40mg/L; Cat: T: 360 min solar (600 nm)	70.56% of MO removal within 360 min using the 1:1 molar ratio of CQDs/TiO$_2$	[108]
CQDs (0.98 wt%)/g-C$_3$N$_4$	Citric acid/urea	bottom-up	–	MB, RhB	$[MB]_0$=$[RhB]_0$: 15 mg/L; Cat: 0.33 g/L; Visible light (320–780 nm)	Complete MB and RhB removal within 20 and 110 min; MB and RhB removal rate constants: 0.214 and 0.048 min^{-1}.	[130]
CQDs/hydrogenated TiO$_2$	Glucose	Oil bath reflux	$d_{CQDs} < 5$ nm $d_{H-TiO2} \approx 22.2$ nm	MO	$[MO]_0$: 20 mg/L; Cat: 1.0 g/L; Visible light (Xe lamp 300 W)	50% MO removal within 25 min under visible light.	[131]

(Continued)

Table 33.2 (Continued)

Nanocomposite	Carbon source	Synthesis method	Properties	Pollutant	Operational condition	Outcome	Ref.
GO/g-C_3N_4/ BiFeO$_3$	Melamine	PVD	-	Cr (VI)	[Cr]$_0$: 5 mg/L; Cat: 2.5 g/L; Visible light (Xe lamp 300 W)	Complete Cr (VI) photoreduction within 90 min.	[132]
rGO/g-C_3N_4 QDs	Flake graphite powder	Hydrothermal	$d_{g-C3N4\ QDs} \approx 4.4$ nm	CIP	[TOC]$_0$: 27.9 mg/L visible light ($\lambda > 420$ nm)	65.9% TOC removal within 150 min, CIP removal rate constant: 0.014 min^{-1}	[133]
Ag/P@g-C_3N_4/ BiVO$_4$	Melamine	photoreduction	-	CIP	[CIP]$_0$: 10 mg/L Cat: 1 g/L; (420–760 nm)	92.6% and 28.5% CIP and TOC removals within 120 min of visible light exposure.	[125]
WO$_3$-TiO$_2$@g-C_3N_4	Melamine	Hydrothermal	10–100 nm 139 m^2/g E_g = 2.27 eV	Aspirin and Caffeine	[ACT]$_0$: 10 mg/L Cat: 1 g/L; Visible light $\lambda > 420$ nm	98% and 97% aspirin and caffeine removals within 90 min.	[134]
Porous g-C_3N_4	dicyandiamide	CNT as hard template	$d_{g-C3N4} \approx 2.0$ nm	RhB	[RhB]$_0$: visible light ($\lambda > 400$ nm)	4.4-fold enhancement in RhB removal rate and enlarged specific surface area (103.3 m^2/g) of porous g-C_3N_4 vs. bulk g-C_3N_4 (10.5 m^2/g).	[135]
Protonated g-C_3N_4/CuS (20%)	Melamine	Electrostatic adsorption	$d_{Cu2O} \approx 400$ nm	*Staphylococcus aureus* and *Escherichia coli* bacteria	Diluted bacterial:160 μL Cat: 40 μL; xenon lamp: 0.2 W/cm^2	99.16% and 98.23% *E. coli* and *S. aureus* removals within 20 min.	[136]
KOH-modified biochar/g-C_3N_4	sunflower straw/ Melamine-cyanuric acid (1:1)	Supramolecular self-assembly/ thermal poly-condensation	-	Phenanthrene (PHT)	[PHT]$_0$: 1 mg/L; Cat: 0.2 g/L; 400–800 nm	76.72% phenanthrene removal within min under visible light at rate constant of 0.355 h^{-1}.	[137]
Mesoporous g-C_3N_4 N-defect	Melamine, SiO$_2$ as a template	Calcination	Specific surface area: 51.34 m^2/g; pore volume: 0.226 cm^3/g	TC	[TC]$_0$: 15 mg/L; Cat: 1.0 g/L; $400 < \lambda < 800$ nm	Complete TC removal within 30 min under visible light	[123]

1) Chemical vapour deposition (CVD);
2) nanoribbons (NR); nanotube (NT); physical vapour deposition (PVD);

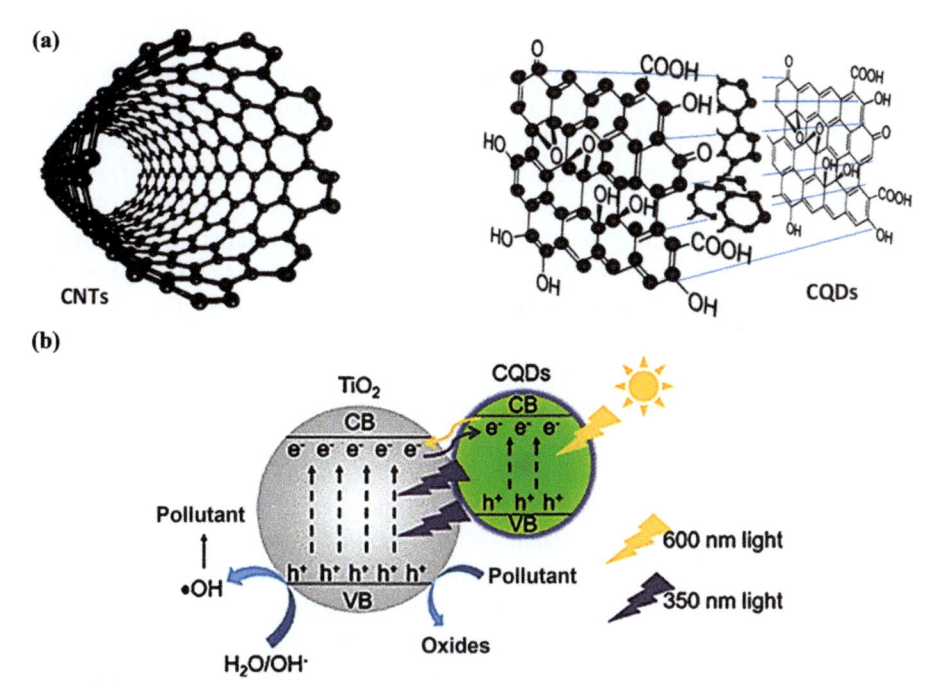

Figure 33.7 (a) The chemical structures of single-walled carbon nanotubes (CNTs) and carbon quantum dots (CQDs). Ref [104] / IOP Publishing. (b) The Vis to UV upconversion phenomenon in CQDs/TiO$_2$ nanocomposite. Ref [108] / IOP Publishing.

photocatalytic action, are in common with other sp^2 hybridized carbon materials, which are rich in electrons. Moreover, they contain sp^3 hybridized carbon atoms, which implies the presence of irregular π–electronic conjugation in its structure [106].

Because more than 50% of sunlight is comprised of infrared radiation, the productive utilization of this portion in a photocatalytic system has long been of interest and valued. From this perspective, the upconversion nanomaterials based on quantum dots (QDs) have impressed researchers for their sustainable applications as highly active solar-light-driven photocatalysts. In this system, the solar photons of long wavelengths (near-infrared photons) are absorbed by QDs and transformed into shorter wavelengths with higher energies (UV-vis radiation) that can stimulate the neighboring semiconductor with wide band energy and thereupon launch photocatalytic reactions. This phenomenon takes place via the process of luminescence resonance energy transfer [107]. The upconversion property of CQDs together with photochemical stability and prominent role in separation and transfer of charge carriers has nominated them as one the most favoured sustainable materials for solar-triggered photocatalytic systems. In a study, the hydrothermal synthesis of anatase TiO$_2$ with m-phenylenediamine and L-glutamic acid, as precursors of CQDs, brought about Vis–UV upconversion CQDs/TiO$_2$ nanocomposite [108]. The resultant photocatalyst could absorb visible light (≈ 600 nm) and convert it into UV photons (≈ 300 to 400 nm) wherein the charge carriers were in turn generated following the TiO$_2$ excitation (Figure 33.7b).

CNTs are among the widely studied carbonaceous materials since their discovery. The intriguing properties of CNTs (i.e. nontoxic, large active surface area, high tensile strength, and advantageous electrical and optical qualities) have made them great candidates for miscellaneous applications [109]. Of the routes established for the synthesis of CNTs-based nanocomposites, the direct one-step approach has received increased attention because of the uniform deposition of the metal/metal oxide nanoparticles over the surface of the oxidized CNTs [110]. Opposing processes have been also applied, wherein CNTs are developed over some other materials (e.g. TiO_2 nanotubes) [111]. In the present instance, an electrochemical anodization approach was used to prepare nanotubular TiO_2 with the aid of Ti foil as the substrate material. Subsequently, a combination of ethanol/toluene was utilized as the carbon precursor and embedded CNTs into the TiO_2 nanotubes through a chemical vapour deposition (CVD) process. The deposition of CNTs inside the nanotubular TiO_2 rather than the exterior wall was mainly attributed to the inner Lewis acid sites [112]. Considering the influence of the electron-electron interactions on the nature of the gap in semiconducting CNTs [113], they demonstrate eminent efficiencies in photocatalytic systems.

$g-C_3N_4$ is a rapidly rising star as a 2D polymeric carbon-based semiconductor for removing aqueous pollutants through photocatalysis [114]. The main features of $g-C_3N_4$ and corresponding synthesis approaches have been detailed in several review articles [107, 115–117]. The striking bandgap (2.7 eV) and surface properties, appealing optical and structural features, high thermochemical stability, and, more importantly, simple synthesis via thermal polymerization have encouraged researchers as ways to satisfy the sustainability criteria while conducting photocatalytic experiments [118]. However, the photocatalytic activity of pristine $g-C_3N_4$ is cramped due to the limited sunlight absorption profile, rapid recombination of charge carriers, and limited electrical conductivity [119]. Wang et al. [120] were the first researchers who reported the photocatalytic application of $g-C_3N_4$, almost 175 years from its first discovery in 1834. Afterwards, thousands of studies on the photocatalytic performance of $g-C_3N_4$ were carried out to enhance its functioning under solar irradiation. The controlled synthesis for attaining a specific morphology [121], defect induction [122, 123], tuning the bandgap through doping [124, 125], and constructing heterojunction/s with some other semiconductor material/s [50, 126] are some strategies to improve the solar photocatalytic degradation of organic molecules by $g-C_3N_4$.

In an effort to fabricate a metal-free and highly efficient visible-light-driven photocatalyst, Cui et al. [121] established a cost-effective and simple approach using an in-air chemical vapour deposition (CVD) procedure to grow $g-C_3N_4$ microstructures in the form of onion rings (Figure 33.8). Despite the conventionally synthesized techniques, a thin layer of highly uniform $g-C_3N_4$ spheres (R-CN) were deposited through this method using a hard template, SiO_2 microspheres of 350 nm in size, and a CVD precursor, melamine. This morphologically modified sample represented an outstanding photocatalytic activity under a broader range of visible light owing to improved bandgap (2.58 eV) and simplified separation of photoinduced charges with expanded lifespans.

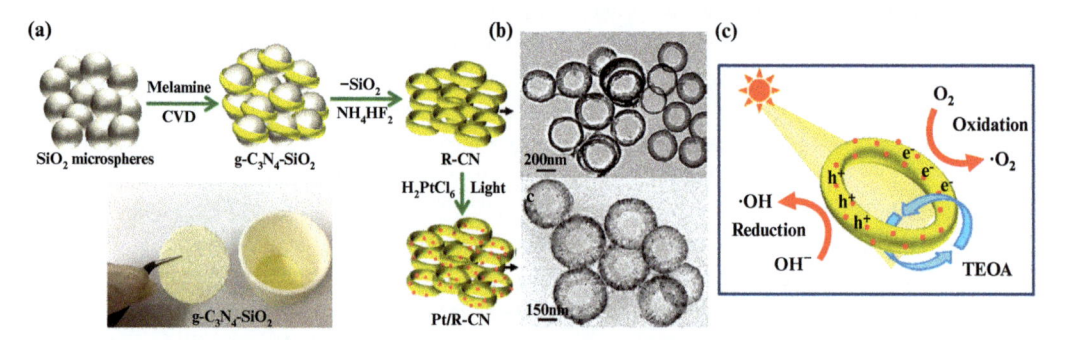

Figure 33.8 (a) Pt-R-CN synthesis procedure. (b) Microscopic images of the samples. (c) Suggested representation for photocatalytic reactions. Reproduced with permission from Ref [121] / American Chemical Society. (Tertiary aliphatic amines [TEOA] as a sacrificial electron donor to stoke up photochemical reactions).

Data on some recent studies of solar-active g-C$_3$N$_4$ nanocomposites are presented in Table 33.2. The information on the operational conditions of the photo-oxidation of various organic pollutants through CQDs, CNTs, and g-C$_3$N$_4$ exhibit the sustainable chemistry practices engaged in the synthesis and photocatalysis processes.

33.4 Remarks and Future Perspectives

Heterogeneous photocatalysis has shown immense potential for sustainable water decontamination. There are wide ranges of ecofriendly materials that comply with sustainable rules. Most metal oxide semiconductors, especially magnetic nanoparticles, and carbon-based semiconductors gave rise to solar-driven photocatalysts that can be synthesized easily using abundant and nature-based materials. However, in most instances, the electrical and optical properties can be modified by conjugation of a wide-bandgap metal oxide with some other semiconductor/s, transition metal/s, or carbon-based materials for powerful solar energy harvesting. In addition, the controlled synthesis procedure of most semiconductors can lead to desired properties with minimum cost and energy input. Nature-derived materials with photocatalytic properties are of great interest as the vast majority are safe, eco-friendly, cheap, and abundant. Additionally, the application of biomass-derived nanomaterials as green alternatives can minimize the application of toxic substances and move closer to sustainability goals. The high porosity, large surface areas, sp^2-hybridized structure, and robust stability of carbon-based composites offer great efficiencies for decontamination of wastewater under solar irradiation. On the other hand, water decontamination at minimal expense and in a sustainable manner can be accomplished via the effective utilization of solar energy. Therefore, the future directions should be to encourage the preparation of photocatalytic nanocomposites that fulfill requirements for easy preparation, high adsorption, wide solar absorption property, and excellent charge separation efficiencies along with facile separation. It must be emphasized that, despite large efforts devoted to developing photocatalysts capable of solar power conversion to chemical energy for water decontamination, there are few studies on some important factors such as photocatalyst content, photoreactor wall materials, temperature control, flow/mixing patterns, and mass transfer efficiency and these should be also taken into account for practical and sustainable implementations.

Acknowledgments

The authors are grateful for the support of the Ardabil University of Medical Sciences (IR.ARUMS. REC.1400.227) and the University of Mohaghegh Ardabili.

References

1. Shiklomanov, I.A. (2000). Appraisal and assessment of world water resources. *Water Int.* 25 (1): 11–32.
2. Cosgrove, W.J. and Loucks, D.P. (2015). *Water Resour. Res.* 51 (6): 4823–4839.
3. Rodriguez, D.J., Serrano, H.A., Delgado, A. et al. (2020). *From Waste to Resource: Shifting Paradigms for Smarter Wastewater Interventions in Latin America and the Caribbean*. Washington, DC: World Bank.

4 Connor, R. (2021). *The United Nations World Water Development Report 2021, VALUING WATER World*, 1–11. Italy: UNESCO World Water Assessment Programme.
5 Shannon, M.A., Bohn, W., Elimelech M. et al. (2008). *Nature* 452 (7185): 301–310.
6 Landrigan, P.J. et al. (2018). *The Lancet* 391 (10119): 462–512.
7 Lema, J.M. and Martinez, S.S. (2017). *Innovative Wastewater Treatment & Resource Recovery Technologies: Impacts on Energy, Economy and Environment*. London: IWA Publishing.
8 Scardozzi, G., Ismaelli, G.T., Leucci, G. et al. (2021). *Archaeol. Prospect.* 28 (2): 137–151.
9 Crini, G. and Lichtfouse, E. (2019). *Environ. Chem. Lett.* 17 (1): 145–155.
10 Rahim Pouran, S., Abdul Aziz, A.R., and Wan Daud, W.M.A. (2015). *J. Ind. Eng. Chem.* 21: 53–69.
11 Keller, N. Ivanez, J., Highfield,J et al. (2021). *Appl. Catal. B Environ.* 296: 120320.
12 Wu, Q. and Zhang, Z. (2019). *Adv. Powder Technol.* 30 (2): 415–422.
13 White, J.L., Baruch, M.F., Pander, J.E. et al. (2015). *Chem. Rev.* 115 (23): 12888–12935.
14 Schneider, J., Matsuoka, M., Takeuchi, M. et al. (2014). *Chem. Rev.* 114 (19): 9919–9986.
15 Pirhashemi, M., Habibi-Yangjeh, A., and Rahim Pouran, S. (2018). *J. Ind. Eng. Chem.* 62: 1–25.
16 Fauzi, A.A., Jalil, A.A., Hassan, N.S. et al. (2022). *Chemosphere* 286: 131651.
17 Ong, W.-J., Tan, L.-I., Ng, Y.H. et al. (2016). *Chem. Rev.* 116 (12): 7159–7329.
18 Singh, P. et al. (2019). *Mater. Today Chem.* 14: 100186.
19 Cao, M., Wang, P., Ao, Y. et al. (2016). *J. Colloid Interface Sci.* 467: 129–139.
20 Deng, J., Xiao, S., Wang, B. et al. (2020). *ACS Appl. Mater. Interfaces* 12 (46): 51537–51545.
21 Gao, Z., Yang, H., Li, J. et al. (2020). *Appl. Catal. B Environ.* 267: 118695.
22 Wang, H., Zhang, R., Yuan, D. et al. (2020). *Adv. Funct. Mater.* 30 (46): 2003995.
23 Zhang, C., Wu, M.-B., Wu, B.-H. et al. (2018). *J. Mater. Chem. A* 6 (19): 8880–8885.
24 Coronado, J.M. (2013). A historical introduction to photocatalysis. In: *Design of Advanced Photocatalytic Materials for Energy and Environmental Applications* (eds. J.M. Coronado, F. Fresno, M. Hernández-Alonso, and R. Portela), 1–5. London: Springer.
25 Li, K., de Rancourt de Mimérand, Y., Jin, X. et al. (2020). *ACS Appl. Nano Mater.* 3 (3): 2830–2845.
26 Mohd Adnan, M.A., Phoon, B.L., and Muhd Julkapli, N. (2020). *J. Clean. Prod.* 261: 121190.
27 Batzill, M. (2011). *Energy Environ. Sci.* 4 (9): 3275–3286.
28 Nunes, D., Pimentel, A., Santos, L. et al. (2019). 3 - Structural, optical, and electronic properties of metal oxide nanostructures. In: *Metal Oxide Nanostructures* (eds. D. Nunes, L. Santos, L. Pereira et al.), 59–102. Elsevier.
29 Pimentel, A., Rodrigues, J., Duarte, P. et al. (2015). *J. Mater. Sci.* 50 (17): 5777–5787.
30 Zhang, H., Zhang, Z., Liu, Y.et al. (2021). *J. Phys. Chem. Lett.* 12 (38): 9188–9196.
31 Raizada, P., Soni, V., Kumit, A. et al. (2021). *J. Materiomics* 7 (2): 388–418.
32 Hassandoost, R., Pouran, S.R., Khataee, A. et al. (2019). *J. Hazard. Mater.* 376: 200–211.
33 Hu, J., Zhao, R., Li, H. et al. (2022). *Appl. Catal. B Environ.* 303: 120869.
34 Kumar, S., Kaushik, R.D., Upadhyay, G.K. et al. (2021). *J. Hazard. Mater.* 406: 124300.
35 Ma, R., Zhang, S., Wen, T. et al. (2019). *Catal. Today* 335: 20–30.
36 Shi, Y., Ma, J., Chen, Y. et al. (2022). *Sci. Total Environ.* 804: 150024.
37 Chen, S., Huang, D., Xu, P. et al. (2020). *ACS Catal.* 10 (2): 1024–1059.
38 Kumar, R., Raizada, P., Verma, N. et al. (2021). *J. Clean. Prod.* 297: 126617.
39 Samuel, O., Othman, M.H.D., Kamaludin, R. et al. (2022). *Ceram. Int.* 48 (5): 5845–5875.
40 Sun, C., Yang, J., Xu, M. et al. (2022). *Chem. Eng. J.* 427: 131564.
41 Wan, X., Wang, Y., Jin, H. et al. (2019). *Ceram. Int.* 45 (17, Part A): 21091–21098.
42 Wei, X., Pan J., Wei, J. et al. (2018). *Photonics Nanostruc. Fundam. Appl.* 30: 20–24.
43 Fujishima, A. and Honda, K. (1972). *Nature* 238 (5358): 37–38.
44 Loeb, S.K., Alvarez, P.J.J., Brame, J.A. et al. (2019). *Environ. Sci. Technol.* 53 (6): 2937–2947.

45 Lee, S.-Y. and Park, S.-J. (2013). *J. Ind. Eng. Chem.* 19 (6): 1761–1769.
46 Al-Hajji, L.A. and Ismail, A.A. (2019). *Superlattices Microstruct.* 129: 259–267.
47 Mancuso, A., Sacco, O., Vaiano, V. et al. (2021). *Catal. Today* 380: 93–104.
48 Jiménez-Salcedo, M., Monge, M., and Tena, M.T. (2022). *Photochem. Photobiol. Sci.* 21: 337–347.
49 Vukoje, I., Kovač, K., Džunuzović, J. et al. (2016). *J. Phys. Chem. C* 120 (33): 18560–18569.
50 Ji, H., Du, P., Zhao, D. et al. (2020). *Appl. Catal. B Environ.* 263: 118357.
51 Kim, M.G., Kang, J.M., Lee, J.E. et al. (2021). *ACS Omega* 6 (16): 10668–10678.
52 Hsien, Y.-H., Chang, C.-F., Chen, Y.-H. et al. (2001). *Appl. Catal. B Environ.* 31 (4): 241–249.
53 He, J., Kumar, A., Khan, M. et al. (2021). *Sci. Total Environ.* 758: 143953.
54 Ding, D., Liu, K., He, S. et al. (2014). *Nano Lett.* 14 (11): 6731–6736.
55 Taylor, C.M., Ramirez-Canon, A., Wenk, J. et al. (2019). *J. Hazard. Mater.* 378: 120799.
56 Wolski, L., Grzelak, K., Muńko, M. et al. (2021). *Appl. Surf. Sci.* 563: 150338.
57 Das, S., Misra, A.J., Habeeb Rahman, A.P. et al. (2019). *Appl. Catal. B Environ.* 259: 118065.
58 Gawande, M.B., Branco, P.S., and Varma, R.S. (2013). *Chem. Soc. Rev.* 42 (8): 3371–3393.
59 Kim, B., Jang, J., and Lee, D.S. (2022). *Chemosphere* 289: 133040.
60 Esquinazi, P., Hergert, W., Spemann, D. et al. (2013). *IEEE Trans. Magn.* 49 (8): 4668–4674.
61 Hernández-Tecorralco, J., Meza-Montes, L., Cifuentes-Quintal, M.E. et al. (2020). *J. Phys. Condens. Matter* 32 (25): 255801.
62 Rahim Pouran, S., Abdul Raman, A.A., and Wan Daud, W.M.A. (2014). *J. Clean. Prod.* 64: 24–35.
63 Ji, W.-C., Hu, P., Wang, X-Y. et al. (2021). *J. Alloys Compd.* 866 (p): 158952.
64 Selvakumar, K., Wang, Y., Lu, Y. et al. (2022). *Appl. Catal. B Environ.* 300: 120740.
65 Brossault, D.F.F., McCoy, T.M., and Routh, A.F. (2021). *J. Colloid Interface Sci.* 584: 779–788.
66 Fazli, A., Khataee, A., Brigante, M.et al. (2021). *Chem. Eng. J.* 404: 126391.
67 Sabri, M., Habibi-Yangjeh, A., Chand, V. et al. (2021). *J. Mater. Sci. Mater. Electron.* 32 (4): 4272–4289.
68 Masudi, A., Harimisa, G.E., Ghafar, N.A. et al. (2020). *Environ. Sci. Pollut. Res. Int.* 27 (5): 4664–4682.
69 Pan, S., Yin, R.-T., Huang, W.T. et al. (2021). *J. Nanosci. Nanotechnol.* 21 (6): 3178–3182.
70 Bhoi, Y.P., Fang, F., Zhou, X. et al. (2020). *Appl. Surf. Sci.* 525: 146571.
71 Byrne, C., Dervin, S., Hermosilla, D. et al. (2021). *Catal. Today* 380: 199–208.
72 Mohan, H., Ramasamy, M., Ramalingam, V. et al. (2021). *J. Hazard. Mater.* 412: 125330.
73 Liu, X., Fang, B., Wang, Z. et al. (2021). *ACS Appl. Nano Mater.* 4 (4): 3742–3749.
74 Mukherjee, I., Cilamkoti, V., and Dutta, R.K. (2021). *ACS Appl. Nano Mater.* 4 (8): 7686–7697.
75 Dehghani, M., Nadeem, H., Singh Raghuwanshi, V. et al. (2020). *ACS Appl. Nano Mater.* 3 (10): 10284–10295.
76 Choudhary, M.K., Kataria, J., and Sharma, S. (2018). *ACS Appl. Nano Mater.* 1 (4): 1870–1878.
77 Chávez, A.M., Quiñones, D.H., Rey, A. et al. (2020). *Chem. Eng. J.* 398: 125642.
78 Sultana, S., Amirbahman, A., and Tripp, C.P. (2020). *Appl. Catal. B Environ.* 273: 118935.
79 Zhu, L., Kong, X., Yang, C. et al. (2020). *J. Hazard. Mater.* 381: 120910.
80 Ahmadpour, N., Sayadi, M.H., Sobhani, S. et al. (2020). *J. Clean. Prod.* 268: 122023.
81 Talukdar, K., Jun, B.-M., Yoon, Y. et al. (2020). *J. Hazard. Mater.* 398: 123025.
82 Chávez, A.M., Solís, R.R., and Beltrán, F.J. (2020). *Appl. Catal. B Environ.* 262: 118275.
83 Padervand, M., Ghasemi, S., Hajiahmadi, S. et al. (2021). *Appl. Surf. Sci.* 544: 148939.
84 He, S., Yan, C., Chen, X.-Z. et al. (2020). *Appl. Catal. B Environ.* 276: 119138.
85 Kumar, J.A., Krithiga, T., Manigandan, S. et al. (2021). *J. Clean. Prod.* 324: 129198.
86 Bayrami, A., Alioghli, S., Rahim Pouran, S. et al. (2019). *Ultrason. Sonochem.* 55: 57–66.
87 Ramanathan, S., Moorthy, S., Ramasundaram, S. et al. (2021). *ACS Omega* 6 (23): 14734–14747.

88 Liu, X. and Dai, L. (2016). *Nat. Rev. Mater* 1 (11): 16064.
89 Gopinath, K.P., Vo, D.-V.N., Gnana Prakash, D. et al. (2021). *Environ. Chem. Lett.* 19 (1): 557–582.
90 Das, G.S., Shim, J.P., Bhatnagar, A. et al. (2019). *Sci. Rep.* 9 (1): 15084.
91 Xie, X., Li, S., Qi, K. et al. (2021). *Chem. Eng. J.* 420: 129705.
92 Faisal, M., Alsaiari, M., Rashed, M.A. et al. (2021). *J. Taiwan Inst. Chem. Eng.* 120: 313–324.
93 Alsaiari, M. (2021). *Arab. J. Chem.* 14 (8): 103258.
94 Bayahia, H. (2022). *J. Saudi Chem. Soc.* 26 (2): 101432.
95 Yu, C., Tang, J., Liu, F. et al. (2021). *Chemosphere* 284: 131237.
96 Zhang, Y., Sun, J., Tan, J. et al. (2021). *Fuel* 305: 121622.
97 Omoriyekomwan, J.E., Tahmasebi, A., Dou, J. et al. (2021). *Fuel Process. Technol.* 214: 106686.
98 Hoang, A.T., Nižetić, S., Cheng, C.K. et al. (2022). *Chemosphere* 287: 131959.
99 Ahuja, V., Bhatt, A.K., Varjani, S. et al. (2022). *Chemosphere* 293: 133564.
100 Xu, X., Ray, R., Gu, Y. et al. (2004). *J. Am. Chem. Soc.* 126 (40): 12736–12737.
101 Hagiwara, K., Horikoshi, S., and Serpone, N. (2021). *Chem. A Eur. J.* 27 (37): 9466–9481.
102 Lim, S.Y., Shen, W., and Gao, Z. (2015). *Chem. Soc. Rev.* 44 (1): 362–381.
103 Pirsaheb, M., Asadi, A., Sillanpää, M. et al. (2018). *J. Mol. Liq.* 271: 857–871.
104 Demchenko, A.P. and Dekaliuk, M.O. (2013). *Methods Appl. Fluoresc.* 1 (4): 042001.
105 Choppadandi, M., Guduru, A.T., Gondaliya, P. et al. (2021). *Mater. Sci. Eng. C* 129: 112366.
106 Zhai, Y., Zhang, B., Shi, R. et al. (2022). *Adv. Energy Mater.* 12 (6): 2103426.
107 Jiang, L., Yang, J., Zhou, S. et al. (2021). *Coord. Chem. Rev.* 439: 213947.
108 Deng, Y., Cheng, M., Cheng, G. et al. (2021). *ACS Omega* 6 (6): 4247–4254.
109 Wu, L., Wu, T., Liu, Z. et al. (2022). *J. Hazard. Mater.* 431: 128536.
110 Chen, W., Pan, X., Willinger, M.-G. et al. (2006). *J. Am. Chem. Soc.* 128 (10): 3136–3137.
111 Alsawat, M., Altalhi, T., Gulati, K. et al. (2015). *ACS Appl. Mater. Interfaces* 7 (51): 28361–28368.
112 Eswaramoorthi, I. and Hwang, L.-P. (2007). *Diam. Relat. Mater.* 16 (8): 1571–1578.
113 Aspitarte, L., McCulley, D.R., Bertoni, A. et al. (2017). *Sci. Rep.* 7 (1): 8828.
114 Mousavi, M., Habibi-Yangjeh, A., and Pouran, S.R. (2018). *J. Mater. Sci. Mater. Electron.* 29 (3): 1719–1747.
115 Zhu, J., Xiao, P., Li, H., et al. (2014). *ACS Appl. Mater. Interfaces* 6 (19): 16449–16465.
116 Wang, L., Wang, K., He, T. et al. (2020). *ACS Sustain. Chem. Eng.* 8 (43): 16048–16085.
117 Huang, R., Wu, J., Zhang, M. et al. (2021). *Mater. Des.* 210: 110040.
118 Zhao, Z., Sun, Y., and Dong, F. (2015). *Nanoscale* 7 (1): 15–37.
119 Ye, C., Li, J.-X., Li, Z.-J. et al. (2015). *ACS Catal.* 5 (11): 6973–6979.
120 Wang, X., Maeda, K., Thomas, A. et al. (2009). *Nat. Mater.* 8 (1): 76–80.
121 Cui, L., Song, J., McGuire, A.F. et al. (2018). *ACS Nano* 12 (6): 5551–5558.
122 Zhang, M., Duan, Y., Jia, H. et al. (2017). *Catal. Sci. Technol.* 7 (2): 452–458.
123 Razavi-Esfali, M., Mahvelati-Shamsabadi, T., Fattahimoghaddam, H. et al. (2021). *Chem. Eng. J.* 419: 129503.
124 Guo, C., Chen, M., Wu, L. et al. (2019). *ACS Appl. Nano Mater.* 2 (5): 2817–2829.
125 Deng, Y., Tang, L., Feng, C. et al. (2018). *J. Hazard. Mater.* 344: 758–769.
126 Asadzadeh-Khaneghah, S. and Habibi-Yangjeh, A. (2020). *J. Clean. Prod.* 276: 124319.
127 Phin, H.-Y., Ong, Y.-T., and Sin, J.-C. (2020). *J. Environ. Chem. Eng.* 8 (3): 103222.
128 Chinh, V.D., Hung, L.X., Di Palma, L. et al. (2019). *Chem. Eng. Technol.* 42 (2): 308–315.
129 Shaban, M., Ashraf, A.M., and Abukhadra, M.R. (2018). *Sci. Rep.* 8 (1): 781.
130 Zhang, L., Zhang, J., Xia, Y. et al. (2020). *Int. J. Mol. Sci.* 21 (3).
131 Tian, J., Leng, Y., Zhao, Z. et al. (2015). *Nano Energy* 11: 419–427.
132 Hu, X., Wang, W., Xie, G. et al. (2019). *Chemosphere* 216: 733–741.

133 Zhao, P., Jin, B., Yan, J. et al. (2021). *RSC Adv.* 11 (56): 35147–35155.
134 Tahir, M.B., Sagir, M., and Shahzad, K. (2019). *J. Hazard. Mater.* 363: 205–213.
135 Chen, X., Sagir, M., and Shahzad, K. (2019). *ChemistrySelect* 4 (20): 6123–6129.
136 Ding, H., Han, D., Han, Y. et al. (2020). *J. Hazard. Mater.* 393: 122423.
137 Lin, M., Li, F., Cheng, W. et al. (2022). *Chemosphere* 288: 132620.

34

Sustainable Homogeneous Catalytic Oxidative Processes for the Desulfurization of Fuels

Federica Sabuzi[1], Giuseppe Pomarico[2,3], Pierluca Galloni[1], and Valeria Conte[1]

[1] *Department of Chemical Science and Technologies, University of Rome Tor Vergata, Via della Ricerca Scientifica snc, Rome, Italy*
[2] *Department of Molecular and Translational Medicine, University of Brescia, Viale Europa, 11, Brescia, Italy*
[3] *CSGI, Research Center for Colloids and Nanoscience, Via della Lastruccia, 3, Sesto Fiorentino, Firenze, Italy*

34.1 Introduction

At present, in addition to CO_2 and nitrogen oxides, sulfide oxides (SO_x) emissions must be strictly controlled, because SO_x are responsible for severe environmental and human health issues. Therefore, there is a strong demand for ultra-low-sulfur fuels to limit the amount of SO_x generated during combustion.

Oxidative fuel desulfurization (ODS) currently represents one of the most valuable approaches to lower S-content in fuel, as it can be carried out in sustainable mild conditions, with simple equipment and low operating costs [1, 2]. In ODS, sulfides are oxidized to the corresponding sulfoxides and/or sulfones by an oxidant, in the presence of a catalyst; oxidation products are easily removed from the fuel by liquid-liquid extraction or adsorption on solid materials.

The major aromatic organosulfur compounds in fuels (usually resistant to other desulfurization methods such as hydrodesulfurization) are benzothiophene (BT), dibenzothiophene (DBT), 4-methyldibenzothiophene (MDBT) and 4,6-dimethyldibenzothiophene (DMDBT) (Figure 34.1).

In this chapter, recent developments in oxidative fuel desulfurization coupled with liquid-liquid extraction methods are reported. In particular, it focuses on the application of homogeneous metal catalysis to obtain fuels with ultra-low content of sulfur. The key features of ODS, i.e. reaction selectivity for sulfides, catalysts and oxidant recovery and reuse, as well as the recyclability of extraction solvent, will be discussed, in order to assess the potential industrial application of the proposed systems.

34.2 Vanadium

Vanadium exerts a chief role in oxidation reactions; next to the conventional commercially available V-catalysts, to date, a plethora of different V-complexes have been synthesized to catalyze the oxidation of alkanes, alkenes, arenes, alcohols, as well as sulfides [3–6].

Catalysis for a Sustainable Environment: Reactions, Processes and Applied Technologies 2V Set, First Edition. Edited by Armando J. L. Pombeiro, Manas Sutradhar, and Elisabete C. B. A. Alegria.
© 2024 John Wiley & Sons Ltd. Published 2024 by John Wiley & Sons Ltd.

Figure 34.1 Principal aromatic organosulfur compounds in fuel.

The first examples of V-catalyzed homogeneous oxidative systems for fuel desulfurization involved the use of vanadyl acetylacetonate (VO(acac)$_2$) and H$_2$O$_2$. Reactions were performed in biphasic system dissolving DBT in a model fuel, while the proper amount of VO(acac)$_2$ and a large excess of H$_2$O$_2$ were dissolved in an immiscible solvent, like acetonitrile. Results showed that DBT was almost completely oxidized in two hours at 40 °C, and oxidation products were extracted in the acetonitrile layer [7]. Later on, eco-friendly alternatives to conventional organic solvents have been proposed as extractants. The ionic liquids (ILs) 1-butyl-3-methylimidazolium tetrafluoroborate ([bmim]BF$_4$) and 1-butyl-3-methylimidazolium bis(trifluoromethanesulfonyl)imide ([bmim]Tf$_2$N) successfully accomplished more than 95% of sulfur removal from model oils, in the presence of 5% mol of VO(acac)$_2$ [8, 9]. Interestingly, catalyst and IL could be recycled without significant loss of activity [9]. Promising results have been obtained applying microwave irradiation (MW) during ODS. Indeed, S-removal could be efficiently achieved using VO(acac)$_2$, H$_2$O$_2$, and N-carboxymethylpyridine hydrosulphate ionic liquid ([CH$_2$CO$_2$HPy]HSO$_4$) with MW power of 500 W in 90 s, at 80 °C [10]. In addition, the system could be improved using a combination of H$_2$O$_2$/H$_2$SO$_4$ [10] or HNO$_3$/H$_2$SO$_4$ with pyridinium phosphate ([HPy]H$_2$PO$_4$) as extractant IL [11]. However, the presence of very strong acids limits the industrial application of such protocol.

The catalytic oxidation of DBT and DMDBT has been accomplished also using more elaborated structures (Figure 34.2). Oxovanadium^{IV} complex **V1** with a tetradentate N$_2$O$_2$-donor ligand has

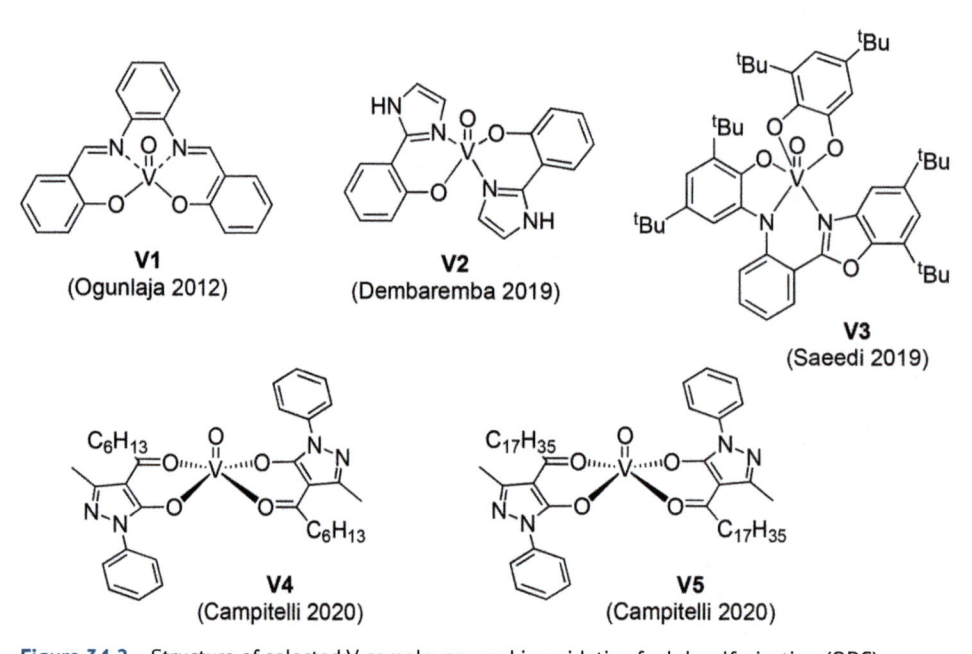

Figure 34.2 Structure of selected V-complexes used in oxidative fuel desulfurization (ODS).

been adopted to perform ODS in mild conditions [5, 12]. Reactions performed on a model oil, prepared dissolving DBT and DMDBT in a mixture toluene–hexane in the presence of an excess of *tert*-butyl hydroperoxide (TBHP), at 40 °C, led to more than 60% of DMBT and c. 87% of DBT oxidation in six hours.

The system was even upgraded, supporting **V1** on a polystyrene-based polymer that improved catalyst stability and reaction selectivity to sulfone. Likewise, preliminary studies on DBT oxidation in acetonitrile have been performed using **V2** and **V3** catalysts [13, 14], with excess of TBHP [13] or H_2O_2 [14] as oxidants. High conversions and good selectivity to sulfone were achieved, but products extraction from the model oil was not reported.

S-removal from a model fuel was also accomplished after DBT and DMDBT oxidation with TBHP, using oxovanadium complexes **V4** and **V5** [15]. Here, the beneficial role of 1-butyl-3-methylimidazolium hexafluorophosphate ([bmim]PF_6) as extractive solvent was highlighted, and recyclability studies confirmed the robustness of such catalysts, which could be recycled with only slight decreasing of activity.

V-oxodiperoxidocomplexes with pyridine, 2- and 4-picoline ligands have been found to be promising catalysts in desulfurization of a diesel fuel in a biphasic system (CH_2Cl_2/H_2O), using cationic surfactants and H_2O_2 as the oxidant [16]. On the contrary, V-peroxido complexes with 2,6-pyridinedicarboxylic acid derivatives as ligands led to unsatisfactory results in diesel desulfurization [17].

Notably, salophen and salen resulted amongst the most promising ligands for V catalyzed ODS [18–20]: in a recent example, a three-phase system (fuel/H_2O_2/IL) was proposed (Figure 34.3), in which DBT was dissolved in a model fuel, while 0.5% of V^V-catalyst (**V6-V8**) was dissolved in an immiscible IL and the primary oxidant (i.e. H_2O_2) resided between the two phases. Importantly, the use of H_2O_2 is preferred to TBHP for improving the atom economy of the process.

Here, **V6** resulted the most active catalyst because it led to almost quantitative DBT conversion under conventional heating or with MW irradiation at 35 W and 100 °C [18, 21]. The successful feature of such a protocol was the remarkable selectivity for sulfides oxidation, because cyclooctene added in the mixture as a competitive substrate was not significantly oxidized [18]. Indeed, it has proved to be a particularly good and economical alternative to hydrodesulfurization methods, being able to strongly reduce S-content in fuels while leaving alkenes amount intact, thus being suitable for industrial purposes.

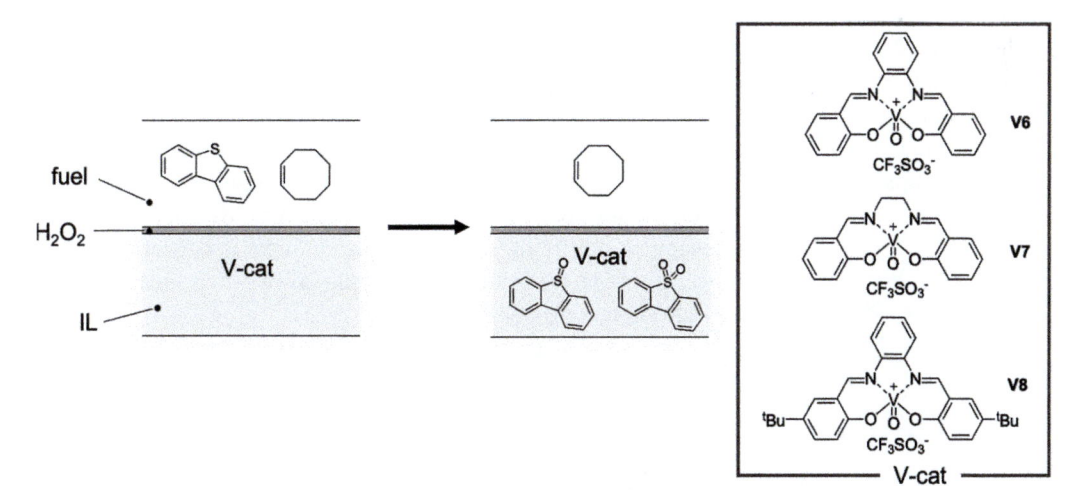

Figure 34.3 Three phase system for oxidative fuel desulfurization (ODS) [18] / Elsevier.

34.3 Manganese

Taking inspiration from oxygenase enzymes, which can convert sulfides in sulfoxides and sulfones under mild conditions, the catalytic activity of Mn-porphyrins has been evaluated in ODS. Desulfurization of a model oil prepared dissolving BT, 2-methyl-BT, 3-methyl-BT and DBT in hexane has been accomplished using [5,10,15,20-tetrakis(2,6-dichlorophenyl)porphyrinato]manganese^{III} chloride as the catalyst and ammonium acetate as co-catalyst, both dissolved in CH_3CN [22, 23]. Reactions performed in the dark, at 22–25 °C, using H_2O_2 as oxidant, led to almost complete substrates oxidation after two hours.

34.4 Iron

Fe-TAML® catalyst (Figure 34.4; TAML: tetraamidomacrocyclic ligand) is a class of iron catalysts with ligand(s) designed to be used under different reaction conditions in term of pH, temperature, and solvents. **Fe1** [24] was investigated both in homogeneous medium and in a two-phase system. Homogeneous reaction was performed in water/*tert*-butyl alcohol, where the latter was chosen to solubilize DBT. The reaction was affected by water/*t*-BuOH ratio, where 7:3 v/v gave the fastest reaction but no conversion was reported with lighter alcohols.

Under these conditions DBT is the most reactive species followed by MDBT and DMDBT. In the two-phase reaction, performed with decane as model for diesel, oxidation requires longer time for completion (three hours vs 30 minutes) mainly because of the oxidant was added in six portions to limit its decomposition, thus extending the overall reaction time.

The properties of μ-oxo dinuclear iron complex in sulfide oxidation were tested and compared to those of the structurally related mononuclear Fe^{III} complex [25]. Dimeric iron-complex dissolved in CH_3CN with H_2O_2 quickly reacts with DBT, as suggested by the blue color disappearance of the peroxido intermediate. DBT can directly react with the peroxido-intermediate or with some decomposition products, yielding DBTO, whereas sulfone derivative appears after the addition of eight equivalents of H_2O_2, ruling out direct formation of $DBTO_2$. Investigation of the reaction mechanism performed with $H_2^{18}O$ and mass analysis suggested the involvement of a $Fe^{IV}(O)$ species formed upon the homolytic cleavage of the peroxido intermediate.

Zhou et al. [26] reported the oxidation of DBT in decalin with O_2 at 1.5 MPa and 160 °C for five hours; oxidation occurred only in the presence of a porphyrin containing catalyst, **Fe2** (Figure 34.5). Results by the product characterization were in agreement with the formation of $DBTO_2$. The molecular structure of the catalyst affects its reactivity that increases with electron-withdrawing groups and decreases with donating substituents. However, the required harsh reaction conditions may lead to catalyst degradation and deactivation. As far as reaction mechanism is concerned, radical process was ruled out, while a two-step nucleophilic addition was proposed.

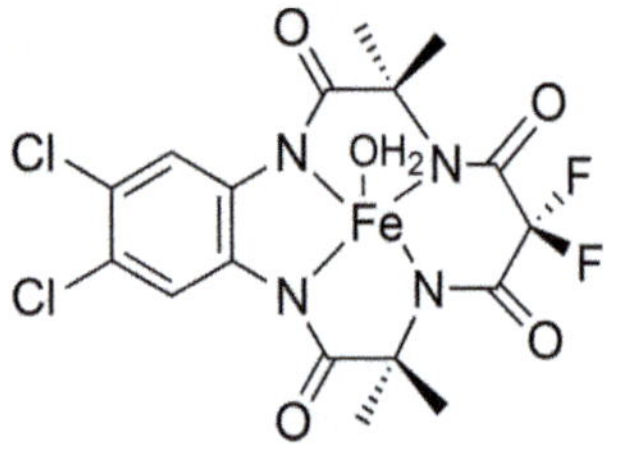

The catalytic activity of iron μ-oxo dimer of **Fe3** was investigated by Aguiar [27] in biphasic system, made by equal volume of the model oil and immiscible extracting solvent. The process was performed in two steps, where the first consists in the extraction of sulfur-based compounds, followed by H_2O_2 addition to start oxidation reaction.

Figure 34.4 **Fe1** structure [24] / Elsevier.

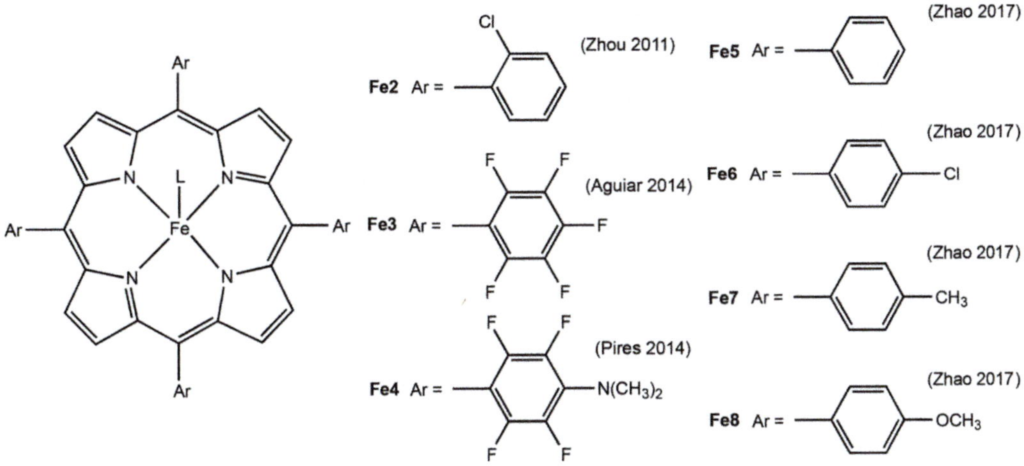

Figure 34.5 Structure of selected iron porphyrins used in oxidative fuel desulfurization (ODS); axial ligand is Cl⁻ unless otherwise stated.

Although highest extraction value was achieved by DMF, no oxidation products were detected in this solvent, while CH_3CN and CH_3OH yielded satisfactory results after two hours. Little loss of activity was reported after three consecutive cycles.

Pires et al. [28] explored the ODS reaction by iron-porphyrin bearing electron-withdrawing group on peripheral phenyl rings (**Fe4**). They focused on the sustainability of the reaction such as the choice of a green solvent (methanol or ethanol) and the ability of Fe-porphyrin to promote the reaction without co-catalyst (only H_2O_2 was used). Once the efficacy toward thioanisole had been proven, this catalyst was applied to more refractory substrates and generally good results were obtained.

The need for the greenest possible systems prompted Zhao et al. [29] to explore ILs as solvent. Fe^{III} porphyrin and H_2O_2 in different ILs as extractants allowed to remove at least more than 97% (complex **Fe5–Fe8** yielded different values) of DBT added to octane as model oil, while lower amount of DBT was removed with no catalyst added. Under optimal reaction conditions, BT removal reached 94.7% with **Fe7** and [bmim]PF_6. High efficiency of this IL could be due to the stabilization effect of both ions toward intermediates and it could be recycled for six times with no loss of catalytic activity. Similar results were obtained on desulfurization of real diesel fuel. The same authors [30] detailed the role of anionic axial ligand of **Fe5** in the catalytic activity enhancement. Native chloride was replaced by PF_6^- or BF_4^- which bind the metal center more loosely, favoring the formation of active intermediates, Fe^{III} hydroperoxides porphyrin complex or high-valent Fe^{IV} oxo-porphyrin cation radical. As a result, better conversion rate of DBT in $DBTO_2$ was obtained.

Fe-phthalocyanine complex **Fe9** (Figure 34.6) was initially investigated by Zhou et al. [31] with O_2 as oxidant in decalin. In the presence of 1% of the catalyst, $DBTO_2$ was produced.

Conversion was enhanced by O_2 pressure (solubility in decalin increases) and temperature increase. Reactivity and stability of phthalocyanines bearing different functional groups was investigated and the general trend is that the lower the electron density, the higher the reactivity. In biphasic water/octane system, Liu et al. [32] investigated oxidation of DBT and related compounds to sulfone with **Fe10**. Under mild conditions (30 °C, 40 equivalents of H_2O_2), DMDBT was more

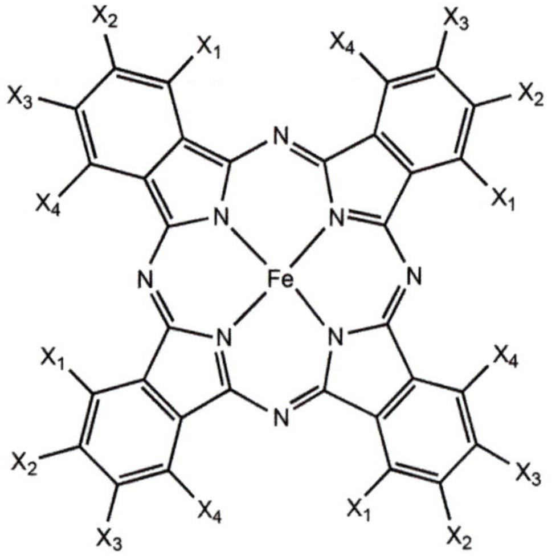

Figure 34.6 Structure of iron phthalocyanines used in oxidative fuel desulfurization (ODS).

Fe9	$X_2 = (NO_2)$; $X_1, X_3, X_4 = H$	(Zhou 2009)
Fe10	$X_1, X_2, X_3, X_4 = Cl$	(Liu 2017)
Fe11	$X_1, X_2, X_3, X_4 = F$	(Fang 2021)

reactive than DBT and BT, despite steric hindrance. Noteworthy, nitrogen-based compounds such as pyridine or quinoline, common contaminants of fuels, facilitate the formation of the active intermediate high-valent iron^{IV} oxo species.

Dimeric O-bridged complex of **Fe11** was chosen by Fang to investigate extractive-oxidative desulfurization in 9:1 mL of ethanol / water mixture and 5 mL of model oil with H_2O_2 as oxidant [33]. Catalyst improved ODS from about 37% to more than 98%. Further improvement was obtained in the presence of 4-mercaptopyridine, by its coordination to iron center. In this case, more hindered DMDBT resulted the species with lowest activity.

34.5 Cobalt

Tripathi et al. [34] elaborated a water-soluble N-benzylated cobalt phthalocyaninetetrasulfonamide, to obtain an oil-soluble derivative, more suitable for oxidative desulfurization. The optimal catalyst/DBT molar ratio was found 1:10; lower amount of catalyst did not allow oxidation of substrate, while larger excess led to its aggregation. Some other parameters were explored, such as H_2O_2/DBT ratio and CH_3CN (extracting solvent)/oil ratio. At the best, 85% of extraction efficiency was reached.

The same author [35] investigated desulfurization by a Co^{II} 5,10,15,20-tetrarylporphyrinate. Reactions were performed in dodecane as model oil, while CH_3CN was used as extracting solvent; 91% desulfurization was achieved with DBT/catalyst 15:1 molar ratio and 20 equivalents (respect to DBT) of H_2O_2 at 50 °C. Porphyrin with different substituents on phenyl rings marginally affected the oxidation process. Properties of three Co^{II} salen complexes were also investigated [36]. By using the same reaction conditions, ligand with 3,5-di-*tert*-butyl groups showed the highest activity. The bulky groups near the metal center enhance catalytic activity and increase solubility in the organic solvent. In general, yield of sulfur oxidation was in the 60–76% range depending on catalyst, H_2O_2 molar ratio, and solvent volume.

34.6 Molybdenum

Mo resulted a promising catalyst in the oxidation of alkyl and aryl sulphides, in homogeneous and heterogeneous processes [37], and thus its activity in fuel desulfurization has been explored (Figure 34.7).

Peroxido-molybdenum amino acid complexes **Mo1–Mo3** have been screened in the oxidation of DBT from a model fuel, with an excess of H_2O_2 [38]. More than 99% of S-removal was achieved with 10% mol of **Mo1**, at 70 °C, using [bmim]BF_4 and [bmim]PF_6 as extractant after oxidation. The same system could be also applied for DMDBT and BT removal from fuel: full elimination of DMDBT was accomplished in three hours, whereas BT removal reached 90%.

The use of dioxomolybdenum^{VI} complexes with 4,4′-di-*tert*-butyl-2,2′-dipyridyl ligand (**Mo4**) [39], *N,N*-dimethylbenzamide (**Mo5**) [40] or *N,N*′-diethyloxamide (**Mo6**) [40], cyclopentadienyl molybdenum tricarbonyl complex (**Mo7**) [41] or indenyl molybdenum tricarbonyl complex (**Mo8**) [41] as pre-catalysts led to remarkable results in term of S-removal from model and real diesels. In particular, preliminary desulfurization studies were performed using octane enriched with BT, DBT, MDBT, and DMDBT as the model oil. Reactions were performed in biphasic system, using acetonitrile or [bmim]BF_4 as extractive solvents, in which an excess of H_2O_2 was dissolved, or in three-phase system (diesel/H_2O_2/IL) using [bmim]PF_6 [39, 41, 43]. In all cases, excellent results were achieved, but [bmim]PF_6 resulted the most promising solvent, because it led to more than 94% of S-removal in two hours at 50 °C with all of the catalysts. Catalysts and IL could be efficiently recycled and reused; furthermore, their application to a commercial diesel confirmed the efficiency of the protocol.

A similar approach has been developed to simultaneously achieve S- and N-removal from fuel [42]. In particular, hybrid Mo^{VI}-bipyridine based catalysts allowed to remove 99.9% of S and 97% of N from a model diesel, using [bmim]PF_6, in the presence of H_2O_2, at 70°C. Reusability and recyclability tests confirmed the possible use of catalysts and IL for consecutive desulfurization and denitrogenation run.

Interestingly, **Mo5** and **Mo6** have been also exploited in ODS using two polyethylene glycol (PEG)-based deep eutectic solvents (DESs) as extraction solvents and reaction media [43]. DESs were obtained by combining PEG as hydrogen bond donor with tetrabutylammonium chloride or choline chloride as hydrogen bond acceptor. Reactions run in biphasic conditions, at 70 °C, led to

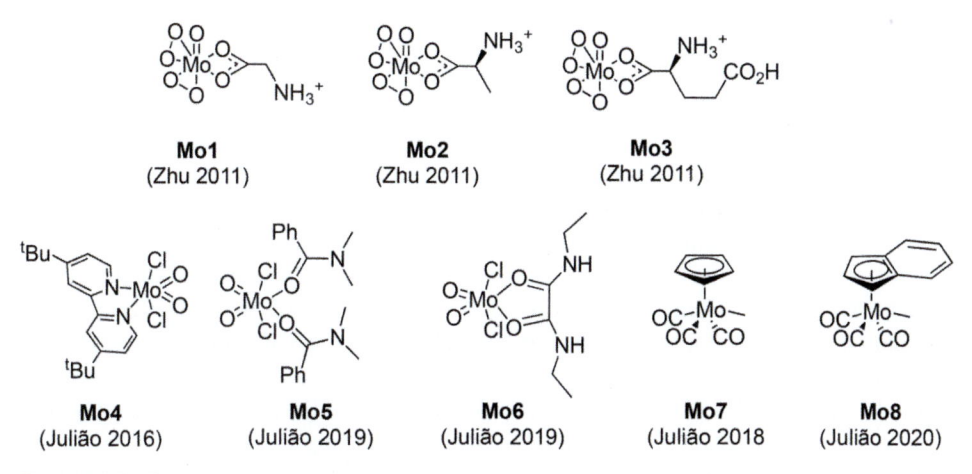

Figure 34.7 Structure of selected Mo-complexes.

S-free diesel after two hours. In addition, 82% desulfurization was achieved after the treatment on a real fuel, demonstrating the applicability of DESs as sustainable extractive solvents in ODS, even though their industrial scalability is still controversial.

34.7 Tungsten

In a recent example, ODS catalyzed by $Na_2WO_4.2H_2O$, in the presence of acetic acid and H_2O_2 as oxidants, has been explored [45]. However, to achieve good conversion results, extraction with methanol and an additional alumina adsorption step were required, leading to approx. 20% diesel loss. Nevertheless, Na_2WO_4 was previously adopted as catalyst in an interesting example of ultra-deep oxidative desulfurization: a model diesel prepared dissolving DBT in tetradecane, was added to a mixture of $Na_2WO_4.2H_2O$ and an excess of H_2O_2 dissolved in 1-(4-sulfonic acid)butyl-3-methylimidazolium p-toluenesulfonate ([$(CH_2)_4SO_3HMIm$]Tos) [46]. The reaction performed at 50 °C for three hours led to > 99% of S-removal, with the formed sulfone extracted in the IL phase. Interestingly, the active catalytic species was a peroxytungstate-ionic liquid complex, that made the system homogenous, and it could be even recycled by adding fresh diesel and H_2O_2. Remarkably, at the working temperature, aromatic compounds, essential components in fuel, were not oxidized.

Encouraging results have been achieved also using hexatungstates (i.e. [$(C_4H_9)_3NCH_3$]$_2W_6O_{19}$, [$(C_8H_{17})_3NCH_3$]$_2W_6O_{19}$, or [$(C_{12}H_{25})_3NCH_3$]$_2W_6O_{19}$) in water-in-IL emulsion systems with H_2O_2 [47]. Specifically, the quaternary ammonium ion of the catalyst and IL cation (1-octyl-3-methylimidazolium) promote the formation of emulsion droplets and extract S-compounds from the model oil to the IL. Here, IL anion (i.e. PF_6^-) promotes the formation of the active W-peroxidocomplex. Indeed, with such system, almost complete DBT and DMDBT removal was achieved at 60 °C in 60 and 80 minutes, respectively, while BT removal was c. 75%; in addition, quite promising results have been achieved on real gasoline, where 94% S-removal was obtained in 4 desulfurization steps.

Interestingly, the catalytic activity of phosphotungstic acid with H_2O_2 for ODS has been evaluated using DESs as extractants. Reactions performed with ChCl/2PEG (obtained by combining choline chloride (ChCl) with 2 eqs. of PEG), led to > 99% of DBT removal at 50 °C within three hours [48].

34.8 Polyoxometalates

Polyoxometalates (POM) are a family of anionic metal-oxide clusters, generally with octahedral structures, exploited in several fields ranging from catalysis to biology to energy and material science [49]. Due to the large selectivity, efficiency, and versatility, POMs have been used in ODS as homogeneous catalysts with H_2O_2 or O_2 as oxidant and organic solvents or ILs as extractants. Two recent reviews detailed desulfurization by using POMs [9, 50, 51].

Even more recently, the so called Venturello compounds $(Bu_4N)_3[PO_4\{MO(O_2)_2\}_4]$, where M is Mo or W, were investigated under different reaction conditions (catalyst and oxidant amount, temperature) and with different extracting solvents [52, 53]. In the case of Mo-derivative, highest extracting ability was achieved for the solvent-free method (respect to CH_3CN or [bmim]PF_6), that actually is a heterogeneous system [53].

Lately, both W and Mo based POMs were compared using CH_3CN, [bmim]PF_6 or DES (tetrabutylammonium/PEG) as extracting systems. After H_2O_2 addition, complete desulfurization within 40 minutes was achieved with peroxidotungstate and [bmim]PF_6. The catalyst has stronger acid sites while [bmim]PF_6 allowed the formation of a triphasic system that in some way protects catalyst from deactivation/decomposition.

Mn polyoxotungstate was investigated by Duarte et al. [54]. With H_2O_2 as oxidant and CH_3CN as extracting solvent, such POM oxidizes sulfur-based substrates into the corresponding sulfones, operating at room temperature with almost 100% conversion with a relatively low S/C molar ratio (150:1). System sustainability was then improved by using ethanol/water mixture to remove sulfone from hexane (chosen as model oil); after 20 minutes, 98 mol% of the organosulfur compounds were removed.

34.9 Ionic Liquids

Ionic liquids, which initially were exploited as non-flammable and less toxic alternative to many organic solvents to extract S-based impurities from fuels, are currently being explored as promising eco-friendly catalysts in ODS. Among the others, the catalytic activity of metal-based ILs is receiving great attention. IL catalytic activity depends by the Lewis-acid character of the organic cation. As general strategy, metals halides are added to ILs leading to the formation of a [organic cation](metal halide anion) moiety, while H_2O_2 is used as oxidant. Accordingly, Lewis acid ionic liquids, with alkylated 1,8-diazabicyclo[5.4.0]undec-7-ene cation and $ZnCl_2$-based complex anion have been explored as catalysts in ODS with H_2O_2 (Figure 34.8) [55].

$ZnCl_2$ coordination to sulfide lone pair promotes ODS, that can be efficiently accomplished, at 50 °C in two hours. IL can be reused for six cycles with only a slight decrease of activity and more than 99% of S-removal can be achieved on a hydrogenated diesel via a one step process. Similarly, a series of metal-based IL having triethylamine hydrochloride cation and a series of anhydrous metal chlorides anions has been explored [56]. Among the others, $[Et_3NH]FeCl_4$ exhibited the best performance, leading to > 99% of S-removal in mild conditions, using O_2 as the oxidant, under UV-light irradiation. To note, such protocol resulted particularly suitable for gasoline with low olefin concentration, since alkenes in solution lowers S-removal.

Jiang et al. [57] reported that the oxidative desulfurization ability for ILs containing different metal ions strongly depends by the nature of the metals. They compared $[(C_8H_{17})_3CH_3N]Cl/MCl_x$ where MCl_x was $CuCl_2$, $SnCl_2$, $ZnCl_2$, or $FeCl_3$; in the absence of H_2O_2, sulfur removal was in the range of 20–30%, whereas the addition of the oxidant allowed an increase in this value close to 98% with 0.5 or 1 equivalent of $FeCl_3$. This system was recycled six times with no significant loss of activity charging fresh H_2O_2 and model oil before starting a new cycle.

1-octyl-3-methylimidazolium tetrachloroferrate ([omim]$FeCl_4$) was found to be a valuable IL for the extraction and catalytic oxidative desulfurization of a model fuel, reaching complete removal of BT, DBT, and DMDBT, with H_2O_2, at 25 °C in 15 minutes [58]. The IL exhibits a beneficial role in promoting sulfides extraction from the oil through the long hydrophobic alkyl chain, π-π interactions between imidazolium and thiophenic ring, and interactions between S lone pairs and Fe^{3+}. In addition, Fe^{3+} catalyzes the formation of reactive oxygen species from H_2O_2, that are involved in sulfides oxidation. Even though [omim]$FeCl_4$ retained its activity when recycled seven times, desulfurization of a real fuel was less effective. Likewise, a magnetic IL $[C_4(mim)_2Cl_2/2FeCl_3]$, Figure 34.9) showed high desulfurization performances, with the advantage that it could be efficiently separated from model oil

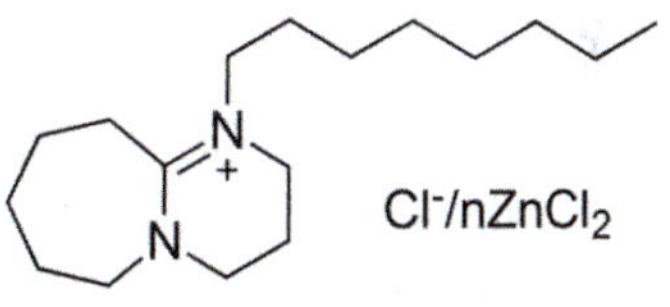

Figure 34.8 Structure of a Zn-based ionic liquid (IL) [55] / Elsevier.

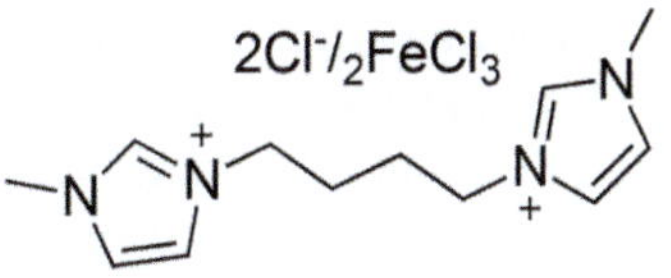

Figure 34.9 Structure of $[C_4(mim)_2]Cl_2/FeCl_3$ [59].

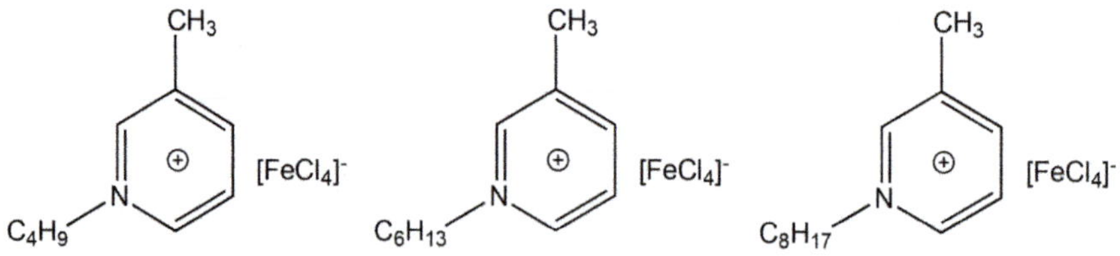

Figure 34.10 Structure of $[C_n{}^3MPy]FeCl_4$ [60] / Elsevier.

after reaction by applying an external magnetic field; thus, recycling could be accomplished with no significant loss of activity [59].

Comparisons of the effects of alkyl chain length in ILs containing of alkylpyridinium cation, like $[C_4{}^3MPy]FeCl_4$, $[C_6{}^3MPy]FeCl_4$, $[C_8{}^3MPy]FeCl_4$ (Figure 34.10), were made by Nie et al. [58]; they reported that the longer the alkyl chain, the higher the sulfur removal, a result owing to better extractive performances.

Addition of H_2O_2 increased the sulfur removal due to the larger efficacy of extraction and oxidation process; $[C_8{}^3MPy]FeCl_4$ removes more than 99% of sulfur in all the H_2O_2/DBT ratio tested. After three cycles, efficacy decreases because of the $DBTO_2$ accumulation in ILs. $[C_6mim]Cl/FeCl_3$ and $[C_6{}^3MPy]_nFeCl_4$ with n = 0.5, 1, 2, 3 were investigated by Dong et al. [61]. With one equivalent of metal halide and H_2O_2/DBT = 4, authors reported a 100% yield. However, to achieve almost complete S-removal from a real gasoline, seven ODS runs were required, with the addition of fresh IL and H_2O_2 after each cycle, raising costs and limiting the industrial applicability of such protocol.

Jiang et al. [62] explored a modified pyridinium based ionic liquids, $[C_4Py]_3Fe(CN)_6$ and $[C_{16}Py]_3Fe(CN)_6$ for the oxidation, coupled with 1-octyl-3-methylimidazolium hexafluorophosphate $[omim]PF_6$ as extractant; these systems removed around 97 and 87% of sulfur respectively, more than sulfur removed by $K_3Fe(CN)_6$ when used alone. Experiments performed on real samples, demonstrated that the presence of aromatics or olefins has a negative impact on DBT oxidation.

Redox couple of hexacyanoferrates were applied in the oxidation of DBT [63] and related derivatives in the form of $[C_nmim]_3Fe(CN)_6$ where n = 2, 4, and 8 as a catalyst and $[bmim]BF_4$ as an extractant. The highest activity was measured for $[bmim]_3Fe(CN)_6$ in the presence of only 7.5% of H_2O_2 (almost 98% of sulfur removal); the radical anion superoxide was identified as the reactive species in the oxidation of sulfide to sulfones.

Li et al. [64] used N-methyl-2-pyrrolidone (NMP) coordinating $FeCl_3$ or $ZnCl_2$; although bare NMP exhibits the highest extractive ability, in agreement with its structural similarity and intermiscibility with DBT, they investigated some derivatives such as $C_5H_9NO \cdot xFeCl_3$ (x = 0.1 or 0.3). As a result, $C_5H_9NO \cdot xFeCl_3$ showed very good catalytic activity (best conditions: 12 equivalents of H_2O_2, 30 °C, sulfur removal from 97 to 94.3% after six cycles) because it decreases the solubility of fuel oil in NMP, and allows DBT to be extracted in the more polar ILs phase.

In the framework of metal-based IL, polyoxometalate-based ionic liquid catalysts are largely explored in ODS, showing excellent results in S-removal under mild conditions, with H_2O_2, in the presence of an extractant solvent, such as acetonitrile [65, 66], a different IL like $[omim]BF_4$ or $[omim]PF_6$ [67, 68], or a deep-eutectic solvent, as the one prepared from benzenesulfonic acid and PEG [69]. The peculiar catalytic activity of ILs having POM anions (i.e. $[PMo_{12}O_{40}]^{3-}$, $[PVMo_{11}O_{40}]^{4-}$, $[PV_2Mo_{10}O_{40}]^{4-}$, $[PW_{12}O_{40}]^{3-}$, $[PMo_6W_6O_{40}]^{3-}$,) is usually ascribed to the formation of metal-peroxidocomplexes, which is pivotal for sulfides oxidation.

Recently, an ultrafast approach for fuel desulfurization has been proposed, using a NaClO oxidant and an ILs-based phase-transfer catalyst (i.e. 1-hexadecyl-3-methylimidazolium phosphomolybdate [C_{16}mim]PMoO) [70]. DBT removal from octane was accomplished in 20 minutes and the authors highlighted the multiple roles of ILs: the C_{16} alkyl chain ensures the proper catalyst dispersibility in the oil phase, the imidazolium participates in DBT extraction from the oil through π-π interactions, and PMoO anion promotes $O_2^{\cdot-}$ and $HO^{\cdot}$ generation from NaClO decomposition and H_2O; such radicals react with the catalyst, leading to Mo-peroxido species, directly involved in DBT oxidation to sulfone. The proposed system was particularly promising, being efficiently applied also on real fuel: 99% desulfurization was obtained in 30 minutes, at 30 °C. Recyclability tests demonstrated the stability and reusability of such IL, which shows that it is suitable for potential industrial application; however, the use of greener oxidant, like H_2O_2 or O_2, led to unsatisfactory results.

34.10 Conclusions

In this chapter, a brief overview of the most efficient homogeneous catalytic systems designed to perform oxidative fuel desulfurization is offered. Several metal complexes have been successfully explored to lower the S-content in fuels in mild conditions, and the most valid protocols to be scaled-up at the industrial level have been highlighted. Of note, the use of ILs as sustainable extractive (and reusable) solvents and, possibly, as eco-friendly catalysts, is widespread and offers new perspectives to make fuels greener.

References

1 Boshagh, F., Rahmani, M., Rostami, K. et al. (2022). *Energy Fuels* 36 (1): 98–132. https://doi.org/10.1021/acs.energyfuels.1c03396.
2 Rajendran, A., Cui, T.-Y., Fan, H.-X. et al. (2020). *J. Mater. Chem. A* 8 (5): 2246–2285. https://doi.org/10.1039/C9TA12555H.
3 Conte, V. and Floris, B. (2010). *Inorganica Chim. Acta* 363 (9): 1935–1946. https://doi.org/10.1016/j.ica.2009.06.056.
4 Langeslay, R.R., Kaphan, D.M., Marshall, C.L. et al. (2019). *Chem. Rev.* 119 (4): 2128–2191. https://doi.org/10.1021/acs.chemrev.8b00245.
5 Sutradhar, M., Martins, L.M.D.R.S., Guedes da Silva, M.F.C. et al. (2015). *Coord. Chem. Rev.* 301-302: 200–239. http://dx.doi.org/10.1016/j.ccr.2015.01.020.
6 Sutradhar, M., Pombeiro, A.J.L., and da Silva, J.A.L. (ed.) (2021). *Vanadium Catalysis*. Croydon: Royal Society of Chemistry. https://doi.org/10.1039/9781839160882.
7 Silva, G., Voth, S., Szymanski, P. et al. (2011). *Fuel Process. Technol.* 92 (8): 1656–1661. https://doi.org/10.1016/j.fuproc.2011.04.014.
8 Mota, A., Butenko, N., Hallett, J.P. et al. (2012). *Catal. Today* 196 (1): 119–125. https://doi.org/10.1016/j.cattod.2012.03.037.
9 Zhu, W., Xu, D., Li, H. et al. (2013). *Pet. Sci. Technol.* 31 (14): 1447–1453. https://doi.org/10.1080/10916466.2010.545790.
10 Mesdour, S., Lekbir, C., Doumandji, L. et al. (2017). *J. Sulphur Chem.* 38 (4): 421–439. http://dx.doi.org/10.1080/17415993.2017.1304550.

11 Benmabrouka, H., Mesdour, S., Boufades, D. et al. (2019). *Pet. Sci. Technol.* 37 (6): 662–670. https://doi.org/10.1080/10916466.2018.1563611.
12 Ogunlaja, A.S., Chidawanyika, W., Antunes, E. et al. (2012). *Dalton Trans.* 41 (42): 13908–13928. https://doi.org/10.1039/C2DT31433A.
13 Dembaremba, T.O., Correia, I., Hosten, E.C. et al. (2019). *Dalton Trans.* 48 (44): 16687–16704. https://doi.org/10.1039/C9DT02505G.
14 Saeedi, R., Safaei, E., Lee, Y.-I. et al. (2019). *Appl. Organomet. Chem.* 33 (3): e4781. https://doi.org/10.1002/aoc.4781.
15 Campitelli, P., Aschi, M., Di Nicola, C. et al. (2020). *Appl. Catal. A: Gen.* 599: 117622. https://doi.org/10.1016/j.apcata.2020.117622.
16 Gobara, H.M., Nessim, M.I., Zaky, M.T. et al. (2014). *Catal. Lett.* 144 (6): 1043–1052. https://doi.org/10.1007/s10562-014-1251-3.
17 Anisimov, A.V., Myltykbaeva, Z.B., Kairbekov, Z. et al. (2017). *Theor. Found. Chem. Eng.* 51 (4): 563–566. https://doi.org/10.1134/S0040579517040029.
18 Coletti, A., Sabuzi, F., Floris, B. et al. (2018). *J. Fuel Chem. Technol.* 46 (9): 1121–1129. https://doi.org/10.1016/S1872-5813(18)30045-8.
19 Floris, B., Sabuzi, F., Coletti, A. et al. (2017). *Catal. Today* 285: 49–56. https://doi.org/10.1016/j.cattod.2016.11.006.
20 Sabuzi, F., Pomarico, G., Conte, V. et al. (2021). Peroxo-vanadium complexes as sustainable catalysts in oxidations, halogenations and other organic transformations. In: *Vanadium Catalysis* (ed. M. Sutradhar, A.J.L. Pombeiro, and J.A.L. da Silva), 97–110. Croydon: Royal Society of Chemistry. https://doi.org/10.1039/9781839160882-00097.
21 Floris, B., Sabuzi, F., Galloni, P. et al. (2017). *Catalysts* 7 (9): 261. https://doi.org/10.3390/catal7090261.
22 Calvete, M.J.F., Piñeiro, M., Dias, L.D. et al. (2018). *ChemCatChem.* 10 (17): 3615–3635. https://doi.org/10.1002/cctc.201800587.
23 Pires, S.M.G., Simões, M.M.Q., Santos, I.C.M.S. et al. (2012). *Appl. Catal. A: Gen.* 439-440: 51–56. https://doi.org/10.1016/j.apcata.2012.06.044.
24 Mondal, S., Hangun-Balkir, Y., Alexandrova, L. et al. (2006). *Catal. Today* 116 (4): 554–561. https://doi.org/10.1016/j.cattod.2006.06.025.
25 Trehoux, A., Roux, Y., Guillot, R. et al. (2015). *J. Mol. Catal. A: Chem.* 396: 40–46. https://doi.org/10.1016/j.molcata.2014.09.030.
26 Zhou, X., Lv, S., Wang, H. et al. (2011). *Appl. Catal. A: Gen.* 396 (1–2): 101–106. https://doi.org/10.1016/j.apcata.2011.01.041.
27 Aguiar, A., Ribeiro, S., Silva, A.M.N. et al. (2014). *Appl. Catal. A: Gen.* 478: 267–274. https://doi.org/10.1016/j.apcata.2014.04.002.
28 Pires, S.M.G., Simões, M.M.Q., Santos, I.C.M.S. et al. (2014). *Appl. Catal. B: Environ.* 160-161: 80–88. https://doi.org/10.1016/j.apcatb.2014.05.003.
29 Zhao, R., Wang, J., Zhang, D. et al. (2017). *Appl. Catal. A: Gen.* 532: 26–31. https://doi.org/10.1016/j.apcata.2016.12.008.
30 Zhao, R., Wang, J., Zhang, D. et al. (2017). *ACS Sustain. Chem. Eng.* 5 (3): 2050–2055. https://doi.org/10.1021/acssuschemeng.6b02916.
31 Zhou, X., Li, J., Wang, X. et al. (2009). *Fuel Process. Technol.* 90 (2): 317–323. https://doi.org/10.1016/j.fuproc.2008.09.002.
32 Liu, H., Bao, S., Cai, Z. et al. (2017). *Chem. Eng. J.* 317: 1092–1098. https://doi.org/10.1016/j.cej.2017.01.086.

33 Fang, Z., Li, N., Zhao, Z. et al. (2022). Bio-inspired strategy to enhance catalytic oxidative desulfurization by *O*-bridged diiron perfluorophthalocyanine axially coordinated with 4-mercaptopyridine. *Chem. Eng. J.* 433 (2): 133569. https://doi.org/10.1016/j.cej.2021.133569.

34 Tripathi, D., Negi, H., Singh, R.K. et al. (2019). *J. Coord. Chem.* 72 (17): 2982–2996. https://doi.org/10.1080/00958972.2019.1683549.

35 Tripathi, D., Inderpal, Y., Negi, H. et al. (2021). *J. Porphyr. Phthalocyanines* 25 (01): 24–30. https://doi.org/10.1142/S1088424620500443.

36 Tripathi, D. and Singh, R.K. (2021). *Catal. Lett.* 151: 713–719. https://doi.org/10.1007/s10562-020-03343-4.

37 Thiruvengetam, P. and Chand, D.K. (2018). *J. Indian Chem. Soc.* 95: 781–788. https://doi.org/10.5281/zenodo.5638611.

38 Zhu, W.S., Li, H., Gu, Q.Q. et al. (2011). *J. Mol. Catal. A: Chem.* 336 (1–2): 16–22. https://doi.org/10.1016/j.molcata.2010.12.003.

39 Julião, D., Gomes, A.C., Pillinger, M. et al. (2016). *Dalton Trans.* 45 (38): 15242–15248. https://doi.org/10.1039/C6DT02065H.

40 Julião, D., Gomes, A.C., Cunha-Silva, L. et al. (2019). *Catal. Commun.* 128: 105704. https://doi.org/10.1016/j.catcom.2019.05.011.

41 Julião, D., Gomes, A.C., Pillinger, M. et al. (2018). *Appl. Catal. B: Environ.* 230: 177–183. https://doi.org/10.1016/j.apcatb.2018.02.036.

42 Julião, D., Gomes, A.C., Pillinger, M. et al. (2020). *Chem. Eng. Technol.* 43 (9): 1774–1783. https://doi.org/10.1002/ceat.201900624.

43 Julião, D., Gomes, A.C., Pillinger, M. et al. (2020). *Appl. Org. Chem.* 34 (4): e5490. https://doi.org/10.1002/aoc.5490.

44 Julião, D., Gomes, A.C., Pillinger, M. et al. (2020). *J. Mol. Liq.* 309: 113093. https://doi.org/10.1016/j.molliq.2020.113093.

45 Bourane, A., Koseoglu, O., Al-Hajji, A. et al. (2019). *React. Kinet. Mech. Catal.* 126: 365–382. https://doi.org/10.1007/s11144-018-1484-z.

46 Liu, D., Gui, J., Ding, J. et al. (2011). *React. Kinet. Mech. Catal.* 104 (1): 111–123. https://doi.org/10.1007/s11144-011-0347-7.

47 Ding, Y., Zhu, W., Li, H. et al. (2011). *Green Chem.* 13 (5): 1210–1216. https://doi.org/10.1039/C0GC00787K.

48 Liu, W., Jiang, W., Zhu, W. et al. (2016). *J. Mol. Catal. A: Chem.* 424: 261–268. https://doi.org/10.1016/j.molcata.2016.08.030.

49 Long, D.L., Tsunashima, R., and Cronin, L. (2010). *Angew. Chem., Int. Ed.* 49 (10): 1736–1758. https://doi.org/10.1002/anie.200902483.

50 Taghizadeh, M., Mehrvarz, E., and Taghipour, A. (2019). Polyoxometalate as an effective catalyst for the oxidative desulfurization of liquid fuels: a critical review. *Rev. Chem. Eng.* 36 (7): 831–858. https://doi.org/10.1515/revce-2018-0058.

51 Ahmadian, M. and Anbia, M. (2021). *Energy Fuels* 35: 10347–10373. https://doi.org/10.1021/acs.energyfuels.1c00862.

52 Gao, Y., Julião, D., Silva, F.L. et al. (2021). *Mol. Catal.* 505: 111515. https://doi.org/10.1016/j.mcat.2021.111515.

53 Julião, D., Gomes, A.C., Cunha-Silva, L. et al. (2020). *Appl. Catal. A, Gen.* 589: 117154. https://doi.org/10.1016/j.apcata.2019.117154.

54 Duarte, T.A.G., Pires, S.M.G., Santos, I.C.M.S. et al. (2016). *Catal. Sci. Technol.* 6: 3271. https://doi.org/10.1039/c5cy01564b.

55 Wang, J., Zhang, L., Sun, Y. et al. (2018). *Fuel Process. Technol.* 177: 81–88. https://doi.org/10.1016/j.fuproc.2018.04.013.
56 Wang, C., Chen, Z., Zhu, W. et al. (2017). *Energy Fuels* 31 (2): 1376–1382. https://doi.org/10.1021/acs.energyfuels.6b02624.
57 Jiang, Y., Zhu, W., Li, H. et al. (2011). *ChemSusChem.* 4: 399–403. https://doi.org/10.1002/cssc.201000251.
58 Andevary, H.H., Akbari, A., and Omidkhah, M. (2019). *Fuel Process. Technol.* 185: 8–17. https://doi.org/10.1016/j.fuproc.2018.11.014.
59 Wang, T., Yu, W.-H., Li, T.-X. et al. (2019). *N. J. Chem.* 43 (48): 19232–19241. https://doi.org/10.1039/C9NJ04015C.
60 Nie, Y., Dong, Y., Lu, B. et al. (2013). *Fuel* 103: 997–1002. https://doi.org/10.1016/j.fuel.2012.07.071.
61 Dong, Y., Nie, Y., and Zhou, Q. (2013). *Chem. Eng. Technol.* 36 (3): 435–442. https://doi.org/10.1002/ceat.201200570.
62 Jiang, W., Li, H., Yin, S. et al. (2016). *Appl. Organomet. Chem.* 30: 753–758. https://doi.org/10.1002/aoc.3500.
63 Jiang, W., Zhu, W., Chang, Y. et al. (2014). *Energy Fuels* 28: 2754–2760. https://doi.org/10.1021/ef500082y.
64 Li, F.-T., Wu, B., Liu, R.-H. et al. (2015). *Chem. Eng. J.* 274: 192–199. https://doi.org/10.1016/j.cej.2015.04.027.
65 Akopyan, A., Eseva, E., Polikarpova, P. et al. (2020). *Molecules* 25 (3): 536. https://doi.org/10.3390/molecules25030536.
66 Li, J., Guo, Y., Tan, J., and Hu, B. (2021). *Catalysts* 11 (3): 356. https://doi.org/10.3390/catal11030356.
67 Hao, L., Sun, L., Su, T. et al. (2019). *Chem. Eng. J.* 358: 419–426. https://doi.org/10.1016/j.cej.2018.10.006.
68 Wang, J., Yang, B., Peng, X. et al. (2022). *Chem. Eng. J.* 429: 132446. https://doi.org/10.1016/j.cej.2021.132446.
69 Chi, M., Su, T., Sun, L. et al. (2020). *Appl. Catal. B: Environ.* 275: 119134. https://doi.org/10.1016/j.apcatb.2020.119134.
70 Li, A., Song, H., Meng, H. et al. (2020). *Chem. Eng. J.* 380: 122453. https://doi.org/10.1016/j.cej.2019.122453.

35

Heterogeneous Catalytic Desulfurization of Liquid Fuels

The Present and the Future

Rui G. Faria, Alexandre Viana, Carlos M. Granadeiro, Luís Cunha-Silva, and Salete S. Balula

LAQV/REQUIMTE & Department of Chemistry and Biochemistry, Faculty of Sciences, University of Porto, Porto

35.1 Introduction

The intensification of worldwide industrial activity has led to an ever-increasing dependence on fossil fuel combustion for energy production. The transition toward cleaner energy production technologies is crucial for ensuring environmental sustainability without significant economic impacts. Fossil fuel combustion still accounts for about 85% of the world's energy production and the prediction for 2040 is a decrease non greater than 20% [1]. Powering civilization with fossil fuels has led to major strides in human development, but we are now aware that this entails severe consequences. Links have been developed between emissions resulting from fossil fuel combustion and severe health and environmental impacts [2]. Discussion around the environmental impacts of these fuels usually circles around the emission of carbon dioxide; however, other pollutants containing heteroatoms such as nitrogen and sulfur deserve just as much attention. The content of sulfur containing compounds (SCCs) is several times higher than that of nitrogen compounds in fossil fuels [1]. Consequently, the sulfur produced during fuel combustion (sulfur oxide[s]) contributes significantly to growing climate instability through the production of deleterious acids, and, when in contact with atmospheric moisture, by producing acid rain. Furthermore, the SSCs in fuels also lead to the corrosion of storage tanks, pipelines, and equipment, as well as the deactivation of catalysts used in refinery processes [3]. Sulfur can also chemisorb on the catalytic coverters of passenger vehicles, promoting deactivation and reducing engine efficiency [3]. These effects highlight the crucial economic and environmental need for the removal of SSCs from fuels.

Strict legislation has been introduced worldwide to drastically reduce the sulfur content of transportation fuels. Since 2009, road fuels in the European Union have been limited to 10 ppm S [4]. Since 2020, sulfur content in marine fuels has been limited to 0,5% of sulfur (0,1% in the Baltic Sea, North Sea, and English Channel) by the International Marine Organization (IMO) [5]. These legislative efforts further pressure the oil processing industry into producing sulfur-free fuels.

Four major groups for the SSCs in fuels can be classified, namely disulfides, mercaptans, sulfides, and thiophenes (Ts). The concentration and the nature of SSCs in fuels change over the

Catalysis for a Sustainable Environment: Reactions, Processes and Applied Technologies Volume 3, First Edition. Edited by Armando J. L. Pombeiro, Manas Sutradhar, and Elisabete C. B. A. Alegria.
© 2024 John Wiley & Sons Ltd. Published 2024 by John Wiley & Sons Ltd.

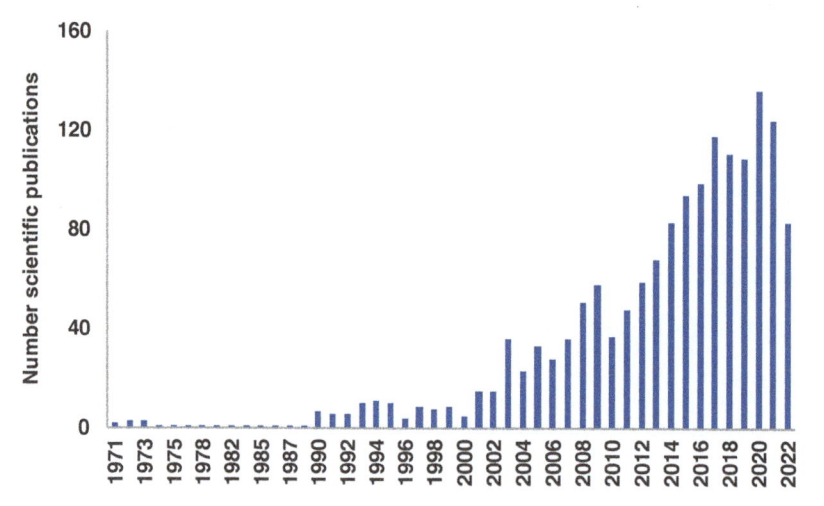

Figure 35.1 Number of scientific publication from 1971 to date found in the Web of Science database, using the keywords "desulfurization" and "fuel".

boiling range during petroleum distillation processes, with the heaviest fractions containing the largest amounts of sulfur species [6].

HDS is the most used method by the petroleum industry to drive down sulfur content in road transportation fuels (diesel and gasoline). This technology requires increased temperatures (300–400 C°), high pressures (3–6 MPa), and a large quantity of hydrogen. The strict regulation imposed for ultra-low sulfur fuel commercialization turn HDS into a costly choice. Large efforts have been done by researchers to develop alternative technologies with higher cost-efficiency and eco-sustainability. A total of 6,551 related scientific papers were published from 1971 to date (Figure 35.1). This high number of the scientific reports presents several possible simple or combined desulfurization technologies able to replace or complement HDS. A large part of these technologies need to use materials as the main key to achieve high desulfurization (5,212 papers found in the Web of Science database in a search incorporating the following three terms: desulfurization, fuel, and materials). These materials integrate mainly novel catalysts (2,474 papers) and/or highly effective and selective absorbent materials.

This chapter presents the most important advances given by various materials in the area of desulfurization of liquid fuels, mainly in the area of HDS, adsorptive desulfurization (ADS), oxidative desulfurization (ODS), and the combination of extractive and oxidative desulfurization (ECODS) achieved by porous powdered and membrane materials. The morphology, porosity, hydrophilicity, and nature of functional groups in the material surfaces are crucial properties in the fuel media and have a remarkable influence in the desulfurization process.

35.2 Hydrodesulfurization

Hydroconversion is a broad term used to refer to the reaction of hydrocarbons with hydrogen during oil processing and includes hydrotreating, hydrocracking, and hydrogenation [7]. HDS is thus the term used to describe the catalytic desulfurization of any hydrocarbon stream during hydrotreating. This technique has been reviewed exhaustively over the past decades [7, 8], and a brief overview is given here for contextualization.

HDS of lower boiling aromatics is carried out in vapor phase, but most other applications of HDS are in liquid phase, in which it is generally more efficient [9]. Oil stocks are desulfurized for many reasons, the main one being to prevent atmospheric pollution that results from the combustion of residual stocks. In mid-distillate treating, desulfurization was originally practiced just to avoid corrosion of heating equipment. In naphtha pre-treating, desulfurization is important to avoid poisoning of precious metal used in its catalytic reforming. Desulfurization is not only practiced on hydrocarbon streams that concern fuel production, as it is also key to assure purity standards in other chemical production processes.

During HDS, sulfur-containing compounds from the feedstock are converted into sulfur-free hydrocarbons at high temperature and pressure conditions and in the presence of hydrogen, producing hydrogen sulfide gas. Although HDS is a well-established industrial process and is being employed for several decades, specific mechanisms of HDS reactions are still a matter of debate. In so far as these reactions are associated with the cleavage of carbon-sulfur bonds, they are categorized as hydrogenolysis reactions. It is understood that there are two possible general pathways for the HDS reaction [10]. The simplest mechanism is referred as direct desulfurization and is related to the substitution of a sulfur atom with a hydrogen atom in the hydrocarbon structure which is initially adsorbed via π-bonding in the active metal site. This happens by direct carbon-sulfur bond cleavage and is carried out without additional hydrogenation of any carbon-carbon double bond. Alternatively, in the hydrogenation route, the aromatic ring adjacent to the sulfur-containing ring is hydrogenated. The destabilization of the aromatic ring leads to the weakening of the carbon-sulfur bond before the substitution of the sulfur atom which is coordinated via π-bonding. Ultimately, hydrogenation of the aromatic ring can also be carried out after removal of the sulfur atom. It is also believed that the neighbouring sulfur-hydrogen bonds enable proton transfer in both pathways. It is accepted that sulfur removal and hydrogenation occur simultaneously on the catalyst surface, while active metal sites for both processes are considered to be the same.

HDS catalysts are available in several different compositions and reactivity. The most widely applied active species are metal-sulfides, the most commonly being MoS_2 or WS_2 [8a]. Catalysts based on supported MoS_2 are extensively used due to their low cost, high stability, and activity. These consist of two-dimensional S–Mo–S layers that are stacked to various degrees and form nano-crystalline structures related to the truncated triangle in which two types of edges are found and termed as the S-edge and the Mo-edge (Figure 35.2) [11]. The edges of these structures play the role of active sites. Because of this, catalytic activity is dependent on the orientation, form, and growth of MoS_2 species over the catalyst support material. Additionally, their activity is significantly increased by the inclusion of transition metals such as cobalt and nickel which can be attached to the S-edges by coordination [12]. The support material provides enhanced surface area and mechanical stability to the catalyst and guaranties dispersion of active sites. Alumina is the most commonly employed support in HDS catalysts, others being silica, titania, or alumina-silica. Commercial HDS catalysts, and hydrotreating catalysts in general, are thus most often a porous alumina matrix impregnated with combinations of cobalt, nickel, molybdenum and tungsten to give composites with surface area between 200 and 300 $m^2.g^{-1}$. Catalytic industrial processes are commercialized by Albemarle, ExxonMobil, Akzo Nobel, IFP, Advanced Refining, RIPP, Haldor Topsoe, Nippon Ketjen, and Criterion.

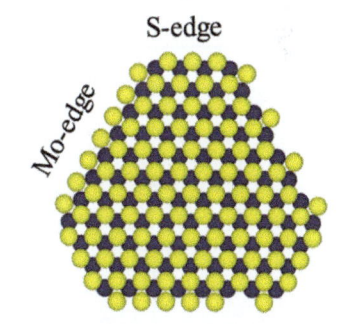

Figure 35.2 Top-view ball model of a MoS_2 nano-crystal. The balls denote the position of Mo (blue) and S (yellow and orange). Reproduced with permission from Ref [13] Springer Nature / CC BY 4.0.

Co–Mo and Ni–Mo based catalysts are the most widely applied catalysts for HDS of most feedstock. However, their catalytic behavior is not exactly the same and differs primarily in their reactivities toward removal of heterocyclic compounds [8, 14]. Co–Mo catalysts act primarily via the direct desulfurization route and prevent the adjacent aromatic ring hydrogenation, whereas Ni–Mo catalysts have a higher selectivity for desulfurization via the hydrogenation route. The extent to which catalysis occurs via one route or the other is determining for considerations on operating hydrogen partial pressure. The two main factors for determining which catalyst should be used are pressure, as high pressure favors Ni–Mo activity, and amount of heterocyclic compounds, as low content of heterocyclic compounds favors Co–Mo catalysts [15]. Co–Mo catalysts are mostly employed in hydrotreating of straight run petroleum fractions. Ni–Mo catalysts are chosen when higher activity is required for the saturation of polynuclear aromatic compounds or for desulfurization of higher amounts of the most refractory sulfur compounds. Additionally, Ni–W catalysts are chosen when very high activity for aromatic saturation is required. State-of-the-art catalytic processes are mostly based in the combination of different catalysts.

A schematic HDS process is shown in Figure 35.3. The working parameters to consider in hydrotreating processes include pressure, temperature, catalyst loading, feed flow rate, and hydrogen partial pressure. The hydrogen partial pressure must be greater than the hydrocarbon partial pressure and increasing it improves both desulfurization and denitrogenation rates. Higher temperatures will also increase the reaction rate constant and improve the kinetics. The overall desulfurization rate is ultimately determined by the rate of removal of the most refractory sulfur-containing compounds. In a standard hydrotreating process [16], the feedstock is pressurized and subsequently mixed with hydrogen gas. The feed oil can be initially preheated by a heat exchanger, but the mixture is always brought to the reaction temperature in a furnace before being fed into the catalytic reactor. The heated mixture is then hydrotreated over a catalyst bed inside the reactor, under a determined flow rate of pressurized hydrogen gas. Depending on the process design, multiple reactors can be used in processes where multi-stage conversion is employed for in-between product separation. Fixed-bed reactors are generally utilized for processing lighter feeds, whereas heavier feedstock can be processed in ebullated-bed reactors or moving-bed reactors. Different reactor systems can be integrated for the processing of complex

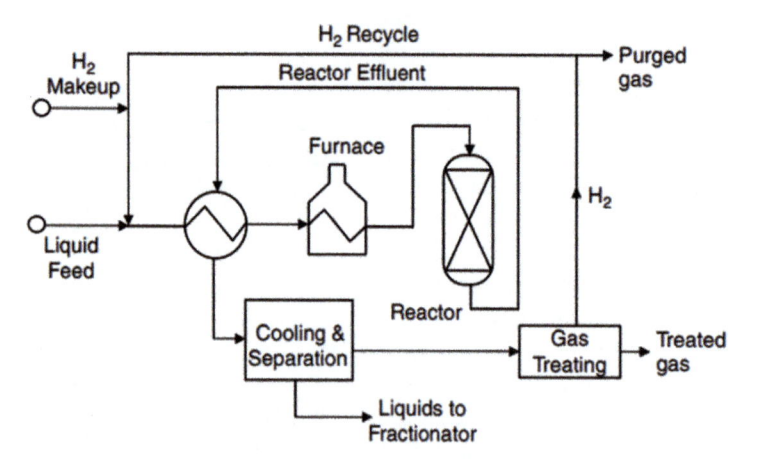

Figure 35.3 Process flow diagram for a hydrodesulfurization (HDS) process. Reproduced with permission from Ref [7] / Elsevier.

feeds and many beds can be packed in a single reactor for different catalyst combinations. Both liquid and gas process fluids are cooled using a heat exchanger and separated in a series of pressure vessels. The liquid products obtained after treatment are separated into desired products in a fractionation column on the basis of their boiling points. The gas-phase mixture obtained is taken into an absorber where hydrogen sulfide gas is separated from hydrogen by using an amine scrubber.[8c]. The treated hydrogen gas is then recycled back into the reactor by using a recycle-compressor. Some of the recycle gas is purged to prevent accumulation of light hydrocarbons and to control hydrogen partial pressure.

HDS presents a huge weakness since this cannot be used to marine fuels. Marine fuels have been prepared based on Heavy Fuel Oils (HFO), which present a whole set of different challenges to conventional desulfurization processes due to their complex matrices and heavy molecular composition (paraffins, cycloparaffins, aromatics, olefins, and asphaltenes, among others), with carbon numbers ranging from approximately C_{20} to $>C_{50}$ depending on the manufacturing processes used [17] (road fuels usually contain C_4 to C_{20} hydrocarbon chains). HDS is still employed to desulfurize HFOs [18]; however, its effectiveness is severely undermined by the aforementioned factors, along with their high metal content, combined with their coking and fouling propensity (which causes catalyst deactivation), and the molecular size and steric protection of sulfur in cyclic SCCs [19]. Other more promising technologies, involving materials as the main key to achieve the success of HFO desulfurization, have been developed, such as adsorptive and oxidative desulfurization.

35.3 Adsorptive Desulfurization

ADS is a cost-effective and energetically sustainable desulfurization methodology, based on the selective removal of organic S-compounds from petroleum fractions by physicochemical adsorption processes over activated/functionalized adsorbents (Figure 35.4) [20]. Generally, ADS does not introduce impurities or affect fuel composition, and its physicochemical properties remain largely unaffected. Adsorption of S-compounds can occur through physical or chemical routes. Physical adsorption occurs essentially by Van der Walls interactions, while chemisorption can

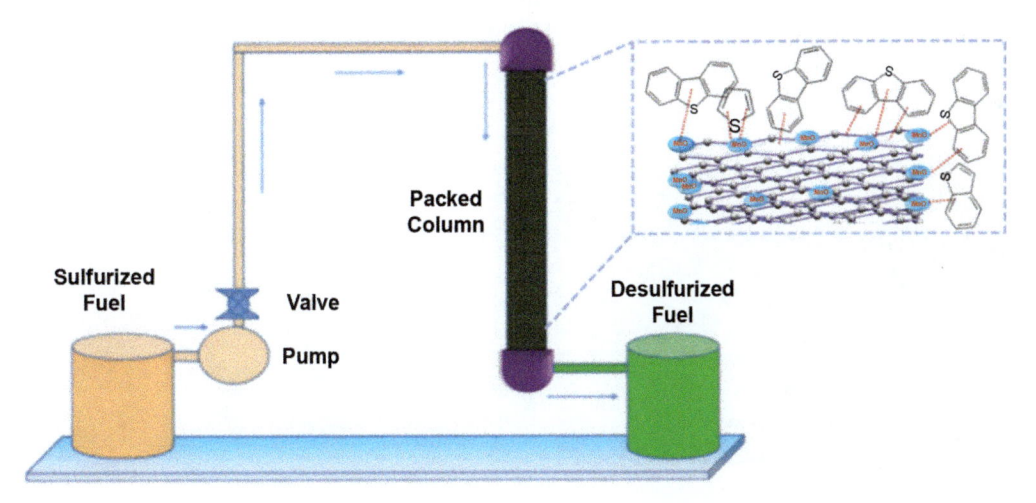

Figure 35.4 Schematic illustration of adsorptive desulfurization (ADS) over carbon-manganese oxide nanocomposite system. Reproduced with permission from Ref [22] / Elsevier.

occur by a single mechanism or by the combination of three mechanisms: π complexation, direct sulfur-metal (S-M) interaction, and acid-base interaction. The desired characteristics for ADS adsorbents are those expected for material in other separation processes: extensive surface area and suitable pore volume, mesoporosity, surface-active sites, structural strength, and stability [21]. This chapter section focuses on the use of carbon-based materials, zeolites, mesoporous silica, and metal-organic frameworks (MOF) for ADS.

35.3.1 ADS with Carbon-based Materials

C-based materials have captured broad scientific interest due to their varying properties caused by different C local bonding environments [23]. Activated carbons (ACs) are microcrystalline materials prepared from cheap and abundant precursors [24]. Processing of these starting materials under adequate conditions produces potential adsorbents with extensive surface areas, suitable porosity and unique surface properties, such as O-containing functional groups that can be modified or tuned [25]. In 1997, ACs were tested for the first time as potential adsorbents for SCCs in naphta solutions, highlighting the effects of porosity features on desulfurization efficiency [26]. More recently, it was reported that pore openings smaller than 1 nm strongly correlate with increased dibenzothiophene (DBT) adsorption capacities [27]. Furthermore, the density of acidic O-containing functional groups on the surface of ACs seems to improve ADS capacity [28]. Other investigations further revealed "pristine" ACs with high S adsorption ability but low selectivity in the presence of N compounds, which occur in real fuel matrices [29]. Thus, solvent treatment and metal loading have been attempted to increase the S adsorption capacity and selectivity of ACs. The treatment of pristine AC with NaOH or nitric acid increased the number of acidic/basic functional groups, and creating extra pores that increased ADS performance [30]. Furthermore, ACs modified with sulfuric acid after steam treatment showed different functional groups onto the AC's surface and revealed DBT removals of over 90% (65% for pristine AC). The adsorption capacity was also correlated with the amount of surface O-containing groups [31].

Metal (M) loading on ACs aims to enhance their adsorption strength by the incorporation of M species into the C matrix. The specific role of metals for S adsorption lies in being high-energy centers for specific removal via strong interactions and π complexation with S in the confined pore space. AC doped with Cu and/or Co were tested in the ADS of model oils, revealed better removal of every SCC, notably for T, benzothiophene (BT) and 4-methylbenzothiophene (MBT), due to the strong S-M interaction (Figure 35.5) [32].

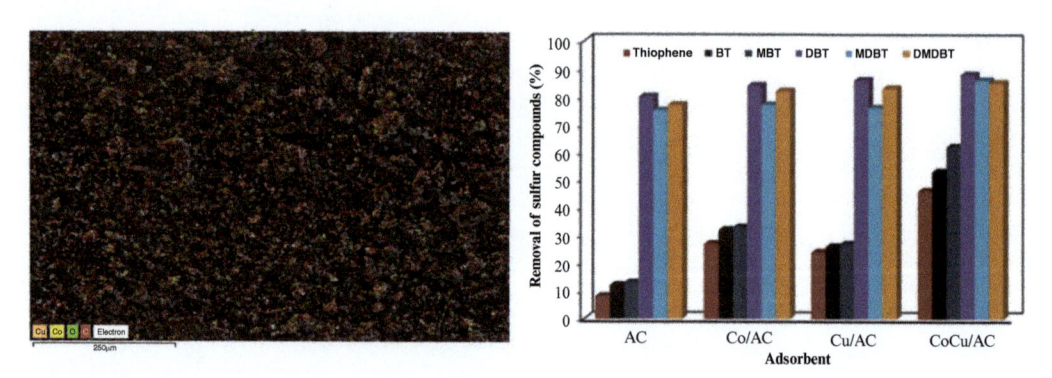

Figure 35.5 Energy dispersive x-ray (EDX) mapping image for CuCo/AC (left). Desulfurization capacity of pristine activated carbon (AC) and metal doped counterparts (right). Reproduced with permission from Ref [32] / Elsevier.

Nanoscale C allotropes have also been explored for ADS as they possess similar surface functionalities to ACs. Graphene and carbon nanotubes (CNTs) have been tested as potential adsorbents of S from petroleum distillates. The unique mechanical properties and extensive surface areas, chemical inertness, and remarkable conductivity of graphene graphene oxide (GO) make it a prime candidate for ADS [33]. A comparative study of the adsorption of DBT on graphene and GO revealed that graphene is able of adsorbing much larger amounts of aromatic S compounds. Graphene's hexagonal ring structure possesses a great density of π-bonds, facilitating π–π interactions with bulky Ts [34]. The density of O-containing functional groups in GO seems to disrupt its capacity to establish π-π interactions with DBT, nearly nullifying its adsorptive capacity and suggesting that S adsorption over graphene is predominantly controlled by π interactions. Despite its shortcomings as S adsorbent in its pristine state, GO has been explored to prepare hybrid nanocomposites with mixed metal oxides (MMOs) to boost its performance. A series of GO supported MMOs (MgAl, CuAl, and CoAl) were prepared and tested in ADS of DBT. The MMO-GO hybrid nanocomposites displayed remarkably superior DBT adsorption capacities, with the introduction of just as little as 5 wt% GO boosting efficiency by 170% [35]. The use of CNTs in ADS is like that of graphene, as both depend on π-π interactions, can be functionalized, and doped with nanoparticles to increase adsorption capacity. It was demonstrated that SCC adsorption over single walled CNTs depends on their diameter. Narrower tubes adsorbed larger amounts of adsorbates than their wider counterparts, as the adsorbate molecules are subjected to a larger interaction potential from the delocalized electrons of the tube, and was also verified that adsorption mainly occurs inside the nanotubes through π-interactions [36]. The comparison of the adsorption capacity of multi-walled carbon nanotubes (MWCNTs), GO, and AC, revealed that AC outperformed the other two adsorbents due to the large differences in their specific surface area (882 m^2/g for AC, 217 m^2/g for MWNTs and 10.8 m^2/g for GO), average pore width (14.5 Å for AC, 73.8 Å for MWNTs, and 68.5 Å for GO), and micropore volume (0.487 ml/g compared to 0.286 ml/g for MWNTs and 0.021 ml/g for GO), further highlighting the importance of adequate textural characteristics for ADS [37].

In general, the C-based materials revealed a high potential for ADS; however, some limitations must still be addressed. The use of C-based adsorbents for desulfurization is extremely limited at the industrial scale due to heterogeneous pore size distributions (reducing reproducibility), poor selectivity, and difficulty of regeneration and recovery (π-interacted S is hard to remove), so further research is need to overcome these difficulties.

35.3.2 ADS with Zeolites

Zeolites are microporous 3D crystalline aluminosilicates with high external surface areas, commonly used as adsorbents and catalysts, and one of the most investigated materials for ADS due to their structural features and ability to perform ion-exchange on surface active sites, which significantly boosts their adsorption capacity and selectivity toward S compounds [38]. ADS with zeolites involves all the previously mentioned adsorption mechanisms (π complexation, S-M, and Lewis acid-base interactions). Clinoptilolite is the most abundant natural zeolite and one of the most studied as an adsorbent for deep ADS. This zeolite was employed for the adsorption of SCCs after dealumination and ion exchange with Ni^{2+}, revealing a strong correlation between Si/Al molar ratio and desulfurization capacity, and suggesting that incorporation of Ni leads to an increase in S removal through π-complexation and S-M bonds. Whereas the pristine Clinoptilolite (Si/Al = 5.65) showed a S removal percentage of 5.4%, this value increased to 68% after dealumination (Si/Al = 10.40).

Synthetic zeolites are also widely employed as adsorbents in ADS. These materials are manufactured by thermal processes, and careful control of the temperature and the composition of its precursors allows close control of the structure and the surface characteristics [39]. Furthermore, post-synthetic treatments to produce mesoporosity and the introduction of metals into the structure are the most reported methodologies to ensure maximum S adsorption. In fact, the Cu-exchanged synthetic zeolite Y selectively adsorbs T through the donation of electron charges from T to the vacant s orbital of metals and, simultaneously, back-donation of electron charges from the d orbitals of metals to T [40]. When zeolite Y was metal-exchanged with both Zn and Cu, high DBT removal efficiencies were achieved, an outcome associated to possible synergistic effects between these two metals. Removal of 4,6-dimethyldibenzothiophene (4,6-DMDBT) on Ag-Y was as effective as that of DBT, suggesting that the predominant interaction between 4,6-DMDBT and the metal-exchanged zeolite is π-complexation, as there was no significant impact from steric hindrance [41].

Modified Y-zeolites have also shown ADS capabilities for real jet fuel, with capacities of 10 mg S/g of adsorbent, at low temperatures (80 °C) [42]. β-zeolites have also been tested as potential ADS adsorbents after solvent post-synthetic treatment and ion-exchange. β-zeolite modified by alkaline treatment at different temperatures, resulted in significantly improved surface areas and pore volumes (Figure 35.6a), and revealed remarkable increases in SCC adsorption relatively to the parent zeolite (Figure 35.6b), as these bulkier species were now able to access the adsorbents' porous structure. In addition, the ion-exchanging with Ce^{3+} on the previously modified β-zeolite, resulted in an adsorbent with superior deep desulfurization, attributed to the introduction of S-M interaction capabilities (Figure 35.6c) [43].

Other zeolite families like Linde type A zeolite (LTA) have been considered as potential adsorbents for ADS; however, this application fell short as LTA-zeolites do not possess adequate pore size (3–4 Å) for accommodating bulky organo-sulfur compounds [44], highlighting the importance of this crucial characteristic. With all these factors in mind, the future of ADS with zeolites is

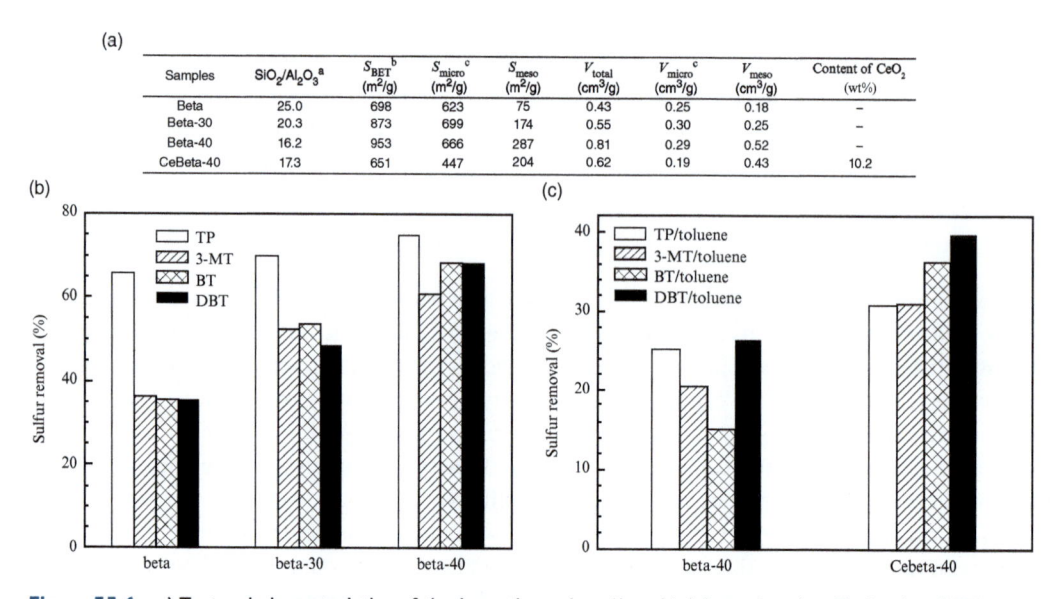

Samples	SiO_2/Al_2O_3 a	S_{BET} b (m^2/g)	S_{micro} c (m^2/g)	S_{meso} (m^2/g)	V_{total} (cm^3/g)	V_{micro} c (cm^3/g)	V_{meso} (cm^3/g)	Content of CeO_2 (wt%)
Beta	25.0	698	623	75	0.43	0.25	0.18	–
Beta-30	20.3	873	699	174	0.55	0.30	0.25	–
Beta-40	16.2	953	666	287	0.81	0.29	0.52	–
CeBeta-40	17.3	651	447	204	0.62	0.19	0.43	10.2

Figure 35.6 a) Textural characteristics of the investigated zeolites. b) Adsorptive desulfurization (ADS) performance of alkaline treated β-zeolites on different sulfur containing compounds (SCCs). c) ADS performance of Ce^{3+} ion exchanged and alkaline treated β-zeolite, compared with alkaline treated β-zeolite. Reproduced with permission from Ref [43] / Elsevier.

strongly dependent on the development of adequate modification methodologies which endow the adsorbent with remarkable textural properties without sacrificing mechanical stability, and on the discovery of the "sweet spot" of metal ion-exchange and acid-base functionality introduction.

35.3.3 ADS with Mesoporous Silica

Mesoporous silica materials (MSM) have been emerging as promising desulfurization adsorbents due to generally large specific surface area (700–1300 m^2/g) and pore volume (0.5–1.2 cm^3/g), tuneable pore size, narrow pore distribution, and possible synthesis in a wide range of morphologies [45]. The application of MSMs in ADS is more recent, and generally display a similar behavior to zeolites. A wide variety of MSM have been applied for ADS, with the most popular ones including Santa Barbara Amorphous-15 (SBA-15) and Mobil Composition of Matter No. 41 (MCM-41).

MCM-41 with different Si/Al ratios and Ni/MCM-41 were applied for the adsorptive removal of DBT from a model oil solution, with the MCM-41 with lower Si/Al ratios revealing higher performance due to a higher density of Lewis acid sites (despite having lower surface areas). The incorporation of Ni (metal centers) into the MCM-41 structure also improved its DBT adsorption due to the creation of strong S-M interactions [46]. Also, the incorporation of distinct amounts of Ni nanoparticles on MCM-41 demonstrated that at 15% Ni loading there is a 206% increase in S adsorption capacity. However, at higher metal loadings, Ni nanoparticle agglomeration reduces the number of available Lewis acid adsorption sites due to channel blockage, reducing desulfurization efficiency. The Cu incorporation into MCM-41 by direct synthesis and incipient wetness impregnation showed that the direct synthesis method allows for better Cu dispersion across the MCM-41 structure, producing a better adsorbent than the wetness impregnation method. In terms of metal loading, it seems there is an intermediate sweet spot that ensures maximum desulfurization efficiency, as both the material with lower and the material with higher Cu loadings performed worse than the MCM-41 with intermediate metal loading [47].

Silver nanoparticles confined within the channels of MCM-41 allow the maintenance of the framework's surface area up to a certain Ag loading, with the maximum S adsorption capacity attained at 20 wt% Ag loading (higher loadings cause silver or silver nitrate aggregates to form in the channels and block adsorption sites). These adsorbents were tested in real jet fuel, achieving S adsorption capacities four times higher than previously reported materials [48]. Mesoporous silica SBA-15 has been the subject of similar modifications. The dispersion of Ni on the hexagonal pore walls of SBA-15, resulted in a significant increase in surface area, pore volume and the number of acid sites available for S adsorption up to 20 wt% Ni loadings, with analogous disabling behavior observed for higher loadings. Similar findings were found with Cu-SBA-15 nanocomposites applied for ADS [49]. As mentioned previously for zeolites, future research into MSM should not only focus on developing novel synthetic/post-synthetic methodologies to improve textural properties, but also the investigation of strategies for the impregnation of metal species into their structures without aggregation or pore blockage.

35.3.4 ADS with Metal-Organic Frameworks

MOFs have porous 3D structures generated from the rational combination of multidentate organic linkers and metal centers, producing multiple topologies, with remarkable surface areas and porosity, as well as nearly à la carte synthetic/post-synthetic tailorability [50]. The first use of a series of MOFs for the adsorptive removal of organosulfur compounds from fuels, attained better performances than those previously reported for Y zeolite [51]. It was verified that larger surface area and pore volume did not necessarily imply better desulfurization performance of MOFs.

MOF-177 had the highest surface area and pore volume of the five prepared materials; however, it adsorbed the least for all three organosulfur compounds, raising the question as to which structural factor influenced S adsorption the most. University of Michigan Crystalline Material-150 (UMCM-150) and MOF-505 were able to adsorb larger amounts of S due to coordinative unsaturated (or open) metal centers, taking advantage of both S-M metal bonding and physicochemical adsorption processes to ensure deep desulfurization, and demonstrating the importance of adequate functionality of the active sites (Figure 35.7) [51].

To further investigate this hypothesis, MOFs Cu-BTC, CPO-27-Ni, and ZIF-76 (ZIF = Zeolitic imidazolate framework) were prepared and tested. Despite having similar textural characteristics, Cu-BTC and CPO-27-Ni performed significantly better than ZIF-76 due to the open metal sites. Moreover, a comparative study of four MOFs, namely, MOF-5, Cu-BTC, MIL-53(Fe), and MIL-101(Cr) (MIL = Matérial Institut Lavoisier), assembled from four different metal ions (Zn^{2+}, Cu^{2+}, Fe^{3+}, and Cr^{3+}) and two different organic ligands, was done to disclose the essential factors influencing adsorption performance. Interestingly, ADS performance followed the Cu-BTC ≈ MOF-5 > MIL-53(Fe) > MIL-101(Cr) order, despite MIL-101(Cr) showing the largest pore size and surface area. In fact, the ADS of aromatic S compounds can be achieved through the interactions of the metal centers in the MOF with the delocalized π electrons of the aromatic rings [52]. On the other hand, the surface functional group introduction has been reported as a strategy to improve desulfurization efficiency of MOFs. Zirconium-based MOF UiO-66 (UiO = Universitetet i Oslo) and two surface-functionalized counterparts (NH_2-UiO-66 and COOH-UiO-66) were prepared and studied [53]. Despite the NH_2-UiO-66 and COOH-UiO-66 showing reduced porosity relatively to UiO-66 due to the introduction of additional groups, these functionalized materials display better adsorption capabilities than the pristine framework. COOH-UiO-66 adsorbs SCCs through acid-base interactions between the carboxylate moieties and basic organosulfurs, while the NH_2-UiO-66 has high density of surface NH_2 groups, which can act as H donor to form hydrogen bonds with T and BT (act as H acceptors). Moreover, NH_2-UiO-66 can be reused for several ADS cycles without significant loss of performance by simple ethanol washing, suggesting no leaching of functional groups.

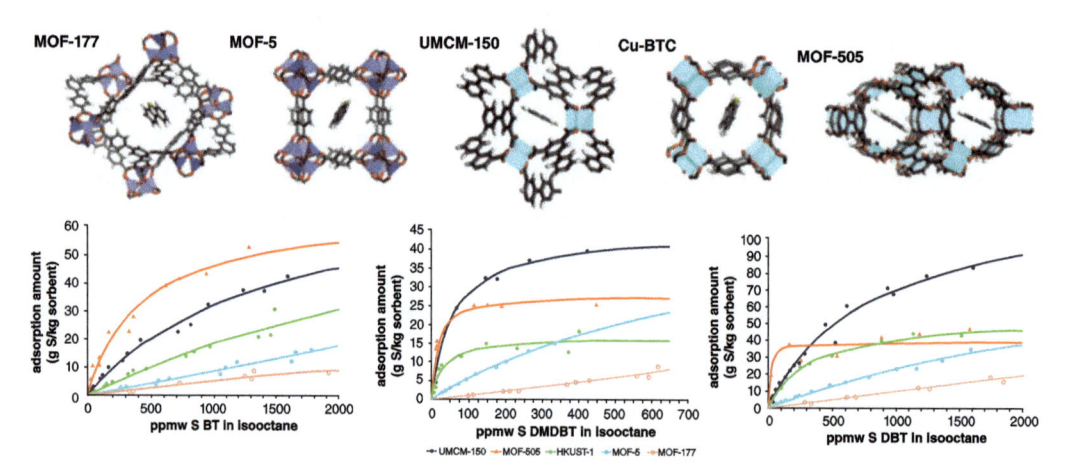

Figure 35.7 Structures of metal-organic framework (MOF)-177, MOF-5, University of Michigan Crystalline Material (UMCM)-150, Copper benzene-1,3,5-tricarboxylate (Cu-BTC), and MOF-505, with one molecule of DBT added in the pore of each MOF to represent scale. Adsorption isotherms for different benzothiophene (BT) (left), dimethyldibenzothiophene (DMDBT) (middle), and dibenzothiophene (DBT) (right) for the different MOFs. Reproduced with permission from Ref [51] American Chemical Society.

Mixed metal loading is also a potential tool to enhance MOF S adsorption capabilities. The CuCl$_2$-loaded MIL-47 showed a similar behavior to that observed for metal-loaded silicas: "intermediate" loading weights provide better ADS performances, which decrease with further loading increase due to the blocking of metal active adsorption sites. Comparing the porosity of pristine MIL-47 and Cu-loaded MIL-47 with their adsorption capacities, it is once again apparent that this is not the only factor impacting performance. In fact, the reduction of Cu^{II} to Cu^{I}, which occurs in the presence of V (the metallic component of MIL-47) produces synergistic effects with the framework itself which facilitate π-complexation with adsorbed SCCs. The incorporation of task-specific ionic liquids (ILs) in MOFs is another interesting approach to ADS with modified MOFs. 1-butyl-3-methylimidazolium chloride was immobilized in the pores of MIL-101(Cr), resulting in a reduction of the pore volume with gradually increasing IL amounts, suggesting they are located inside the porous framework [54]. The adsorbed S quantity by IL/MIL-101 increased up to 33% IL/MIL-101 and decreased for 50% IL/MIL-101 due to a high degree of pore blockage with excess amounts of ILs. ILs supported on MIL-101 act as acidic centers toward the basic S species, resulting in a favorable interaction between IL/MIL-101 and increasing the adsorption capacity by 71% in relation to pristine MIL-101. This aproach opens interesting possibilities into the development of ADS efficient MOFs, as there are hundreds of reported MOFs structures and ILs, thus, many MOF/IL combinations can be prepared with interesting properties favoring effective S/adsorbent interactions. Future research on MOFs for ADS should focus on the development of structures with adequate porous systems, which allow for selective S removal, as well as further chemical functionalization strategies which can significantly enhance MOF–S interactions and allow for ultra-low sulfur diesel (ULSD) production.

35.4 Oxidative Desulfurization

ODS is an energetically sustainable deep desulfurization methodology that involves a chemical reaction between an appropriate oxidant and the sulfur compounds (SCCs). This process can be performed under mild reaction conditions of temperature and pressure and using environmentally benign oxidants. The oxidation of SCCs present in the fuel matrix occurs by the presence of an oxidant that is activated by an appropriate catalyst. The reactivity of SCCs depends on their electron density and steric accessibility. Comparing the oxidative facility of T, BT, DBT, and 4,6-DMDBT under the same conditions, this follows the order DBT > 4,6-DMDBT > BT > T. The electron densities on the sulfur atom from 4,6-DMDBT, DBT, BT and T are 5.760, 5.758, 5.739 and 5.696, respectively. Therefore, compounds with higher electron densities around the S atom are more readily oxidized into their respective sulfoxides and/or sulfones [55]. The DBT is more easily oxidized than the 4,6-DMDBT despite having lower electron density. However, 4,6-DMDBT presents a spatial steric hindrance around the S atom caused by alkyl substitution, obstructing the formation of new S=O bonds [56].

The oxidative catalytic reaction can be performed in the presence or the absence of an extraction solvent that need to be immiscible with the treated liquid fuel. When the extraction solvent is used, ODS occurs simultaneously with the extractive desulfurization (EDS). This last bifunctional extractive and catalytic oxidative desulfurization process (ECODS) promotes a deep desulfurization, where the SCCs are extracted from the liquid fuel due to the difference of polarity and the higher affinity of SCCs by the higher polar extractive phase. In the latter occurs the oxidation of the SCCs to produce the respective sulfoxide and/or sulfone compounds [57].

35.4.1 Oxidants for (EC)ODS

The choice of an adequate oxidant is crucial for the success of ODS. The first patent in the area of ODS, published in 1974, reported the use of nitrogen dioxide (NO_2) as an oxidant followed by extraction with an organic solvent to remove oxidized SCCs from petroleum distillates [58]. Air has been successfully used as oxidant to convert SCCs from gasoline and diesel fuels to sulfones, in the presence of γ-butyrolactone as an oxygen transfer agent, at 140 °C, affording high conversion rates and selectivities [59]. Sundararaman et al. also proposed using air as an oxidant for the desulfurization of a commercial jet propellant-8 (JP-8) jet fuel and a commercial diesel fuel in the presence of CuO catalyst, which promote in situ generation of peroxides [60]. Other option, is the use of ozone, since this is a strong oxidant. Ma et al, used ozone produced by dielectric barrier discharge plasma combined with IL extraction [61]. The desulfurization efficiency for TS and BT was investigated and both were removed to achieve under 0.1% of S in the presence of O_3/[BMIM]Ac [61]. Wang et al. used ozone and hydrogen peroxide and 99.1% of desulfurization was achieved for DBT. Using this combination of oxidants guaranteed that the DBT could be oxidized in both fuel and solvent extraction phases; while aqueous H_2O_2 ensure a complete oxidation in the polar extraction phase [62]. The use of *tert*-butyl hydroperoxide (TBHP) as an oxidant brings some disadvantageous, such as the high price and the production of t-butylalcohol as by-product. The presence of peracetic acid in ODS is easily obtained in situ from the reaction between H_2O_2 and carboxylic acids [63]. Dehkordi et al. reported the desulfurization of kerosone using a combination of H_2O_2 and acetic acid, observing that the desulfurization efficiency increased when the ratio acid/S increased, achieving as the best performance to 83.3% of S removals [64].

The solo H_2O_2 is often cited as the best oxidant for ODS, not only due to its high amount of active oxygen by mass unit, but also due to its superior sustainability claims, since the only by-product is water [65]. The first report on the use of H_2O_2 in ODS was published by Shiraishi et al. in 1999 and showed that the sulfur content in real fuel was reduced from 0.2 to 0.05 wt% even requiring 48 h of reaction [66].

35.4.2 Heterogeneous Catalysts for (EC)ODS

The main goals of heterogeneous catalysis research are the development of remarkably active, selective, and recoverable/reusable catalysts, tailored to the desired reactions. Textural properties play a major role toward these goals, as different types of porosity tend to enhance/hinder different types of interactions with substrates, thus, it is desirable for heterogeneous catalysts to exhibit hierarchical porosity (i.e. combine micro/mesoporosity to increase surface area and macroporosity to enhance substrate transport capabilities) [67]. Combining supporting materials that exhibit these properties with catalytically active species as active centers allows for the production of heterogeneous catalysts that also act as molecular sieves, enhancing selectivity, with the porous system acting as micro-reactors.

35.4.2.1 (EC)ODS with Zeolites

Zeolites have recently received attention as potential heterogeneous desulfurization catalysts due to their desirable combination of micro- and mesoporosity along with excellent thermal and chemical stabilities. Titanium silicalite-1 (TS-1) is one of the most important and widely used zeolite as heterogeneous catalyst in several important reactions, including ODS, due to its remarkable efficiency and high selectivity under mild conditions [68]. Nevertheless, the microporous framework with narrow and tortuous channels strongly limits its application in catalytic reactions involving large-dimension reactants and products [69]. For that reason, several efforts have been made to develop hierarchically structured TS-1 zeolites combining the typical micropores with additional mesoporosity [70]. Du et al. prepared titanium silicalite-1 (TS-1) zeolite using polyvinyl butyral gel as the mesopore template, producing optimized TS-1 structures that combine meso- (30 nm), micro- (0.55 nm) and super-microporosity (1.6 nm) with catalytically active tetrahedral Ti species. Commercial TS-1 reached 22% conversion of DBT after 40 minutes, due to is exclusively microporous structure which caused steric hindrance for DBT access to active sites. TS-1 derived structures obtained from mesopore templating displayed much higher catalytic activities, reaching over 95% DBT conversions in under 20 minutes, due to the introduction of abundant meso- and super-micropores in the rugged surface and interior of the zeolite, enhancing its DBT transport capabilities [71]. Kong et al. have also reported enhanced desulfurization performance of acid-treated TS-1 in the oxidation of T using H_2O_2 as oxidant. In fact, the HCl treated TS-1 samples were able to reach 79.6% conversion in 30 minutes while the pristine zeolite was only able to convert 37.1% of T. The superior performance of the treated samples is attributed to the removal of non-coordinated Ti from the framework which enhanced porosity and enabled substrate access to catalytically active sites [72]. Yu et al. have proposed an alternative method for the preparation of hierarchical TS-1 using a green surfactant as mesoporous template [73]. Materials with uniform intracrystalline mesopores were prepared using Triton X-100 as assisted surfactant during hydrothermal synthesis. The hierarchical TS-1 materials were evaluated as heterogeneous catalysts in the ODS of T, BT, DBT and 4,6-DMDBT using H_2O_2 or TBHP as oxidant. The materials exhibited enhanced catalytic activity when compared with conventional TS-1 and were able of being recycled for at least four consecutive cycles.

35.4.2.2 (EC)ODS with Metal-organic Frameworks

Owing to their aforementioned properties of excellent stability, porosity and tailorability (adjustable acidity/basicity, introduction of functional groups, etc.), it comes as no surprise that MOFs have played a key role in the research of sustainable catalytic oxidation desulfurization. MOFs exhibit 3 important components which can grant them catalytic importance: the metallic components, the organic linkers, and the porous framework structure. In the case of intrinsically catalytically active MOFs, the first two components are of utmost importance. Metallic centers in MOFs can directly coordinate with substrates or oxidants to form reactive oxygen species through the expansion of their coordination sphere, or the reversible delocalization of labile linkers [74]. The organic linkers can provide catalytic activity to a MOF through the presence of adequate functional groups [75]. While examples using V [76], Cr [77] or Co [78] MOFs have been reported, the majority of literature regarding intrinsically active MOFs for ODS is based on metals from the IV (Ti, Hf, Zr) group, due to their exceptional stability and remarkable catalytic activity. The first report of an intrinsically active Zr MOF for application in ODS was published by Granadeiro et al., using the porous Zr(IV) terephthalate UiO-66 for the desulfurization of model oil containing DBT, 4-MDBT and 4,6-DMDBT. The authors suggest that the catalytic activity of UiO-66 must be related to the formation of Zr^{IV}-peroxo groups on the surface of the material by the interaction with H_2O_2, with this MOF achieving complete desulfurization of the model oil after just 30 minutes. Another important result of this work was the correlation of catalyst performance with the degree of crystallinity of the prepared UiO-66 samples, with the least crystalline structures displaying enhanced catalytic activity due to a higher number of coordinative unsaturated zirconium sites, facilitating the formation of active oxygen species [79]. Following this study, Viana et al. compared the catalytic activity of UiO-66 samples prepared under different methodologies (solvothermal and microwave assisted synthesis), noting that faster post-synthetic cooling promotes a higher incorporation of Cl anions (arising from the zinc precursor) in the UiO-66(Zr) framework, which can be correlated with a higher number of defects in the structure and a higher catalytic activity [80]. Viana et al. also proposed a straightforward activation of low-defect pristine UiO-66 (Zr and Hf), using chloride-based salts, aiming to introduce catalytically active defects into the frameworks. The activation method performed on UiO-66(Hf) by the insertion of chloride into the framework structure led to an increase of the desulfurization performance from 48% to 91% (Figure 35.8) [81].

Another Zr-MOF obtained using trimesic acid as organic linker, MOF-808, has also been extensively reported in ODS related literature. Similarly to UiO-66, pristine MOF-808 is less active in oxidation reactions than its defect-containing analogues [82]. Gu et al. reported the acid-treatment of pristine MOF-808 aiming to create abundant defect sites in its structure by the partial removal of organic linkers. The HCl treatment did not alter the MOF-808 morphology or particle size, but caused slight improvements to its textural properties which, associated with available active metal sites from linker removal, ensured this defective MOF-808 to achieve near complete desulfurization of model oil in five minutes, a significant improvement from the performance of the pristine MOF (60% desulfurization) [83]. Similar results were reported for the preparation of defective engineered MOF-808 with enhanced catalytic activity in ODS processes [84] The methods typically involve the use of formic acid as modulator agent during synthesis followed by the post-synthetic removal of formate with the consequent generation of open metal sites in the MOF structure. Zheng et al. compared a series of Zr-based MOFs (UiO-66, UiO-67, NU-1000, and MOF-808) with intrinsic peroxidase-like activity for ultradeep ODS. All the selected Zr-MOFs possessed intrinsic peroxidase-like activity and exhibited catalytic activity in ODS with H_2O_2 as oxidant. Nevertheless, MOF-808 could completely remove DBT within five minutes while UiO-66, UiO-67, and NU-1000, only achieved 8.8%, 20.8%, and 11.1% respectively [85].

Ti-based MOFs such as Ti terephtalate MIL-125 and Ti-carboxylate COK-47 structures have also been reported as active desulfurization catalysts, with their activity strongly influenced by the density of structural defects. Li et al. prepared size-modulated MIL-125, inducing hierarchical porosity and structural defects on the nanocrystals. Increasing amounts of modulator led to a crystal size reduction from c. 1500 nm to 50 nm, with a significant surface area increase of the MOF due to the downsizing of the nanocrystals and the prevalence of missing-linker defects. The "defective" MIL-125 achieved complete desulfurization of model oil in 30 minutes, while pristine MIL-125 was only capable of removing 80% of DBT and 45% of 4,6-DMDBT [86]. Smolders et al. recently reported a new Ti-based MOF comprised of a layer of TiO_6 octahedra forming a three-dimensional framework through the connection of $bpdc^{2-}$ linkers. This MOF, baptized

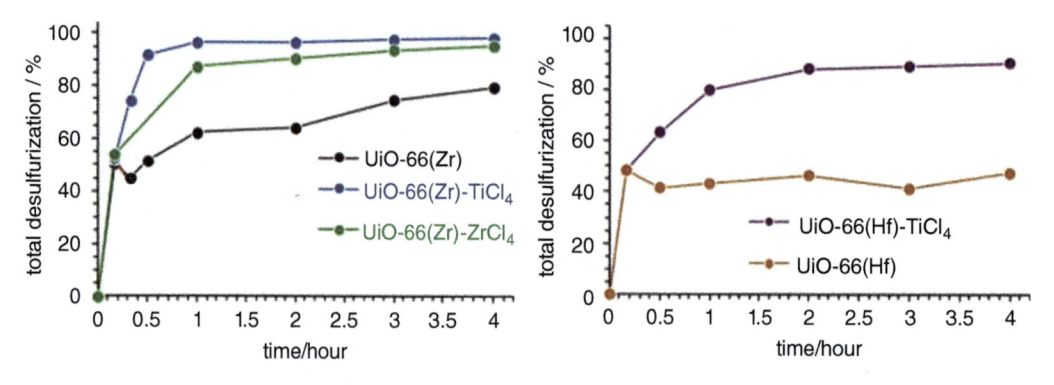

Figure 35.8 Desulfurization profiles of defect free UiO-66(Zr) and activated counterparts (left). Desulfurization performance of pristine UiO-66(Hf) and activated counterpart (right). Reproduced with permission from Ref [81] / Elsevier.

COK-47, possesses a remarkable density of methoxy groups (Me-O-Ti) and structural defects producing open Ti-sites, allowing for the efficient generation of reactive oxygen species when in contact with TBHP and achieving 99% DBT oxidation in 120 minutes [87].

Despite the previously described works, intrinsically active ODS MOF catalysts are, however, rare, and most literature regarding the use of MOFs for the preparation of heterogeneous desulfurization catalysts deals with their use as solid supports for catalytically active homogeneous species.

35.4.2.3 (EC)ODS with Carbon-based Materials

Haw et al. prepared a series of functionalized ACs with micro/mesoporosity features. Functionalization was achieved with phosphoric acid activation which converts carbon surface hydroxyl groups into oxygen-containing acidic moieties. H_2O_2/acetic acid mixture was used as the oxidant, producing peroxyacetic acid which was then decomposed by the ACs, forming •OOH free radicals, which were the main oxidant for the reaction. H_2O_2 can also be directly decomposed by the functionalized ACs by transfer of an electron from the reducing site of the AC to form •OH radicals which can also oxidize SCCs. The prepared functionalized-ACs can thus act both as catalysts and adsorbents in this process, providing the adsorption surface for the organosulfur compounds and oxidation reaction to take place, while an acetonitrile extraction phase ensures the continuous transfer of SCCs from the oil phase allowing their removal. The functionalized-ACs were able to reduce the sulfur content of a commercial diesel fuel from 2,189 ppm to 190 ppm of sulfur after three cycles [88].

Kampouraki et al. performed chemical modification of ACs via treatment with two different acids (HNO_3 and H_2SO_4) aiming to introduce additional functional groups onto their surfaces. The surface chemistry of carbon materials, expressed by the density of the acidic functional groups, was found to be a critical parameter with regards to both adsorption and catalytic oxidative performance. The authors also noted that among the several tested ACs, materials with higher abundance of smaller micropores were more efficient due to increased confinement effects caused by size similarity with 4,6-DMDBT. A significant increase of the removal efficiency via catalytic oxidation can be observed for both acid-treated AC samples, reaching complete sulfur removal, further suggesting the positive effect of surface acidity on the catalytic oxidation (Figure 35.9). These surface groups can also be responsible for the strong retention of the generated oxidation products (sulfoxide and sulfones) [89].

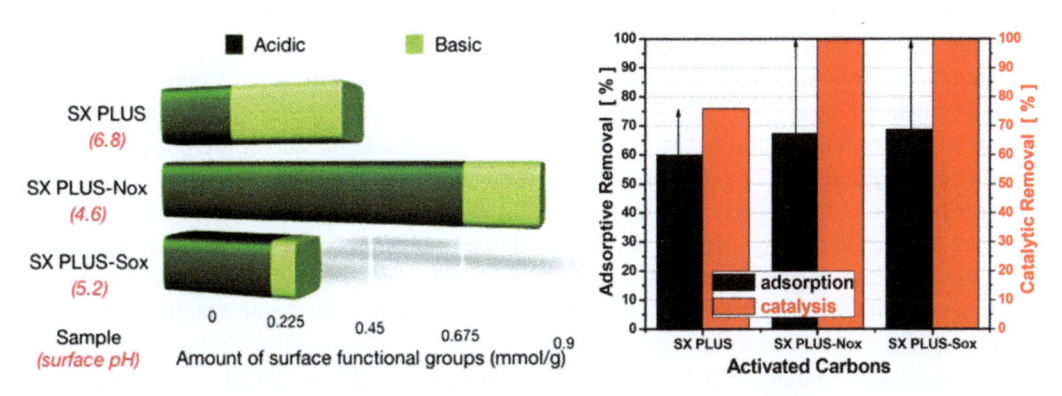

Figure 35.9 Quantification of acidic and basic surface functional groups of the parent activated carbon (AC) and its counterparts oxidized by HNO_3 and H_2SO_4 (left). Comparison of 4,6-dimethyldibenzothiophene (4,6-DMDBT) adsorptive and catalytic oxidation removal of the parent activated carbon (AC) and its counterparts oxidized by HNO_3 and H_2SO_4 (right). Ref [89] / Royal Society of Chemistry / CC BY 3.0.

Nanoscaled carbon-based materials have also showed promising results as heterogeneous catalysts for combined adsorption/oxidation/extraction sustainable desulfurization processes. Gu et al. reported the use of metal-free reduced graphene oxide for the oxidative removal of SCCs, due to presence of chemically active defects in the graphene oxide structure which generate carbonyl groups in situ [90]. Oxygen molecules could interact with the carbon atoms adjacent to carbonyl groups to form super oxygen anion radicals, which then oxidize the adsorbed organosulfurs. The reduced graphene oxide material has proved to be a highly efficient heterogeneous catalyst for aerobic ODS reaching 96.1%, 90.5%, 100%, and 97.7% of sulfur removal for 3-methylbenzothiophene (3-MBT), BT, DBT, and 4,6-DMDBT, respectively.

Recently, a new class of highly ordered porous materials has been reported using MOFs as self-sacrificial templates, the so-called MOF-Derived Porous Carbons (MDPCs) [91]. MDPCs are obtained by carbonization of MOFs under inert atmosphere resulting in uniform metal/metal oxide nanoparticles distributed in the ligand-derived carbon matrix [92]. The high specific surface area, tunable porosity, non-toxicity, lightweight and simple synthetic procedures of MDPCs makes them highly suitable for a wide range of application perspectives [93].

Hicks et al. reported a nanoporous carbon-based material obtained by pyrolysis of titanium-modified zinc-containing IRMOF-3 [94]. The pyrolyzed material exhibited higher porosity, surface area and thermal stability than the corresponding pristine MOF. Moreover, its catalytic activity in the oxidation of DBT using TBHP as oxidant was significantly enhanced as well as its recycling ability in consecutive catalytic cycles. The MIL-47(V) has also been used as template for the preparation of MDPCs for application in the oxidation of DBT [95]. The authors demonstrated that the pyrolysis temperature influences the V phase (V_2O_5, V_2O_3, VO_2, V_8C_7) present in the final carbon material. The best desulfurization performance was attained with V carbide carbon materials showing increased catalytic activity, enhanced chemical resistance to the oxidant (TBHP) and reduced leaching of active species when compared with the other prepared heterogeneous catalysts.

Another report deals with the production of novel MDPCs obtained by pyrolysis at different temperatures of hierarchical microporous/mesoporous MIL-125 frameworks prepared by a vapor-assisted technique [95]. The pyrolyzed mesoporous MIL-125-derived materials were evaluated as heterogeneous catalysts in the oxidation of DBT using TBHP and their activity compared with pyrolyzed microporous MIL-125. The results reveal a direct relationship between the catalytic activity and the pyrolysis temperature as a result of the increasing Ti content. The authors suggest that the enhanced desulfurization performance can be attributed to the higher amount of mesopores and smaller Ti particle size present in the mesoporous-derived materials. Jhung et al. described the pyrolysis of Ti-loaded zinc-containing hydrophobic MAF-6 and hydrophilic MOF-74 [96]. The adequate combination of the type of MOF and solvent (hydrophobic/hydrophilic) allowed to control the location of the loaded Ti-precursors (inside/outside of the porous framework). The catalytic activity of the pyrolyzed carbon materials on the oxidation of DBT seems to be favored when the initial Ti-precursors are mainly located inside the porous framework (prior to pyrolysis) as it leads to a smaller size of TiO_2 particles after the thermal treatment. In another report by Jhung et al., mesoporous MDPC materials were prepared by pyrolysis under inert atmosphere of bimetallic Zn/Ni or Zn/Mn MOF-74-type structures [97]. Several MOF precursors with different metallic ratios were prepared and its influence on the properties of the final carbonized materials was evaluated. The increasing Zn content in the MOF precursors lead to MDPCs with higher extent of mesopores and better catalytic activity in DBT oxidation using H_2O_2 (>95% of 1000 ppm S in 120 minutes). A similar strategy was employed for the preparation of Co-supported on nitrogen-doped porous carbon by pyrolysis of bimetallic Zn/Co MAF-6 with different metal ratios [98]. The pyrolyzed carbonaceous materials exhibited increased porosity and enhanced catalytic activity in the ECODS removal of SCCs from model fuel compared with the parent MOFs. In particular, the MDC-6(75Zn25Co)-900 sample revealed promising results reaching 93.6% of DBT conversion after 120 minutes using H_2O_2 as oxidant and acetonitrile with good recyclability for five consecutive cycles.

35.4.2.4 (EC)ODS with Mesoporous Silicas

The large and accessible porous framework of mesoporous silica allows the facile diffusion of bulky reactants and products in oxidative reactions. However, the ODS activity of pristine mesoporous silicas is practically inexistent as a result of the amorphous nature of their inner channels [99]. To overcome this limitation, several efforts have been made to prepare Ti-functionalized mesoporous silica for application in the removal of SCCs [100]. By doing so, researchers aimed to combine the exceptional ODS activity of Ti-species with the channel-type porous structure of mesoporous ordered silica for fast diffusion of reactants and products. Several reports can be found in the literature dealing with the preparation of heterogeneous ODS catalysts based on Ti-functionalized hexagonal mesoporous silica (HMS), SBA-15, SBA-16, and MCM-41 [101]. Lee et al. reported a Ti-SBA-15 catalyst obtained through post-synthetic grafting of Ti using tetrabutyl orthotitanate (TBOT) as source [101c]. The Ti-SBA-15 material was tested as heterogeneous catalyst in the ODS of model feed and real light cycle oil (LCO) with TBHP as oxidant. The effect of nitrogen containing compounds (NCCs), aromatic and aprotic solvents on the catalytic activity of Ti-SBA-15 was evaluated. The results reveal that the presence of NCCs decreases the ODS activity, although the addition of aromatic (tetralin or 1-methylnaphthalene) or aprotic (acetonitrile) solvents allows to recover the catalytic activity. In these conditions, the catalyst exhibited a remarkable desulfurization performance by attaining complete ODS conversion of real LCO with 3,700 ppm S.

Beltramone et al. prepared heterogeneous catalysts based on titanium-modified SBA-16 with tetrahedral or octahedral Ti coordination and compared their ODS activity in the oxidation of DBT using H_2O_2 as oxidant [100]. The results showed superior ODS performance of octahedral coordinated Ti-SBA-16 reaching 90% of DBT (2000 ppm) conversion after 60 minutes without loss of activity in four cycles. An alternative heterogeneous ODS catalyst based on mesoporous silica has been proposed by Plikarpova et al. [102]. Instead of the typical transition metal-functionalized silicas, the authors developed a cost-effective non-metal ODS catalyst by chemical immobilization of sulfonic groups on the surface of mesoporous MCM-41. MCM-SO_3H proved to be a highly efficient catalyst reaching complete DBT and 4-MDBT conversion (500 ppm) in 120 minutes with H_2O_2 as an oxidant. Moreover, the chemical coordination of sulfonic groups avoids leaching of the active species and allows the catalyst to retain its catalytic activity for 10 consecutive ODS cycles.

35.4.2.5 (EC)ODS with Titanate Nanotubes

Recently, titanate nanotubes (TiNTs) have emerged as an alternative type of Ti-containing mesoporous materials for ODS applications. TiNTs have attracted the interest of the scientific community owing to their high surface area, mesoporous structure, morphology, low cost and possibility to scale-up [103]. The majority of reports deals with the use of titanate nanotubes as supports for catalytic active species, while only a few works investigated their intrinsic catalytic properties. Regarding ODS application, Lorençon et al. described the alkaline treatment of TiO_2 anatase to produce sodium (Na-TiNTs) and hydrogen-titanate nanotubes (H-TiNTs) [104]. Catalytic studies revealed that the counter-ion has significant influence on the catalytic activity of TiNTs. In fact, Na-TiNTs showed no catalytic activity in the DBT conversion with H_2O_2 as oxidant while H-TiNTs exhibited high efficiency by achieving complete DBT removal (500 ppm) after just 60 minutes and could be recycled for five cycles without significant loss of activity. The authors propose that the mechanism involves the interaction of superficial Ti(IV) sites with the oxidant to form superoxide radicals. These radical active species are able to oxidize DBT into the corresponding sulfones. Similar results have been reported by Yao et al. by also studying protonated titanate nanotubes (H-TiNTs) as heterogeneous catalysts for desulfurization of fuels [105]. Cedeño-Caero et al. have compared the catalytic activity of TiNTs (pristine and calcined) and Ti-substituted mesoporous SBA-15 in the ECODS of DBT, 4-MDBT and 4,6-DMDBT with H_2O_2 as oxidant [106]. The catalytic results showed a superior performance of TiNTs over Ti-SBA-15 materials which, according to the authors, might be attributed to the higher exposure of superficial Ti-species in TiNTs. Interestingly, calcined TiNTs, despite having smaller surface area than the uncalcined material, exhibited superior desulfurization performance. In fact, calcined TiNTs reached a remarkable 99% of DBT conversion in just 60 minutes while the uncalcined material converted 90.5% for the same reaction time.

35.4.3 (EC)ODS Catalyzed by Heterogeneous Polyoxometalates

Transition metal compounds have been receiving considerable attention as catalytic active centers for ODS. Some of these compounds containing V, Mo, W, Nb, and Cr form peroxo-compounds by the interaction with H_2O_2. A particular class of transition metal compounds largely used in ODS are the polyoxometalates (POMs). These are anionic metal oxides, comprised of $[MO_x]_n$ (M = Mo, W, V, Nb; x = 4–7) basic units, producing a variety of possible structures with interesting physico-chemical properties [107]. The use of POMs in ODS is centered on their ability to form easily active peroxo-compounds, increasing the electrophilic character of peroxidic oxygens [108]. One of the most active POMs derived from Keggin structure $[XM_{12}O_{40}]^{n-}$, forming easily peroxo-POM species [109]. Among the different transition metals that can incorporate POM structures, Mo^{VI} and W^{VI} stand out due to a favorable combination of ionic radius, charge and accessibility of empty d orbitals for metal-oxygen π bonds [110]. In 1997, Collins published the first report of using POMs in ODS, using the phosphotungstic acid (PW_{12}) for the single oxidation of DBT with H_2O_2 [111]. Collins suggested that there is a fast reaction between H_2O_2 and the catalyst to produce active species which ensures the oxidation of DBT [111]. Wang et al. noted the influence of the Brønsted acidity and the cation in the ODS performance of POMs [112].

The efficiency of POMs as catalysts for ODS systems to treat fuels led to more than 200 papers to date, in which various reaction conditions, extraction solvents, oxidants, fuel feedstock were attempted. More recent studies are dedicated to the preparation of POMs composites (i.e. their heterogenization using various support materials). This is a clever procedure to use active catalytic centers able to be recycled. Several successful examples of POMs composites for ODS using model and real fuels are reported.

35.4.3.1 Carbonaceous Composites

Recently, several commercial ACs differing in acid-base properties were tested as supports for active POM catalysts. Various Keggin-type POMs structures decomposed due to their interaction with more basic ACs, forming monomeric and/or oligomeric M^{VI} oxo species on the carbon surface, which showed to enhance POM-AC composite activity, allowing 100% of DBT oxidation in just 30 minutes [113]. Further, these catalysts showed to be stable and recyclable for the oxidation of DBT [113]. Previously, Dizaji et al. took a different approach and immobilized different POMs on graphene oxide, using these for ECODS processes to treat multicomponent model fuel (BT, DBT and 4,6-DMDBT) [114]. SCCs were continuously extracted from model fuel and adsorbed into the graphene oxide surface. SCCs were then oxidized and desorbed from the graphene oxide surface, freeing the active sites for further reactions. Complete desulfurization was achieved after 30 minutes [114]. In the beginning of last decade, Wang et al. prepared a Multi-walled carbon nanotube composite $Cs_{2.5}H_{0.5}PW_{12}O_{40}$/MWNT which was found to be very effective for the oxidative removal of DBT, with a desulfurization efficiency of up to 100%. The catalyst was reused without significant loss of activity; however, lower efficiencies were found for the most difficult to oxidize SSCs [115].

35.4.3.2 MOF Composites

The encapsulation of homogeneous active POMs into porous MOFs has been forming high performance heterogeneous catalysts for ODS. In 2013, Ribeiro et al. presented the first published work using POM@MOF catalysts, in which the tetrabutylammonium salt $TBA_3PW_{12}O_{40}$ (TBA = Tetra-n-butylammonium) was encapsulated in the chromium terephthalate metal-organic framework MIL-101(Cr), following an impregnation method [116]. This new heterogeneous catalyst showed to have a similar activity than the homogeneous POM (i.e. near complete desulfurization obtained after three hours) [116]. The composite $TBA_3PW_{12}O_{40}$@MIL-101(Cr) was reused for three cycles with minimal loss of activity, attributed to the leaching of the active center [116]. The absence of POM leaching was found some years later by Wang et al. using an amine-modified NH_2-MIL-101 support [117]. The presence of amine-functional group in the MOF structure showed to be essential to prevent POM leaching via electrostatic interaction. More recently, Wang et al. presented a correlation between desulfurization efficiency and the pore and the window size of different MOF structures: MIL-100(Fe), UiO-66, and ZIF-8 [118]. These were used as support materials for the encapsulation of $[PW_{12}O_{40}]^{3-}$ active centers and these solid catalysts were used to treat model and real gasolines. From these MOFs, the POM size was smaller than the MOF nanocage sizes and larger than MOFs window sizes, preventing POM leaching. However, the catalytic performance in these cases was lower probably due to the mass transfer limitations [118]. The best catalytic performance was obtained for POM@MIL-100(Fe) (8.6 × 8.6 Å) achieving a desulfurization efficiency of 92% to treat a multicomponent model fuel (BT, DBT, and 4,6-DMDBT) using H_2O_2 as oxidant; while UiO-66 (6.0 × 6.0 Å) displayed high desulfurization efficiency only for the smaller size sulfur compounds (i.e. BT and DBT), because these are more easily diffused into the inner pores of UiO-66. The low activity of ZIF-8 was ascribed to its narrow pores (3,4 × 3,4 Å), which limit the sulfur compounds diffusion [118].

35.4.3.3 Zeolite Composites

Zeolites have also been modified via the introduction of POMs into their structures to produce active catalysts for organosulfur oxidation. One of the most recent works was published by Wang et al. that dispersed PW_{12} on the MWW zeolite, which consists of two ranges of independent porous structures (sinusoidal channels and supercages), and tested the composite catalyst for the desulfurization of model oil, straight-run gasoline, and fluid catalytic cracking gasoline [119]. The introduction of 38 wt% of W into the zeolite boosted its desulfurization efficiency from 35 to 99%. The prepared catalyst was reused for three consecutive cycles with minimal loss of activity (91.4% oxidation on the third cycle) due to the adsorption of oxidation products. These were easily removed by washing with ethanol and the composite was once again reused, achieving 98% S removal. The POM-MWW catalyst was also able to remove 98% of S content from straight-run gasoline, and 47% of organosulfurs from fluid catalytic cracking gasoline, due to its high olefin content, which hinders efficiency [119]. One previous work was reported by Zhang et al. that prepared cup-like hollow Zeolite Socony Mobil–5s (ZSM-5s) which were then used as a support for a POM@MOF composite of phosphomolybdic acid (PMo_{12}) in MOF-199 [119]. Each ZSM-5 cup acts as a micro-reactor allowing for efficient formation of reactive oxygen species by the POM@MOF composite and subsequent adsorption of DBT, leading to its oxidation. The introduction of the catalytically active centers achieved 91% of desulfurization, a significant increase from the pristine zeolite, 21% [120]. One of the first works using POM@zeolite to (EC)ODS first was presented by Wang et al. in 2011 and used a ship in the bottle methodology to encapsulate PMO_{12} in Y-type zeolites [121]. Using H_3PO_4 and MoO_3 as the P and Mo precursors, respectively, this preparation method allows for in situ POM formation in the zeolite's surface and pores. Surface POMs are then washed away with water, unblocking substrate and oxidant access to the porous structure's microreactors. The Y-zeolite with 10% Mo loading achieved 85% DBT oxidation in 20 minutes and was recycled for more three cycles with just 0.7% loss of activity [121].

35.4.3.4 Mesoporous Silica Composites

To guarantee an effective immobilization of active POMs on the surface of mesoporous silica supports, these hydrophilic materials need to be pre-functionalized (i.e. their surface needs to be modified by the incorporation of organic functional groups). The presence of strategic functional groups will promote an interaction via electrostatic or covalent with the POM and it will prevent leaching occurrence. In 2019, Julião et al. presented successful work that modified SBA-15 by functionalization with propyltrimethylammonium groups (TMA) to immobilize tetranuclear peroxotungstate $TBA_3\{PO_4[WO(O_2)_2]_4\}$ (PW_4) [122]. The composite PW_4@TMA-SBA-15 achieved a desulfurization efficiency similar to the homogeneous PW_4 to treat model oil and real fuels (Figure 35.10). This heterogeneous catalyst could be recovered and reused for at least 10 consecutive cycles with the first to eighth cycle achieving complete desulfurization and minimal losses of activity between the eighth and tenth cycle [122].

In the same year, Ribeiro et al. performed structural modification on a Keggin-type POM, obtaining a catalytically active lacunar structure which was then supported on amine-functionalized SBA-15. Surface functionalizing with amines minimizes leaching owing to the strong interaction between the POM via dative bonding. The catalytic activity of the PW_{11}@aptesSBA-15 composite was similar to that of the homogeneous PW_{11}, with the heterogeneous catalyst displaying remarkable activity for the ODS of real diesel, achieving 83.4% desulfurization when applied in a tandem EDS/ODS system. The recycling capacity of the composite was confirmed for eight cycles [123]. Ribeiro et al. also tested an ethylene-bridged periodic mesoporous organosilica (PMO), functionalized with a cationic group from N-trimethoxysilylpropyl-N, N, N-trimethylammonium (TMA), for the immobilization of the same PW_{11}. The same functionalization and immobilization was

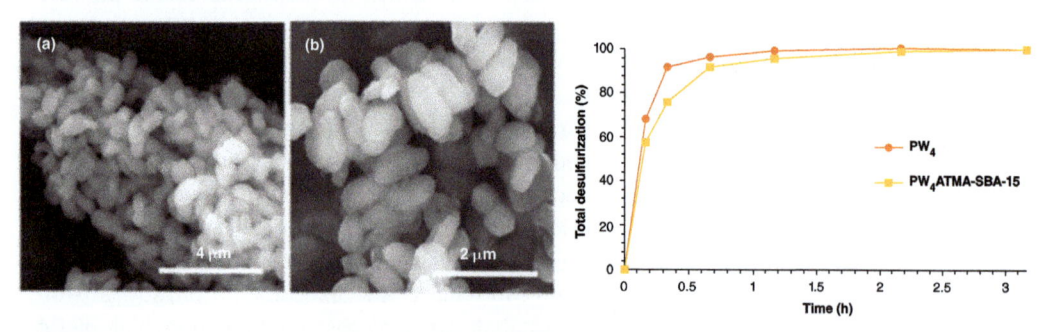

Figure 35.10 Scanning electron microscopy (SEM) micrographs of the solid support TMA-SBA-15 (a) and the composite material PW_4@TMA-SBA-15 (b) (left, center). Comparison of the desulfurization profiles of the homogeneous and heterogeneous catalysts (right). Reproduced with permission from Ref [122] / Elsevier.

performed using SBA-15 support. The PW$_{11}$@TMA-SBA-15 and PW$_{11}$@TMA-PMOE composites were tested as heterogeneous catalysts in the ODS of both model and real fuels. The PW$_{11}$@TMA-SBA-15 catalyst was able to achieve complete desulfurization for DBT, 4-MDBT and 4,6-DMDBT, and 93.9% for 1-BT, after 30 min, while PW$_{11}$@TMA-PMOE reached, 92.8% for 1-BT, 98.2% for DBT, 99.0% for 4-MDBT and 99.3% for 4,6-DMDBT after 60 minutes of catalytic oxidation, resulting in a total desulfurization of 96.9%. PW$_{11}$@TMA-SBA-15 retained its activity for six consecutive cycles, at which point some loss of performance was observed, ascribed to active site deactivation by the presence of strongly adsorbed sulfones. This composite was also used in the desulfurization of real untreated diesel, achieving 93% S removal [124].

35.5 (EC)ODS Catalyzed by Membranes

Applying membrane technology to the desulfurization of industrial fuels can be one of the most efficient approaches to mitigate the frequently reported catalyst mass losses during recycling and deactivation due to active species leaching and sintering [125]. This approach may be the defining step toward making other desulfurization technologies, such as ADS and/or (EC)ODS as true alternatives for application at an industrial level, as it dramatically enhances catalyst handling and may lower operating and energy costs, allowing for process scale-up. In fact, several reports can be find in the literature preparing mixed-matrix membranes (MMM) for desulfurization related applications, mainly for sulfur extraction. These works are mostly focused on using polymeric membranes due to their wide availability, low cost and ease of preparation [126]. These MMMs are based on the combination of organic polymers with some of the materials previously discussed in this chapter (carbon-based materials, silicas, and MOFs) and the reports mostly focus on their applications for selective adsorptive or pervaporative desulfurization. Recently, Peng et al. supported graphene oxide in a polyurethane membrane through a simple methodology of mixing each of the MMM components, followed by solvent evaporation [127]. The prepared MMMs were thiophene-permselective, displaying a significantly better separation performance over the non-modified membrane [127]. The decade before, Lin et al. used a similar approach to prepare a CuY-zeolite@polyethylene glycol (PEG) MMM [128], and also Cao et al. to obtain a polydimethylsiloxane MMM enhanced with silver/silica core–shell microspheres, noting that the composite membrane displayed more desirable characteristics than the non-modified membrane [129]. A couple of years ago, MOFs have also been used for the enhancement of polymeric membranes. Copper benzene-1,3,5-tricarboxylate (Cu-BTC) was incorporated into PEG, producing a MMM with a permeation flux that was 100% higher than a pristine PEG membrane prepared under the same conditions [130]. MOF-505 was inserted into the same polymeric matrix by Shi et al., resulting in a 158% increase of permeation flux over the corresponding pristine membrane [131]. None of these examples used the oxidative catalytic potential that can have these membranes. In fact, a gap in the literature is found for the application of MMMs that can combine oxidative catalytic activity and suitable sulfur extraction. One of the two published work was presented by Vigolo et al. in 2016, reinforcing a poly(methylmetacrylate) (PMMA) membrane by crosslinking with different molar ratios of Zr or Hf oxoclusters [132]. The different prepared MMMs were tested for the oxidation of DBT in a biphasic n-octane/acetonitrile system, with H_2O_2 as the oxidant. The best performance was obtained using a 1:50 Hf$_4$:polymer monomer ratio, reducing sulfur content from 300 to 25 ppm in 24 hours, a 16% higher than the obtained using the solo oxocluster as homogeneous catalyst. This increase in activity was noted for all seven prepared MMMs, highlighting the significant role of the polymeric matrix in shaping a suitable catalytic environment. More recently, a second report on the preparation of oxidative catalytic MMM for desulfurization was published by Mirante et al. [133]. The researchers reported on the development and application of a membrane-supported layered coordination polymer as an advanced sustainable catalyst for desulfurization by the immobilization of a previously reported active powdered coordination polymer catalyst [45a] in the PMMA matrix [133]. This membrane was then tested as an oxidative catalyst membrane for desulfurization of a model diesel containing four SCCs (1-BT, DBT, 4-MDBT and 4,6-DMDBT), with a BMIM based IL as extraction solvent and H_2O_2 as the oxidant, achieving complete desulfurization after four hours of reaction [133]. The powdered coordination polymer achieved a similar result after just two hours of reaction; however, the significantly enhanced handling of the membrane catalyst, added to its ease recovery and reusing (six ECODS cycles with no loss of activity), can open a new generation of advanced catalysts with viability to the industrial application of ODS processes.

35.6 Future Perspectives

The need of sustainable alternatives for clean energy production should be one of the drivers of scientific innovation for the next century. However, energy transition to clean energy future is a long-term process. In opposite, the Organization of the Petroleum Exporting Countries predicts a global economic growth doubling from today to 2040, as the number of people on the planet expands by 1.7 billion [1]. Significantly for the oil sector, which is transportation driven, another 1.2 billion people will be behind the wheel of an automobile. Commercial vehicles on the road will double, while air travel will soar. Even with a massive effort to decrease the usage of fossil fuels (mainly in occidental countries), it is reported that the use of fossil fuels could still represent 60% of primary energy consumption by 2040. Therefore, the combustion of fossil fuels still represents an enormous fraction of the planet's energy production. According to this scenario, the investment of developing technologies to reduce the harmful environmental impacts of fossil fuels is a priority. Desulfurization processes focus on limiting the eventual consequences of the release of sulfur into the atmosphere. The enhancement of the current industrial desulfurization process (HDS) by its combination with more efficient and cost-effective processes is crucial to provide clean fuel oil. Throughout this chapter, ADS and ODS, and the tandem use of ODS and EDS through (EC)ODS, were presented as potential sustainable alternatives/complementary processes to ensure deep desulfurization; however, their potential for industrial application is still in a research and development stage for real transportation fuels. Future advances in the actual industrial desulfurization technology will pass by strategically combining the well stablished HDS with other technology that is highly efficient for desulfurizing heavy molecular aromatic sulfur compounds. This latter will principally rely on materials with capacity to combine catalytic oxidation and extraction, resulting from its adsorptive or selective porosity. A promising class of materials able to incorporate all these functionalities are the membranes. Active and robust MOF or silica bulk porous materials that integrate oxidative catalytic and adsorptive capacities need to be incorporated in membranes containing selective permeability to heavy molecular aromatic sulfur compounds present in fuels. The combination of these desulfurization technologies integrating suitable materials will open the possibility to desulfurize a large number of liquid fuels, including HFO, saving time, costs, and material loss and deactivation.

Acknowledgments

The work developed in our research group received financial support from Portuguese national funds (FCT/MCTES, Fundação para a Ciência e a Tecnologia and Ministério da Ciência, Tecnologia e Ensino Superior) through the strategic project UIDB/50006/2020 (for LAQV / REQUIMTE). Carlos Granadeiro, Luís Cunha-Silva and Salete S. Balula thank FCT/MCTES for funding through the Individual Call to Scientific Employment Stimulus (Ref. 2022.02651.CEECIND/CP1724/CT001, CEECIND/00793/2018 and Ref. CEECIND/03877/2018, respectively). Alexandre Viana and Rui Faria thank FCT/MCTES and ESF (European Social Fund) through POCH (Programa Operacional Capital Humano) for their PhD grants (Refs. SFRH/BD/150659/2020 and UI/BD/151277/2021, respectively).

References

1. Johnsson, F., Kjärstad, J., and Rootzén, J. (2019). *Clim. Policy* 19: 258–274.
2. Lelieveld, J., Klingmüller, K., Pozzer, A. et al. (2019). *Proc. Natl. Acad. Sci.* 116: 7192.
3. Sikarwar, P., Gosu, V., and Subbaramaiah, V. (2019). *Rev. Chem. Eng.* 35: 669–705.
4. Sefoka, R.E. and Mulopo, J. (2017). *Int. J. Ind. Chem.* 8: 373–381.
5. (2020). *Oil Energy Trends* 45: 3–5.
6. Singh, D., Chopra, A., Mahendra, P.K. et al. (2016). *Pet. Sci. Technol.* 34: 1248–1254.
7. Fahim, M.A., Alsahhaf, T.A., and Elkilani, A. (2010). *Fundamentals of Petroleum Refining* (ed. M.A. Fahim, T.A. Alsahhaf, and A. Elkilani), 153–198. Amsterdam: Elsevier.
8. (a) Grange, P. (1980). *Catal. Rev.* 21: 135–181; (b) Babich, I.V. and Moulijn, J.A. (2003). *Fuel* 82: 607–631; (c) Shafiq, I., Shafique, S., Akhter, P. et al. (2022). *Catal. Rev.* 64: 1–86.
9. Schuman, S.C. and Shalit, H. (1971). *Catal. Rev.* 4: 245–318.
10. a. Duayne Whitehurst, D., Isoda, T., and Mochida, I. (1998). In *Advances in Catalysis*, 42. (eds. D.D. Eley, W.O. Haag, B. Gates, and H. Knözinger), Academic Press, pp. 345–471.; b. Song, C. and Ma, X. (2003). *Appl. Cataly. B Environ.* 41: 207–238.
11. Hansen, L.P., Ramasse, Q.M., Kisielowski, C. et al. (2011). *Angew. Chem. Int. Ed.* 50: 10153–10156.
12. Zhu, Y., Ramasse, Q.M., Brorson, M. et al. (2014). *Angew. Chem. Int. Ed.* 53: 10723–10727.
13. Salazar, N., Rangarajan, S., Rodríguez-Fernández, J. et al. (2020). *Nat. Commun.* 11: 4369.
14. Ma, X., Sakanishi, K., and Mochida, I. (1996). *Ind. Eng. Chem. Res.* 35: 2487–2494.
15. Knudsen, K.G., Cooper, B.H., and Topsøe, H. (1999). *Appl. Catal. A: Gen.* 189: 205–215.
16. Song, C. (2003). *Catal. Today* 86: 211–263.
17. McKee, R.H., Reitman, F., Schreiner, C. et al. (2013). *Int. J. Toxicol.* 33: 95S–109S.
18. Rana, M.S., Sámano, V., Ancheyta, J., and Diaz, J.A.I. (2007). *Fuel* 86: 1216–1231.
19. Anthony, E.J., Talbot, R.E., Jia, L., and Granatstein, D.L. (2000). *Energy Fuels* 14: 1021–1027.
20. Dias da Silva, P., Samaniego Andrade, S.K., Zygourakis, K., and Wong, M.S. (2019). *Ind. Eng. Chem. Res.* 58: 19623–19632.
21. Linares, N., Silvestre-Albero, A.M., Serrano, E. et al. (2014). *Chem. Soc. Rev.* 43: 7681–7717.
22. Saleh, T.A., Sulaiman, K.O., Al-Hammadi, S.A. et al. (2017). *J. Clean. Prod.* 154: 401–412.
23. Dinadayalane, T.C. and Leszczynski, J. (2010). *Struct. Chem.* 21: 1155–1169.
24. Saleem, J., Shahid, U.B., Hijab, M. et al. (2019). *Biomass Convers. Biorefin.* 9: 775–802.
25. Sevilla, M. and Mokaya, R. (2014). *Energy Environ. Sci.* 7: 1250–1280.

26 Salem, A.B.S.H. and Hamid, H.S. (1997). *Chem. Eng. Technol.* 20: 342–347.
27 Shi, Y., Liu, G., Wang, L., and Zhang, X. (2015). *Chem. Eng. J.* 259: 771–778.
28 Ania, C.O. and Bandosz, T.J. (2005). *Langmuir* 21: 7752–7759.
29 a. Moreira, A.M., Brandão, H.L., Hackbarth, F.V. et al. (2017). *Chem. Eng. Sci.* 172: 23–31.; b. Yu, C., Qiu, J. S., Sun, Y. F. et al. (2008). *J. Porous Mater.* 15: 151–157.
30 Jung, B.K. and Jhung, S.H. (2015). *Fuel* 145: 249–255.
31 Deng, L., Lu, B., Li, J. et al. (2017). *Fuel* 200: 54–61.
32 Saleh, T.A. (2018). *J. Clean. Prod.* 172: 2123–2132.
33 a. Novoselov, K.S., Fal'ko, V.I., Colombo, L. et al. (2012). *Nature* 490: 192–200. b. Chen, D., Feng, H., and Li, J. (2012). *Chem. Rev.* 112: 6027–6053.
34 Song, H.S., Ko, C.H., Ahn, W. et al. (2012). *Ind. Eng. Chem. Res.* 51: 10259–10264.
35 Menzel, R., Iruretagoyena, D., Wang, Y. et al. (2016). *Fuel* 181: 531–536.
36 Crespo, D. and Yang, R.T. (2006). *Ind. Eng. Chem. Res.* 45: 5524–5530.
37 Khaled, M. (2015). *Res. Chem. Intermed.* 41: 9817–9833.
38 Jiang, K., Li, Z., Zheng, Z. et al. (2021). *Environ. Sci. Atmos.* 1: 569–576.
39 Zhao, X. (2010). *Materials for Energy Efficiency and Thermal Comfort in Buildings* (ed. M.R. Hall), 399–426. Sawston: Woodhead Publishing.
40 Yang Ralph, T., Hernández-Maldonado Arturo, J., and Yang Frances, H. (2003). *Science* 301: 79–81.
41 Zhang, Z.Y., Shi, T.B., Jia, C.Z. et al. (2008). *Appl. Catal. B: Environ.* 82: 1–10.
42 Neubauer, R., Husmann, M., Weinlaender, C. et al. (2017). *Chem. Eng. J.* 309: 840–849.
43 Tian, F., Yang, X., Shi, Y. et al. (2012). *J. Nat. Gas Chem.* 21: 647–652.
44 Zhou, Y., Dong, Z., Terasaki, O., and Ma, Y. (2022). *Acc. Mater. Res.* 3: 110–121.
45 a Mendes, R.F., Antunes, M.M., Silva, P. et al. (2016). *Chem. Eur. J.* 22: 13136–13146; b Narayan, R., Nayak, U.Y., Raichur, A.M., and Garg, S. (2018). *Pharmaceutics* 10: 118.
46 Alvarado-Perea, L., Colín-Luna, J.A., López-Gaona, A. et al. (2020). *Catal. Today* 353: 26–38.
47 Song, L., Chen, J., Bian, Y. et al. (2015). *J. Porous Mater.* 22: 379–385.
48 Palomino, J.M., Tran, D.T., Hauser, J.L. et al. (2014). *J. Mater. Chem. A* 2: 14890–14895.
49 Tian, W.H., Sun, L.B., Song, X.L. et al. (2010). *Langmuir* 26: 17398–17404.
50 Zhou, H.-C., Long, J.R., and Yagh, O.M. (2012). *Chem. Rev.* 112: 673–674.
51 Cychosz, K.A., Wong-Foy, A.G., and Matzger, A.J. (2008). *J. Am. Chem. Soc.* 130: 6938–6939.
52 Li, Y.-X., Jiang, W.-J., Tan, P. et al. (2015). *J. Phys. Chem. C* 119: 21969–21977.
53 Zhang, X.-F., Wang, Z., Feng, Y. et al. (2018). *Fuel* 234: 256–262.
54 Khan, N.A., Hasan, Z., and Jhung, S.H. (2014). *Chem. Eur. J.* 20: 376–380.
55 Otsuki, S., Nonaka, T., Takashima, N. et al. (2000). *Energy Fuels* 14: 1232–1239.
56 Zhang, M., Zhu, W., Xun, S. et al. (2013). *Chem. Eng. J.* 220: 328–336.
57 Zhao, H. and Baker, G.A. (2015). *Front. Chem. Sci. Eng.* 9: 262–279.
58 Guth, A.D.E., KVB Inc. (1974). *United States Patent 1191*. 1451 Nov. 12, Method for removing sulfur and nitrogen in petroleum oils.
59 Xu, X., Moulijn, J.A., Ito, E. et al. (2008). *ChemSusChem.* 1: 817–819.
60 Sundararaman, R., Ma, X., and Song, C. (2010). *Ind. Eng. Chem. Res.* 49: 5561–5568.
61 Ma, C., Dai, B., Liu, P. et al. (2014). *J. Ind. Eng. Chem.* 20: 2769–2774.
62 Wang, J., Zhao, D., and Li, K. (2010). *Energy Fuels* 24: 2527–2529.
63 Kim, J. and Huang, C.-H. (2021). *ACS ES&T Water* 1: 15–33.
64 Dehkordi, A.M., Sobati, M.A., and Nazem, M.A. (2009). *Chin. J. Chem. Eng.* 17: 869–874.
65 Akopyan, A., Eseva, E., Polikarpova, P. et al. (2020). *Molecules (Basel, Switzerland)* 25: 536.
66 Shiraishi, Y., Hara, H., Hirai, T., and Komasawa, I. (1999). *Ind. Eng. Chem. Res.* 38: 1589–1595.

67 Isaacs, M.A., Robinson, N., Barbero, B. et al. (2019). *J. Mater. Chem. A* 7: 11814–11825.
68 (a) Napanang, T. and Sooknoi, T. (2009). *Catal. Commun.* 11: 1–6; (b) Shen, C., Wang, Y.J., Xu, J.H., and Luo, G.S. (2015). *Chem. Eng. J.* 259: 552–561.
69 Lv, Q., Li, G., and Sun, H. (2014). *Fuel* 130: 70–75.
70 a Ding, Y., Ke, Q., Liu, T. et al. (2014). *Ind. Eng. Chem. Res.* 53: 13903–13909; b Li, L., Wang, W., Huang, J. et al. (2022). *Appl. Cataly. A General* 630: 118466.
71 Du, S., Chen, H.-M., Shen, H.-X. et al. (2020). *ACS Appl. Nano Mater.* 3: 9393–9400.
72 Kong, L., Li, G., and Wang, X. (2004). *Catal. Today* 93-95: 341–345.
73 Du, S., Li, F., Sun, Q. et al. (2016). *Chem. Commun.* 52: 3368–3371.
74 Hall, J.N. and Bollini, P. (2019). *React. Chem. Eng.* 4: 207–222.
75 Llabrés I Xamena, F.X., Luz, I., and Cirujano, F.G. (2013). *Metal Organic Frameworks as Heterogeneous Catalysts*. 237–267. The Royal Society of Chemistry.
76 a McNamara, N.D., Neumann, G.T., Masko, E.T. et al. (2013). *J. Catal.* 305: 217–226; b Li, X., Gu, Y., Chu, H. et al. (2019). *Appl. Cataly. A General* 584: 117152.
77 Hwang, Y.K., Hong, D.-Y., Chang, J.-S. et al. (2009). *Appl. Catal. A: Gen.* 358: 249–253.
78 Masoomi, M.Y., Bagheri, M., and Morsali, A. (2015). *Inorg. Chem.* 54: 11269–11275.
79 Granadeiro, C.M., Ribeiro, S.O., Karmaoui, M. et al. (2015). *Chem. Commun.* 51: 13818–13821.
80 Viana, A.M., Ribeiro, S.O., Castro, B. et al. (2019). *Materials (Basel)* 12: 3009.
81 Viana, A.M., Julião, D., Mirante, F. et al. (2021). *Catal. Today* 362: 28–34.
82 Dissegna, S., Epp, K., Heinz, W.R. et al. (2018). *Adv. Mater.* 30: 1704501.
83 Gu, Y., Xu, W., and Sun, Y. (2021). *Catal. Today* 377: 213–220.
84 (a) Gu, Y., Ye, G., Xu, W. et al. (2020). *ChemistrySelect* 5: 244–251; (b) Fu, G., Bueken, B., and De Vos, D. (2018). *Small Methods* 2: 1800203.
85 Zheng, H.-Q., Zeng, Y.-N., Chen, J. et al. (2019). *Inorg. Chem.* 58: 6983–6992.
86 Li, N., Zhang, Z.-W., Zhang, J.-N. et al. (2021). *Dalton Trans.* 50: 6506–6511.
87 Smolders, S., Willhammar, T., Krajnc, A. et al. (2019). *Angew. Chem. Int. Ed. Engl.* 58: 9160–9165.
88 Haw, K.-G., Bakar, W.A.W.A., Ali, R. et al. (2010). *Fuel Process. Technol.* 91: 1105–1112.
89 Kampouraki, Z.C., Giannakoudakis, D.A., Triantafyllidis, K.S., and Deliyanni, E.A. (2019). *Green Chem.* 21: 6685–6698.
90 Gu, Q., Wen, G., Ding, Y. et al. (2017). *Green Chem.* 19: 1175–1181.
91 Shen, K., Chen, X., Chen, J., and Li, Y. (2016). *ACS Catal.* 6: 5887–5903.
92 Yap, M.H., Fow, K.L., and Chen, G.Z. (2017). *Green Energy Environ.* 2: 218–245.
93 (a) Marpaung, F., Kim, M., Khan, J.H. et al. (2019) *Chem. Asian J.* 14: 1331–1343; (b) Chen, Y.-Z., Zhang, R., Jiao, L., and Jiang, H.-L. (2018). *Coordinat. Chem. Rev.* 362: 1–23.
94 Kim, J., McNamara, N.D., Her, T.H., and Hicks, J.C. (2013). *ACS Appl. Mater. Interfaces* 5: 11479–11487.
95 Kim, J., McNamara, N.D., and Hicks, J.C. (2016). *Appl. Catal. A: Gen.* 517: 141–150.
96 Sarker, M., Bhadra, B.N., Shin, S., and Jhung, S.H. (2019). *ACS Appl. Nano Mater.* 2: 191–201.
97 Bhadra, B.N. and Jhung, S.H. (2018). *Nanoscale* 10 (31): 15035–15047.
98 Bhadra, B.N., Khan, N.A., and Jhung, S.H. (2019). *J. Mater. Chem. A* 7: 17823–17833.
99 Capel-Sanchez, M.C., Campos-Martin, J.M., and Fierro, J.L.G. (2010). *Energy Environ. Sci.* 3: 328–333.
100 Rivoira, L.P., Vallés, V.A., Ledesma, B.C. et al. (2016). *Catal. Today* 271: 102–113.
101 (a) Cui, S., Ma, F., and Wang, Y. (2007) *React. Kinet. Catal. Lett.* 92: 155–163; (b) Kim, T.-W., Kim, M.-J., Kleitz, F. et al. (2012). *ChemCatChem.* 4: 687–697; (c) Cho, K.-S. and Lee, Y.-K. (2014). *Appl. Cataly. B Environ.* 147: 35–42; (d) Shah, A.T., Li, B., and Ali Abdalla, Z.E. (2009). *J. Colloid*

Interface Sci. 336: 707–711; (e) Chica, A., Corma, A., & Dómine, M.E. (2006). *J. Catalysis* 242: 299–308.

102 Polikarpova, P., Akopyan, A., Shlenova, A., and Anisimov, A. (2020). *Catal. Commun.* 146: 106123.

103 (a) Kasuga, T., Hiramatsu, M., Hoson, A. et al. (1998). *Langmuir* 14: 3160–3163; (b) Bavykin, D.V., Friedrich, J.M., and Walsh, F.C. (2006). Adv. Mater. 18: 2807–2824.

104 Lorençon, E., Alves, D.C.B., Krambrock, K. et al. (2014). *Fuel* 132: 53–61.

105 Lu, S.-X., Zhong, H., Mo, D.-M. et al. (2017). *Green Chem.* 19: 1371–1377.

106 Cedeño-Caero, L., Ramos-Luna, M., Méndez-Cruz, M., and Ramírez-Solís, J. (2011). *Catal. Today* 172: 189–194.

107 Rao, C., Müller, A., and Cheetham, A. (2004). *The Chemistry of Nanomaterials: Synthesis, Properties and Applications*, 2. (ed. C.N.R. Rao, A. Müller, and A.K. Cheetham), 761. Wiley-VCH. ISBN 3-527-30686-2 March 2004, -1.

108 Trakarnpruk, W. and Rujiraworawut, K. (2009). *Fuel Process. Technol.* 90: 411–414.

109 Te, M., Fairbridge, C., and Ring, Z. (2001). *Appl. Catal. A: Gen.* 219: 267–280.

110 Ivanova, S. (2014). *ISRN Chem. Eng.* 2014: 963792.

111 Collins, F.M., Lucy, A.R., and Sharp, C. (1997). *J. Mol. Catal. A Chem.* 117: 397–403.

112 (a) Wang, R., Zhang, G., and Zhao, H. (2010). *Catal. Today* 149: 117–121; (b) Misra, A., Kozma, K., Streb, C., and Nyman, M. (2020). *Angewandte Chemie Int. Ed.* 59: 596–612.

113 Ghubayra, R., Yahya, R., Kozhevnikova, E.F., and Kozhevnikov, I.V. (2021). *Fuel* 301: 121083.

114 Khodadadi Dizaji, A., Mortaheb, H.R., and Mokhtarani, B. (2019). *Catal. Lett.* 149: 259–271.

115 Wang, R., Yu, F., Zhang, G., and Zhao, H. (2010). *Catal. Today* 150: 37–41.

116 Ribeiro, S., Barbosa, A.D.S., Gomes, A.C. et al. (2013). *Fuel Process. Technol.* 116: 350–357.

117 Wang, X.S., Huang, Y.B., Lin, Z.J., and Cao, R. (2014). *Dalton Trans.* 43: 11950–11958.

118 Wang, X.-S., Li, L., Liang, J. et al. (2017). *ChemCatChem.* 9: 971–979.

119 Wang, H., Jibrin, I., and Zeng, X. (2020). *Front. Chem. Sci. Eng.* 14: 546–560.

120 Zhang, Y., Zhang, W., Zhang, J. et al. (2018). *RSC Adv.* 8: 31979–31983.

121 Wang, H. and Wang, R. (2011). *Collect. Czechoslov. Chem. Commun.* 76: 1595–1605.

122 Julião, D., Mirante, F., Ribeiro, S.O. et al. (2019). *Fuel* 241: 616–624.

123 Ribeiro, S.O., Granadeiro, C.M., Almeida, P.L. et al. (2019). *Catal. Today* 333: 226–236.

124 Ribeiro, S.O., Granadeiro, C.M., Corvo, M.C. et al. (2019). *Front. Chem.* 7: 756.

125 Argyle, M.D. and Bartholomew, C.H. (2015). *Catalysts* 5: 949–954.

126 Fihri, A., Mahfouz, R., Shahrani, A. et al. (2016). *Chem. Eng. Process.- Process Intensif.* 107: 94–105.

127 Peng, P., Lan, Y., Zhang, Q., and Luo, J. (2022). *J. Appl. Polym. Sci.* 139: 51514.

128 Lin, L., Zhang, Y., and Li, H. (2010). *J. Colloid. Interface Sci.* 350: 355–360.

129 Cao, R., Zhang, X., Wu, H. et al. (2011). *J. Hazard. Mater.* 187: 324–332.

130 Cai, C., Fan, X., Han, X. et al. (2020). *Polymers* 12: 414.

131 Shi, W., Han, X., Bai, F. et al. (2021). *Sep. Purif. Technol.* 272: 118924.

132 Vigolo, M., Borsacchi, S., Sorarù, A. et al. (2016). *Appl. Catal. B: Environ.* 182: 636–644.

133 Mirante, F., Mendes, R.F., Faria, R.G. et al. (2021). *Molecules* 26: 2404.

Part VII

Hydrogen Formation, Storage, and Utilization

36

Paraformaldehyde

Opportunities as a C1-Building Block and H_2 Source for Sustainable Organic Synthesis

Ana Maria Faísca Phillips[1], Maximilian N. Kopylovich[1], Leandro Helgueira de Andrade[2], and Martin H.G. Prechtl[1,3]

[1] Centro de Química Estrutural, Institute of Molecular Sciences, Instituto Superior Técnico, Universidade de Lisboa, Lisbon, Portugal
[2] Departamento de Química Fundamental, Instituto de Química, Universidade de São Paulo, São Paulo, Brazil
[3] Department of Synthesis and Analysis, Albert Hofmann Institute for Physiochemical Sustainability, Vlotho, Germany

36.1 Introduction

In recent years, environmental concerns related to the production of fine chemicals has become a major issue, because hazardous reagents are commonly applied in many industrial processes and this results in large amounts of polluted waste and requires expensive reaction setups for safe processing [1–3]. A modernization of traditional reaction schemes considers the use of catalysts for higher selectivitis, shorter reaction times, and lower reaction temperatures combined with the use of less toxic and renewable solvents, as well as reagents which can be derived from renewable sources. Another consideration is related to the need to minimize the risk related to the use and handling of hydrogen gas and CO under elevated pressure and/or at high concentrations. To overcome these issues for synthetic setups, liquid/solid H_2 and CO sources are attractive alternatives [1–6]. Among these are small molecules that can be derived from CH_4 and CO_2 with a relatively high hydrogen content (in wt-%), such as C1-molecules like methanol (12.5 wt-%), formaldehyde (6.7 wt-%), methanediol (8.4 wt-%), or formic acid (4.3 wt-%), which are catalytically convertible to H_2 and CO_2, H_2 and CO or CO, and H_2O (Figure 36.1) [1, 4, 7].

One of these C1-molecules, formaldehyde, is known to form oligomers and polymers of different chain lengths in the absence of water and paraformaldehyde (PFA) can act as a source of formaldehyde and methanediol, respectively (Figure 36.1) [1–3, 5, 8–10]. Formaldehyde can be synthesized by the oxidation/dehydrogenation of methanol via the Formox process at elevated temperature, applying a transition metal catalyst [3, 11, 12], or by the conversion of syngas ($CO:H_2 = 1:2$) in the aqueous phase with metal catalysts [3, 11, 13]. PFA [$(CH_2O)_n$], the corresponding colorless oligomer/polymer solid powder (n = 8–100), is formed by dehydration of aqueous formaldehyde and, respectively, methanediol.

In this chapter, we give a brief overview about selected examples of homogeneous metal-catalyzed reactions with PFA which can be used as a source of hydrogen gas and C1-building blocks in

Catalysis for a Sustainable Environment: Reactions, Processes and Applied Technologies Volume 3, First Edition. Edited by Armando J. L. Pombeiro, Manas Sutradhar, and Elisabete C. B. A. Alegria.
© 2024 John Wiley & Sons Ltd. Published 2024 by John Wiley & Sons Ltd.

addition to its potential use as a CO surrogate, depending on the reaction conditions. In particular, PFA in water can be used as source of H_2 for transfer hydrogenation reactions such as for the reduction of C=C, C=O, –CC–, and –CN bonds (Figure 36.2a) [9, 10, 14, 15]. Underlining that water acts as an H_2 source as well because it reacts with PFA to yeild methanediol, which is subsequently dehydrogenated to yield two eq. of H_2 and one eq. of CO_2 (Figure 36.3) [1, 8]. PFA itself is produced from methanol, as by-product of formaldehyde [3, 11, 12], which can be produced from CO_2 [16–20], and therefore PFA could become a renewable molecule in the future as well. Moreover, PFA is a convenient CO-surrogate and C1 building block; therefore by the use of appropriate metal complex catalysts, it can be applied in hydroformylation [6], N-formylation [5], carbonylation [21], methylation [2], hydroxymethylation [22], or C–H activation [22–24], among many other reactions which will be outlined further in this chapter (Figure 36.2b).

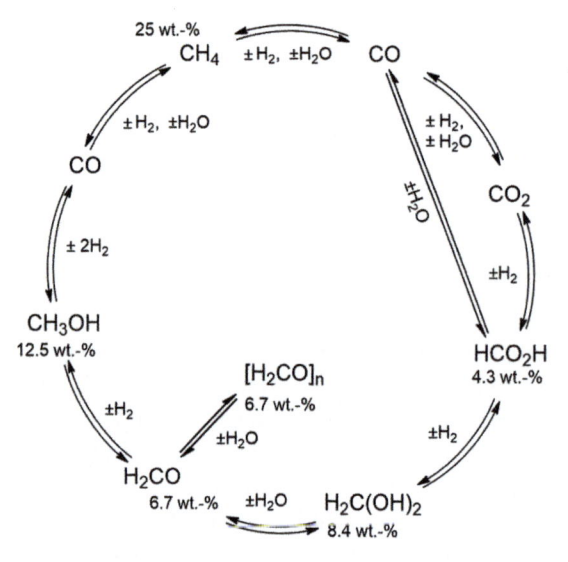

Figure 36.1 Interconversion of C1-molecules by shuttling hydrogen and oxygen in the presence of catalysts.

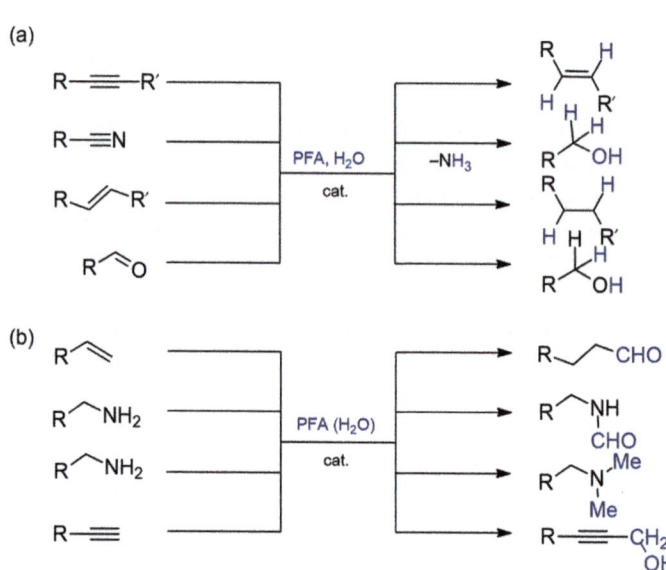

Figure 36.2 a) Metal-catalyzed transfer-hydrogenation of multiple-bonds with paraformaldehyde (PFA) as an H_2 source [9, 10, 14, 15]. b) Metal-catalyzed N-formylation and (hydroxyl)methylation [2, 5, 6, 22].

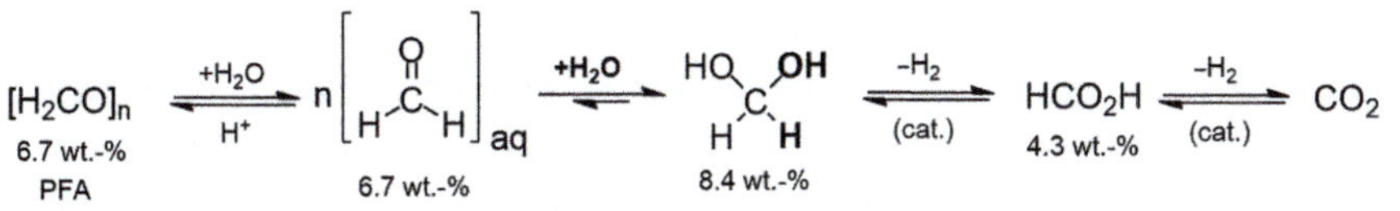

$$[H_2CO]_n \underset{6.7 \text{ wt.-\%}}{\overset{+H_2O}{\underset{H^+}{\rightleftarrows}}} n\left[H\overset{O}{\underset{H}{C}}H\right]_{aq} \overset{+H_2O}{\rightleftarrows} \underset{8.4 \text{ wt.-\%}}{HO\overset{OH}{\underset{H}{C}}H} \underset{(cat.)}{\overset{-H_2}{\rightleftarrows}} \underset{4.3 \text{ wt.-\%}}{HCO_2H} \underset{(cat.)}{\overset{-H_2}{\rightleftarrows}} CO_2$$

6.7 wt.-% PFA   6.7 wt.-%

Figure 36.3 Transition metal catalyzed hydrogen evolution from paraformaldehyde (PFA) via methanediol [1, 3, 12, 25, 26].

Gaseous reagents such as dihydrogen, syngas, and CO require the use of more sophisticated and expensive laboratory infrastructure to handle volatile, hazardous, explosive, and/or toxic gases. In contrast, the solid form and simple use of PFA shows advantages for application in research labs [2, 5, 9, 10]. The adjustment of the required quantity of the H_2 carrier, respectively C1-surrogate and water, can be easily conducted gravimetrically and these reactions can be run in simple glassware suitable for low pressure reactions (e.g. headspace vials). Expensive stainless steel pressure reactors are not required. In addition, in several cases air and moisture stable catalysts and reagents are used, thus expensive and inert conditions, and likewise glovebox or Schlenk techniques, are not necessary. The robustness of the catalysts, the use of more environmentally benign solvents, and PFA and water as hydrogen carriers, give access to more sustainable synthetic methodologies.

36.2 Carbonylation and Related Reactions

The incorporation of a C=O unit into a molecule, known as carbonylation, provides access to a variety of useful carbonyl-containing compounds, including aldehydes, esters, acids, or amides [27]. These products may themselves be subsequently transformed into a whole range of fine chemicals, due to the high synthetic versatility of the carbonyl function. Carbon monoxide (CO) gas is the cheapest source of this C1 unit and carbonylation reactions are amongst the largest applications of homogeneous catalysis in industry [27]. An example is the synthesis of propanoic acid, produced on a multi-ton scale by the hydroxycarbonylation of ethene, as an intermediate for the production of plastics, pharmaceuticals, and pesticides [28]. Methyl methacrylate (MMA), produced by the palladium-catalyzed methoxycarbonylation of ethene (the Lucite process), followed by reaction of the methyl propanoate obtained with formaldehyde and elimination, is another example. Methyl methacrylate is used to prepare poly(methyl methacrylate), a transparent and strong polymer with many industrial applications. The carbonylation of alkenes is also highly important in the pharmaceutical industry [6, 29] and it can even be used to convert the very inert alkanes from the petrochemical and natural gas industries into value-added products [6, 30]. Although CO is usually used for carbonylation, it is flammable and highly toxic and therefore many attempts are underway to find surrogates that are easier to handle and safer, without compromising the efficiency of the reactions, particularly (but not only) for laboratory-scale applications [28]. A few organic and inorganic chemical compounds have been utilized to perform this role, although these compounds have other drawbacks, related either to their atom efficiency, reactivity, toxicity, or price. For example, metal carbonyls such as $Mo(CO)_6$ may require stoichiometric amounts of other transition metals and harsh conditions for the release of CO. Organic reagents include pivaloyl chloride, silicon-based carboxylic acids, *N*-formylsaccharin, formic acid, and formamides such as dimethylformamide (DMF). Waste generation, harsh reaction conditions, poor selectivity in the presence of other nucleophiles, low atom efficiency, and even the need for external CO pressure in some cases, can be some of their disadvantages, which make the search for other CO sources a subject of continued interest [28]. Formaldehyde is the most economic surrogate; it has suitable reactivity and the added advantage that the percentage by molecular weight of the CO unit is 93%, the highest of all surrogates used so far. High yields and high levels of selectivity have been obtained with its solid form, paraformaldehyde, in the alkoxycarbonylation of alkenes, in N-carbonylation, in the carbonylation of aryl halides, and even in cascade reactions, to name a few, as described next.

36.2.1 Alkoxycarbonylation of Olefins

The alkoxycarbonylation of olefins allows the synthesis of carboxylic acid esters from readily available olefins, alcohols, and CO [31, 32]. The main issues to overcome are i) the regioselectivity (i.e. whether linear or branched products are formed), ii) the activity of the catalyst, and iii) the rate of depolymerization of PFA. At present, a number of catalysts are available to promote this transformation, based on Ru or Pd, and the nature of the ligand used has a strong impact on activity and selectivity. An early example, in which PFA was used as the CO source [33], relied on the utilization of triruthenium dodecacarbonyl [$Ru_3(CO)_{12}$] as catalyst, and an electron-rich monodentate phosphine ligand, PCy_3 (**L1**, Cy=cyclohexyl), in order for high yields to be obtained. The utilization of 1-butyl-3-methylimidazolium chloride (BMIMCl) as an additive helped to improve conversion and selectivity. In the Lucite process for the synthesis of MMA, CO gas is used, and methane sulfonic acid is required as an additive to ensure high catalyst activity. The new PFA-based method by-passed the need for high pressure reactors, as well as that of acid, which may cause reactor vessel corrosion problems. As the examples in Table 36.1 show, the Ru/**L1** system afforded carboxylate esters from a variety of terminal and internal alkenes. Methyl propionate, the

Table 36.1 The alkoxycarbonylation of olefins.

No	Alkene	Major product	Metal/ligand[1]	Yield [%] (linear:branched selectivity)	Ref
1	=	CO₂Me	[Ru]/**L1**[2]	51	[33]
			[Pd]/**L2**	99	[28]
2	Ph⏜	Ph⏜CO₂Me	[Ru]/**L1**	51 (70:30)	[33]
			[Pd]/**L2**	90 (78:22)	[28]
3	Ph⇃	Ph⇃CO₂Me	[Ru]/**L1**	89 (> 99:1)	[33]
			[Pd]/**L3**	97 (> 99:1)	[28]

Table 36.1 (Continued)

#	Substrate	Product	Catalyst	Yield (selectivity)	Ref
4	1-(4-chlorophenyl)ethylene (α-methylstyrene, 4-Cl)	methyl 3-(4-chlorophenyl)butanoate	[Ru]/L1	68	[33]
			[Pd]/L2	43 (>99:1)	[28]
			[Pd]/L3	92 (>99:1)	[34]
5	methyl 2-phenylacrylate	dimethyl 2-phenylsuccinate	[Ru]/L1	43	[33]
			[Pd]/L2	89 (>99:1)	[28]
6	2-vinylnaphthalene	methyl 3-(naphthalen-2-yl)propanoate	[Ru]/L1	53	[33]
			[Pd]/L3	89 (>99:1)	[34]
7	cyclohexene	methyl cyclohexanecarboxylate	[Ru]/L1	88	[33]
			[Pd]/L3	94	[34]
8	cyclopentene	methyl cyclopentanecarboxylate	[Ru]/L1	90	[33]
			[Pd]/L3	91	[34]
9	norbornene	methyl norbornane-2-carboxylate	[Ru]/L1	82	[28]
10	C_6H_{13}–CH=CH$_2$ (1-octene)	C_6H_{13}-CH$_2$CH$_2$CO$_2$Me	[Ru]/L1	74 (51:49)	[33]
				93 (95:5)	[28]
11	C_5H_{11}-CH=CH-CH$_3$	C_6H_{13}-CH$_2$CH$_2$CO$_2$Me	[Pd]/L2	93 (94:6)	[28]
12	internal octene	C_6H_{13}-CH$_2$CH$_2$CO$_2$Me	[Pd]/L2	75 (89:11)	[28]
13	3,3-dimethyl-1-butene	methyl 4,4-dimethylpentanoate	[Pd]/L2	85 (>99:1)	[28]
14	methyl acrylate derivative (CO$_2$Me$_2$)	dimethyl succinate derivative	[Pd]/L2	92 (>99:1)	[28]
15	methyl methacrylate derivative	dimethyl 2-methylsuccinate	[Pd]/L2	89 (>99:1)	[28]
16	2,3,3-trimethyl-1-butene	methyl 3,4,4-trimethylpentanoate	[Pd]/L3	93 (>99:1)	[34]
17	indene	methyl indane-1-carboxylate	[Pd]/L3	42 (89:11)	[34]
18	2,4,4-trimethyl-2-pentene	methyl 3,5,5-trimethylhexanoate	[Pd]/L3	93 (>99:1)	[34]
19	2,3-dimethyl-2-butene	methyl 3,4-dimethylpentanoate	[Pd]/L3	93 (>99:1)	[34]
20	(E)-3,4-bis(4-hydroxyphenyl)hex-3-ene (diethylstilbestrol)	corresponding hydroxymethoxycarbonyl product	[Pd]/L3	42	[34]

1) Conditions: **[Ru]/L1**: [Ru$_3$(CO)$_{12}$] (1.5 mol%), PCy$_3$ (4.5 mol%), NMP, BMIMCl (2 equiv), 130 °C; **[Pd]/L2**: [Pd(OAc)$_2$] (1 mol%), d^{t}bpx (4 mol%), PTSA (5 mol%), 100 °C; **[Pd]/L3**: 1.0 mmol substrate, 1.0 mol% Pd(OAc)$_2$, 4.0 mol% L1, 5.0 mol% PTSA·H2O, 200 mg (CH$_2$O)n, 2.0 mL MeOH, 120 °C, 72 h; NMP = N-methylpyrrolidone.
2) [Ru$_3$(CO)$_{12}$] 0.7 mol%.

intermediate in the synthesis of methyl methacrylate, could be obtained in 51% yield from ethylene, with an even lower catalyst loading of 0.7 mol%. With styrene derivatives, terminal aliphatic esters were obtained (entries 3 and 4), and the use of terminal and internal aliphatic alkenes produced a mixture of branched and linear esters with almost the same yields and selectivities (e.g. entry 10). The major side reaction observed was alkene hydrogenation, but, in the reactions of aliphatic alkenes, hydrogenated and isomerized byproducts (n-octane and iso-octenes) were also observed. In this study, other alcohols (e.g. ethanol, butanol, benzyl alcohol) were also reacted with cyclohexene, and the corresponding esters were obtained in moderate to excellent yields, which decreased as the length of the carbon chain increased. An isotope-labeling experiment with ^{13}C-labeled methanol and cyclohexene showed that paraformaldehyde was the single source of CO in the reaction, because ^{13}C only appeared in the methyl ester portion of the product, not in the CO group.

In addition, palladium catalysis appeared to be suitable for this type of conversion [28], and the use of palladium acetate in combination with α,α′-bis(di-tertbutylphosphino)-o-xylene (**L2**), a more hindered bidentate diphosphine ligand, provided very high linear selectivities (up to >99:1). In general, linear selective (anti-Markovnikov) reactions are preferred, because linear carboxylate esters are favored by industry. Terminal aliphatic olefins yielded esters in very high yields and n-selectivity (e.g. entry 10, Table 36.1). Aliphatic internal olefins were converted into the corresponding linear esters with high yields, by double bond isomerization followed by alkoxycarbonylation (entries 11 and 12). This isomerization is a useful feature of this process, because it allows the use of mixtures of double bond isomers as starting materials to obtain a single product, which can be of interest for industrial applications because olefin mixtures are usually less expensive than pure isomers. Fatty acids could also be functionalized to yield linear diesters (entries 14 and 15). This palladium-catalyzed process required an acid additive (p-toluene sulfonic acid [PTSA] or methanesulfonic acid [MSA]) for high yields to be obtained. With other acids (e.g. H_2SO_4, acetic acid [AcOH], trifluoroacetic acid [TFA]), there was no product formation. In this reaction, ^{13}C isotope labeling experiments showed different results from those obtained with the Ru-catalyzed process. In this case, both ^{13}C-labeled paraformaldehyde and the independent use of $^{13}CH_3OH$ as the solvent produced a ^{13}C labeled methyl nonanoate, suggesting that both formaldehyde and methanol act as the carbonyl source in this process simultaneously.

Despite the good results obtained with the Pd/**L2** system, tri- and tetra-substituted olefins did not react under these conditions. However, an optimized system to convert these highly substituted olefins into the corresponding products could be realized [34]. The key to success was the development of the ligand **L3**, which incorporated pyridyl substituents. The success of this ligand is related to the ability of the amphoteric group to act as a proton shuttle to form the active palladium hydride, as well as for enabling an N-assisted alcoholysis step. The pyridyl substituents also helped to improve olefin isomerization. The desired methyl esters could be obtained with very good yields and selectivitites, as shown in Table 36.1, even when tetrasubstituted olefins were reacted (entries 19 and 20).

(2-Pyridylmethylene)cyclobutanes were subjected to rhodium-catalyzed carbonylation with gaseous CO in xylene [35]. CO insertion in the presence of the Rh(I) catalyst proceeded via pyridine-directed C–C oxidative addition. The overall result is cyclobutane to cyclopentanone ring expansion. Good to high yields of products (55–95%) and selectivities in the range 50:50–26:74 were obtained with the internal double bonded products predominating over those with an exocyclic double bond.

36.2.2 Carbonylation of Aryl Halides

The reaction of aryl halides with CO gas has been known for a while, but the utilization of paraformaldehyde as a CO surrogate has been much less explored [6]. N-(o-bromoaryl)amides were used as substrates for the preparation of substituted benzoxazinones heterocycles (36–86% yield) by a simple palladium-catalyzed carbonylation with PFA (Figure 36.4a) [36]. The presence of electron-withdrawing substituents on the amidoaryl moiety had a detrimental effect on the yield. 4,5-Bis(diphenylphosphino)-9,9-dimethylxanthene (Xantphos) was found to be the best ligand for this process. An experiment with ^{13}C-labeled paraformaldehyde yielded a 4-^{13}C-labeled benzoxazinone derivative that has many important pharmaceutical and biological applications.

In addition 2-bromobiphenyls could be transformed into a range of fluoren-9-one derivatives in good to high yields, via palladium catalysis (Figure 36.4b) [37]. The fluoren-9-one structure is found in many optoelectronically and biologically active compounds, as well as in key synthetic intermediates. In this process, there is not only aryl halide activation, but also the much more difficult to achieve direct C–H carbonylation by cleavage of a C–H bond. The presence of a base

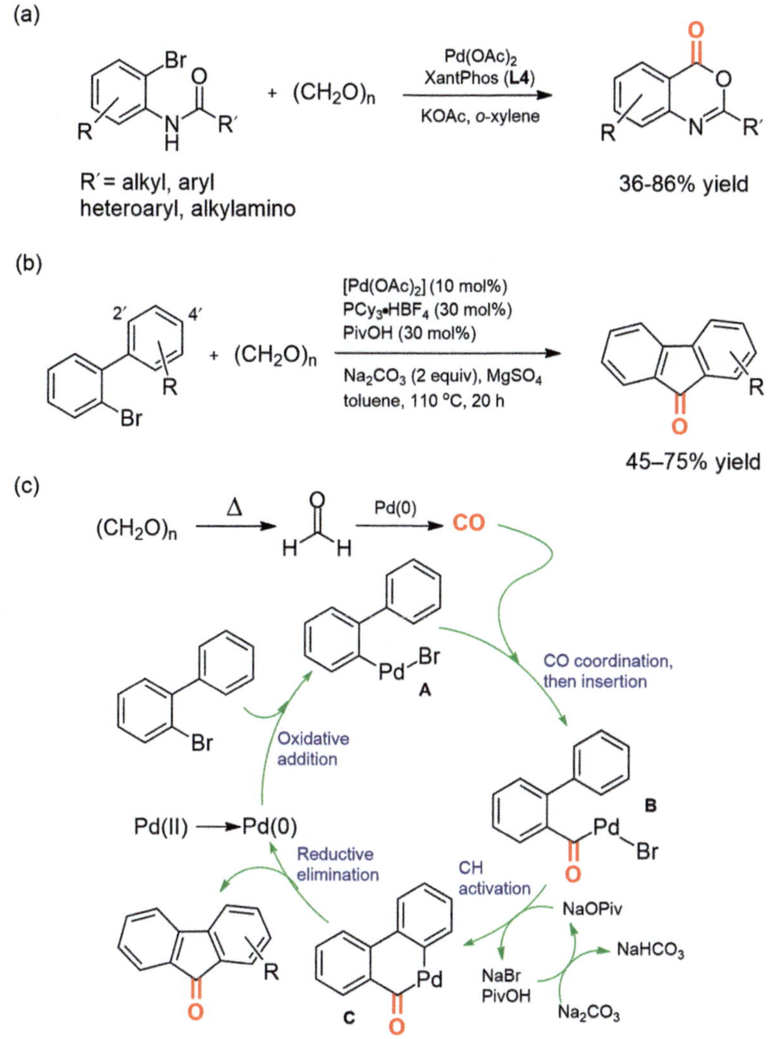

Figure 36.4 The carbonylation of aryl halides. (a) Synthesis of benzoxazinones by Pd-catalyzed carbonylation with paraformaldehyde (PFA) [36]. (b) The synthesis of fluorinones [37]. (c) The reaction mechanism proposed for the synthesis of fluorinones [37].

(Na_2CO_3) and a dehydrating agent ($MgSO_4$) are important in this process, the latter for suppressing the reductive hydrodebromination of the aryl halide. The mechanism postulated is shown in Figure 36.4c [37]. Oxidative addition of the aryl bromide to Pd(0) obtained from Pd(II) gives rise to complex **A**. CO formed by decarbonylation of formaldehyde under Pd(0) catalysis coordinates to **A**, then undergoes migratory insertion giving rise to acyl palladium species **B**. C–H bond cleavage can occur in one of two ways: either by electrophilic aromatic substitution or via concerted metal-lation-deprotonation to yield palladacycle **C**. Reductive elimination releases the product and Pd(0) for another catalytic cycle. The cyclocarbonylation of 2-bromo-1,1,2-triphenylethylene was possible under the same conditions, to afford 2,3-diphenylinden-1-one in 52% yield.

Rhodium catalysis can be used for the aryloxycarbonylation of aryl iodides with PFA and phenols [38]. However, in this case, good yields are only obtained with phenyl iodide, although several substituents may be present in the phenolic reaction partner. The corresponding reaction performed with free CO was less sensitive to phenol substitution and good yields could be obtained (Table 36.2). The reaction conditions had to be modified for the paraformaldehyde reaction. The addition of ethyl acetate as co-solvent improved the solubility of the in situ formed formaldehyde, although the catalyst activity dropped substantially if too much was added. Side products were

Table 36.2 The Rh-catalyzed aryloxycarbonylation of aryl iodides with PFA.[1]

Conditions A: CO (120 bar), [Rh(CO)$_2$(acac)], Xantphos, Et$_3$N
Conditions B: (CH$_2$O)$_n$, [Rh(nbd)Cl]$_2$, DPPP, EtOAc, Na$_2$CO$_3$, 110 °C

Entry	R^1	R^2	Conditions	Conv. [%]	Yield [%]
1	F	H	A	80	72
2			B	21	19
3	Cl	H	A	95	88
4			B	67	3
5	Me	H	A	97	88
6			B	43	35
7	H	F	A	>99	94
8			B	>99	86
9	H	iPr	A	98	95
10			B	65	61
	H	CF$_3$	A	98	93
			B	49	43

[1] Acac = acetylacetone; xantphos = 4,5-Bis(diphenylphosphino)-9,9-dimethylxanthene; nbd = 7-nitrobenzo-2-oxa-1,3-diazole; DPPP = 1,3-Bis(diphenylphosphino)propane.

were observed in all of the reactions to a small extent, resulting from reduction of the iodobenzene, which led to benzene formation or ester hydrolysis leading to carboxylic acid formation.

36.2.3 Cascade C–H Activation/carbonylation/cyclization Reactions and Related Processes: The Synthesis of Heterocycles

The synthesis of quinolines and their derivatives has been recently achieved via cascade processes involving carbonylation with paraformaldehyde using rhodium catalysis [23]. In this strategy, unprotected anilines are reacted with electron-deficient alkynes to yield initially C–C bonded products instead of the more common C–N bonded species often used in the synthesis of these compounds. For example, anilines usually undergo Michael-type addition with electron-deficient alkynes, leading to C–N bonded adducts (Figure 36.5a) [23]. A C–N bond is also formed in

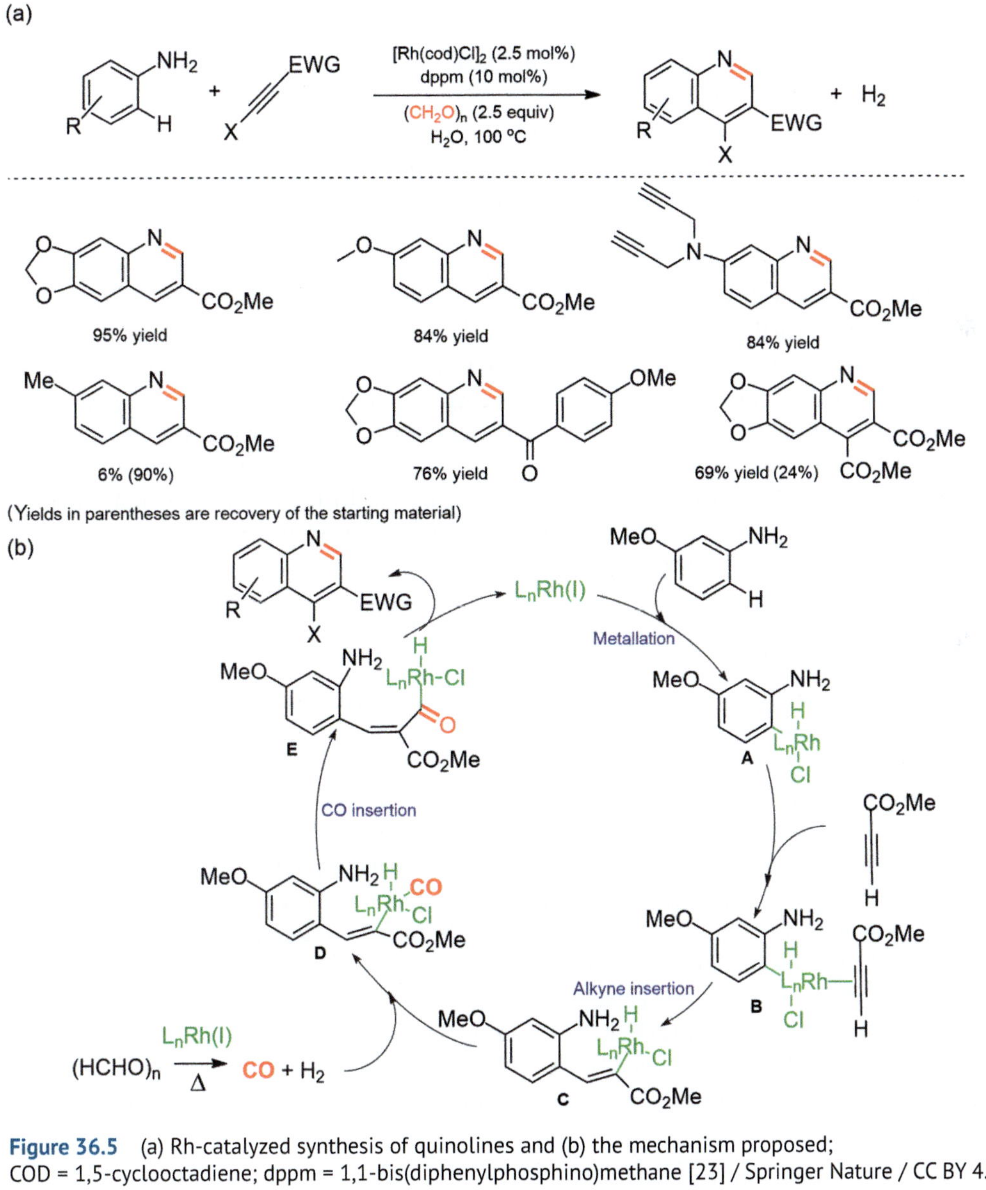

Figure 36.5 (a) Rh-catalyzed synthesis of quinolines and (b) the mechanism proposed; COD = 1,5-cyclooctadiene; dppm = 1,1-bis(diphenylphosphino)methane [23] / Springer Nature / CC BY 4.0.

TM-catalyzed hydroamination of alkynes [39]. This reversal in reactivity, in which there is TM-catalyzed C–H bond activation of unprotected or directing group-free amines, is very rare, and this appears to be the first example of non-directed catalytic C–H bond activation of unprotected primary anilines.

The synthesis of these carboannulated products could also be achieved using pre-prepared ortho-vinylanilines, PFA, and the same rhodium catalyst (Figure 36.5a). In that case, 1 mol% of catalyst was enough for high yields to be obtained (up to 86% of isolated compound). 3-Methylaniline showed poor reactivity (6% product yield after 24 hours), which was attributed to a lower arene electron density: less η^2-coordination and greater Lewis basic coordination to the metal center. The fact that water can be used as a solvent adds to the environmental friendliness of the procedure. Labelling experiments with ^{13}C-labelled PFA showed unambiguously that CO originates from the PFA. A mechanism was proposed for the reaction (Figure 36.5b) [23]. Non-directed C–H activation by the rhodium catalyst results in C–C coupling after coordination of the alkyne to rhodium activated species **A** and migratory insertion. **C**, the product of these initial steps, reacts with CO obtained from PFA. CO insertion can lead to intermediate E via D. Protodemetalation of intermediate E and intramolecular imination affords the desired quinoline.

The synthesis of quinolines from anilines has also been made possible by the reaction with ketones and PFA by Yi et al. using cobalt catalysis [40]. A C–H activation/carbonylation/cyclization reaction takes place to give products in good to very high yields (Figure 36.6). Exclusive site- or/and region-selectivity was also observed when meta-substituted anilines were reacted with unsymmetrical ketones, and a large range of substrates were compatible with the reaction conditions. As in the previous example, no directing groups were required. The only by-products in this atom and step-economical synthesis are H_2O and H_2. It was postulated that $AgNTf_2$ has the dual role of activating the catalyst (i.e. from $Cp*Co(CO)I_2$ to $Cp*Co(CO)NTf_2$) and the ketone enolate, and that C–H bond activation probably occurs via an internal electrophilic substitution (IES)-type mechanism. This rules out the probability of there being ring closure through an electrophilic aromatic substitution (EAS) pathway. Palladium-catalyzed carbonylative C–H activation of arenes in the presence of norbornene has been used to obtain various 5-(pyridin-2-yl)-hexahydro-7,10-methanophenanthridin-6(5H)-ones in moderate yields (Figure 36.6b) [24]. Although molybdenum hexacarbonyl was the main source of CO, PFA could also be utilized under similar conditions to provide a product in moderate yield (Figure 36.6).

36.2.4 Hydroformylation of Alkenes

The hydroformylation of alkenes with syn gas is widely known as an important industrial process [6, 41]. It normally requires a high pressure of H_2/CO and a metal catalyst, of which Rh, Co or Ru are commonly used. The interest to find a replacement that can work under pressure free conditions, namely paraformaldehyde, has also attracted much interest. PFA can sometimes provide complementary regioselectivity to the gaseous hydroformylation reactions [6].

Hydroformylation performed with PFA was already reported in 1982 [42], where 0.5 mol% $RhH_2(O_2COH)[P(i-Pr)_3]_2$ catalyzed the conversion of olefins to aldehydes, with the later being obtained in low to good yields (11–67%) (Table 36.3). With α,β-unsaturated alkenes, C=C bond reduction products predominated. It was found that β-oxygen-substituted olefins (i.e. allyl alcohol and methyl acrylate) were more reactive and yielded higher n/iso ratio of products than non-oxygenated alkenes when $RhH(CO)(PPh_2)_2/PPh_3$ was used as catalyst, although the highest yield obtained was only 23% [43]. Employing $[Rh(CO)_2(acac)]/2dppe$ as the catalytic system, obtained a modest *n/iso* ratio with 1-hexane (*n/iso* = 2, turnover number [TON] = 200), whereas a high

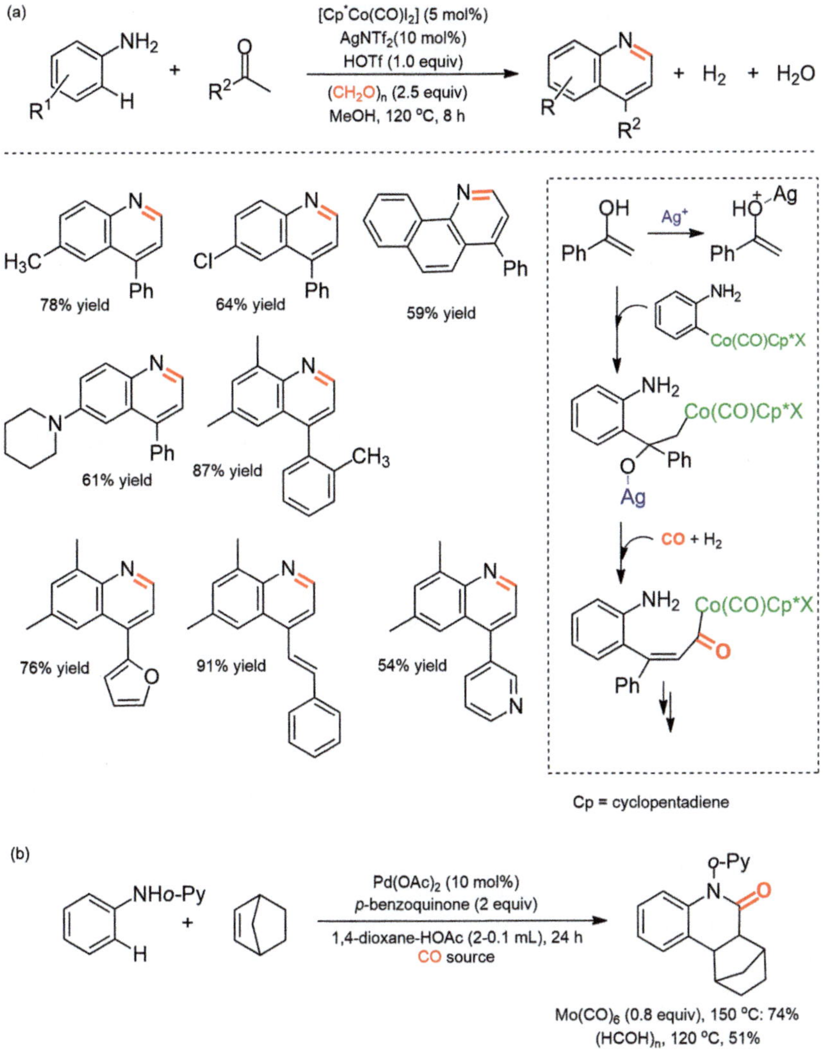

Figure 36.6 (a) Co-catalyzed synthesis of quinolines from anilines, ketones and PFA and the mechanism proposed [40]. (b) Pd-catalyzed carbonylative activation of arenes and norbornene [24].

iso/n ratio was observed for allyl alcohol (*iso/n* = 21, TON = 129) [44]. A highly efficient (up to 95%) and regioselective (l/b=up to 98/2) hydroformylation of 1-alkenes can be performed by using a combination of two ligands (2,2′-bis(diphenylphosphino)-1,1′-biphenyl [BIPHEP] and Nixantphos [4,6-bis(diphenylphosphino)phenoxazine]) for a Rh-catalyzed process that operated under mild conditions [45]. The yields obtained under similar conditions using a formalin solution instead of PFA were generally 20% higher, except for entry 3. The same process could be applied under microwave conditions in domino processes involving hydroformylation [46]. An asymmetric hydoformylation process using the same metal complex, PFA, and the chiral phosphine ligand (*S,S*)-Ph-BPE can be realized [47]. Good enantio- and regioselectivity was obtained in the hydroformylation of a range of styrenes, with branched products being obtained preferentially. However, the yields obtained with a formalin solution were >20% higher, except for entry 7.

Table 36.3 The hydroformylation of alkenes with paraformaldehyde (PFA) as a CO source.

Entry	R	Conditions	Selectivity	Yield of the major product [%][1]; %ee	Ref
1	n-C$_6$H$_{13}$	[Rh(COD)Cl]$_2$ (1 mol%), BIPHEP, Nixantphos (2 mol% each), toluene, 90 °C	97:3 (l:b)	72	[31]
2	BnO(CH$_2$)$_6$		96:4 (l:b)	47	[31]
3	PivO(CH$_2$)$_8$		98:2 (l:b)	73	[31]
4	Ph	[Rh(COD)Cl]$_2$ (0.5 mol%), [(S,S)-Ph-BPE] (1.2 mol%), tolune, 80 °C	93:7 (b:l)	92%; 90% ee	[33]
5	4-OMe-C$_6$H$_4$		96:4	63%; 95% ee	[33]
6	4-F-C$_6$H$_4$		96:4	69%; 95% ee	[33]
7	4-CF$_3$-C$_6$H$_4$		88:12	96%; 67% ee	[33]
8	Ph	[Rh(COD)Cl]$_2$ (0.1 mol%), DPPP (0.4 mol%), toluene, 120 °C, H$_2$ (10 bar)	3:5	83	[35]
9	4-OMe-C$_6$H$_4$		2:5	74	[35]
10	t-Bu		22:1	88	[35]
11	CO$_2$Bu		3.7:1	17	[35]

[1]Conversions are shown, not yields; BIPHEP=2,2′-bis(diphenylphosphino)-1,1′-biphenyl; Nixantphos = 4,6-bis(diphenylphosphino)phenoxazine; COD = 1,5-cyclooctadiene.

The same Rh catalyst with dppp as ligand in the hydroformylation of styrenes with PFA [48], required higher operating temperature of 120 °C and the use of gaseous H$_2$ to provide a positive pressure that allowed higher yields to be obtained. In this case, the linear products were favored. Aliphatic alkenes (e.g. cyclooctene, butylacrylate) provided a linear product in a highly regioselective and efficient manner.

The enantioselective hydroformylation of Z-alkenes using PFA as CO source [49] uses [Rh(acac)(CO)$_2$] (2 mol%) in combination with 3 mol% Ph-BPE. The aldehyde products could be obtained in good yields, as a 65:35 mixture of regioisomers, ees between 90–98%, from unsymmetrically substituted silbenes were reported. Using a dual catalyst system (i.e. catalysts A, [RhCl(cod)]$_2$ with (S,S)-Ph-BPE) was shown by separate experiments to be responsible for syngas formation and catalyst B, [Rh(acac)(CO)$_2$] with (S$_{ax}$,S,S)-BOBPHOS as a ligand produced the best regioselectivity (69:31) and a good ee (Figure 36.7). With this system, there was a slow release of syngas with catalyst A,

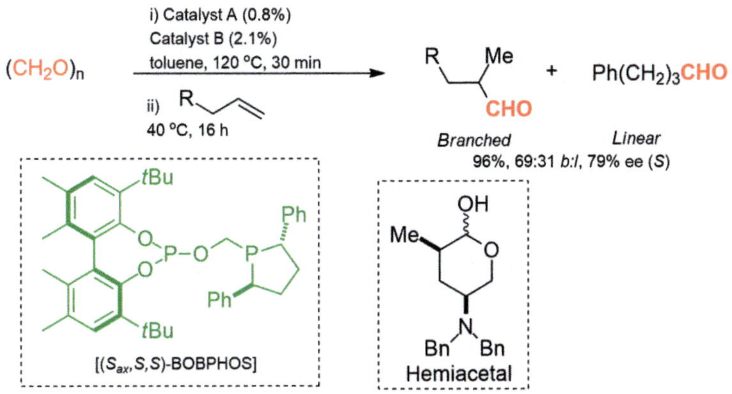

Figure 36.7 The Rh-catalyzed hydroformylation of allylbenzene and the synthesis of cyclic hemiacetal [49]. Catalyst A = [RhCl(cod)]$_2$ with (S,S)-Ph-BPE; catalyst B = [Rh(acac)(CO)$_2$] with (S$_{ax}$,S,S)-BOBPHOS.

whereas catalyst B promoted the asymmetric hydroformylation reaction. The same catalyst combination was used with allylbenzene, which is a challenging example because isomerisation to the internal conjugated alkene, prop-1-en-yl benzene, is thermodynamically favored. Most catalysts give either no selectivity or favor the linear aldehyde in hydroformylation of alkenes of this type. Catalysts with the BOBPHOS ligand (Figure 36.7) are exceptions and 75–80% selectivity can be observed. With the catalysts A and B in combination, a 69:31 b:l ratio could be obtained at high conversion with a high ee of 79% (Figure 36.7). The same strategy and reaction conditions were used to convert (S)-N,N-dibenzyl allylglycinol into a cyclic hemiacetal also with very high conversion (>99%), good regioselectivity (b : l, 82 : 14) and d.r (86 : 14). It was found that if the alkene and catalyst B were joined to the solution of A and paraformaldehyde after a certain time (30 minutes), presumably enough for CO and H$_2$ formation, the b:l ratio improved.

While the search for novel applications of PFA in synthesis continues, other studies have provided important clues that may help with future developments. One of them was the use of in situ Raman spectroscopy to follow paraformaldehyde depolymerization at different temperatures [50]. In these experiments the concentration of H$_2$C(OMe)$_2$, the product of the paraformaldehyde depolymerization in methanol, was monitored by the change in intensity of the ν_s(OC–O) vibration at 913 cm^{-1}. It was found that the presence of even small (catalytic) amounts (0.15 mol%) of the common carbonylation catalyst [Pd(dtbpx)(MeOH)$_2$]$^{2+}$, when present in solution, can slow down the depolymerization process. [Pd(dtbpx)]$^{2+}$ fragments formed from [Pd(dtbpx)(MeOH)$_2$]$^{2+}$ act as inhibitors in this reaction. They change the electron density at its reactive sites. The development of ways to control this inhibitory effect could help to develop better processes. More recent studies using in situ nuclear magnetic resonance (NMR) spectroscopy and quantum mechanical calculations were used to unravel the mechanism of the alkoxycarbonylation of alkenes with different surrogate molecules, including PFA [21]. Interestingly, no free CO could be found in these studies, although free CO resulting from the depolymerization of (CHO)$_n$ and its interaction with the metal catalysts, is postulated to be formed in many of these reactions. The authors found that, instead, the reaction proceeds via the C–H activation of in situ generated methyl formate, the key intermediate. The authors proposed a reaction mechanism based on the fact that Pd-hydride, Pd-formyl, and Pd-acyl species could be observed by NMR, the last one for the first time. Also of interest in the field of carbonylation are recent developments on the application of [^{13}C]PFA for the introduction of isotope labels [51].

36.2.5 N-formylation

Formamides represent a specific class of amides that are classically synthesized by using activated carboxylic acid derivatives or more recently using methanol under dehydrogenative or oxidative conditions in presence of amines [5, 52]. An early report for the utilization of PFA for the N-formylation of primary and secondary (a)cyclic amines uses [(Cp*IrI$_2$)$_2$] (1 mol%) as a catalyst in water at elevated temperature for 5–10 hours, giving the corresponding formamides in moderate to good yields (Figure 36.8; 13 examples, yields 41–95%) [53]. One of the latest findings demonstrated that PFA, as well as methanol, undergoes N-formylation with amines in presence of a bicatalytic system consisting of Cu(I)/TEMPO (2,2,6,6-tetramethylpiperidine 1-oxyl radical) catalysts that are well known for the selective oxidation of alcohols to aldehydes with oxygen as terminal oxidant (Figure 36.8) [5]. Interestingly, the reaction proceeds at low temperature and selectively to form N-formamides using PFA or methanol for the formylation. This observation is specific for these reagents and stays in vast contrast to the conversion other alcohols or aldehydes in presence of amines that form selectively imines using Cu/TEMPO [5]. Primary, secondary, (a)cyclic, aliphatic, and benzylic amines react to formamides in low to good yields (17 examples; yields 12–97%). In contrast, simple aniline is not suitable for the conversion to formamides because it undergoes homo-coupling under these conditions to form azobenzene. However, other aromatic amines can be N-formylated. Furthermore, in another study it has been demonstrated that catalytic

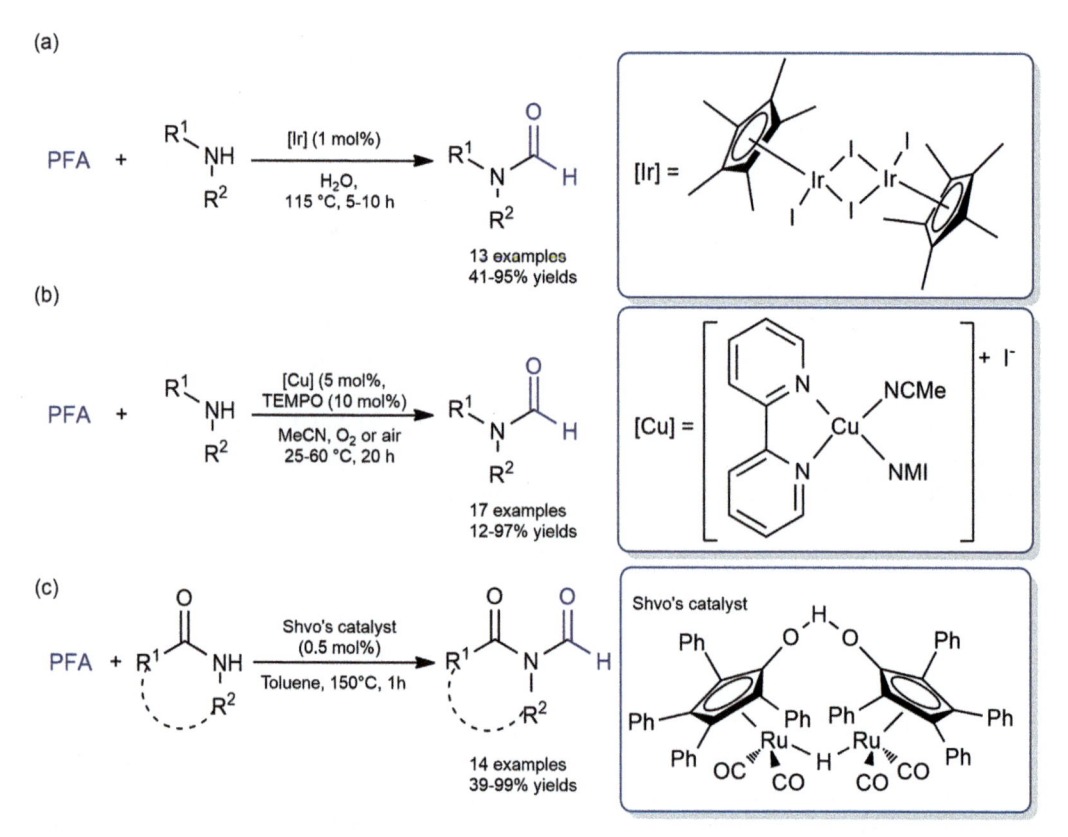

Figure 36.8 a) N-formylation of amines with paraformaldehyde (PFA) with an iridium catalyst. b) N-formylation of amines with PFA with Cu/TEMPO as catalysts. c) N-formylation of lactams with Shvo's catalysts yielding N-formylimides under hydrogen evolution.

N-formylation of lactams (14 examples; yields 39–99%) is possible with PFA by the utilization of Shvo's catalyst at elevated temperature (150 °C) within 1h (Figure 36.8) [54]. Various lactams were smoothly converted into *N*-formylimides without the use of a stoichiometric activating reagent and hydrogen as the only byproduct.

36.3 Methylation and Related Reactions

For a long time, formalin and PFA has been applied to insert methyl (-CH$_3$), methylene (-CH$_2$-) and hydroxymethyl (-CH$_2$OH) moieties into various substrates (Figure 36.9) [6, 27]. Thus, aldol condensation, Baeyer diarylmethane synthesis, Blanc and Mannich reactions, Petasis, Prins-Kriewitz, Tiffeneau, and calixarenes syntheses were widely used to introduce C1 terminal or bridging block into numerous scaffolds (Figure 36.9) [27]. For example, in the classical Mannich

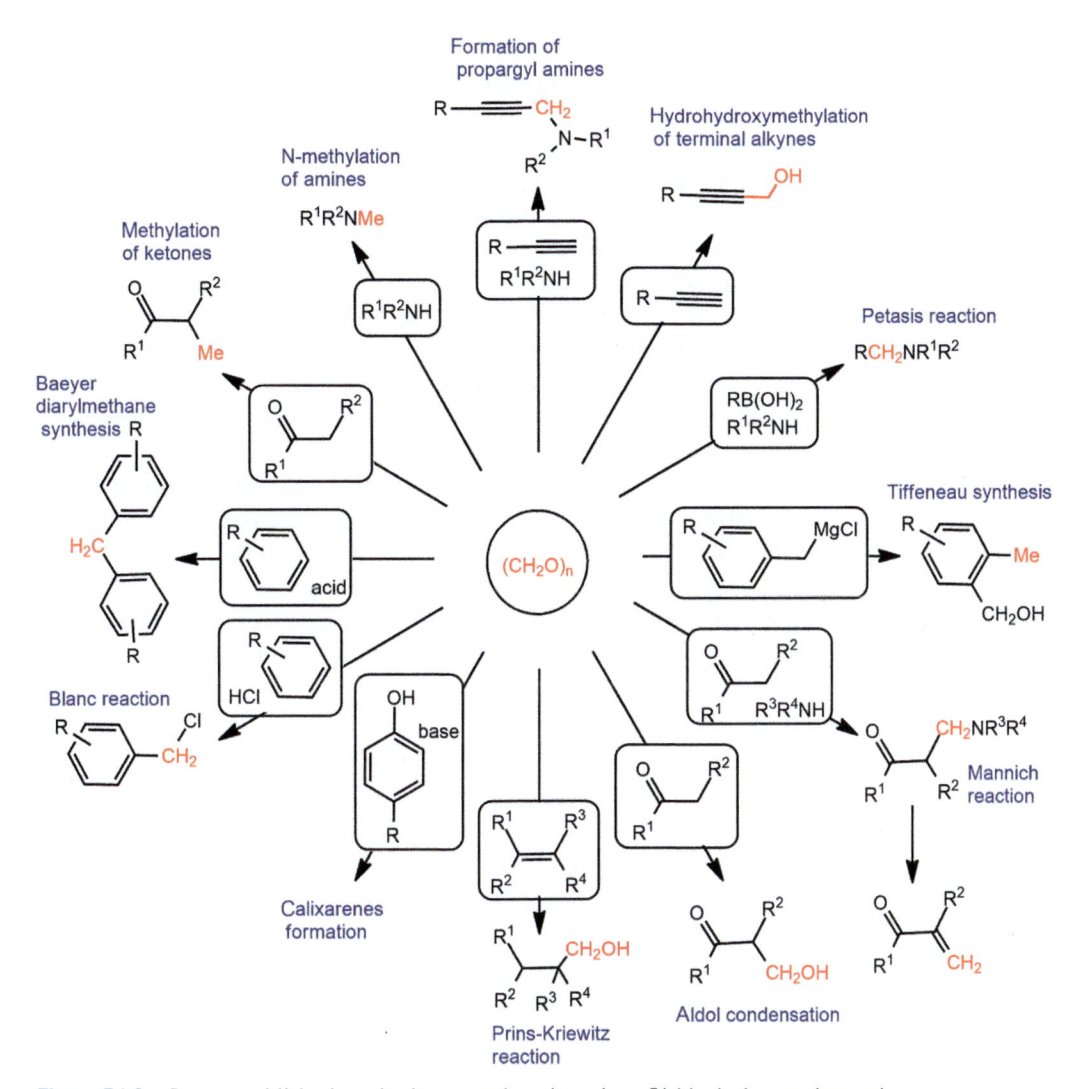

Figure 36.9 Some established synthetic protocols to introduce C1 blocks into various substrates.

reactions of formalin, an amine and an enolizable carbonyl compound propargyl amines can be obtained (Figure 36.9) [55]. To expand the substrate scope and improve yields, many metal catalyst-based systems have been developed [56–59]. The hydrohydroxymethylation of terminal alkynes is an important reaction for the synthesis of propargylic alcohols (Figure 36.9). Thus, CuI-catalyzed alkynylation of formaldehyde with formation of primary propargyl alcohols has been developed [60]. Compared to the former procedures, gaseous formaldehyde and moisture sensitive Grignard reagents were omitted, while the yields were higher and the reaction proceeded at a faster rate. As early as 1979, the first example of a copper bromide-catalyzed synthesis of terminal allenes from alkynes, PFA and diisopropylamine was reported [61, 62], and this direction has been greatly developed since then [63, 64].

PFA is also known as a methylating agent in a Rh-catalyzed methylation of ketones with CO/formaldehyde/water system (Figure 36.9) [65]. Under similar reaction conditions, the N-methylation of amines can be also performed. For example, morpholine can be methylated with 92% yield using [Rh] as a catalyst [66], while under high CO pressure PFA was employed as the formylation reagent for amines in the presence of a cobalt complex [67]. In this case, the imine was first formed and then reduced by hydrogen from the water-gas shift in the presence of a cobalt catalyst. In a more recent study, it has been demonstrated that a redox self-sufficient reductive amination occurs employing PFA both as a carbon source and reducing agent [2]. The methylation of primary and secondary amines is possible in the presence of 0.5 mol% of various ruthenium p-cymene dimers, giving the best results with the [(Ru(p-cymene)I$_2$)$_2$], and with excellent yields for many substrates after only two hours (Figure 36.10). In the case of primary amines, the dimethylated products have been formed. In addition, biphasic reductive N-methylation and catalyst recycling has been successfully demonstrated. The reductive methylation of amines is also related to transfer-hydrogenation, respectively, to the concept of the borrowing hydrogen principle, because the hydrogen is generated from one of the substrates which are incorporated into the target product [68–70].

In the traditional Baeyer diarylmethane synthesis (Figure 36.9), strong acids are usually needed in large amounts. However, in the presence of an [In] catalyst, the reactions of electron-rich arenes proceeded smoothly with trioxane to give the corresponding diarylmethanes in good yields [71]. And a selective palladium-catalyzed cross-coupling reaction between arylalkenes and aminals, which were generated in situ from amines and paraformaldehyde (Figure 36.11), can be realized as well [72].

PFA can be used for regioselective metal-catalyzed reductive coupling with 1,3-dienes, allenes, and alkynes [6]. The products of these reactions cannot be formed in a selective manner by

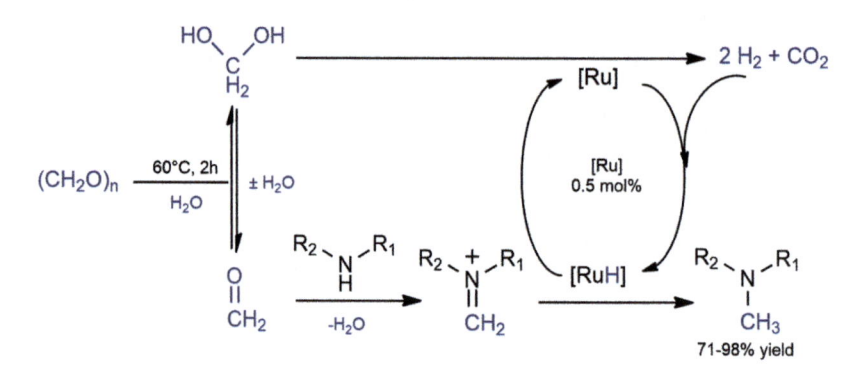

Figure 36.10 Redox self-sufficient reductive N-methylation.

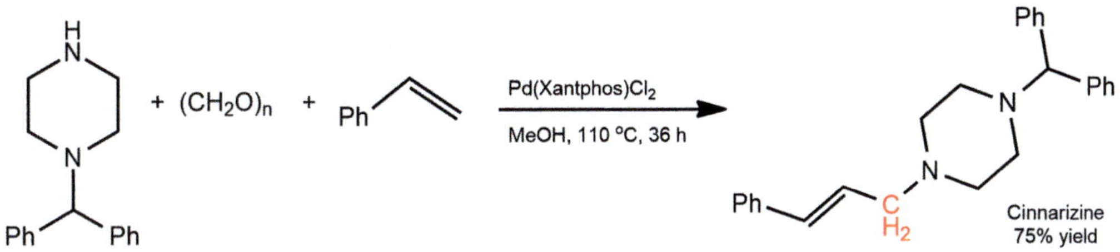

Figure 36.11 Pd-catalyzed synthesis of cinnarizine via a one-pot vinylation of amines via aminals to allylic amines.

conventional hydroformylation. The complete inversion of the regioselectivity is currently possible in the reactions of dienes and of alkynes. Examples are shown in Figure 36.12. Coupling of PFA to 2-substituted dienes can occur at the C1, C2, C3 or even the C4 carbon atoms in a regioselective manner, by a variation in catalysts and reaction conditions (Figure 36.12a). Nickel and ruthenium-based catalysts have been found to work well for C1 coupling, a process involving Ni(0) catalysis [73]. Formaldehyde plays the role of carbonyl electrophile and terminal reductant. Continuing studies on the reactions of 2-substituted dienes with higher aldehydes using silanes, boranes, and organozinc reagents as the terminal reductants [74], isolated the metallacycles involved and showed that metallacycle formation was reversible, whereas the formation of the C1 adducts, by oxidative coupling, was kinetically preferred [75]. Such oxidative pathways also appear to take place in the corresponding diene–PFA reductive couplings, with the observed C1 regioselectivity suggesting a kinetically controlled process (Figure 36.12, Eq 1.). Accordingly, it is expectable that 2-substituents on the diene capable of weakening the newly formed C–C bond of the C1 adduct might enable equilibration between π-allyl A and π-allyl B. This would lead to the formation of C4 adducts, e.g. eqs 1 and 4. Trialkylsilyl or trialkylstannyl substituents at position 2, in which hyperconjugation of the C–Si or C–Sn σ-bond with the σ-antibonding orbital of the newly formed C–C bond at the C1 position can occur, would be candidates for this type of coupling, which was indeed observed with Ni(0) catalysis. A neutral ruthenium catalyst allows the synthesis of C3 coupled products [76], whereas cationic Ru catalysts hydrometalate reversibly at all diene positions, providing access to the C2 coupling products bearing all-carbon quaternary centers (Eqs 2 and 3, Figure 36.12b) [77]. In the ruthenium-catalyzed reactions to form the C3 and C2 coupling products, isopropanol and PFA both contribute as terminal reductants. Alkyne hydroxymethylation has been achieved with PFA and formic acid as the terminal reductant [78]. Primary allylic alcohols were obtained with good to complete levels of regioselectivity (Figure 36.12c). The use of nickel-catalysis under conditions of reductive coupling provides the isomeric allylic alcohols with good to complete levels of regioselectivity. The initial products are formate esters, which are hydrolyzed during isolation. 1,1-Disubstituted allenes and PFA can react under conditions of reductive coupling with ruthenium catalysis and isopropanol as terminal reductant, to provide the primary neopentyl alcohols with complete levels of branch regioselectivity [79]. When CF_3-substituted allenes were employed, alcohols containing CF_3-bearing all-carbon quaternary stereocenters could be obtained (Figure 36.12d) [80]. These transformations provide a greener alternative to conventional methods employed in alkyne functionalization that require metallic, pyrophoric, or highly mass intensive terminal reductants (e.g. ZnR_2, BEt_3, $HSiR_3$) [6].

Cycloaminomethylation of pyrrole or indole involving their CH- and NH-reactive sites can be accomplished with a 1:2 mixture of primary alkyl(phenyl)amine and PFA in the presence of 5 mol% zirconium or nickel catalysts (Figure 36.13) [81]. Thus, performing the reaction in the

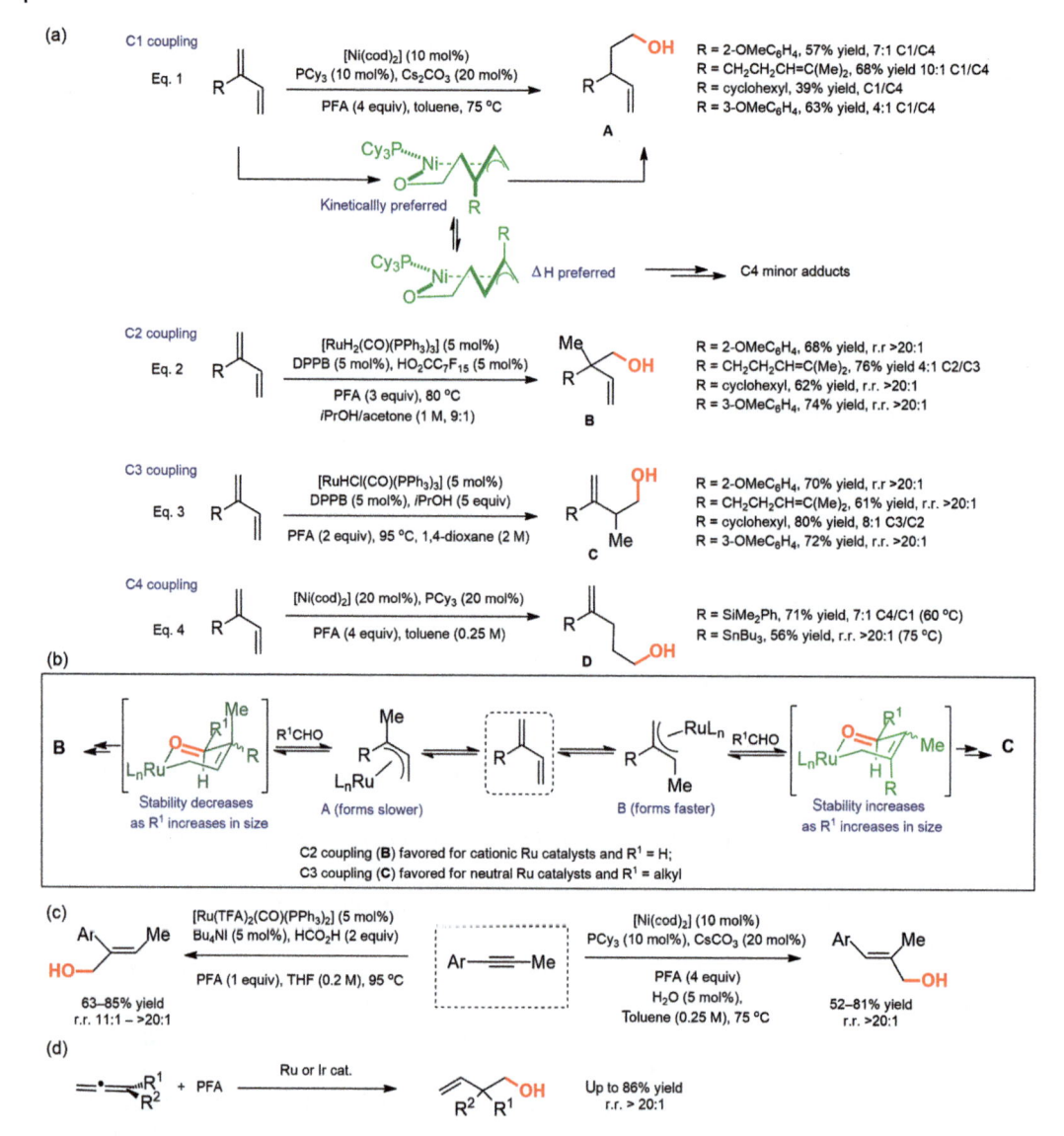

Figure 36.12 (a) Hydroxymethylation of dienes. (b) C2 vs. C3 regioselectivity in the ruthenium-catalyzed reductive coupling of 2-substituted dienes with paraformaldehyde (PFA). (c) Hydroxymethylation of alkynes; (d) Hydroxymethylation of allenes.

presence of $ZrOCl_2 \cdot 8H_2O$ catalyst resulted in a mixture of aminomethylated products with up to 86% combined yield (Figure 36.13a), whereas cycloaminomethylation of pyrrole with $NiCl_2 \cdot 6H_2O$ or $[Ni(Py)_4Cl_2] \cdot 0.76H_2O$ catalysts proceeds at positions 2 and 5 of the pyrrole ring, leading to the formation of piperazinopyrroles reaching maximum of 48% yield (Figure 36.13b). Cycloaminomethylation of indole in the presence of $ZrOCl_2 \cdot 8H_2O$ catalyst was accomplished effectively at positions 1 and 3, giving rise to 3-alkyl(phenyl)-3,4-dihydro-2H-1,5-(metheno)[1,3]benzodiazepines (Figure 36.13c).

Addressing the essential need in a sustainable strategy for the synthesis of quinazoline compounds from aniline derivatives and PFA, a recyclable GO@Fe$_3$O$_4$@APTES@FeL (GOTESFe)

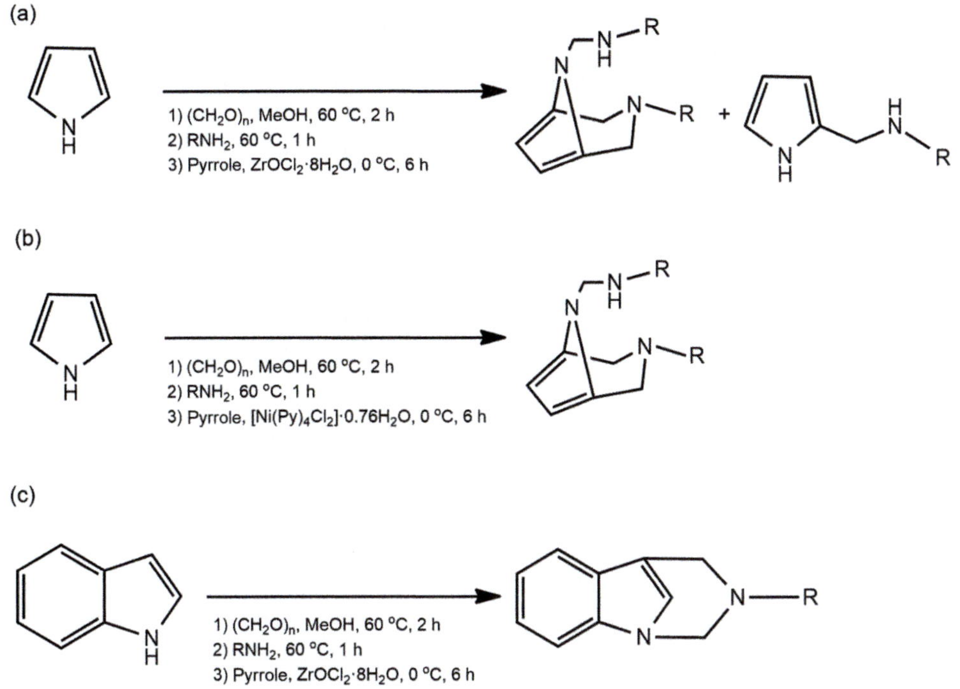

Figure 36.13 Cycloaminomethylation of pyrrole (a, b) and indole (c) (R = n-Pr, n-Bu, t-Bu, Hex, Ph).

catalytic composite was developed by grafting the Fe(III)–Schiff base complex (FeL) (L = 2,2-dimethylpropane-1,3-diyl)bis(azanylylidene)bis-(methanylylidene)bis(2,4-Cl-phenol) onto 3-aminopropyltrie-thoxysilane (APTES)-coated Fe_3O_4 nanoparticle-decorated graphene oxide (GO). The catalytic composite was exploited for the synthesis of dihydroquinazoline-based compounds with acetonitrile/tetrahydrofuran (THF) as the solvent and paraformaldehyde as the carbon source under mild and acid free conditions (Figure 36.14) [82]. The magnetic property of the GOTESFe composite enables the catalyst to be recycled and reused up to five times without a visible loss in catalytic activity.

The synthetically useful catalytic preparation of primary propargylic alcohols from terminal alkynes and formaldehyde can be realized with a nanosilver catalyst and PFA (Figure 36.15) [83]. To devise such catalytic particles, titanium oxide, and nanosilver were dispersed on polymethylhydrosiloxane-based cross-linked semi-interpenetrating networks (PMHS-SIPNs) to form supported nanoAg@TiO_2@PMHSIPN composite. Although the aliphatic and silicon-linked alkynes were not suitable substrates in this reaction, this procedure deserves consideration for the preparation of aromatic propargylic alcohols due to its practicality and simplicity of operation.

The modification of the the Mannich condensation of α,ω-diacetylenes with secondary diamines and aldehydes has been found as an efficient approach to diazaalkatetraynes and tetraazatetraacetylenic macrocycles [84]. In this study, the reaction of α,ω-diacetylenes with PFA and acyclic and cyclic diamines (2 : 2 : 1) in the presence of CuCl catalyst (10 mol.%) in toluene at 100 °C within eight hours produces diazatetraacetylenic compounds in 29–68% yields (Figure 36.16a). The catalytic cyclocondensation of diaza tetraacetylenes with formaldehyde and piperazine (1 : 2 : 1 molar ratio) in dioxane in the presence of CuCl affords tetraazatetraacetylenic macrocycles in 34–39% yields (Figure 36.16b).

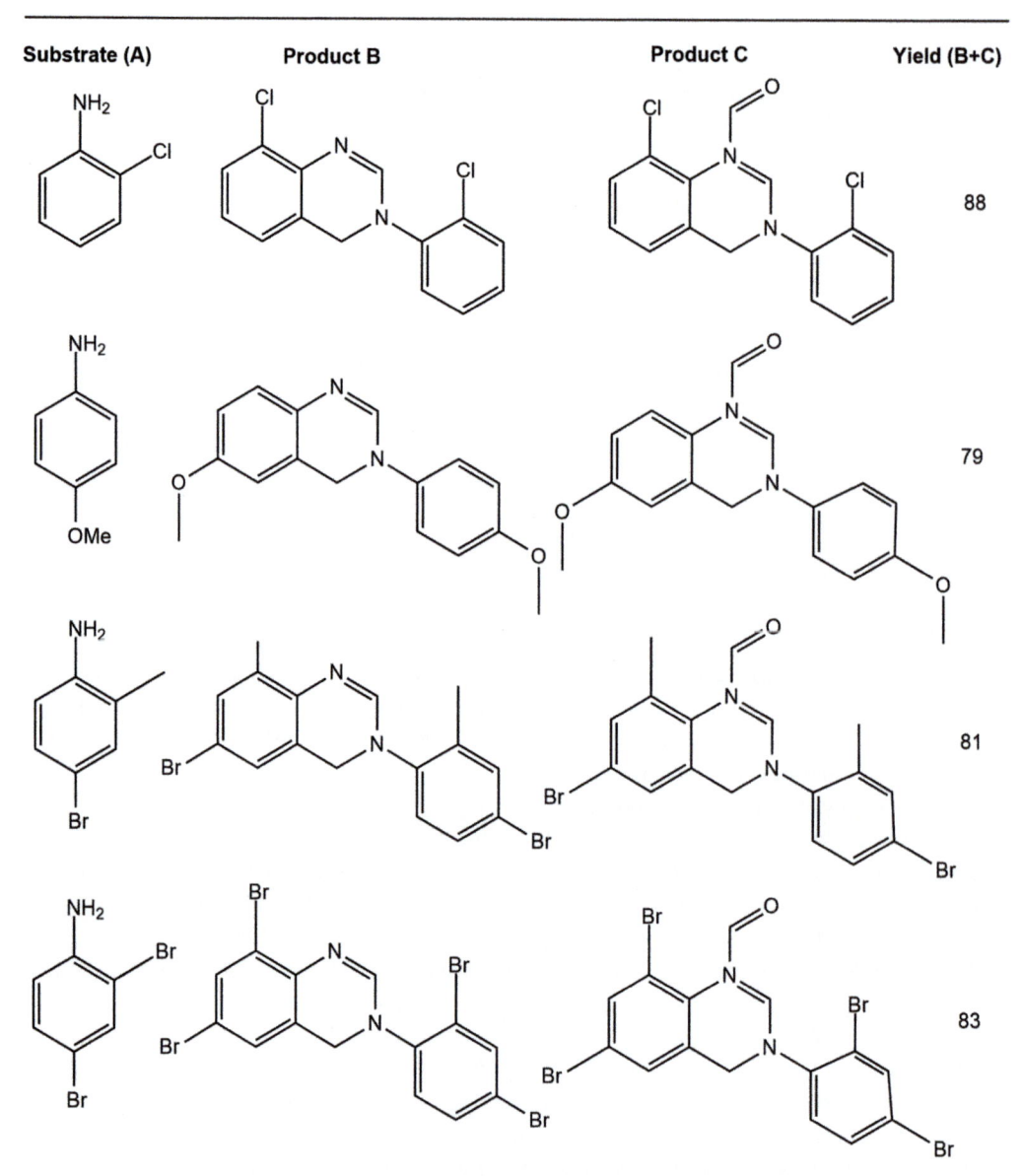

Figure 36.14 Synthesis of substituted dihydroquinazolines from aniline derivatives and paraformaldehyde (PFA) with GO@Fe$_3$O$_4$@APTES@FeL (GOTESFe) composite as a catalyst.

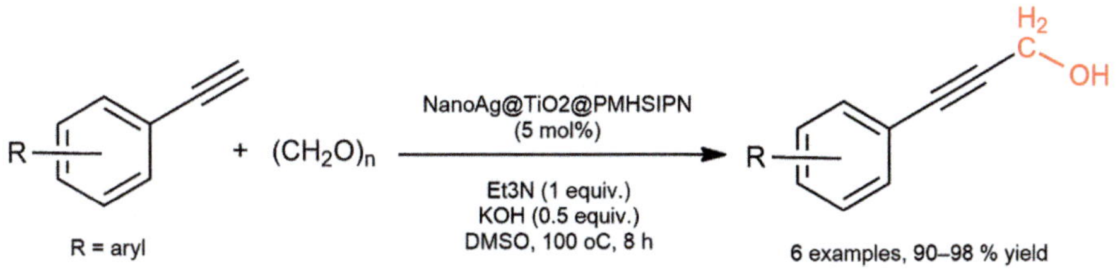

Figure 36.15 Synthesis of primary propargylic alcohols from terminal alkyne and formaldehyde catalyzed with nanosilver Ag@TiO$_2$@PMHSIPN composite.

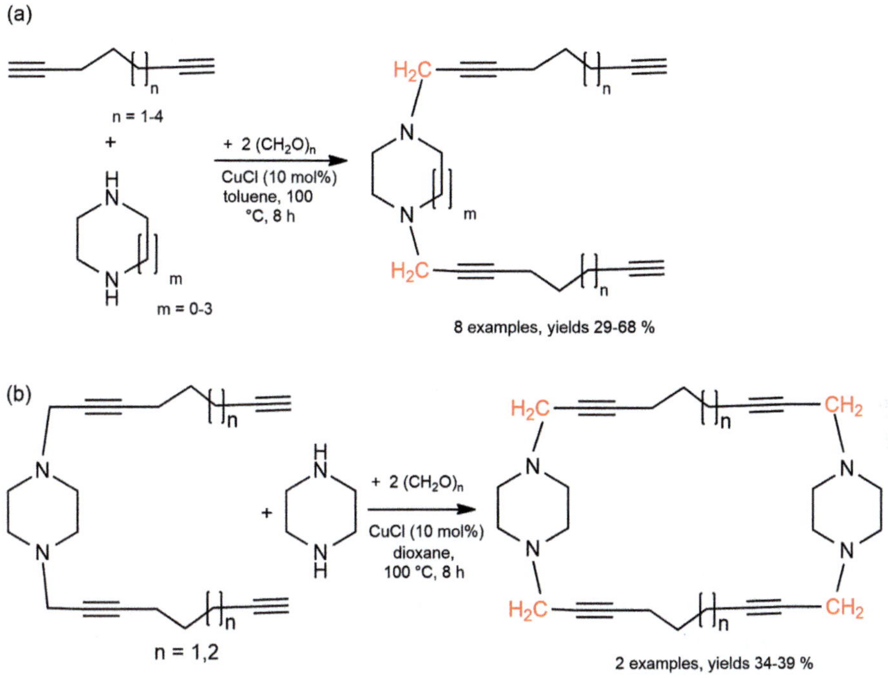

Figure 36.16 Catalytic aminomethylation of diacetylenes with secondary diamines and paraformaldehyde (PFA) to give diazaalkatetraynes (a) and tetraazatetraacetylenic macrocycles (b).

Medicinally relevant and biologically active chiral tertiary alcohols can be prepared by asymmetric decarboxylative cycloaddition of vinylethylene carbonates (VECs) with formaldehyde under the cooperative catalysis of achiral palladium complex and chiral squaramide [85]. With combination of a palladium complex generated in situ from Pd$_2$(dba)$_3$·CHCl$_3$ (2.5 mol%), an achiral phosphine ligand (10 mol%) and chiral squaramide **OC** (25 mol%; OC = organocatalyst) as cooperative catalysts, the reaction of VECs with PFA (10 equiv.) proceeded smoothly to give desired tertiary alcohol derivatives (acetal protected diols) in good yields (51%–65%) with moderate enantioselectivities (62%–79% ee) (Figure 36.17). The reaction conditions are also suitable for the reaction of VECs with electronic deficient arylaldehydes to afford desired products in high yields with good enantioselectivities, although the catalytic system is less effective for the control of the selectivities.

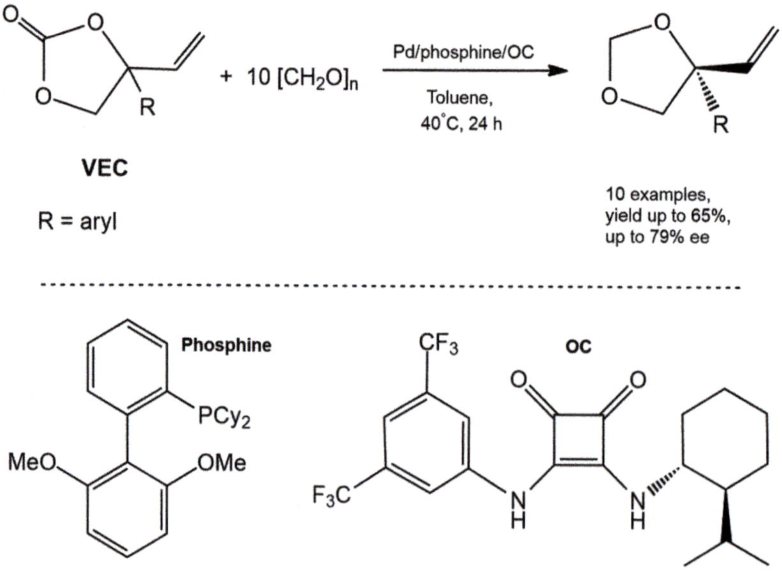

Figure 36.17 Asymmetric decarboxylative cycloaddition of vinylethylene carbonates (VECs) with formaldehyde for the construction of acetal protected diols with a chiral tertiary alcohol.

The azaindole ring system is one of the most valuable heterocyclic moieties; however, only limited methods have been developed for functionalization of 7-azaindoles. Recently a facile synthesis of hydroxymethylated N-aryl-azaindoles was developed via a Ru(II)-catalyzed regioselective C-H addition to PFA (Figure 36.18) [22]. The reaction is compatible with air, shows high functional group tolerance and regioselectivity, and is an environmentally benign method without any undesired byproduct. The C-H hydroxymethylation proceeded efficiently over a broad range of substrates irrespective of their electronic nature. A variety of substrates containing electron-donating and electron-withdrawing groups at the *meta-* or *para-* position of the phenyl ring were proved to be productive substrates for this coupling reaction, affording the corresponding products in moderate to good yields. Synthetically important functional groups, such as halogen, methoxy, ester, and ketone groups were well tolerated, enabling further functionalization. Moreover, for substrate bearing *meta*-Me group, the C-H hydroxymethylation exhibited excellent regioselectivity in favor of the sterically more accessible C-H bond. Several aryl and aliphatic aldehydes were also examined by using the current catalytic system.

A Rh(II)-catalyzed multicomponent reaction is able to trap the in situ generated α-imino enols, and rapidly affords α-amino-β-indole ketones [86]. In particular, 1-sulfonyl-1,2,3-triazoles can be transformed via α-imino metal carbene species by vinylimine ions using C(2)-substituted indoles and PFA as precursors in the presence of a Rh(II) catalyst (Figure 36.19). A broad range of triazoles provided the corresponding products in moderate to good yields. 1-Mesyl–substituted triazoles bearing a bromide or other electron-rich groups at the C-4′ of the phenyl ring produced good yields of the corresponding products. 1-Mesyl–substituted triazole featuring a heterocyclic group also produced a good yield of the desired product. Bulky triazoles were found to be tolerated despite the steric effect and can equally furnish the corresponding products. This transformation was also applicable to a range of C(2)-substituted indoles. An evaluation of the substituents on the indolic nitrogen revealed that N-H and N-Bn indoles were both compatible with the reaction conditions to

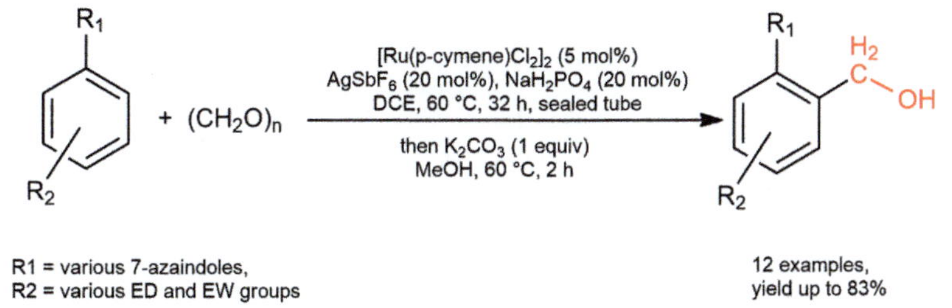

Figure 36.18 Ru(II)-catalyzed C–H hydroxymethylation of 7-azaindoles.

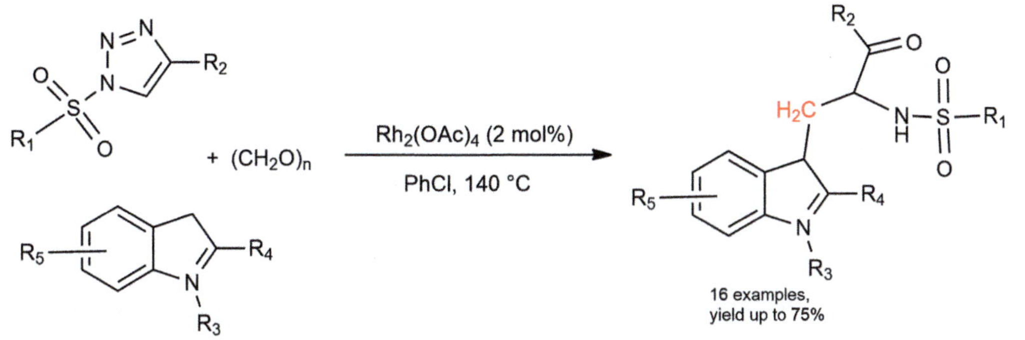

R1 = various 7-azaindoles,
R2 = various ED and EW groups

12 examples, yield up to 83%

Reaction conditions: [Ru(p-cymene)Cl$_2$]$_2$ (5 mol%), AgSbF$_6$ (20 mol%), NaH$_2$PO$_4$ (20 mol%), DCE, 60 °C, 32 h, sealed tube; then K$_2$CO$_3$ (1 equiv), MeOH, 60 °C, 2 h.

Figure 36.19 Synthesis of α-amino-β-indole ketones from 1-sulfonyl-1,2,3-triazoles, C(2)-substituted indoles and paraformaldehyde in the presence of a rhodium(II) catalyst.

16 examples, yield up to 75%

Reaction conditions: Rh$_2$(OAc)$_4$ (2 mol%), PhCl, 140 °C.

provide the products in 55 and 62% yield, respectively. The resulting products and density functional theory (DFT) calculations indicated that the enolic carbon had a stronger nucleophilicity than the traditional enamic carbon in the trapping process.

Dearomatization reactions have garnered considerable interest because they represent an useful way to prepare complex three-dimensional molecular scaffolds from readily available aromatic compounds. In particular, dearomative methods that form both C–H and C–C bonds represent an important class of reaction that are particularly versatile and efficient. In this view, the recently developed single point activation of pyridines, using an electron-deficient benzyl group, facilitates the ruthenium-catalyzed dearomative functionalization of a range of electronically diverse pyridine derivatives [87]. The key aspect involves the use of a ruthenium catalyst in conjunction with a highly electron-deficient 2,4-bistrifluoromethylbenzyl activating group (Figure 36.20). This tailored activating group can be readily removed to furnish the free amine. Ranging from 4-arylpyridines to the electron-rich 4-methoxypyridine, the hydroxymethylation delivers hydroxymethylated piperidines in good yields. A noteworthy feature of this work is that paraformaldehyde acts as both a hydride donor and an electrophile in the reaction, enabling the use of cheap and readily available feedstock chemicals. Removal of the activating group can be achieved readily, furnishing the free NH compound in only two steps. Mechanistic work has shown that the metal catalyst oxidizes formaldehyde in methanol to methyl formate, forming a metal hydride in the process. The synthetic utility of the method was illustrated in the synthesis of paroxetine.

Since the last century, the challenge has remained to produce the valuable ethylene glycol (EG) directly from the C1 building block formaldehyde in a single step. In the classical systems, the reaction conditions were very harsh, often with pressures above 400 bar. However, under milder

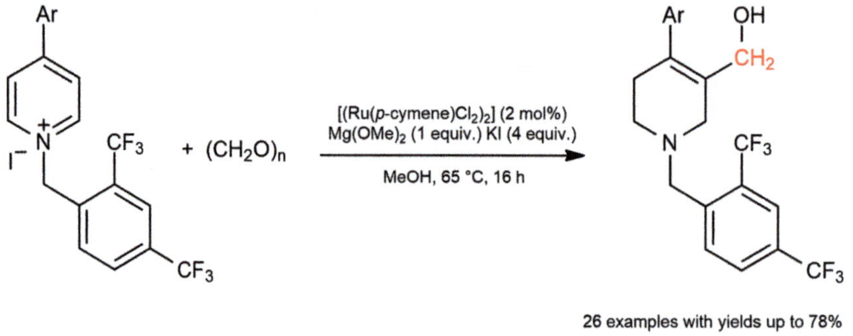

Figure 36.20 Ruthenium-catalyzed dearomative hydroxymethylation of pyridines toward hydroxymethylated piperidines.

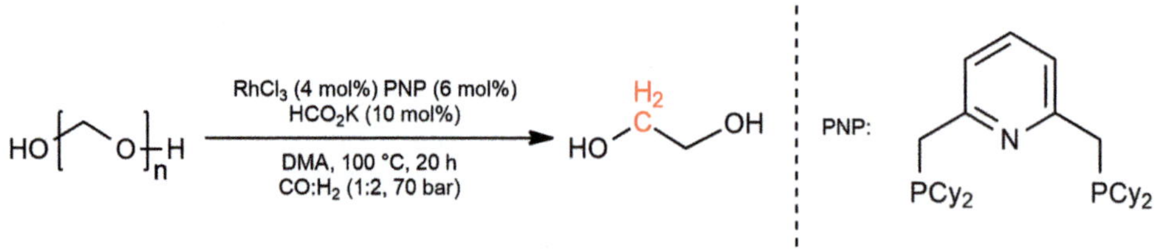

Figure 36.21 [Rh-(pincer ligand)] catalyzed carbonylation of paraformaldehyde (PFA) toward ethylene glycol.

conditions, the selectivity was on the side of glycol aldehyde (GA) and the hydrogenation product methanol. However, Rh-catalyzed method for the carbonylation of PFA, which allows the direct one pot synthesis of EG from PFA at relatively mild conditions (70 bar, 100 °C) with yields up to 40% has been described (Figure 36.21) [88]. Application of a symmetrical pyridine-based PNP ligand was essential.

Moreover, a straightforward and convenient ruthenium(II)-catalyzed synthesis of 3-unsubstituted phthalides from aryl amides and PFA via C–H activation can be realized (Figure 36.22) [89]. The reaction proceeds through tandem ortho-hydroxymethylation of aryl amide and subsequent intramolecular lactonization.

An one-pot copper(I)-catalyzed synthesis of multi-substituted 2-azolylimidazole derivatives is possible with N-propargylcarbodiimides, azoles, PFA and secondary amines through a domino addition/A3 coupling/ cyclization process (Figure 36.23) [90]. The N-propargyl-N'-arylcarbodiimides bearing different substituents underwent the reaction smoothly. Reactions of the N-propargylcarbodiimides containing an electron-donating or electron-withdrawing phenyl group proceeded well to afford the corresponding 2-imidazolylimidazoles in moderate to good yields. The N-propargylcarbodiimides with a naphthyl group could also successfully participate in the one-pot reaction to deliver the desired products. The substrates containing a pendant o-tolyl, o-iodophenyl or α-naphthyl group were found to have some influence on the outcome of the reaction, probably due to the steric hindrance. The reaction of an N-propargyl-N'-alkylcarbodiimide gave only a 37% yield of the desired 2-imidazolylimidazole. Then the one-pot reactions of the substrates bearing different branched chains (such as n-hexyl, benzyl, β-phenylethyl, and c-hexyl chains) at the α-position of the propargyl group were also tested, and moderate to good yields of the desired multi-substituted 2-imidazolylimidazole derivatives were isolated. A variety of secondary amines including linear amines [such as $NH(iPr)_2$ and $NH(nBu)_2$] and cyclic amines

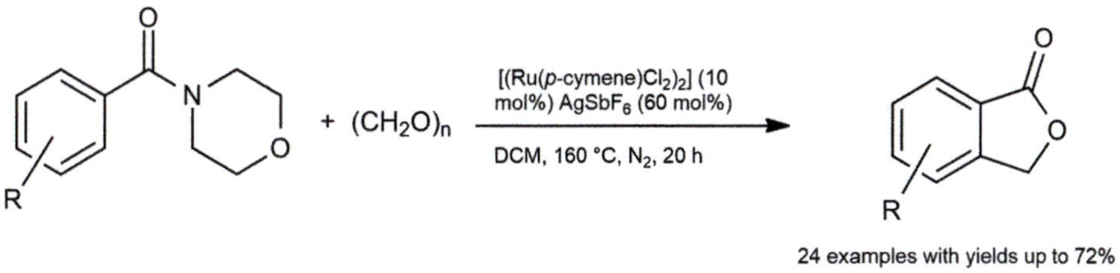

Figure 36.22 Ru (II)-catalyzed synthesis of 3-unsubstituted phthalides from aryl amides and paraformaldehyde (PFA).

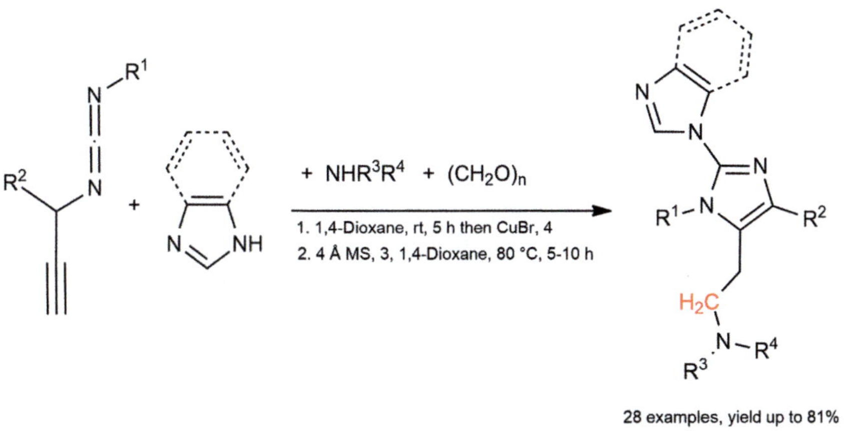

Figure 36.23 Synthesis of multi-substituted 2-azolylimidazole derivatives from N-propargylcarbodiimides, azoles, paraformaldehyde (PFA), and secondary amines.

(such as piperidine, morpholine, N-methylpiperazine, and 1,2,3,4-tetrahydroiso-quinoline) were shown to be good partners for this one-pot reaction. The reactions with substituted imidazoles and benzimidazoles also furnished the desired 2-azolylimidazole derivatives successfully. However, the use of other azoles such as pyrrole, indole, 1,2,3-triazole, and 1,2,4-triazole failed to afford the desired 2-azolylimidazoles, probably because of their weaker nucleophilicity.

The carbon-carbon bond-forming reaction between two carbonyl-containing substances to yield an α-hydroxyketone or derivative thereof is of great synthetic value. Product dehydration, self-condensation, or polymerization are potential problems. The aldol reaction can be catalyzed either by acid or base. In the Mukaiyama-aldol version a silyl enol ether (or a chemical equivalent) is used. Formaldehyde is the smallest aldehyde that can be employed in the aldol reaction. An obvious advantage is that only the other reaction partner can be enolized and act as the nucleophile, thus limiting the number of possible products that can be obtained. Its very high reactivity can, however, be an obstacle, that the use of PFA can help. In this case the issue becomes how to generate the reactive monomer efficiently. Both metal- and organocatalyzed processes have been developed with the aid of PFA, including asymmetric reactions [91].

PFA is also an efficient hydroxymethylating agent in an enantioselective synthesis of chiral 4,5-dihydrooxazoles from α-isocyanocarboxylates catalyzed by a complex formed in situ between bis(cyclohexylisocyanide)gold (I) tetrafluoroborate and a diphosphine ligand (Figure 36.24a) [92]. This reaction proceeded in dichloromethane at room temperature, and ees of 81% could be

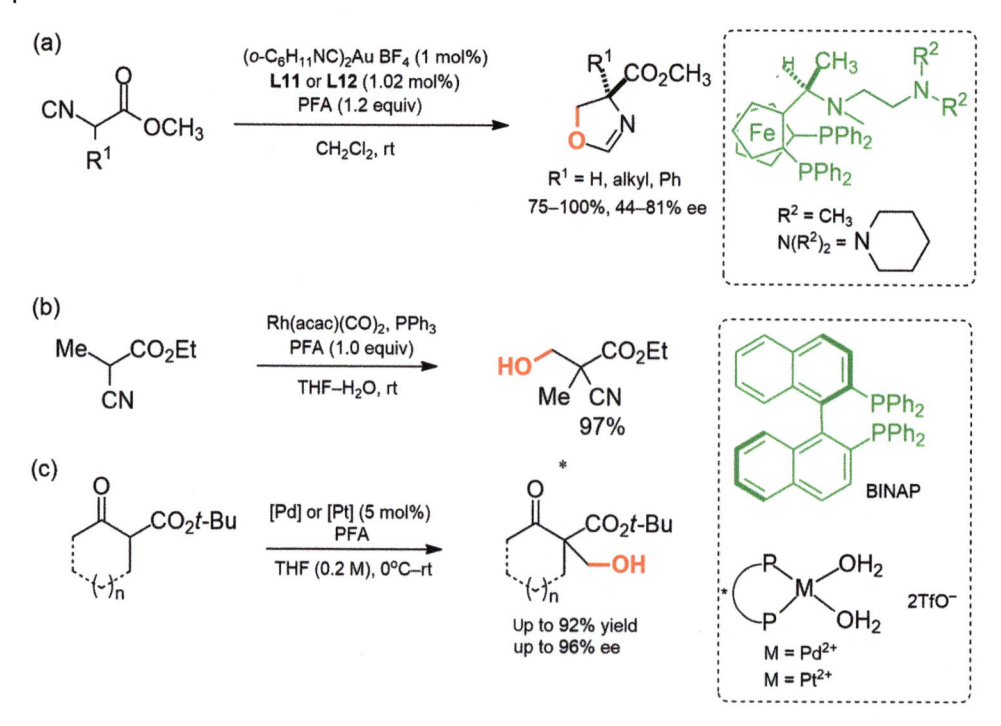

Figure 36.24 (a) Asymmetric aldol reaction of α-isocyanocarboxylates with paraformaldehyde (PFA). (b) Rh-catalyzed aldol reaction of methyl 2-cyanopropionate with PFA. (c) Pd-catalyzed hydroxymethylation of α-substituted β-ketoesters.

obtained. The initially obtained 4,5-dihydrooxazoles were subsequently hydrolyzed to α-alkyl serines. Moreover, methyl 2-cyanopropionate could also be reacted with PFA via Rh(I) catalysis, provided by 1 mol% of a rhodium(I) complex generated in situ from Rh(acac)(CO)$_2$ and triphenylphosphine to yield aldol product at rt in 1 h, in 97% yield (Figure 36.24b) [93]. A catalytic asymmetric hydroxymethylation of α-substituted β-ketoesters is possible with chiral Pd(II) and Pt(II)-(R)-BINAP (BINAP = 2,2-bis(diphenylphosphino)-1,1-binaphthyl) [94]. Aldol products bearing a chiral quaternary center were obtained in high yields and ees (Figure 36.24c). It is suggested that a nucleophilic chiral square planar M(II)-enolate intermediate was the active species in this process, which did not require protection from oxygen (air) or moisture.

36.4 Hydrogen Generation and Transfer-hydrogenation Reactions

In addition to the use of PFA as CO-surrogate and C1-building block (see previous sub-chapter) [6, 27], the application of PFA, respectively aq. formaldehyde for homogenously catalyzed hydrogen generation is another field of application (Figure 36.25a) [1, 95]. [(Ru(p-cymene)Cl$_2$)$_2$] has been identified as catalyst precursor (13 mol%) at 95 °C, under acidic conditions for PFA depolymerisation, resulting in 75% conversion within 74 minutes following the sequence depicted in Figure 36.3. In addition to the application of PFA, the use of aq. formaldehyde solution resulted in a TON of 700 and turnover frequency (TOF) >3000 h^{-1} [96]. The catalytically active species formed *in situ* has been identified as [(Ru(p-cymene))$_2$μ-H(μ-HCO$_2$)μ-Cl]$^+$ by NMR [1] and MS with multiple isotope-labelling using ^{2}H, ^{13}C labelled PFA respectively ^{2}H and ^{18}O labelled water [1, 9, 10]. Subsequently, more transition metal catalysts have been developed and identified for the hydrogen generation from PFA and/or aq. formaldehyde [25, 26, 97–99]. Other examples of active catalysts include cationic IrCp* complexes with cooperative ligand site which were used under basic conditions (pH 11) with moderate activity for the hydrogen evolution in the aqueous phase with formalin, respectively PFA (Figure 25b-c) [26, 100]. A promising activity at low temperature (60 °C) has been revealed using an anionic ruthenium complex known to be active for methanol dehydrogenation [12, 25, 101]; with a TON >1700 and initial TOFs of >20000 h^{-1} (Figure 36.25d). However, this

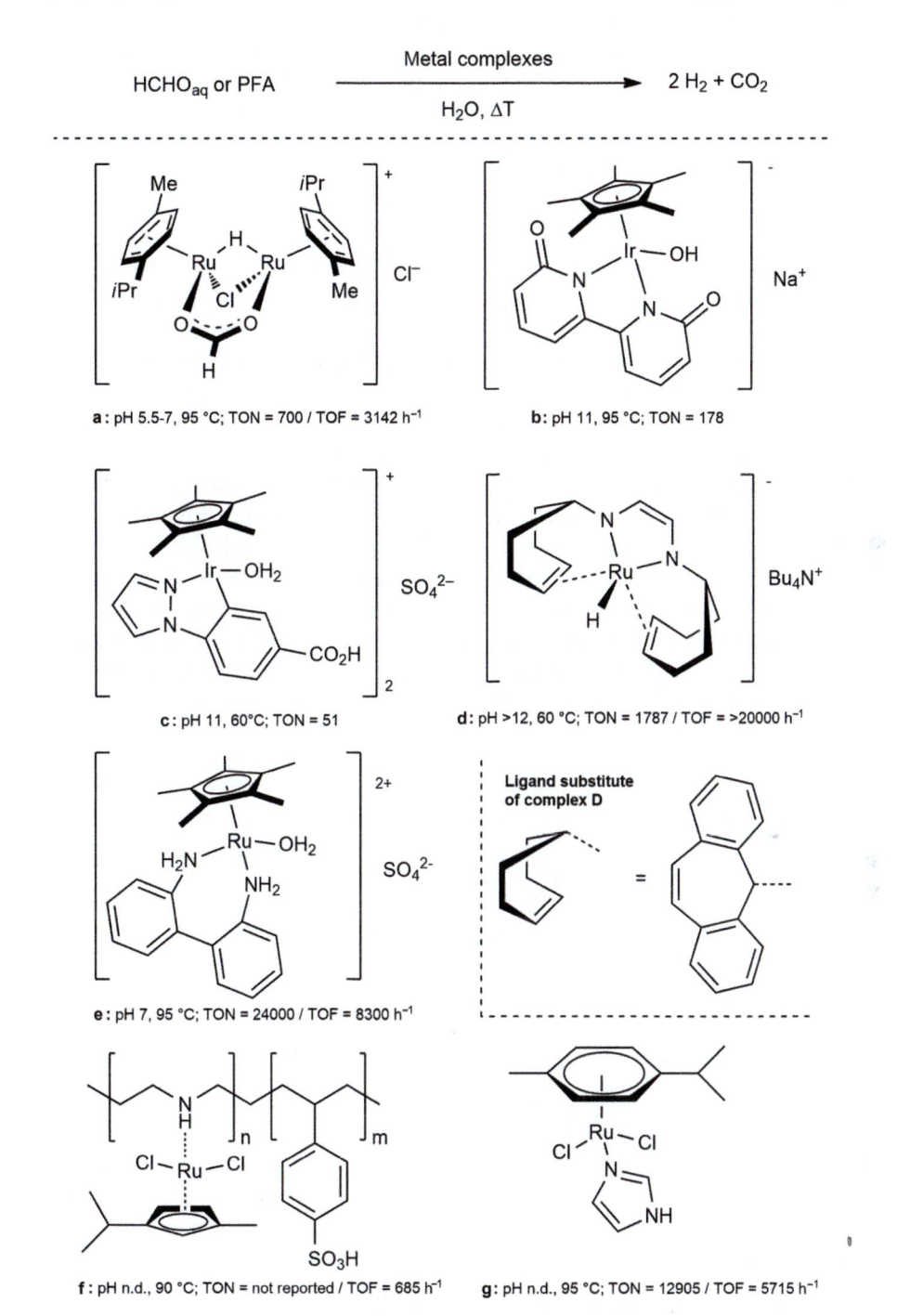

Figure 36.25 Examples of metal complexes suitable for H$_2$ generation in aq. formalin and aq. paraformaldehyde (PFA) solutions [1, 25, 26, 96–100].

setup requires strong basic conditions of pH >12. The highest conversion has been observed by the application of a water-soluble biphenyldiamine ruthenium complex for hydrogen production from aq. formaldehyde under additive- and base-free conditions at pH 7 and 95 °C (Figure 36.25e) [97]. A maximum TON of 24, 000 was observed for the catalytic system after 100 hours. Moreover, the dimeric [(Ru(p-cymene)Cl$_2$)$_2$] complex which is frequently used as precursor for the synthesis of monomeric ruthenium amine complexes, forms by treatment with imidazole the corresponding monomeric complex (Figure 36.25f) [98]. This complex is active for hydrogen production from formaldehyde and paraformaldehyde in water with no external base at 95 °C. The catalytic system demonstrated high activities with TONs >12000 and TOFs 5175 h^{-1}. Also, recyclability experiments of the catalyst exhibited high activity after eight consecutive cycles [98]. Interestingly, immobilized heterogeneous bifunctional catalyst materials have been realized based on coordination polymers (Figure 36.25g). The most active bifunctional catalyst contains two catalytically active sites, a sulfonic acid group and an organometallic ruthenium entity. The sulfonic acid is responsible to catalyze the hydrolysis of paraformaldehyde into formaldehyde. Then, the ruthenium catalyst unit can catalyze the formaldehyde-water shift reaction generating hydrogen and formic acid which is then decomposed into hydrogen and carbon dioxide. The TOF is up to 685 h^{-1} at 363 K [99].

Following the demonstrations of acceptorless metal-catalyzed PFA degradation to H$_2$ and CO$_2$, such approaches have been further developed for transfer-hydrogenation (including isotopelabelling with deuterated PFA and D$_2$O) reactions for the selective conversion of enones [14], aldehydes [15], alkynes [10], and nitriles [9] and for the formation of methanol [8].

A commercially available ruthenium complex as catalyst [RuCl$_2$(PPh$_3$)$_3$] is suitable for transfer-hydrogenation with PFA in water to reduce C=C bond of α,β-enones in a chemoselective fashion (Figure 36.26). The optimized conditions required 2 mol% of catalyst, five equivalents of PFA and K$_2$CO$_3$ as base for 18 hours at 110 °C [14]. Similarly transfer-hydrogenation with PFA in aqueous dimethyl sulfoxide (DMSO) can reduce aromatic aldehydes with an iron complex as catalyst (Figure 36.26). The optimized conditions required 3 mol% of the iron-catalyst, 10 eq. PFA and Na$_2$CO$_3$ as base for 24 hours at 120 °C [15]. The performance of the reaction in a two-chamber system to understand the nature of the production of H$_2$ from PFA and water confirmed the in situ

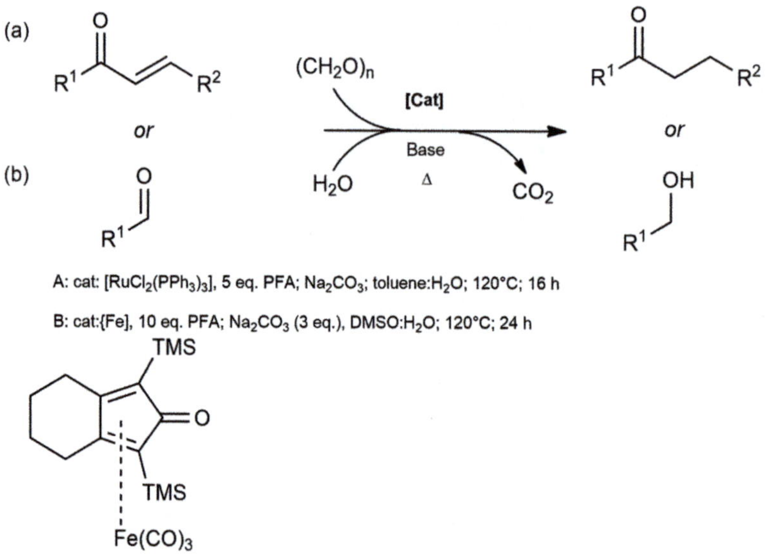

Figure 36.26 Transfer-hydrogenation of enones and aldehydes [14, 15].

formation of formic acid [1] previous to the decomposition to CO_2 and H_2, which is the reducing agent to transform aromatic aldehydes to alcohols. Several aromatic aldehydes were applied as substrates, yielding the corresponding alcohols in good yields (up to 89%).

Applying PFA for alkyne transfer-hydrogenation, it has been demonstrated that E-alkenes are formed with up to >99% E/Z selectivity via Ru-catalyzed partial hydrogenation of different aliphatic and aromatic alkynes (Figure 36.27). Best results were obtained with [Ru(p-cymene)Cl$_2$]$_2$ complex as pre-catalyst in combination with BINAP as a ligand (1:1 ratio per Ru monomer to ligand). Mechanistic investigations indicate that the E-selectivity is due to the fast Z to E isomerization of alkenes. This method is a complementary procedure to the well-known Z-selective Lindlar reduction in late-stage syntheses and also suitable for the production of deuterated alkenes simply using d$_2$-paraformaldehyde and D_2O mixtures. The use of pFA and the base-free conditions results in a high functional group tolerance and many different substrates can be converted into E-alkenes with this protocol.

In the context of transfer hydrogenation reactions applying PFA in aqueous media, it turned out that a very mild method for the selective reductive deamination of nitriles to primary alcohols under very mild conditions can be realized (Figure 36.28) [9]. This reaction represents an artificial pathway which goes even further than nitrilases go during the natural detoxification of nitriles by the conversion to carboxylic acids and NH_3 [102–104]. Usually, in synthetic systems, nitriles are reduced to primary amines with secondary/tertiary amines as side-products. Additionally, nitriles are often hydrolyzed to amides or even carboxylic acids in the presence of water. In contrast, the present protocol reduces nitriles to imines that react in situ to the corresponding aminals in presence of water, and under ammonia evolution, the corresponding aldehydes are formed in situ that are then further reduced to the corresponding alcohols (Figure 36.28). The protocol is applicable for a wide range of nitriles (20 examples) and neither expensive nor air or moisture sensitive chemicals are required for the conversion of nitriles to alcohols. A broad substrate scope showed very good to excellent yields under the optimised conditions. [Ru(p-cymene)Cl$_2$]$_2$ acts as the catalyst precursor in the presence of pFA resulting in its degradation to CO_2 and H_2. Nitriles play a dual role as substrate and as ligand where the binuclear catalyst structure converts to monomeric ones upon coordination of nitrile molecules.

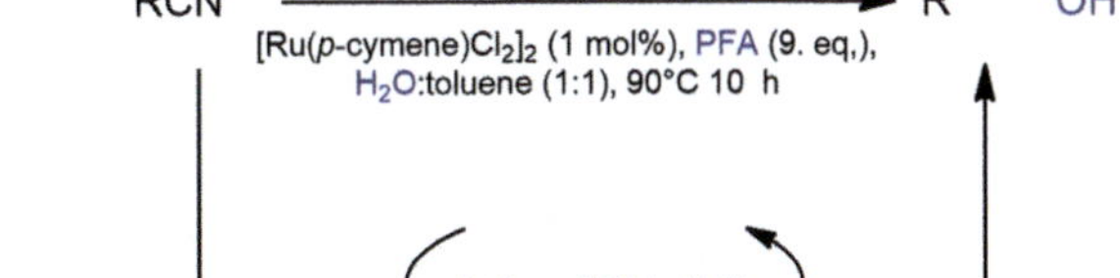

Figure 36.27 E-selective hydrogenation of alkynes with paraformaldehyde (PFA) in presence of water.

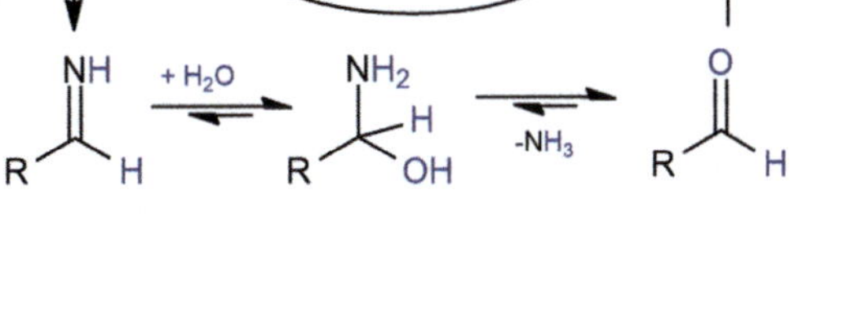

Figure 36.28 Catalytic deaminative conversion of nitriles to alcohols with paraformaldehyde (PFA) in water as hydrogen source.

Inspired by nature, it has been shown that in addition to formaldehyde dehydrogenase mimicry, it is possible to mimic formaldehyde metabolism with a second pathway using *dismutases*, independent of external sacrificial redox partners (Figure 36.29, right) [8]. On one side, these formalin disproportionating enzymes exhibit a considerable structural and functional resemblance to the glutathione- independent zinc-containing dehydrogenases, with a sequence similarity greater 70%. Considering the analogy of NADH as biological hydrogen carrier, and our lately described bioinspired process featuring acceptorless H_2 liberation [25], it was proposed that modification of the organometallic species and/or the reaction environment will lead to a novel homogeneously catalyzed formaldehyde-to-methanol converting system, (Figure 36.29, right), as yet unprecedented in the context of abiotic C_1-valorization pathways [8]. Indeed, it was possible by slight modification of the catalyst backbone and the addition of larger amounts of phosphate buffer [8]. To exclude further reaction pathways that would lead to methanol formation, the pH-dependency and base-catalyzed disproportionation (Cannizzaro reaction), and isotope-labelling experiments (^2H and $^{13}CO_2$) with labelled PFA were investigated to exclude CO_2-coupled or H_2-coupled disproportionation (Figure 36.29). The variation of the pH showed that a Cannizzaro reaction occurs only at a pH >9.5 and, since no $^{12}C/^{13}C$- and $^1H/^2H$-scrambeling occurred during the reaction, we could exclude CO_2-coupled or H_2-coupled disproportionation. A dismutase pathway would require a H_2-decoupled disproportionation, or, in other words, a direct hydride transfer would imitate the enzymatic dismutation. Under optimized conditions, a dismutase mimic reached >90% selectivity with an imidazolium-tagged ruthenium complex [8]. In a similar study, an IrCp* complex was evaluated for the PFA conversion to methanol (Figure 36.29) [105]. The cationic iridium complex, Cp*IrL(OH$_2$)$^{2+}$ (Cp* = pentamethylcyclopentadienyl, L = 2,2′,6,6′-tetrahydroxy-4,4′-bipyrimidine), is very active and highly selective for PFA to methanol conversion at 25 °C with a TOF of 4120 h^{-1} yielding methanol in high amounts (93%) with a high TON of 18,200. In the Ir-catalyzed reaction, the base-catalyzed Cannizzaro reaction does not play a role.

In addition to the PFA-to-methanol conversion (see Figure 36.29), it has been demonstrated that the conversion of PFA with CO_2 and H_2 to ethanol can be catalyzed by a mixture of Ru(acac)$_3$ and CoBr$_2$ in 1,3-dimethyl-2-imidazolidinone (DMI) as the solvent with lithium iodide (LiI) employed as promoter (Figure 36.30) [106]. The optimized conditions and recycling tests used 2.3 mol% Ru(acac)$_3$ and 14 mol% μmol CoBr$_2$, 0.9 equivalent LiI, 2 mL DMI, and 3.2 mmol PFA under a pressure of 3:5

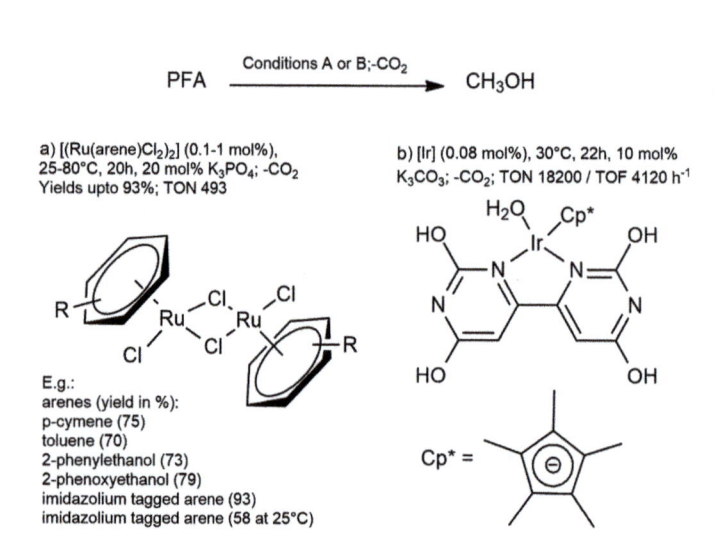

Figure 36.29 Investigations toward the realization of an organometallic dismutase mimic.

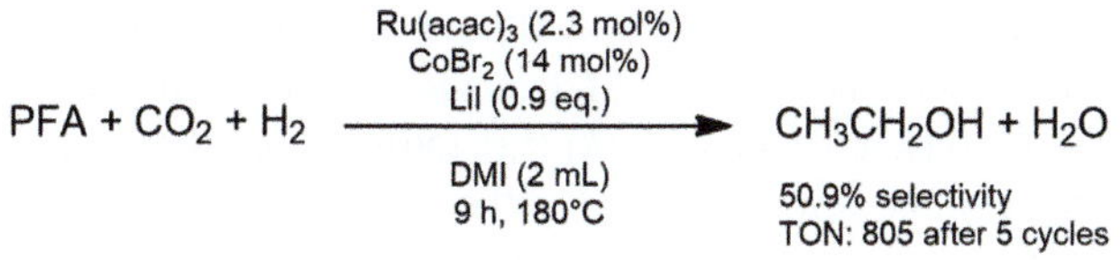

Figure 36.30 Conversion of paraformaldehyde (PFA) to ethanol under CO_2 (3 MPa) and H_2 (5 MPa) pressure.

$$PFA + CO_2 + H_2 \xrightarrow[\substack{DMI\ (2\ mL) \\ 9\ h,\ 180°C}]{\substack{Ru(acac)_3\ (2.3\ mol\%) \\ CoBr_2\ (14\ mol\%) \\ LiI\ (0.9\ eq.)}} CH_3CH_2OH + H_2O$$

50.9% selectivity
TON: 805 after 5 cycles

MPa (CO_2:H_2) during nine hours at 180 °C, resulting in a TOF of 17.9 h^{-1} for the conversion to ethanol with a product selectivity of 50.9% and a total TON of 805 after five cycles [106].

36.5 Summary and Outlook

In summary, the described efforts to use paraformaldehyde as a versatile reagent in the field of transition metal catalyzed organic synthesis are very promising for future applications in a wide range of applications. Paraformaldehyde can be used under mild conditions for i) transfer-hydrogenation reactions for the conversion of several functional groups acting as hydrogen source, ii) carbonylation reactions in the function of a CO surrogate and iii) as a C1-building block to incorporate methyl, methylene, and other groups. Moreover, the methods are suitable for the synthesis of isotope-labelled compounds with hydrogen and carbon isotopes. The mild conditions such as low-temperature, robustness of the catalytic systems (non-inert conditions under air in presence of moisture or even in the aqueous phase), and high chemo-selectivities are often accompanied with a good functional group tolerance, which are promising for further applications. For the future, one can expect that more base-metal catalysts will be identified for PFA activation in the so far precious metal dominated field. Moreover, the future synthesis of PFA from renewable sources and from carbon dioxide will open further possibilities for the application in organic synthesis and as energy storage molecule for the generation of hydrogen for paraformaldehyde reforming fuel cell applications.

Acknowledgement

The authors acknowledge the Fundação para Ciência e Tecnologia (FCT, EXPL/QUI-QOR/1079/2021; PTDC/QUI-QOR/02069/2022, UIDB/00100/2020, UIDP/00100/2020, LA/P/0056/2020; AMFP, MNK, and MHGP), the Deutsche Forschungsgemeinschaft (DFG 411475421; MHGP) the Centro de Química Estrutural / Institute of Molecular Sciences (AMFP, MNK, MHGP), Instituto de Química da Universidade de São Paulo (IQ-USP; LHA), Fundação de Amparo à Pesquisa do Estado de São Paulo (FAPES; 2019/10762-1; LHA), and Conselho Nacional de Desenvolvimento Científico e Tecnológico / Conselho Nacional de Pesquisas(CNPq, 312751/2018-4; LHA). Moreover acknowledgement goes to the Insituto Politécnico de Lisboa for funding the IPL/IDI&CA2023/SMARTCAT_ISEL project (MHGP).

References

1 Heim, L.E., Schlorer, N.E., Choi, J.H., and Prechtl, M.H.G. (2014). *Nat. Commun.* 5: 3621.
2 van der Waals, D., Heim, L.E., Gedig, C. et al. (2016). *Chemsuschem* 9: 2343–2347.
3 Heim, L.E., Konnerth, H., and Prechtl, M.H.G. (2017). *Green Chem.* 19: 2347–2355.
4 Heim, L.E., Thiel, D., Gedig, C. et al. (2015). *Angew. Chem. Int. Edit.* 54: 10308–10312.
5 Pichardo, M.C., Tavakoli, G., Armstrong, J.E. et al. (2020). *Chemsuschem* 13: 882–887.
6 Sam, B., Breit, B., and Krische, M.J. (2015). *Angew. Chem. Int. Edit.* 54: 3267–3274.

7 Scholten, J.D., Prechtl, M.H.G., and Dupont, J. (2010). *Chemcatchem* 2: 1265–1270.
8 van der Waals, D., Heim, L.E., Vallazza, S. et al. (2016). *Chem. Eur. J.* 22: 11568–11573.
9 Tavakoli, G. and Prechtl, M.H.G. (2019). *Catal. Sci. Technol.* 9: 6092–6101.
10 Fetzer, M.N.A., Tavakoli, G., Klein, A., and Prechtl, M.H.G. (2021). *Chemcatchem* 13: 1317–1325.
11 Heim, L.E., Konnerth, H., and Prechtl, M.H.G. (2016). *Chemsuschem* 9: 2905–2907.
12 Trincado, M., Grutzmacher, H., and Prechtl, M.H.G. (2018). *Phys. Sci. Rev.* 3: psr-2017-0013.
13 Bahmanpour, A.M., Hoadley, A., and Tanksale, A. (2015). *Green Chem.* 17: 3500–3507.
14 Li, W.F. and Wu, X.F. (2015). *Eur. J. Org. Chem.* 2015: 331–335.
15 Natte, K., Li, W.F., Zhou, S.L. et al. (2015). *Tetrahedron Lett.* 56: 1118–1121.
16 Leopold, M., Siebert, M., Siegle, A.F., and Trapp, O. (2021). *Chemcatchem* 13: 2807–2814.
17 Seibicke, M., Siebert, M., Siegle, A.F. et al. (2019). *Organometallics* 38: 1809–1814.
18 Siebert, M., Seibicke, M., Siegle, A.F. et al. (2019). *J. Amer. Chem. Soc.* 141: 334–341.
19 Schieweck, B.G. and Klankermayer, J. (2017). *Angew. Chem. Int. Edit.* 56: 10854–10857.
20 Thenert, K., Beydoun, K., Wiesenthal, J. et al. (2016). *Angew. Chem. Int. Edit.* 55: 12266–12269.
21 Geitner, R., Gurinov, A., Huang, T.B. et al. (2021). *Angew. Chem. Int. Edit.* 60: 3422–3427.
22 Li, S.Q., Yu, Y., Yang, Y.X., and Zhou, B. (2020). *Heterocycles* 100: 934–945.
23 Midya, S.P., Sahoo, M.K., Landge, V.G. et al. (2015). *Nat. Commun.* 6.
24 Chen, J.B., Natte, K., and Wu, X.F. (2016). *J. Organom. Chem.* 803: 9–12.
25 Trincado, M., Sinha, V., Rodriguez-Lugo, R.E. et al. (2017). *Nat. Commun.* 8: 14990.
26 Suenobu, T., Isaka, Y., Shibata, S., and Fukuzumi, S. (2015). *Chem. Commun.* 51: 1670–1672.
27 Lia, W.F. and Wu, X.F. (2015). *Adv. Synth. Catal.* 357: 3393–3418.
28 Liu, Q., Yuan, K.D., Arockiam, P.B. et al. (2015). *Angew. Chem. Int. Edit.* 54: 4493–4497.
29 Martins, L.M.D.R.S., Phillips, A.M.F., and Pombeiro, A.J.L. (2018). In: *Sustainable Synthesis of Pharmaceuticals: Using Transition Metals as Catalysts* (eds. M.M. Pereira and M.J.F. Calvete), 193–229. Oxford: Royal Society of Chemistry.
30 Faisca Phillips, A.M. and Pombeiro, A.J.L. (2019). In: *Alkane Functionalization* (ed. A.J.L. Pombeiro and M.F.C.G. da Silva), 476–513. Chichester: John Wiley & Sons.
31 Wu, L.P., Liu, Q., Jackstell, R., and Beller, M. (2014). *Angew. Chem. Int. Edit.* 53: 6310–6320.
32 Sang, R., Hu, Y.Y., Razzaq, R. et al. (2021). *Org. Chem. Front.* 8: 799–811.
33 Liu, Q., Wu, L.P., Jackstell, R., and Beller, M. (2014). *Chemcatchem* 6: 2805–2809.
34 Sang, R., Schneider, C., Razzaq, R. et al. (2020). *Org. Chem. Front.* 7: 3681–3685.
35 Matsuda, T., Fukuhara, K., Yonekubo, N., and Oyama, S. (2017). *Chem. Lett.* 46: 1721–1723.
36 Li, W.F. and Wu, X.F. (2014). *J. Org. Chem.* 79: 10410–10416.
37 Furusawa, T., Morimoto, T., Oka, N. et al. (2016). *Chem. Lett.* 45: 406–408.
38 Abu Seni, A., Kollar, L., Mika, L.T., and Pongracz, P. (2018). *Mol. Catal.* 457: 67–73.
39 Martins, L.M.D.R.S., Faisca Phillips, A.M.M.M., and Pombeiro, A.J.L. (2020). In: *Synthetic Approaches to Nonaromatic Nitrogen Heterocycles* (ed. A.M.M.M.F. Phillips), 119–160. Chichester: John Wiley & Sons.
40 Xu, X.F., Yang, Y.R., Chen, X. et al. (2017). *Org. Biomol. Chem.* 15: 9061–9065.
41 Chakrabortty, S., Almasalma, A.A., and de Vries, J.G. (2021). *Catal. Sci. Technol.* 11: 5388–5411.
42 Okano, T., Kobayashi, T., Konishi, H., and Kiji, J. (1982). *Tetrahedron Lett.* 23: 4967–4968.
43 Ahn, H.S., Han, S.H., Uhm, S.J. et al. (1999). *J. Mol. Catal. A - Chem.* 144: 295–306.
44 Rosales, M., Gonzalez, A., Gonzalez, B. et al. (2005). *J. Organomet. Chem.* 690: 3095–3098.
45 Makado, G., Morimoto, T., Sugimoto, Y. et al. (2010). *Adv. Synth. Catal.* 352: 299–304.
46 Cini, E., Airiau, E., Girard, N. et al. (2011). *Synlett* 42: 199–202.
47 Morimoto, T., Fujii, T., Miyoshi, K. et al. (2015). *Org. Biomol. Chem.* 13: 4632–4636.
48 Uhlemann, M., Doerfelt, S., and Borner, A. (2013). *Tetrahedron Lett.* 54: 2209–2211.

49 Pittaway, R., Dingwall, P., Fuentes, J.A., and Clarke, M.L. (2019). *Adv. Synth. Catal.* 361: 4334–4341.
50 Geitner, R. and Weckhuysen, B.M. (2020). *Chem. Eur. J.* 26: 5297–5302.
51 Nielsen, D.U., Neumann, K.T., Lindhardt, A.T., and Skrydstrup, T. (2018). *J. Labelled Comp. Radiopharma.* 61: 949–987.
52 Ortega, N., Richter, C., and Glorius, F. (2013). *Org. Lett.* 15: 1776–1779.
53 Saidi, O., Bamford, M.J., Blacker, A.J. et al. (2010). *Tetrahedron Lett.* 51: 5804–5806.
54 Lee, H., Kang, B., Lee, S.I., and Hong, S.H. (2015). *Synlett* 26: 1077–1080.
55 Tramontini, M. (1973). *Synthesis* 1973 (12): 703–775.
56 Kabalka, G.W., Zhou, L.L., Wang, L., and Pagni, R.M. (2006). *Tetrahedron* 62: 857–867.
57 Li, P.H. and Wang, L. (2007). *Tetrahedron* 63: 5455–5459.
58 Ohta, Y., Oishi, S., Fujii, N., and Ohno, H. (2008). *Chem. Commun.* 835–837.
59 Feng, H.D., Jia, H.H., and Sun, Z.H. (2015). *Adv. Synth. Catal.* 357: 2447–2452.
60 Kundu, S.K., Mitra, K., and Majee, A. (2015). *RSC Adv.* 5: 13220–13223.
61 Crabbe, P., Fillion, H., Andre, D., and Luche, J.L. (1979). *J. Chem. Soc. - Chem. Commun.* 859–860.
62 Searles, S., Li, Y., Nassim, B. et al. (1984). *J. Chem. Soc. - Perkin Trans.* 1: 747–751.
63 Yu, S.C. and Ma, S.M. (2012). *Angew. Chem. Int. Edit.* 51: 3074–3112.
64 Luo, H.W. and Ma, S.M. (2013). *Eur. J. Org. Chem.* 2013: 3041–3048.
65 Watanabe, Y., Shimizu, Y., Takatsuki, K., and Takegami, Y. (1978). *Chem. Lett.* 215–216.
66 Watanabe, Y., Yamamoto, M., Mitsudo, T., and Takegami, Y. (1978). *Tetrahedron Lett.* 1289–1290.
67 Sugi, Y., Matsuda, A., Bando, K.I., and Murata, K. (1979). *Chem. Lett.* 363–364.
68 Watson, A.J.A. and Williams, J.M.J. (2010). *Science* 329: 635–636.
69 Edwards, M.G., Jazzar, R.F.R., Paine, B.M. et al. (2004). *Chem. Commun.* 90–91.
70 Edwards, M.G. and Williams, J.M.J. (2002). *Angew. Chem. Int. Ed.* 41: 4740–4743.
71 Sun, H.B., Hua, R.M., and Yin, Y.W. (2006). *Tetrahedron Lett.* 47: 2291–2294.
72 Xie, Y.J., Hu, J.H., Wang, Y.Y. et al. (2012). *J. Amer. Chem. Soc.* 134: 20613–20616.
73 Kopfer, A., Sam, B., Breit, B., and Krische, M.J. (2013). *Chem. Sci.* 4: 1876–1880.
74 Kimura, M. and Tamaru, Y. (2007). *Top. Curr. Chem.* 279: 173–207.
75 Ogoshi, S., Tonomori, K., Oka, M., and Kurosawa, H. (2006). *J. Amer. Chem. Soc.* 128: 7077–7086.
76 Smejkal, T., Han, H., Breit, B., and Krische, M.J. (2009). *J. Amer. Chem. Soc* 131: 10366–10367.
77 Han, H. and Krische, M.J. (2010). *Org. Lett.* 12: 2844–2846.
78 Bausch, C.C., Patman, R.L., Breit, B., and Krische, M.J. (2011). *Angew. Chem. Int. Edit.* 50: 5686–5689.
79 Ngai, M.Y., Rucas, E., and Krische, M.J. (2008). *Org. Lett.* 10: 2705–2708.
80 Sam, B., Montgomery, T.P., and Krische, M.J. (2013). *Org. Lett.* 15: 3790–3793.
81 Akhmetova, V.R., Bikbulatova, E.M., Akhmadiev, N.S. et al. (2018). *Chem. Heterocycl. Comp.* 54: 520–527.
82 Chakraborty, A., Chowdhury, T., Menendez, M.I., and Chattopadhyay, T. (2020). *ACS Appl. Mater. Interf.* 12: 38530–38545.
83 Dong, X.Y., Gao, L.X., Zhang, W.Q. et al. (2016). *Chemistryselect* 1: 4034–4043.
84 Khabibullina, G.R., Zaynullina, F.T., Tyumkina, T.V. et al. (2019). *Russ. Chem. Bull.* 68: 1407–1413.
85 Khan, I., Li, H.F., Wu, X., and Zhang, Y.J. (2018). *Acta Chim. Sin.* 76: 874–877.
86 Liu, S.Y., Yao, W.F., Liu, Y. et al. (2017). *Sci. Adv.* 3.
87 Marinic, B., Hepburn, H.B., Grozavu, A. et al. (2021). *Chem. Sci.* 12: 742–746.
88 Meyer, T., Konrath, R., Kamer, P.C.J., and Wu, X.F. (2021). *Asian J. Org. Chem.* 10: 245–250.
89 Zhou, C., Zhao, J.Q., Chen, W.K. et al. (2020). *Eur. J. Org. Chem.* 2020: 6485–6488.

90 Lin, Y.F., Li, E.F., Wu, X. et al. (2020). *Org. Biomol. Chem.* 18: 1476–1486.
91 Meninno, S. and Lattanzi, A. (2016). *Chem. Rec.* 16: 2016–2030.
92 Ito, Y., Sawamura, M., Kobayashi, M., and Hayashi, T. (1988). *Tetrahedron Lett.* 29: 6321–6324.
93 Kuwano, R., Miyazaki, H., and Ito, Y. (2000). *J. Organomet. Chem.* 603: 18–29.
94 Fukuchi, I., Hamashima, Y., and Sodeoka, M. (2007). *Adv. Synth. Catal.* 349: 509–512.
95 Prechtl, M.H.G., Heim, L.E., and Schloerer, N.E. (2013). H2-Produktion—Verwendung von Formaldehyd, Paraformaldehyd, Wasser und Methandiol (Formaldehydhydrat) zur Wasserstoffproduktion (H2, HD, D2, T2) und Energiespeicherung. German patent 10 2013 011 379 2 (2013).
96 Heim, L.E., Vallazza, S., van der Waals, D., and Prechtl, M.H.G. (2016). *Green Chem.* 18: 1469–1474.
97 Wang, L., Ertem, M.Z., Kanega, R. et al. (2018). *ACS Catal.* 8: 8600–8605.
98 Awasthi, M.K. and Singh, S.K. (2021). *Sustain. Energy Fuels* 5: 549–555.
99 Shen, Y.B., Bai, C., Zhan, Y.L. et al. (2020). *Chempluschem* 85: 1646–1654.
100 Fujita, K., Kawahara, R., Aikawa, T., and Yamaguchi, R. (2015). *Angew. Chem. Int. Edit.* 54: 9057–9060.
101 Rodriguez-Lugo, R.E., Trincado, M., Vogt, M. et al. (2013). *Nat. Chem.* 5: 342–347.
102 Gong, J.S., Lu, Z.M., Li, H. et al. (2012). *Microb. Cell Fact.* 11: 142.
103 Pawar, S.V. and Yadav, G.D. (2014). *Ind. Eng. Chem. Res.* 53: 7986–7991.
104 Sugai, T., Yamazaki, T., Yokoyama, M., and Ohta, H. (1997). *Biosci. Biotech. Bioch.* 61: 1419–1427.
105 Wang, L., Ertem, M.Z., Murata, K. et al. (2018). *ACS Catal.* 8: 5233–5239.
106 Zhang, J.J., Qian, Q.L., Cui, M. et al. (2017). *Green Chem.* 19: 4396–4401.

37

Hydrogen Storage and Recovery with the Use of Chemical Batteries

Henrietta Horváth, Gábor Papp, Ágnes Kathó, and Ferenc Joó

Department of Physical Chemistry, University of Debrecen, P.O. Box 400, Debrecen, Hungary

37.1 Introduction

Except for nuclear and tidal energy, the ultimate source of all energy presently available on Earth is the Sun. The energy of sunshine which made possible animal and plant life on Earth many million years ago and has been preserved for us in the so-called fossilic energy sources (coal, oil, and gas). However, whatever rich deposits of such fossilic energy carriers are available, their total amount is limited (and unevenly distributed). An even greater problem of our times is that burning (or other chemical use) of coal, oil, and natural gas severely contributes to disrupting the fine equilibrium between the absorption (fixation) and emission of atmospheric carbon dioxide by the global ecosystem, including human activity. Transportation and various industrial processes (such as the production of portland cement) yield huge amounts of carbon dioxide. In May 2022, the CO_2 concentration in the atmosphere passed over 420 ppm (Mauna Loa Observatory, Hawaii, USA), in contrast to 280 ppm prior to the Industrial Revolution. Such a huge increase in the concentration of this greenhouse gas has already brought about unfavorable climate changes (commonly referred to as global warming), and a continuing exponential accumulation of CO_2 in the atmosphere ultimately may lead to global climate catastrophes with severe detrimental consequences on civilization.

An ideal solution to this problem would be the capture and utilization of at least as much atmospheric CO_2 during a given time as is released to the atmosphere in the same time period. Note that such a steady-state situation would only stabilize the atmospheric CO_2 level at the present concentration and would not result in elimination of the climate problems caused so far by global human activity. It is brutally shown by the data of environmental monitoring stations, that despite international efforts, even this steady-state of CO_2 emission/fixation could not be achieved in the last few decades. More, and more effective, measures have to be taken!

Science has long realized the need for procedures for decreasing CO_2 emissions to the atmosphere and several strategic directions have been identified. One of those is the replacement of fossil fuels by renewable ones. Coal, oil, and natural gas store and concentrate the energy of the early Sun. Apart from their other uses as chemical raw materials, these concentrated energy sources can be conveniently stored, transported, and used as fuels, too, according to demand. Harnessing the Sun today is still a possibility either directly in photovoltaic devices, or indirectly (e.g. by means of wind, gravitational, hydroelectric, tide, or underwater current power plants). However, the

---

Catalysis for a Sustainable Environment: Reactions, Processes and Applied Technologies Volume 3, First Edition. Edited by Armando J. L. Pombeiro, Manas Sutradhar, and Elisabete C. B. A. Alegria.
© 2024 John Wiley & Sons Ltd. Published 2024 by John Wiley & Sons Ltd.

production of such renewable forms of energy fluctuates over time and needs concentration and storage in forms (preferably as liquids or gases) that can be used in a manner adapted to the actual energy demand. The term renewable is somewhat misleading, because nothing is renewed here; instead, the million years old fossilic storage materials of solar energy are replaced by new storage forms of recently captured energy of unceasing sunlight. Nevertheless, this approach is, indeed, suitable for mitigation of atmospheric CO_2 emisssions, provided that high capacity storage solutions and specific devices are available both for local and mobile applications. It should also be borne in mind that the capture and storage of renewable energy has its price (which also can be expressed in quantities of CO_2); obviously, this price must compare favorably to the obviated CO_2 emission. It is also worth mentioning here that although energy storage may critically contribute to the success of atmospheric CO_2 depletion via replacement of fossil fuels, its role is much more diverse and includes, for example, the synthesis of high-energy materials, construction of small and large electric batteries for myriad uses, and other applications, some of which may be even accompanied by zero or negative net carbon dioxide emission.

In the complex system of questions and answers for replacing fossil fuels with renewable ones, one of the key problems is energy storage. Energy can be stored in several forms (e.g. electric, magnetic, gravitational), including storage in suitable chemicals (i.e. as the energy of chemical bonds). The interested reader can find a large number of books and reviews describing various aspects of this important topic [1–5]. In this chapter, we treat only a small section of the field, namely chemical energy storage with the use of hydrogen batteries.

37.2 Hydrogen as an Energy Storage Material

At present, hydrogen is used in several large-cale industrial processes (e.g. steel, petrochemical, fertilizer industries). For practical reasons, 95% of industrial H_2 is produced from fossil raw materials, mainly by decomposition of methane (natural gas) in the presence of steam, a process which results in formation of carbon dioxide, too.

Hydrogen has long been advocated as a general energy carrier of the future, playing a central role in the so-called Hydrogen Economy. This suggestion is based on two very important properties of hydrogen. First, it can be produced by electrolysis of water (or direct photochemical water fission, not yet realized on the technological scale). Second, its direct oxidation in fuel cells produces electricity accompanied by formation of water as the sole product [6]. Provided that cheap electricity is available (from renewable, or nuclear/geothermal sources), electrolysis of water results in storage of electric energy as the energy content of the generated hydrogen. The energy can be liberated in fuel cells based upon the oxidation of H_2 by O_2 with no obvious harm to the environment since only water is produced as unique product. This way the overall process converts electric energy first to H_2 and, after some time of storage, back to electric energy (Figure 37.1). Efficient methods of storage are needed for mitigation of daily or seasonal fluctuations of H_2 availability from renewable sources, or for transportation of energy in another form than electricity.

In addition, the $H_2O \rightarrow H_2 \rightarrow H_2O$ cycle does not involve carbon-containing compounds and does not contribute to the increase of atmospheric CO_2 level (i.e. it is carbon neutral). Not to forget, though, that production and operation of the various elements of the process (electrolyzers, fuel cels, energy production facilities) have their carbon footsteps. The produced hydrogen can be used in many applications, and, beause the location and time of electrolysis and the end use of H_2 can be separated, in principle, hydrogen can be a major chemical for energy storage.

At 1 bar pressure, hydrogen is a diatomic gas with a molecular mass $M_w = 2.016 \times 10^{-3}$ kg (2.016 g) and a molar volume $V_m = 22.41 \times 10^{-3}$ m^3 (22.41 L) at 0 °C, and $V_m = 24.05 \times 10^{-3}$ m^3 (at 20 °C).

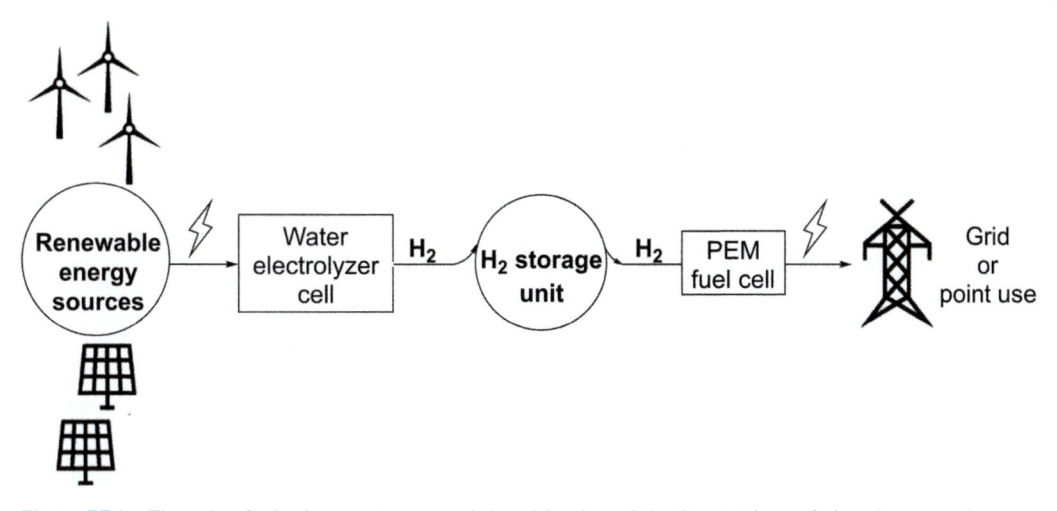

Figure 37.1 The role of a hydrogen storage unit in mitigation of the fluctuations of electric energy from renewable sources.

Because the heat of combustion of H_2 is −285.82 kJ/mol (−141.80 MJ/kg), burning of 1 L H_2 in a fuel cell supplies 11.88 kJ/mol heat energy, as a theoretical maximum [7]. The energy need of a 60 kW compact car is 216 MJ/h, which would require the burning of 756 mol (1.523 kg) i.e. 18.1 m^3 (!) of H_2 stored at room temperature. In comparison, the mass and volume of gasoline with the same energy content (heat of combustion of 47.31 MJ/kg, average density 0.75 L/kg) are only 4.567 kg or 6.1 L, respectively [8]. These data show clearly that H_2 at atmospheric pressure is unsuitable for powering transportation vehicles and that for fuel purposes H_2 must be stored in more condensed form.

Storage of hydrogen is a complex problem that, however, can be solved on many ways [2, 3]. Hydrogen can be stored with the use of physical methods (as a liquid or compressed gas, adsorbed on the surface of various solids, including cryoadsorption, or in interstitial hydrides). Chemical methods include the use of hydrolyzable hydrides, hydrogenation-dehydrogenation of hydrocarbons or nitroaromatic compounds, alcohols, formic acid, and others. Solutions of solids such as formate salts, or carbohydrates are also suitable for hydrogen storage and generation. A few biological methods are also known which apply enzymes or whole cells for catalysis of H_2 fixation in suitable compounds. The advantages and disadvantages of the individual methods should be carefully analyzed not only with respect to storage economics, but also with consideration of the requirements of the actual application. A very large field of (conceived) practical use concerns on-board generation of H_2 from suitable storage systems, while a similarly important approach aims at temporary storage of electric energy available from renewable sources (such as solar, wind, and other power plants). Typically, such stationary H_2 storage/delivery facilities have to deal with enormous amounts of energy storage materials. As will be described later, several reactions, applicable for chemical storage of hydrogen have been considered as basis of hydrogen batteries.

37.3 Chemical Hydrogen Storage

In its simplest definition, chemical energy storage is a chemical reaction in which one chemical (the storage material, H_2 acceptor A) is transformed into a product (AH_2) of higher energy content (Eq. 37.1).

$$A + H_2 \rightleftharpoons AH_2 \tag{37.1}$$

In many cases, the product is used as a fuel (i.e. its energy content is liberated in a combustion process, and then the amount of stored energy is equal to the Gibbs free energy difference ($\Delta\Delta G_r$) of combustion of the product and the storage material). In favorable cases the process results in a product with physical properties which facilitate its use in already existing and proven devices, such as gas burners, fuel cells, or internal combustion engines.

In open-end hydrogen storage and utilization processes, in general, the hydrogenated storage material is used as fuel; however, collection and reuse of the products are not a major concern. Examples are the methanation of carbon dioxide (Eq. 37.2) or the hydrogenation of CO_2 to methanol (Eq. 37.3).

$$CO_2 + 4H_2 = CH_4 + 2H_2O \tag{37.2}$$

$$CO_2 + 3H_2 = CH_3OH + H_2O \tag{37.3}$$

Combustion of CH_4 or CH_3OH results in CO_2 and H_2O. Methanol can be utilized in direct methanol fuel cells, too. The resulting carbon dioxide can be released to the atmosphere, which in this case serves as a CO_2 reservoir. Alternatively, CO_2 can be directly captured from flue gases of power stations or other point sources with concentrated CO_2 emission. What is important, though, is that the used CO_2 originates from the atmosphere and finally returns to the atmosphere, so that in principle these processes are carbon neutral, i.e. they do not result in a net increase of the amount of atmospheric CO_2 (except the quantities produced in the technical implementation). Methane and methanol can be efficiently transported in pipelines or tanker ships, so it is possible to synthesize these hydrogen carriers at large-capacity central facilities and distribute them geographically if required. A practical realization of this concept, hydrogenation of atmospheric carbon dioxide is carried out at the George Olah Renewable Methanol plant at Svartsengi, Iceland, where H_2 production (water electrolysis) and general operation of the plant is made possible at the expense of energy from the Svartsengi geothermal power plant. In fact, methanol was suggested by Olah as the central platform chemical of the so-called Methanol Economy [9]. A very recent review of Onishi and Himeda describes the state-of-art of homogeneous catalysis for CO_2 hydrogenation to methanol and methanol dehydrogenation to hydrogen generation [10]. Several homogeneous catalytic processes for aqueous reforming of methanol resulting in H_2/CO_2 mixtures have been developed recently with Ru(II)-complexes **1–5** (Figure 37.4) [11–14]. These processes were not applied in hydrogen batteries, although the reversibility of the hydrogenation/dehydrogenation steps were demonstrated in some cases [15]. An efficient homogeneous catalytic acceptorless dehydrogenation of methanol selectively to formaldehyde has not been disclosed, yet. Formaldehyde is a commodity chemical which can be dehydrogenated in aqueous systems with metal-complex catalysts under reasonably mild conditions, so it can be used as an important storage material for H_2 [16]. Nevertheless, processes involving formaldehyde dehydrogenation have not been applied in hydrogen batteries, yet.

37.4 Liquid Organic Hydrogen Carriers

Hydrogen is often stored chemically in liquid organic compounds (LOHCs) [17–23]. In principle, an LOHC is an organic liquid of high hydrogen content. From such a compound H_2 can be made free by catalytic dehydrogenation. The liberated hydrogen is utilized for a given purpose. It is usually converted into electricity in a proton exchange membrane (PEM) fuel cell (FC), often described as a polymer electrolyte membrane fuel cell (PEMFC) [6]. The organic product of this decomposition can be hydrogenated back to the original LOHC, exemplified here with the interconversion of toluene/methylcyclohexane (Figure 37.2, Table 37.1). In principle, organic hydrogen carriers could be solids or gases, as well. However, liquids are preferred, since they are easier to handle than solids, and offer higher hydrogen storage capacity than gases. Most often, both the hydrogenation and dehydrogenation processes are catalyzed by heterogeneous catalysts, and require rather harsh conditions (elevated temperatures, especially for dehydrogenation reactions) [18, 19]. The possibilities of H_2-storage offered by homogeneous catalysis are also actively studied recently [17, 22]. Regeneration of the liquid organic hydrogen carriers (LOHCs; hydrogenation of the decomposition products) is usually done in separate devices and at different times than their actual application as fuel. This resembles the present utilization of fossil-derived fuels for vehicles or for other purposes.

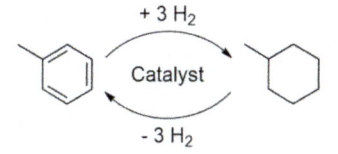

Figure 37.2 Reversible hydrogen storage in the toluene – methylcyclohexane interconversion.

Typical liquid organic hydrogen carriers include aromatic hydrocarbons, N-heterocyclic or other N-containing compounds, methanol, formic acid and its salts (Table 37.1), and several others [18, 19].

The capacity of hydrogen carriers (storage materials) is characterized by their gravimetric and volumetric hydrogen densities (capacities). Gravimetric hydrogen density (capacity) is defined as the mass of available hydrogen in unit mass of the given hydrogen carrier (wt. %). Similarly, the volumetric hydrogen density means the amount of available hydrogen in the decomposition of a hydrogen storage material of unit volume. The amount of available H_2 is often given as its volume under standard conditions, but can also be in units of mass, moles, or even as the energy (kJ) of its combustion. These definitions refer to single compounds, such as an alcohol or a hydrocarbon, and can be regarded as the theoretical maximum of hydrogen storage capacities. Nevertheless, solid hydrogen storage materials can also be used in solutions (these must be made with a chemically stable solvent). When working with solutions, the solubility of the hydrogenated storage material limits the theoretical amount of deliverable H_2 (deliverable capacity). Conversely, when the solubility of the dehydrogenated storage material is lower than that of the hydrogenated form (this is the case with the $MHCO_3/MHCO_2$ pairs, M = alkali metal ions) exploitation of only a part (i.e. the usable capacity) of the theoretically deliverable hydrogen is possible under homogeneous conditions.

In addition to its hydrogen capacity, the choice of a given LOHC depends on many other factors, such as chemical stability and side reactions, reversibility of hydrogenation/dehydrogenation, rate of charge/discharge, solvent (if any), operational temperature, corrosion, toxicity, investment and operational costs, and other technical considerations.

Table 37.1 Hydrogen storage materials and their theoretical gravimetric hydrogen capacity (wt. %).

CO_2/Formic acid	4.4
$CsHCO_3$ /$CsHCO_2$ (70% aq. soln)	0.8
CO_2/Methanol	12.6
Toluene/Methylcyclohexane	6.2
Benzyltoluene/Perhydrobenzyltoluene	6.1
Dibenzyltoluene/Perhydrodibenzyltoluene	6.2
Quinoline/1,2,3,4-Tetrahydroquinoline	3.0
N-Ethylcarbazole/Perhydro-N-ethylcarbazole	5.8

The use of formic acid (FA) as a LOHC deserves a special mention here. Formic acid has a gravimetric hydrogen density of 4.4 wt. %, whereas its volumetric H_2 density is 53 g H_2/L, and this makes it an attractive hydrogen storage material. Extensive research over the last 40 years has identified extremely active and stable homogeneous and heterogeneous catalysts for the decomposition of FA to H_2 and CO_2 (Eq. 37.4). In fact, a large number of homogeneous catalysts have been designed and studied in detail for FA decomposition, which resulted in an enormous contribution to the general field of chemical hydrogen storage.

$$HCO_2H \rightleftharpoons H_2 + CO_2 \tag{37.4}$$

It has been demonstrated unambiguously that this decomposition can supply $H_2 + CO_2$ gas mixtures suitable for use in PMEFCs with no prior separation of CO_2 from H_2. Because carbon monoxide is a poison for catalysts used in the fuel cells, it is important that such CO_2/H_2 mixtures contain less, than 10 ppm CO. The pressure of the generated gas mixture can reach as high as 120–140 bar [24–26], so that the fuel cells can operate under their optimal pressure with high efficiency. Furthermore, direct formic acid fuel cells may be available in the foreseeable future. One major drawback of using formic acid as an H_2 storage material is that at temperatures, required for the practical delivery of the stored hydrogen, FA is highly corrosive. Besides, despite promising recent results, direct hydrogenation of CO_2 to FA is not possible on an industrial scale without the stoichiometric use of suitable bases, such as amines or alkalies, although electrolysis of aqueous electrolytes under CO_2 pressure may produce FA solutions suitable as H_2 storage media.

37.5 Definitions and Fundamental Questions

Because this chapter deals with hydrogen storage in so-called hydrogen batteries, it is time to define what this term covers. A hydrogen battery is a device in which H_2 can be absorbed at a given temperature and H_2 pressure and which is suitable for delivery of the stored hydrogen upon the sole modification of the temperature and/or H_2 pressure. This definition requires that in case the hydrogen storage and delivery are catalytic processes with both directions efficiently catalyzed with the same catalytically active compound (heterogeneous or homogeneous catalyst). Furthermore, the cyclic H_2 charge and discharge of a H_2-battery do not require continuous or cyclic addition of any auxiliary chemicals (e.g. acids or bases). Altogether, the operation of a H_2-battery is much similar to an electric battery. This is exemplified in Figure 37.3.

This definition has rather strict requirements regarding the chemical process of hydrogen storage. First, the reaction (Eq. 37.1) should be fully reversible, with no degradation of the storage material during repeated cycles. Second, the eqilibrium between the hydrogenated and dehydrogenated form of the storage material must be mobile (kinetic requirement) and (under the technically feasible pressure and temperature conditions) must not be in extreme favor of one or the other (thermodynamic requirement). Any required catalyst, too, should be chemically stable during the repeated cycles of hydrogen storage and delivery. These requirements highly limit the number of chemical reactions which can serve as basis for hydrogen batteries which truly conform to this definition.

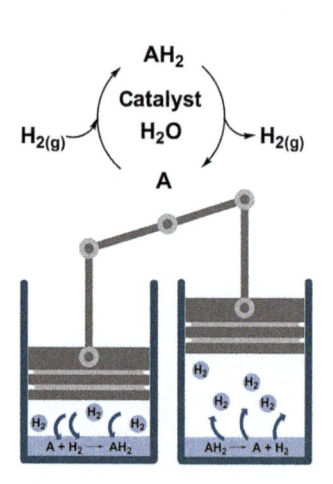

Figure 37.3 General scheme of a hydrogen battery. Reproduced with permission from Ref [27] / John Wiley & Sons.

A hydrogen battery is not a closed system thermodynamically, because there is cyclic in/out mass transfer of H_2. In the literature, though, one may find mention of so-called open hydrogen storage systems. This description is used for systems composed of two processes, in which the hydrogen storage (half cycle) occurs with one catalyst, and the hydrogen delivery (the other half-cycle) proceeds with another catalyst, usually after some kind of physical separation of the storage material. In several cases, the corresponding reactions of the hydrogen storage material have been demonstrated, but the two half cycles were not coupled to result in a H_2-battery that could satisfy the definition.

These considerations lead us to the question: What are H_2-batteries good for? Their most obvious feature is simplicity, in that a single reactor serves both for the charge and for the discharge of H_2 upon only reasonable changes in pressure and temperature. This may be advantageous at remote or hardly accessible locations, where renewable solar, wind, and other energy is available in abundance but with fluctuating intensity, which prevents direct supply of the generated electricity to the grid. In these cases, application of a charge/discharge H_2-buffer device can be useful to achieve the required constant electrical parameters. Hydrogen batteries may also be practical, when the charged storage material produced by the individual wind, solar, or other generators can be collected in large reservoirs and processed (dehydrogenated) in central units for feeding H_2 to fuel cells. With processes for sufficiently fast hydrogenation of the oxidized (depleted) storage materials, even mobile applications of hydrogen batteries can be envisaged, provided that H_2 becomes available at "gas" stations of the future. Alternatively, freshly charged containers of H_2-storage materials could simply replace the exhausted ones (which may be recharged at the filling station). Obviously, in the latter cases the theoretical mass or volumetric H_2-density of the storage material is of paramount importance (together with the efficiency of the discharge process). These considerations are valid for all types of H_2 storage, but of course, some particular properties of a storage process may contribute with different weight than others to the overall efficiency (and applicability) of the actual method of hydrogen storage.

For energy requirements, the bottom line is that the energy consumed for H_2 storage must be only a small fraction of the utilizable energy of the stored hydrogen. This is why storage of liquid hydrogen is not preferred (high costs of liquification and cooling in addition to safety concerns), and this is also the reason for which an ideally functioning storage device should make possible the delivery of the total (theoretically available) amount of stored H_2.

In chemical hydrogen storage, the storage step is the hydrogenation of a suitable substrate. This reaction is usually an exothermic process at the mild temperature and pressure conditions of H_2 entry into storage. (Elevated hydrogen pressure may be used for speeding up the charging of the storage device.) Conversely, H_2 removal from storage generally requires high temperature and low H_2 pressure. In case both processes result in high conversions, the thermal efficiency (so the incurred costs) of the total charge/discharge process mostly depends on the thermodynamic parameters of the reaction partners (H_2 and the storage material). The process parameters, such as the temperature difference between the charge and discharge steps (heating and cooling requirements), and the need for compression/decompression of H_2, together with the general costs of operation (e,g, those of moving the liquid storage solution, automatization, continuous monitoring of the process, safety precautions) must also be taken into account. Consequently, for the practical utilization of chemical hydrogen storage devices, heat management is of utmost importance (e.g. use of proper insulation, heat-exchangers) [18, 28, 29]. It should also be mentioned that when the stored hydrogen is converted to electricity in fuel cells, the heat generated by H_2 oxidation can be utilized to cover (at least in part) the heat requirement of the discharge process [29]. In general, these considerations on heat management are applicable to both stationary and mobile equipments which use fuel cells and stored H_2.

37.6 Catalysts Applied in Hydrogen Batteries

Reduction of oxidized (dehydrogenated) storage materials with H_2 can be carried out both with heterogeneous and homogeneous catalysts and also with biocatalysts (with the use of isolated enzymes or whole-cell systems). The catalysts applied in hydrogen batteries can be chosen from any of the categories described in this section. Heterogeneous catalysts [30] include the well-known Pd/C [31], but may be much more complex (e.g. metal nanoparticles, Pd nanoclusters supported on graphitic carbon nitride [GCN]) [32, 33] or a bimetallic single atom catalyst supported on reduced graphite oxide (rGO) [34]. In hydrogenation of CO_2 to formate in the presence of 1,8-diazabicyclo[5.4.0]undec-1-ene (DBU), Ir-complexes with polymerized cyclic (alkyl)(amino)carbene ligands (PCAAC-Ir), **6**, were found more effective than their un-polymerized analogs [35]. Homogeneous catalysts are exemplified by well-known transition metal complexes, such as $[RuCl_2(PPh_3)_3]$ (PPh_3 = triphenylphosphine) or the similar, but water-soluble, dimeric $[\{RuCl_2(mtppms-Na)_2\}_2]$ ($mtppms$-Na = *meta*-monosulfonated triphenylphosphine, sodium salt; **7**). Mixed ligand phosphine-NHC (NHC = N-heterocyclic carbene) complexes, such as [Ir(cod)(emim)(*mtppms*)] (cod = 1,5-cyclooctadiene, emim = 1-ethyl-3-methyl-imidazole-2-ylidene) also proved excellent catalysts both for hydrogenation and dehydrogenation reactions. So far, only one hydrogen battery is known which is based on biocatalysis.

When required, heterogeneous (contact) catalysts are easily removed from the liquid hydrogen storage medium (e.g. by stopping the flow through the catalyst bed). This is an obvious advantage when charged batteries are transported between locations, or when the supply of H_2 as fuel has to be discontinued. Conversely, tailoring the properties of a heterogeneous catalyst to the specific chemical process and reaction parameters is usually more difficult than in the case of soluble metal complexes. Dissolved metal complex catalysts come with the advantage of all metal ions being available for catalysis in contrast to the availability of active catalytic centers (atoms or their assemblies) only on the surface of a heterogeneous catalyst. However, a homogeneous catalyst resides in the same solution as the storage material, consequently, triggering and stopping the hydrogen evolution must be controlled by means other than physical separation. In general, this can be achieved by setting the optimum temperature both for the storage and delivery steps. An universal strict requirement of applicability of a catalyst (either homogeneous or heterogeneous) is its high chemical stability under operational conditions of the battery. By definition, there is no possibility to replace or reactivate a spent catalyst within the battery without removing the liquid storage material (not even mentioning the costs of frequent catalyst regeneration). Oxidative deactivation of catalysts is usually not a real danger in the reductive atmosphere of a battery, however, other types of reactions leading to loss of activity are known from long-run industrial processes. For example, hydrogenolytic splitting of phenyl groups of tertiary phosphine ligands was detected in hydroformylation reactions [36]. However, similar stability studies under real life conditions and long time-on-stream periods are scarcely found in the literature on hydrogen batteries. High chemical stability of the storage material and auxiliaries (if any) is an important requirement per se, but also because the decomposition products may poison the catalysts. In addition to their effect on catalytic efficiency, multidentate ligands, such as $(^iPr)_2PCH_2CH_2P(^iPr)_2$ [37], $P(CH_2CH_2PPh_2)_3$ [38, 39] or chelating *bis*-NHC-s, **10, 11** [40, 41] provide increased stability to the complexes. For the same reason pincer-type complexes (several examples on Figure 37.4) are also among the preferred catalysts. Examples of these catalysts are shown in Figure 37.4.

Chemical hydrogen storage, including the use of hydrogen batteries, too, is envisaged from small to very large scale applications. Consequently, the price and availability of the catalysts are of crucial concern. So far, the most investigated and best performing catalysts contain precious metals

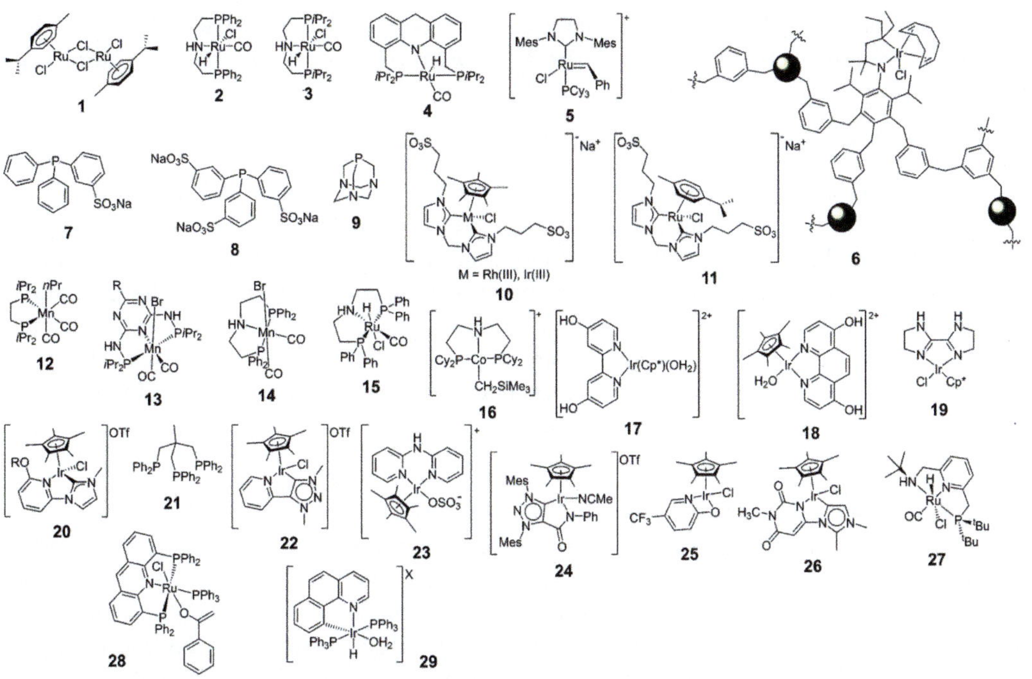

Figure 37.4 Selected ligands and catalysts for chemical hydrogen storage.

(Pd, Ru, Rh, Ir). However, these metals are not only expensive, but all are among the "endangered elements" in the category of "rising threat from increased use" [42]. Intensive research is being devoted to development of catalysts based on more abundant and cheaper metals [43], such as Fe [38, 44], Mn **12**, **13**, **14** [37, 45, 46], and Co **16** [47]. Very recently, a Mn(I)-pincer complex, **13**, has been described with appropriate catalytic properties for use in hydrogen batteries (Figure 37.4) [45].

37.7 Formic Acid and Formate Salts as Storage Materials in Hydrogen Batteries

It is hard to distinguish precisely the chemical hydrogen batteries based on formate/bicarbonate equilibrium in aqueous solution from those operating on the principle of HCOOH decomposition to $H_2 + CO_2$. Although neat formic acid can be used for this purpose, in most cases the storage medium is an aqueous solution of FA, so the reactions take place in the presence of water. The rate of decomposition (gas evolution) as a function of the pH goes through a sharp maximum around the pK_a of formic acid, so much, that several catalysts show greatly reduced activity in very acidic FA solutions. For this reason, some formate or bicarbonate salt or alkali hydroxide (e.g. $KHCO_2$ or KOH, respectively) are given to the storage medium [39, 48, 49] (e.g. in a $HCO_2H:HCO_2Na = 9:1$ ratio) [48]. With the progress of formic acid decomposition, the reaction mixture becomes more and more basic due to the formate additive dissolved in the residual water content of the used FA, which again results in diminished activity of the applied catalyst. Correspondingly, in the known systems of catalytic FA decomposition with formate additives, the added formate salt is found unchanged after the complete dehydrogenation of formic acid. Such problems are not encountered in catalytic dehydrogenation of aqueous formate to bicarbonate, because the pH of the reaction mixture can change only between the pH of the aqueous solutions of the formate and bicarbonate salts, respectively (e.g. between pH 8.9 and 10.9 in case of the HCO_2Na/HCO_3Na pair at 1 M, 25 °C).

37.7.1 Formic Acid as a Hydrogen Storage Material

In the gas phase, hydrogenation of CO_2 to formic acid is thermodynamically unfavored with a $\Delta G_{r,298 K} = +32.9$ kJ/mol. In an aqueous solution, however, hydration of the reactant gases and the product formic acid makes the reaction thermodynamically favorable with $\Delta G_{r,298 K} = -4$ kJ/mol (Eq. 37.5). Still, under a gas phase with partial pressures of $P(H_2) = P(CO_2) = 1$ atm, the calculated equilibrium concentration of formic acid is only $[HCO_2H]_{aq} = 1.33 \times 10^{-4}$ M [50]. Indeed, attempts of FA synthesis in pure water produced only very low amounts of formic acid even under elevated gas pressures [51–53]. Clearly, this reaction is not practical for large scale synthesis of formic acid.

$$CO_{2(aq)} + H_{2(aq)} \rightleftharpoons HCO_2H_{(aq)} \tag{37.5}$$

All processes described so far utilize some kind of additives, such as amines or other bases [4, 54–57], or water-organic solvent mixtures that include co-solvents such as DMSO [28, 53], THF [55, 58, 59], toluene [60], and ionic liquids [61–63] that facilitate dissolution of H_2 and CO_2 and may stabilize the product formic acid by hydrogen bond formation. Nevertheless, the highest FA concentrations achieved in these systems were typically around 2 M (well below of the concentration of neat formic acid; [FA] = 26.5 M). Aqueous-organic biphasic mixtures were also applied for continuous separation of formic acid from the reaction mixture to prevent its decomposition [64]. In a flow system, separation of formic acid from the purely aqueous reaction mixture was also achieved by electrodialysis, resulting in FA concentrations up to 2 M [65]. Interestingly, in aqueous solutions, addition of some cations (most often applied as formate salts) also resulted in some increase of the final FA concentration, up to 0.13 M [66–68]. This list of research results on formic acid synthesis is far from being comprehensive, because the main focus of this chapter is on hydrogen batteries.

For the reasons outlined in the previous paragraph, formic acid is not suitable for use in hydrogen batteries in the strictest sense. Conversely, it may serve as an important hydrogen storage material provided it is synthesized in separate chemical or electrochemical processes. Hence, a very brief outline of dehydrogenation processes is presented here. Again, the list of examples is non-exhaustive.

Decomposition of formic acid catalyzed by heterogeneous catalysts [69] (e.g. Pd/C) has long been known [30, 31]. Recently, active research is directed to the use of supported Pd nanoparticles [30, 32], as well as to mono- or bimetallic single atom catalysts on graphitic supports [33, 34]. Soluble catalysts (Figure 37.4) were first studied by Coffey, with the use of FA solutions in acetic acid, and of various Pt(II)-, Ru(II)-, and Ir(III)-complexes; $[IrH_2Cl(PPh_3)_3]$ was found the most active [70]. In their landmark studies [48], Fellay, Laurenczy, and Dyson first applied Ru(II) complexes prepared from $[Ru(H_2O)_6](tos)_2$ (tos = tosylate) and various water-soluble tertiary phosphines (such as $m$tppts-Na$_3$ = *meta*-trisulfonated triphenylphosphine Na-salt, **8**, or pta = 1,3,5-triaza-7-phosphaadamantane, **9**) as catalysts. Hydrogen was generated from aqueous formic acid solutions containing approximagely 10% Na-formate. Beller et al. studied Ru(II)-based catalysts with several monodentate and bidentate tertiary phosphine ligands, but the most active and stable catalysts were pincer-type complexes [such as e.g. **2** or **13**]. Louloudi et al employed a triglyme/water mixed solvent to study hydrogen storage with a Ru(II)–P(CH$_2$CH$_2$PPh$_2$)$_3$ (Ru-PP$_3$) catalyst [39]. At the beginning of the process, 20% of the used formic acid was converted to formate with addition of KOH. After decomposition of FA (but not of HCO_2K), the resulting solution could be stored with no protection from air and light and applied again for decomposition of a new batch of FA the next day (hence the term Use-Store-Reuse) and altogether for 30 days. Himeda [71–73] and Li [74] introduced Ir(III)-complexes, such as **17, 18** and **19**, respectively, with pH-sensitive multidentate N-donor ligands as catalysts of FA decomposition. Similarly, pH-responsive Ir(III) and Ru(II) complexes with pyridin-2-ol-based ligands, **20**, were synthesized and studied by Papish et al [75]. Papp et al described a straightforward synthesis of *cis,mer*-[IrH$_2$Cl(*mtppms*-Na)$_3$], which proved to be an extremely active catalyst for decomposition of formic acid [26]. The latter catalyst generated virtually CO-free (< 10 ppm) hydrogen, and was characterized by a turnover number (TON) of 674,000 (mol H$_2$) × (mol catalyst)$^{-1}$ and turnover frequencies (TOF) up to 298 000 (mol H$_2$) × (mol catalyst)$^{-1}$ × h^{-1}. Such catalytic activites may allow practical applications; however, other reqirements, such as catalyst stability, corrosion, availability of formic acid, should also be considered.

37.7.2 Formate Salts as Hydrogen Storage Materials

As defined in Section 37.5, a hydrogen battery should involve a reversible and mobile hydrogenation/dehydrogenation reaction in which the chemical equilibrium is not in extreme favor of any of the reactants. The bicarbonate/formate equilibrium in aqueous solution (Eq. 37.6) is well suited for this purpose [27, 50–52, 76–78].

$$HCO_3^-{}_{(aq)} + H_{2(aq)} \rightleftharpoons HCO_2^-{}_{(aq)} + H_2O_{(l)} \tag{37.6}$$

Russo et al. made a rigorous thermodynamic analysis of this equilibrium (Eq. 37.6) in the presence of various cations (H$^+$, Na$^+$, K$^+$, NH$_4^+$) [78]. They considered several parameters which may influence the position of the equilibrium, such as the nonlinear effects on the activity coefficients and operating conditions (concentration of solutes, pressure of hydrogen, temperature). Their paper gives a wealth of information on the thermodynamic properties of the compounds (ions) involved in the equilibria of the mentioned formate salts (and formic acid) Moreover, the values of equilibrium constants are tabulated for the 20–100 °C temperature range in 2.5 °C steps. This largely helps the reader in determining the available maximum conversions for both directions, depending on the operational mode of the battery. For example, for the dehydrogenation process this may be a continuous hydrogen discharge against a constant outside H$_2$ pressure, or hydrogen evolution in a closed vessel with increasing inside H$_2$ pressure until equilibrium is reached. Results of the thermodynamic calculations were compared to experimental findings from many sources (mostly observed with heterogeneous catalysts). It should be emphasized, that in lack of thermodynamic data for non-aqueous or mixed solvent systems, such calculations are presently possible only for fully aqueous reaction mixtures. Nevertheless, in those cases the calculated data can be used as benchmark for comparison of the various catalytic systems studied under widely different operational conditions [78]. One of the main conclusion of this work is that despite the relatively low hydrogen capacity of aqueous alkali formate salt solutions, development of such hydrogen storage systems is strongly advised due to the minimal safety risks in comparison to other methods of hydrogen storage.

During the operation of a hydrogen battery based on the bicarbonate-formate interconversion, reaction of bicarbonate may yield CO$_2$ according to Eq. 37.7.

$$2HCO_3^- \rightleftharpoons CO_3^{2-} + CO_2 + H_2O \tag{37.7}$$

The extent of this side reaction strongly depends on the concentration and the cation of the bicarbonate salt, the catalyst used for hydrogen storage, and the reaction temperature. However, in unfavorable cases concentration of CO_2 may reach as high as 30 vol % of the gas flow delivered by the battery [79]. This leads to loss of the storage material and also may hamper the use of the stored hydrogen in fuel cells. Obviously, generation of CO_2 during discharge of the battery must be minimized (eliminated) with the proper choice of the catalyst, the formate salt and the operating conditions of the battery. Autrey et al. analyzed chemical and technological (e.g. heat management) requirements of using bicarbonate/formate mixtures in aqueous systems in much detail [77].

Dehydrogenation of formate salts in aqueous systems was first studied with the use of Pd/C by Wrighton et al [31], and later by Sasson et al [80]. Generation of hydrogen with negligible CO content was also observed to accompany the transfer hydrogenation of aldehydes from aqueous formate catalyzed by [{$RuCl_2$(*mtppms*-Na)$_2$}$_2$] [81]. Similarly, the reversibility of hydrogenation of $Ca(HCO_3)_2$ was clearly demonstrated in aqueous solution with the use of [RhCl(*mtppms*-Na)$_3$] as the catalyst [67]. The first deliberate studies on using water-soluble Rh(I)-, Ru(II)-, and Ir(I)-complexes as catalysts for bicarbonate hydrogenation in aqueous solution (without additives such as e.g. amines) were published in 1999 [52]. Homogeneous catalysis of hydrogenation of a bicarbonate slurry to formate in a phase-transfer catalyzed toluene/water biphasic reaction system has also been described [82]. Very recently, Beller described a Mn(I)-pincer complex, **13**, which, in the presence of the potassium salt of lysine (LysK), was suitable for both CO_2 hydrogenation and formate dehydrogenation in H_2O/THF solutions (see also Section 39.9) [45]. Large-scale use of aqueous solutions of $KHCO_2$ as a hydrogen storage material was suggested by Sasson et al already in 1986 [83] and recent studies by Autrey et al [77], and Russo et al [78] clarified many of the requirements that should be fulfilled for a successful practical application of the bicarbonate/formate equilibrium in hydrogen storage. Excellent reviews of the field are also available [50, 76, 84].

An important aspect of the work with aqueous solutions of alkali bicarbonates and formates is the solubility of these salts (Table 37.2) [50, 76, 78, 85]. In general, with the same cation, at room temperature formates dissolve better in water than bicarbonates. Consequently, under fully homogeneous conditions, only a fraction of hydrogen present as aqueous formate can be utilized (deliverable capacity), since hydrogen availability (usable capacity) is limited by the lower solubility of the corresponding bicarbonate. It should be mentioned, though, that only H_2 enters/leaves the battery in a bicarbonate-formate hydrogen storage system. Solubility of alkali formates and bicarbonates is also influenced by the temperatures at which the hydrogenation and dehydrogenation processes take place.

Table 37.2 Solubility of formate and corresponding bicarbonate salts.

	Solubility, wt.%	T, °C	Solubility, mol/kg
HCO_2Na	44.8	22	6.59
HCO_3Na	8.72	20	1.04
HCO_2K	76.8	18	9.13
HCO_3K	25	20	2.5
HCO_2Cs	81.7	21	4.59
HCO_3Cs	67.8	20	3.5

Data from Linke, W.F. and Seidell, A. (1965). *Solubilities: Inorganic and Metal-organic Compounds*, 4th edition. Washington, D.C.: American Chemical Society.

37.8 Catalysts and Reaction Conditions Potentially Applicable in Hydrogen Batteries Based on the Formate-Bicarbonate Equilibrium

Next, we describe a few systems in which the applicability of the same catalyst for both the hydrogenation of bicarbonate and dehydrogenation of formate was demonstrated; however, the two half-cycles were not coupled and operated as a hydrogen battery.

Olah et al. used Ru(II)-PNP pincer complex **2** (Figure 37.4) for both hydrogenation of HCO_3Na and dehydrogenation of HCO_2Na in H_2O/THF (50 vol.%) solutions [86]. Dehydrogenations proceeded relatively slowly, but with no siginificant loss of activity in time. The total turnover number (TON) after six cycles of hydrogenation/dehydrogenation reached 11,500. In principle, construction of a hydrogen battery with this catalyst is feasible; however, such attempts were not described.

Beller et al. reported that a catalyst system comprising [{$RuCl_2$(benzene)}$_2$] and 1,2-bis-(diphenylphosphino)methane (dppm) was able to catalyze both hydrogenation of various alkali and alkaline earth bicarbonates in H_2O/THF (5/1 V/V) and dehydrogenation of the corresponding formates in DMF/H_2O (4/1 V/V) mixed solvents [79, 87]. Hydrogenation reactions typically proceeded at 70 °C under 50–80 bar H_2 pressure without or with CO_2 in the gas phase (up to 30 bar) with turnover numbers in the 320–1,754 in two hours. Hydrogen generation from the corresponding formates with the same catalyst was studied at 60 °C, and in a three hour reaction time 889–2000 TON was observed. HCO_2NH_4 reacted sluggishly with a TON of only 93. Following dehydrogenation of HCO_2Na with 90% conversion, the resulting HCO_3Na/HCO_2Na solution was evaporated to dryness and the solid residue was subjected to hydrogenation (with no further addition of catalyst), resulting in HCO_2Na in 80% overall yield. The dehydrogenation–hydrogenation (and vice versa) reactions were succesfully carried out with commercially available HCO_2Na and HCO_3Na, too. Based on these proof-of-concept findings, the system was suggested for construction of hydrogen batteries, although in the two half-cycles different reaction conditions (solvent mixtures, temperature, reaction time) were employed.

In an other publication [88], Xin et al disclosed important findings on the effect of Lewis-acid co-catalysts on decomposition of formic acid, and also on hydrogenation/dehydrogenation of bicarbonate/formate salts catalyzed by complexes formed in situ from [Ru(2-Me-allyl)$_2$(cod)], triphos (**21**) and Al(OTf)$_3$ in 50% aqueous dioxane. The reaction was investigated in a sealed NMR tube in three consequtive runs. At 90 °C and under 70 bar H_2 pressure, in all three cycles HCO_3Na was hydrogenated with approximately 30% conversion within 10–14 hours, whereas dehydrogenation of HCO_2Na proceeded with about the same conversions but required longer reaction times (10–30 hours). These experiments demonstrated that the catalyst system may allow construction of a hydrogen battery without any other additives.

Sasson made extensive studies on the decomposition of aqueous formate to H_2 and bicarbonate both with heterogeneous and homogeneous catalysts. In fact, he was the first to realize the possible industrial importance of this reaction. Already in 1986, his seminal paper [83] gave an account of the methods of hydrogen storage already known; furthermore, the technological aspects of the use of the formate-bicarbonate equilibrium for this purpose were also analyzed. In addition, Wiener, Sasson, and Blum studied the kinetic and mechanistic details of formate dehydrogenation with 10% Pd/C catalyst [80] during which <5% evolution of CO_2 was also observed. On the basis of the results of kinetic experiments (including also deuterium labeling), they suggested the following reaction mechanism, which described well the experimental findings.

$$Pd + HCO_2^- \rightleftharpoons \left[HPd(CO_2)\right]^- \qquad (37.8)$$

$$[HPd(CO_2)]^- + H_2O \rightleftharpoons [HPd(H_2O)(CO_2)]^- \quad (37.9)$$

$$[HPd(H_2O)(CO_2)]^- \rightleftharpoons [H_2Pd(HCO_3)]^- \quad (37.10)$$

$$[H_2Pd(HCO_3)]^- \rightleftharpoons Pd + H_2 + HCO_3^- \quad (37.11)$$

Recently, the synthesis of Pd/graphitic carbon nitride catalysts and their application for H_2 generation from aqueous formate solutions have been also studied by Sasson et al [33]. Nevertheless, a functional H_2 battery was not constructed with the use of these heterogeneous Pd catalysts.

Treigerman and Sasson applied a Ru(II)-pincer catalyst (Ru-MACHO, **2**) for the hydrogenation of sodium bicarbonate to formate in isopropanol-water 1:1 mixtures at 70 °C under 20 bar H_2 pressure [89]. Under such conditions, with a substrate/catalyst ratio, [S]/[C] = 1,600, a TON = 1083 was achieved in seven hours corresponding to 68% Na-bicarbonate conversion. The kinetic profile of the reaction (change of the pressure in the autoclave in time) referred to a typical second-order reaction. Interestingly, both isopropanol and water were required for the hydrogenation reaction; in the absence of water no H_2 uptake was observed, whereas in neat water the reaction was very slow (TON = 30 with [S]/[C]= 800). This may be due solely to the different solubilities of H_2 and HCO_3Na in the respective solvents. However, transfer hydrogenation of CO_2 is known [58], so a chemical effect can also be expected in this case. The stability of the catalyst was tested in five cycles. After completion of the reaction, the autoclave was cooled to room temperature, and the gas phase was vented. Following an overnight "rest" of the used catalyst solution under N_2, a new batch of HCO_3Na was added together with fresh isopropanol and water, then the autoclave was heated to 70 °C and repressurized with addition of 70 bar H_2. For each cycle, TON > 610 was determined.

With the aim of constructing a hydrogen battery, Treigerman and Sasson designed a separation method of formate from the reaction mixtures obtained by hydrogenation of bicarbonate as described previously [90]. The separation process used a strongly basic anion exchanger, Dowex-Cl, and approximately 70% of formate could be separated from dilute HCO_2Na solutions (0.05 M–0.50 M) in two steps. The anion-exchanger was regenerated with the use of 1 M HCl, leaving behind formic acid in the filtered solution. Because several efficient methods are known for decomposition of formic acid to H_2 and CO_2, the separation of formate as HCO_2H during acid regeneration of the anion-exchanger may be regarded as the final stage of a formic acid cycle. Such a cycle could start with CO_2, converted with a base to CO_3^{2-} and HCO_3^-, which are then hydrogenated to HCO_2^-, recovered as HCO_2H by ion-exchange and catalytically decomposed to H_2 and CO_2. Although such a reaction sequence is certainly possible, unfortunately each cycle (together with regeneration of the ion-exchanger) consumes 1–1 mol of a base and acid, respectively, which does not make the process practical for large scale storage and regeneration of H_2.

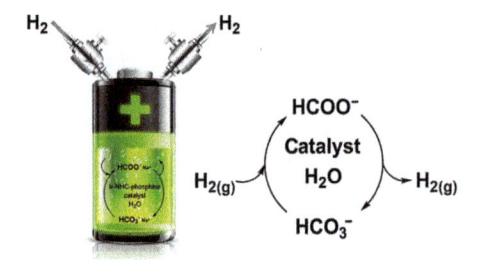

Figure 37.5 General scheme of a hydrogen battery based on the formate-bicarbonate equilibrium.

37.9 Functional Hydrogen Batteries

In the following sections, the term functional hydrogen battery is applied to systems or devices that were designed with the purpose of reversible H_2 uptake and delivery, with the corresponding chemical reactions catalyzed by the same catalyst in both directions, and the charge and discharge of the battery regulated only by the hydrogen pressure and/or the temperature.

37.9.1 Hydrogen Batteries Based on CO_2–Formic Acid Cycles

Plietker et al. designed a hydrogen battery with the use of the [RuCl(PNNP)(acetonitrile)]PF_6 complex (Figure 37.6) as the catalyst in the presence of DBU as a basic additive [91], In a typical experiment, 20 g dry ice, i.e. 455 mmol CO_2 and 65.7 mmol DBU reacted under 70 bar H_2 pressure and 100 °C to yield DBU formate salt with TON values around 2000 with 0.075 mol% catalyst. Addition of toluene ensured homogeneous conditions and increased the rate of CO_2 hydrogenation to formate. When the reaction stopped (indicated by no further pressure change), the reactor was cooled to room temperature and flushed with N_2. The resulting solution (DBU formate in toluene) could be stored at room temperature for days, and could be decomposed to yield a $H_2 + CO_2$ gas mixture by heating to 100 °C at ambient pressure. After cooling, and addition of fresh CO_2 and H_2 to the reactor, the hydrogenation and dehydrogenation steps could be repeated with no need of replenishing the catalyst, DBU, or toluene. This hydrogen storage/delivery cycle was repeated five times with no significant change of the rate of the discharge process and the amount of the delivered H_2. Furthermore, it was demonstrated that decomposition of DBU formate could supply a $H_2 + CO_2$ mixture of 22 bar total pressure which is advantageous for efficient use of the stored H_2 in PEM fuel cells. Gas evolution could also be regulated by changes in the reactor temperature, since no decomposition of DBU formate was detected at room temperature.

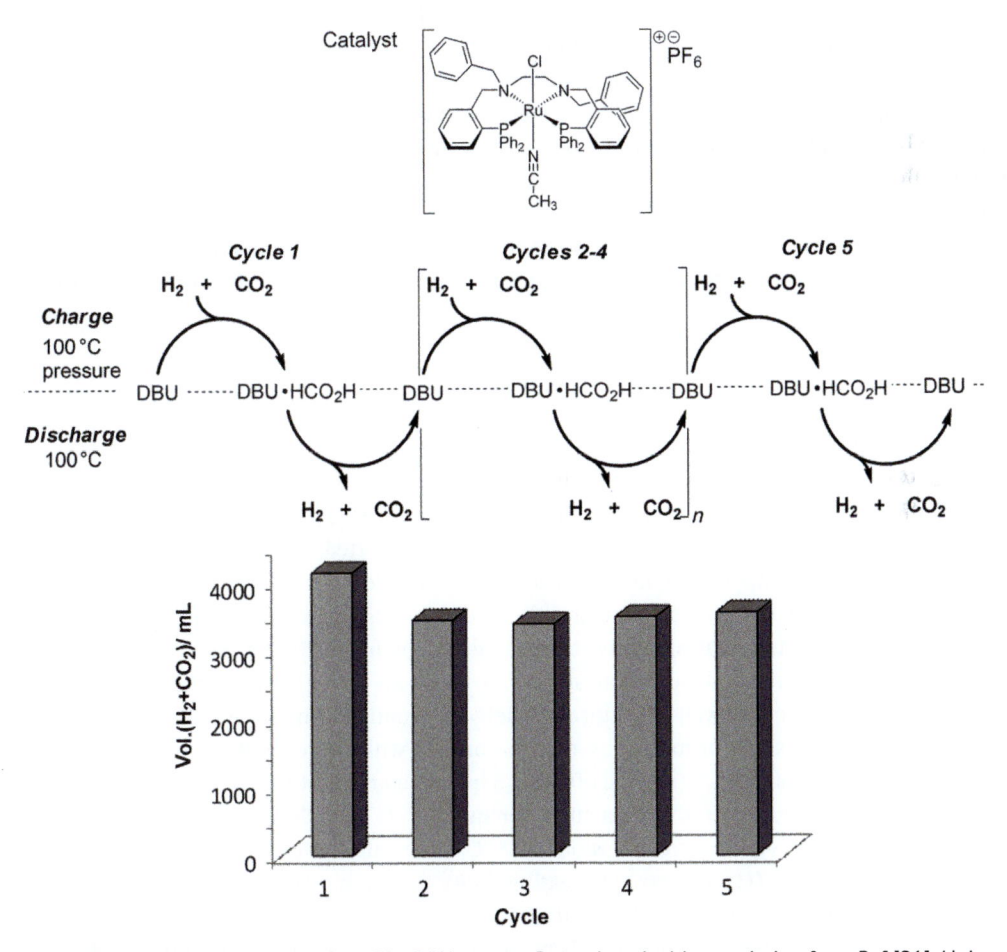

Figure 37.6 Hydrogen storage in a CO_2-DBU system. Reproduced with permission from Ref [91] / John Wiley & Sons.

As described in the paper of Plietker et al., this hydrogen battery requires addition of not only H_2 but also CO_2 to start each cycle. Furthermore, due to the high CO_2 excess, only about 15% of the used CO_2 is converted to DBU formate and the rest is vented at the end of the hydrogenation step. Also, in a complete H_2 storage/delivery cycle, the reaction mixture should be twice heated to 100 °C and cooled to room temperature to start a new cycle. Nevertheless, most of these problems may likely be eliminated by proper engineering solutions, and therefore the system has potential for practical applications.

Beller et al. made a very important step forward in the quest to develop viable hydrogen batteries based on reversible hydrogenation of CO_2 to formic acid. They have found that in water/THF mixtures, several [MnBr(CO)$_2$(PNP)] complexes were active both in hydrogenation of CO_2 and in decomposition of FA. The most promising results were obtained with the use of complex **13** (Figure 37.4). In the presence of the potassium salt of lysine, LysK, total turnover numbers of 2,000,000 for CO_2 hydrogenation and 600,000 in FA dehydrogenation were observed in studies of catalyst stability through several cycles. Dehydrogenations were achieved at 90 °C under ambient pressure conditions, whereas hydrogenations were run at 85 °C under 80 bar initial H_2 pressure. Cycling of the battery started with dehydrogenation of formic acid in the presence of equivalent amount of LysK. At the end of this step, the gas phase was vented and analyzed, while the solution phase was subjected to hydrogenation. LysK made possible the retention of >99.9% of the CO_2 formed in the dehydrogenation of FA (internal carbon dioxide capture); i.e. the evolved gas contained only less than 0.01% CO_2. At the same time, the CO content was less than 10 ppm (the detection limit with the GC method used). Consequently, in contrast to Plietker's system, the hydrogen storage and delivery steps coud be repeated without replenishing CO_2 (or other components of the battery) between the cycles. The operation of this battery requires equivalent amounts of lysine to the FA used in the first dehydrogenation step (or to the desired FA concentration to be obtained in the hydrogenation step). However, lysine is an industrial product produced in large quantities via microbial fermentation so its availability may not limit the practical application of the H_2 storage process. The battery was operated for 10 charge/discharge cycles on a 90.0 mmol scale, which shows the process to be suitable for scale-up and gives good chances for large-scale applications [45].

37.9.2 Hydrogen Batteries Based on Formate–Bicarbonate Cycles

Cao et al. developed a fully aqueous hydrogen battery based on heterogeneous catalysis of HCO_2K/HCO_3K interconversion under pressurized H_2 and pressure-free conditions, respectively [92]. They synthesized a catalyst comprised of Pd nanoparticles supported on reduced graphene oxide (rGO). The strong metal-support interaction resulted in distortion of the crystal structure of the Pd-NP-s (leading to so-called lattice microstrain), which, in turn, largely increased the catalytic activity of the rGO-suported catalyst relative to Pd catalysts obtained with the use of other carbon-based supports. This heterogeneous catalyst proved to be an active and durable catalyst for bicarbonate hydrogenation as well as of formate dehydrogenation. Among the studied formate salts, HCO_2K was by far superior to Na-, Li- or NH_4-formate with regard to its reactivity and stability. In hydrogenation of HCO_3K (storage of H_2) of 5 mL aqueous 4.8 M HCO_3K solution containing 9.6 μmol Pd under $P(H_2)$ = 40 bar, a turnover number TON = 7088 was achieved (94.5% bicarbonate conversion). After decompression at 25 °C, the resulting formate solution was subjected to dehydrogenation (H_2 delivery) by heating to 80 °C, yielding H_2 gas by nearly complete conversion of formate in 40 minutes. This charge/discharge cycle was repeated six times with no impairment of the battery performance. Furthermore, when a charged hydrogen storage solution was stored for four months under ambient conditions, subsequent heating to 80 °C led to hydrogen evolution at the same rate and amount than in the previous cycles. These observations unambiguously showed the stability of the catalyst. It is important to note, that even at high HCO_2K concentrations (>8 M) and temperatures (>150 °C) only a very small amount of CO_2 (<0.05 vol%) was generated. The discharge process was very sensitive to the temperature with apparently no reaction at or below 30 °C, which allowed regulation of the rate of H_2 generation by simply changing the temperature of the reaction mixture. This hydrogen battery has the advantageous properties of being simple, fully aqueous, with only inorganic hydrogen storage materials and no need for other (organic) additives. Cao et al. estimated that under the best conditions of their studies, 5 L of 4.8 M HCO_2K solution could give off enough H_2 to fuel a 1 kW fuel cell, provided a continuous regeneration or replacement of the energy-rich formate solutions could be performed [92].

During our studies on the hydrogenation of bicarbonate [51, 52, 67, 93] and transfer hydrogenation of aldehydes catalyzed by [{$RuCl_2$(mtppms-Na)$_2$}$_2$] [81], we noticed the evolution of hydrogen gas from aqueous formate solutions. Closer examination of this phenomenon revealed, that [{$RuCl_2$(mtppms-Na)$_2$}$_2$] was a potent catalyst for hydrogen storage and delivery in the bicarbonate–formate equilibrium (Eq. 6), which allowed for the construction of a hydrogen battery [27]. The charge and discharge processes were studied both in medium pressure glass reactors and in high pressure sapphire NMR tubes. In a closed glass reactor, at a temperature of 80 °C, [Ru] = 2 mM and an initial $[HCO_2]^- $ = 2.5 M, the amount of liberated hydrogen (as indicated by the increasing pressure in the reactor) followed a saturation curve. Under these conditions, the formate/bicarbonate equilibrium was attained at 6.2 bar H_2 pressure in 250 minutes (32% conversion of formate). Then the pressure of the gas phase was adjusted to 2.0 bar, which initiated further dehydrogenation of formate leading to an equilibrium pressure of 3 bar H_2 in 250 minutes. Hydrogenation of bicarbonate proceeded with 90% conversion at 83 °C and $P(H_2)$ = 100 bar reaching saturation concentration of formate in 200 minutes. These experiments showed, that, indeed, the bicarbonate-formate equilibrium could be reached reasonably quickly from both sides with the use of the [{$RuCl_2$(mtppms-Na)$_2$}$_2$] catalyst. The hydrogenation and dehydrogenation reactions of bicarbonate and formate, respectively, were succesfully coupled into a cyclic charge/discharge operation of a hydrogen battery. This was demonstrated by the use of a high pressure NMR tube in the reaction of [13] C-labeled bicarbonate/formate. As seen in Figure 37.7, hydrogenation of bicarbonate was fast at 83 °C and $P(H_2)$ = 100 bar, however, dehydrogenation at the same temperature was relatively slow (initial $P(H_2)$ = 1 bar), so that a complete charge/discharge cycle required about 1200 minutes. Again, in the closed NMR tube, dehydrogenation stopped about 45% conversion of formate to bicarbonate. These conversions observed in our measurements agree well with the value calculated by Russo et al recently [78]. However, in practical application of a hydrogen battery the

Figure 37.7 Comparison of the catalytic activity of the [{$RuCl_2$(mtppms-Na)$_2$}$_2$] and Na_2[Ir(cod)(NHC)(mtppts)] catalysts in hydrogen storage in aqueous solution of HCO_2Na and HCO_3Na, respectively. Reproduced with permission from Ref [95] / John Wiley & Sons.

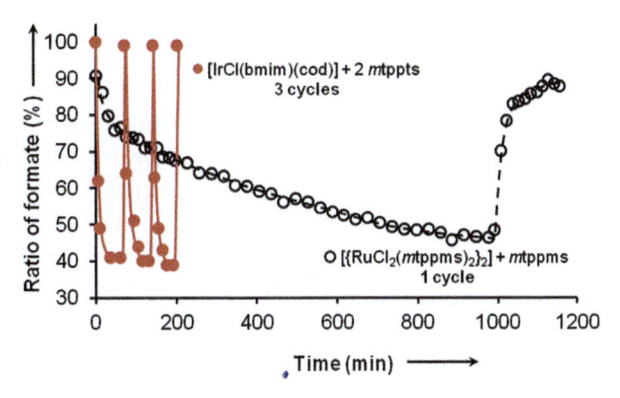

evolved hydrogen is continuously released from the reactor wich allows close-to-complete delivery of the stored amount of H_2. Importantly, the hydrogen obtained in the decomposition of aqueous formate contained ≤3 ppm CO. This was the first hydrogen battery, which operated in a purely aqueous bicarbonate/formate soution without the need of any additives, and in which the charge/discharge process was regulated solely by the hydrogen pressure.

In our investigations into aqueous organometallic catalysis, we have synthesized a series of water-soluble [IrCl(cod)(NHC)] (Figure 37.9) and Ir(I)-NHC-phosphine complexes with the general formula [Ir(cod)(NHC)(*mtppms*)] or Na_2[Ir(cod)(NHC)(*mtppts*)] [94]. Several of these complexes showed high catalytic activity in both the hydrogenation of bicarbonate and dehydrogenation of aqueous formate [95]. For example, in aqueous solutions [Ir(cod)(bmim)(*mtppms*)] (bmim = 1-butyl-3-methylimidazole-2-ylidene) reduced HCO_2Na at 80 °C and 4 mol% catalyst loading with a turnover frequency TOF = 310 h^{-1}. Very importantly, with these Ir(I)-NHC-phosphine complexes, dehydrogenation was also unprecedently fast, exemplified by the turnover frequency TOF = 15 110 h^{-1} achieved with the [Ir(cod)(emim)(*mtppms*)] (emim = 1-ethyl-3-methylimidazole-2-ylidene) catalyst at 80 °C. This catalyst also tolerated very high substrate/catalyst ratios (≥ 10 000), an important feature for construction of high capacity hydrogen batteries.

The reversible hydrogen storage was studied in a high pressure NMR tube with the use of [Ir(cod)(emim)(*mtppms*)] as the catalyst, and $H^{13}CO_2Na$ in aqueous solutions at 80 °C. The results revealed that a full charge/discharge cycle required only 60 minutes, in sharp contrast to 1,200 minutes needed in the case of the [{$RuCl_2$(*mtppms*-Na$)_2$}$_2$] catalyst (Figure 37.7). The catalyst showed remarkable stability. When the reaction mixture was stored under 100 bar H_2 for 71 days at room temperature, and then subjected to the same cycling procedure as before, several new charge/discharge cycles could be observed with unchanged rates of the hydrogenation and dehydrogenation steps. The hydrogen storage was also studied an a flow system [96]. The formate/bicarbonate storage solution was continuously pumped through a heated tubular reactor and the generated hydrogen was collected in a coupled atmospheric gas burette. In this experimental setup, we could not vary the pressure; however, the rate of hydrogen evolution could be regulated by changes in the temperature. No reaction was found at 25 °C, but a fast H_2 evolution was observed with the use of aqueous solutions of cesium formate (c = 81 mM) and [Ir(cod)(emim)(*mtppms*)] (c = 0.1 mM) at elevated temperatures. Hydrogen generation could be triggered and stopped by switching the temperature in the reactor back and forth between 100 °C and 25 °C. The pronounced sensitivity of the reaction rates of both in the hydrogen uptake and delivery steps, offers a possibility to construct a hydrogen battery working at constant H_2 pressure.

37.9.3 Hydrogen Batteries Based on N-heterocyclic Compounds

Iridium complexes, such as e.g. **22, 23, 25** (Figure 37.4), play a prominent role in homogeneous catalysis of hydrogenation and dehydrogenation of N-heterocyclic compounds [17, 97, 98]. Several triazolylidene-ligated Ir-NHC-complexes (e.g. **22**) were used as catalysts for this purpose in aqueous systems [98, 99].

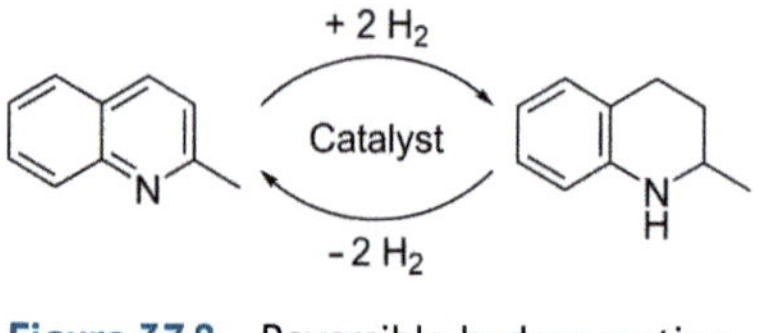

Dehydrogenation of the reduced hydrogen storage materials (such as e.g. 1,2,3,4-tetrahydro-2-methylquinoline, (TH-2-MeQ), Figure 37.8) is a thermodynamically uphill reaction, therefore it requires high energy input to achieve reasonable reaction rates. As an example, a series of N-heterocycles were dehydrogenated in aqueous reaction mixtures with the use of Ir-dipyridylamine complexes

Figure 37.8 Reversible hydrogenation of 2-methylquinoline (2-MeQ) to 1,2,3,4-tetrahydro-2-methylquinoline (TH-2-MeQ).

(e.g. **23**) as catalysts at reflux temperature for 30 hours, resulting in 67–93% isolated yields [99]. It is therefore of high interest to find photochemical procedures for speeding up such dehydrogenations [100]. Indeed, Mata et al have found that an Ir(III)-complex containing mesoionic triazolylidene, pentamethylcyclopentadienyl (Cp*), and acetonitrile ligands (**24**) catalyzed the dehydrogenation of various N-heterocycles at room temperature under blue light irradiation (LED, 455 nm) [101].

In their ground-breaking research, Yamaguchi et al showed that it is possible to construct a hydrogen battery based on the interconversion of N-heterocycles and their hydrogenated products [97]. With the use of the Cp*-Ir(III)-5-trifluoromethylpyridonate complex, **25**, they have dehydrogenated 1,2,3,4-tetrahydro-2-methylquinoline (TH-2-MeQ) with 100% conversion to 2-MeQ in refluxing p-xylene (b.p. 138 °C) under Ar for 20 hours. Then the mixed H_2/Ar atmosphere in the flask was replaced with 1 bar H_2, and the solution was stirred at 110 °C for 20 h during which the 2-MeQ product of the dehydrogenation step was converted back to TH-2-MeQ, again with 100% conversion. This reaction sequence was repeated for a total of five cycles, with a conversion of 98% even in the fifth cycle in both directions. Although the reactions were rather slow and the conditions rather harsh for practical applications, only a single catalyst was needed, and the H_2 storage-release process was regulated mainly by changes in hydrogen pressure (and to a smaller extent by the change in the temperature).

Choudhury et al. developed Ir-NHC complexes for the bidirectional catalysis of hydrogenation/dehydrogenation of various N-heterocycles [102]. Specifically, an abnormal Cp*Ir(III)-NHC complex, **26**, with 1,2-dimethyl-3-uracilyl-4-imidazolylidene as the NHC ligand, proved suitable for construction of a hydrogen battery [103]. This complex could be obtained with the synthesis of the respective Ag(I)-NHC compound with Ag_2O, followed by transmetallation with [{$IrCl_2Cp^*$}$_2$]. Because the catalyst dissolves well in water, hydrogenation of various quinolines and quinoxalines, as well as dehydrogenation of the corresponding reduced products, were studied in aqueous reaction mixtures. Up to 98% yields were obtained in hydrogenations at 50 °C with a reaction time of four hours with 1 mol % catalyst. Dehydrogenations with the same catalyst loading required a higher temperature (100 °C) and longer reaction times (12–36 hours); however, yields up to 98% could be achieved under such conditions. The excellent water-solubility of the catalyst allowed its recycling after removing the organic products from the aqueous phase by extraction with ethyl acetate. With this method, the catalyst-containing aqueous phase was used in five consequtive runs with only 8% loss of the catalytic activity (which, at least partially, could be caused by phase separations between runs). In a proof-of-concept experiment, the H_2 liberated by dehydrogenation of 1,2,3,4-tetrahydroquinoxaline in one flask, was utilized for hydrogenation of phenyl-benzylimine (with the same catalyst) in an other container directly connected to the dehydrogenation flask. Although this desing of hydrogen storage and delivery does not match the strict definition of a hydrogen battery, its elements are, indeed, suitable for construction of a battery. However, the largely different reaction rates of hydrogenation and dehydrogenation (charged/discharge) may render this approach impractical.

37.9.4 Hydrogen Batteries Based on Alcohols

As discussed in Section 39.2, methanol is a suitable hydrogen storage material which may serve as platform chemical in a so-called "Methanol Economy". Other alcohols, di- and polyols (including bio-based alcohols, too) were also studied for hydrogen storage and generation [19, 104, 105], and the term "Alcohol-Based Hydrogen Economy" has been coined to draw attention to this class of possible H_2-storage materials [22]. Unfortunately, the hydrogen capacity of alcohols sharply decreases with increasing molecular weight (carbon content) of the alcohol in question, and becomes impractically low with aromatic substrates. Furthermore, catalytic dehydrogenation of alcohols usually requires high temperature and in many cases leads to a mixture of products which cannot be converted cleanly back to the hydrogen rich storage material. A notable exception is the system of Milstein et al. who reported the ruthenium-catalyzed reversible dehydrogenation of 2-aminoethanol with formation of a mixture of piperazine-2,5-dione (glycine anhydride) and a linear oligopeptide. The use of [RuHCl(CO)(PNN)], **27**, as the catalyst (0.5 mol%) resulted in 85% conversion of 2-aminoethanol (other conditions: 4 mL dioxane solvent, 1.2 mol% KOtBu 135 °C, 12 hours, Ar) [15]. The product mixture could be hydrogenated back to 2-aminoethanol with the same catalyst resulting in 85% yield under 50 bar H_2 pressure at 110 °C. The dehydrogenation/hydrogenation cycle coud be repeated several times with no need of replenishing the catalyst, demonstrating the applicability of this system for construction of a hydrogen battery.

Despite the many achievements in this field [22, 105], we are not aware of publications on functional hydrogen batteries based upon this concept.

We have found, that the [IrCl(cod)(NHC)] complexes, (NHC: emim; Bnmim=1-benzyl-3-methylimidazole-2-ylidene; mmim=1,3-dimethylimidazole-2-ylidene) and their phosphine-substituted derivatives, [Ir(cod)(NHC)(*mtppms*)] catalyzed the hydrogenation and transfer hydrogenation of aromatic ketones selectively to the corresponding alcohols, such as e.g. acetophenone to 1-phenylethanol [106]. We were pleased to observe, that the same catalysts were also active in the dehydrogenation of 1-phenylethanol to acetophenone upon lowering the hydrogen pressure and with no change in the temperature [107]. Hydrogenations were carried out in toluene solution at 95 °C, with a substrate:catalyst ratio 50:1 in the presence of t-BuOK under 10 bar H_2 pressure. Dehydrogenation of 1-phenylethanol was carried out under the same conditions, except that the reaction flask was continuously purged with argon at ambient pressure. Despite the mild reaction conditions, with [IrCl(cod)(mmim)] (Figure 37.9) as the catalyst, a TOF = 490 h^{-1} was achieved in hydrogenation, and TOF = 11 h^{-1} in dehydrogenation. Based on these findings, a

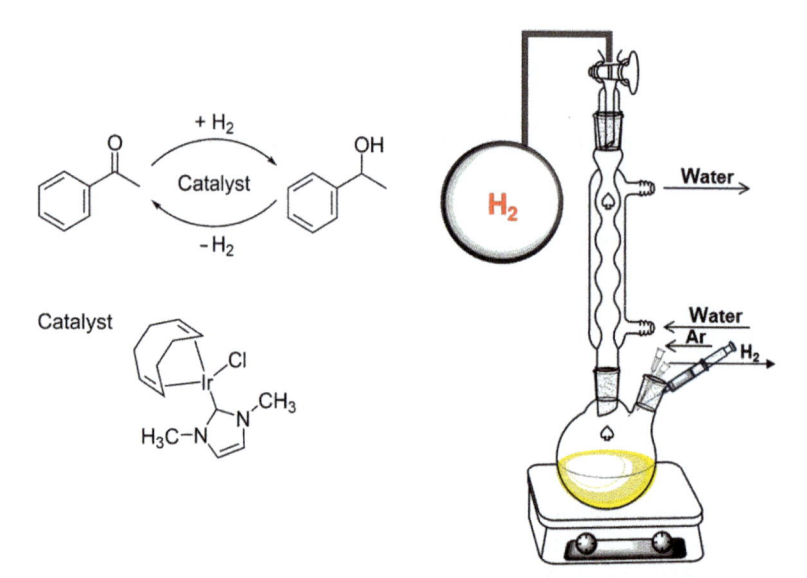

Figure 37.9 Reversible hydrogenation-dehydrogenation of acetophenone and 2-phenylethanol, respectively.

hydrogen battery was designed, in which both the hydrogenation and dehydrogenation were carried out under the stated reaction conditions but at 1 bar H_2 pressure in both directions. H_2 for the hydrogenation step was provided from a balloon. In the dehydrogenation step, the reaction flask and the ballon was first flushed with argon, and the evolved hydrogen was collected in the balloon (Figure 37.9). Progress of the reaction was followed by GC analysis of the reaction mixture.

Starting the reaction sequence with the dehydrogenation of 1-phenylethanol, 35% conversion into acetophenone was observed in 120 minutes, whereas in the following 60 minutes hydrogenation regime, the ketone concentration decreased to 5%. In the next dehydrogenation step (120 minutes) the concentration of acetophenone increased to 25%, followed by a decrease to 8% in the subsequent hydrogenation step (60 minutes). The data obtained with this very simple experimental setup demonstrate the applicability of the acetophenone-phenylethanol H_2 storage material and the [IrCl(cod)(NHC)] catalysts for construction of functional H_2 batteries, with an obvious need for further optimization [107].

37.9.5 Hydrogen Batteries Based on Whole-cell Biocatalysis

Acetogenic bacteria reduce CO_2 to acetic acid with intermediate formation of formic acid (FA). However, further reaction of HCO_2H to CH_3CO_2H can be blocked with the use of specific ionophores, such as monensin. Recently a hydrogen-dependent CO_2 reductase (HDCR) enzyme complex was identified in *Acetobacterium woodii* which requires H_2 for the production of formic acid, while in a H_2-free anaerobic atmosphere it dehydrogenates HCO_2H into a mixture of H_2 and CO_2 [108]. The isolated enzyme showed high catalytic activity at 30 °C both for production (TOF = 101600 h^{-1}) and decomposition of FA (141960 h^{-1}), which were substantially higher activities than those achieved previously with the use of synthetic homogeneous catalysts at such a low temperature. It was also disclosed, that K_2CO_3 too, could be reduced to formic acid with this enzyme [108].

Based on these findings, Müller et al. designed a bio-catalytic hydrogen battery [109]. Importantly, instead of the isolated HDCR enzyme, this battery used a resting *A. woodii* cell culture. Resting cells retain their full metabolic activity, however, cell-division does not occur due to the lack of essential nutrients. The working hours of the battery mimicked the daily periods of supposed photovoltaic electricity generation (i.e. eight hours of H_2 uptake [storage; day period] and 16 hours of H_2 release [night period]). Both FA generation and H_2 release proceeded in the same bioreactor, and the direction of the process was regulated by switching the purge gas concentration from 45% H_2, 45% CO_2 and 10% N_2 (day period) to pure 100 % N_2 (night period). In a buffered cell suspension with 1 mg/mL protein concentration, formic acid was produced at 30 °C with a rate of 2.3 ($\pm$ 0.8) mmol $\times g^{-1} \times h^{-1}$. During the night period (16 hours), the formic acid concentration decreased from 28 to 4.9 mM. This H_2 storage/release cycle was repeated for 15 days (360 hours total). During such cycling with the wild type of *A. woodii*, the available FA concentration showed a slow decrease in time from cycle to cycle, despite the presence of monensin, and after 72 hours, formation of acetic acid was detected in large quantities. This could be mostly eliminated by genetic engineering of the bacterium. The resulting mutant showed approximately the same catalytic activity than the wild type *A. woodii*, producing 23 mM FA in eight hours, and lowering the FA concentration to 5.4 mM during 16 hours of dehydrogenation. Under optimized conditions, formate was produced with a specific rate of 1.7 ($\pm$ 0.2) mmol $\times g^{-1} \times h^{-1}$. These proof-of-concept results are encouraging for development of a practical battery for H_2 storage and delivery, directly using CO_2 for generation of formic acid [110]. It should be mentioned that other microorganisms, such as *Escherichia coli* [111] and

Desulfovibrio desulfuricans [112] can also be used for whole-cell biocatalysis of both CO_2 reduction to FA, and for the oxidation of the latter to H_2 and CO_2. However, these were not studied in a single bioreactor (i.e. in a working model of a hydrogen battery).

37.10 Summary and Conclusions

Chemical hydrogen batteries are a special class of chemical hydrogen storage devices in which a hydrogen-lean storage material is catalytically hydrogenated to a hydrogen-rich compound (charge step), which can be dehydrogenated (discharge step) with lowering the hydrogen pressure and/or raising the temperature in the device. The storage materials can be one-component liquids (e.g. LOHCs: alcohols, N-heterocyclic compounds, formic acid) or solutions of solids, such as formate salts in water. Hydrogen batteries stand out of the rest of hydrogen storage devices due to their operational simplicity. Chemically stable and easily available storage materials, solvents (if any) and catalysts are inevitable for efficient hydrogen storage processes for long-term and large-scale applications.

Intensive research into various methods of chemical hydrogen storage identified several reversible hydrogenation-dehydrogenation processes together with both heterogeneous, homogeneous, and biocatalysts (applied as whole-cell catalysts) that may allow construction of hydrogen batteries. Nevertheless, the number of known functional hydrogen batteries is still low, because in most cases the hydrogenation and dehydrogenation steps require largely different conditions. The most promising of these batteries are based on the carbon dioxide-formic acid interconversion, the reversible hydrogenation-dehydrogenation of N-heterocyclic compounds and alcohols, as well as the interconversion of alkali formate salts and the corresponding bicarbonates in aqueous solution. All known examples have their weak and strong points in the following properties: deliverable hydrogen capacity, rates of both the charge and discharge steps, chemical stability of the components (including solvents, too), opportunities of heat management. When all of these requirements are considered together, aqueous formate equilibria deserve special attention. In addition to the previously mentioned chemical characteristics, such all-aqueous systems with stable inorganic storage materials provide the highest process safety in comparison to other methods of hydrogen storage.

Climate and energy crisis put research into energy storage under very high pressure. There is no question, that chemical hydrogen storage, including that in hydrogen batteries, will further play a most important role in solving these problems of our time.

Acknowledgements

This research was supported by the EU and co-financed by the European Regional Development Fund under the projects GINOP-2.3.3-15-2016-00004 and GINOP 2.3.2-15-2016-00008. Project TKP2020-NKA-04 has been implemented with support provided from the National Research, Development and Innovation Fund of Hungary, financed under the 2020-4.1.1-TKP2020 funding scheme. Financial supports from the Hungarian National Research, Development and Innovation Office (FK-128333), and from the National Laboratory for Renewable Energy, project No. RRF-2.3.1-21-2022-00009, are gratefully acknowledged.

References

1 Cokoja, M., Bruckmeier, C., Rieger, B. et al. (2011). *Angew. Chem. Int. Ed.* 50 (37): 8510–8537. doi: 10.1002/anie.201102010.
2 Eberle, U., Felderhoff, M., and Schüth, F. (2009). *Angew. Chem. Int. Ed.* 48 (36): 6608–6630. doi: 10.1002/anie.200806293.
3 Hauer, A. (2022). *Advances in Energy Storage*. ZAE-Bayern Garching, Germany: John Wiley & Sons Ltd.
4 Leitner, W. (1995). *Angew. Chem. Int. Ed. Engl.* 34 (20): 2207–2221. doi: 10.1002/anie.199522071.
5 Monteiro, J. and Roussanaly, S. (2022). *J. CO2 Util.* 61: 102015. doi: 10.1016/j.jcou.2022.102015.
6 Wang, Y., Pang, Y., Xu, H. et al. (2022). *Energy Environ. Sci.* 15 (6): 2288–2328. doi: 10.1039/D2EE00790H.
7 Haynes, W.M. (2014). *CRC Handbook of Chemistry and Physics*, 95e. Boca Raton: CRC Press, LLC.
8 U.S. Department of Energy *Alternative Fuels Data Center*. https://afdc.energy.gov/fuels/properties (accessed 4 July 2023).
9 Olah, G.A., Goeppert, A., and Prakash, G.K.S. (2018). *The Methanol Economy*, 3e. Weinheim, Germany: Wiley-VCH.
10 Onishi, N. and Himeda, Y. (2022). *Coord. Chem. Rev.* 472: 214767. doi: 10.1016/j.ccr.2022.214767.
11 Heim, L.E., Thiel, D., Gedig, C. et al. (2015). *Angew. Chem. Int. Ed.* 54 (35): 10308–10312. doi: 10.1002/anie.201503737.
12 Nielsen, M., Alberico, E., Baumann, W. et al. (2013). *Nature* 495 (7439): 85–89. doi: 10.1038/nature11891.
13 Luo, J., Kar, S., Rauch, M. et al. (2021). *J. Am. Chem. Soc.* 143 (41): 17284–17291. doi: 10.1021/jacs.1c09007.
14 Wang, Q., Lan, J., Liang, R. et al. (2022). *ACS Catal.* 12 (4): 2212–2222. doi: 10.1021/acscatal.1c05369.
15 Hu, P., Fogler, E., Diskin-Posner, Y. et al. (2015). *Nat. Commun.* 6 (1): 6859. doi: 10.1038/ncomms7859.
16 Heim, L.E., Konnerth, H., and Prechtl, M.H.G. (2017). *Green Chem.* 19 (10): 2347–2355. doi: 10.1039/C6GC03093A.
17 Manas, M.G., Sharninghausen, L.S., Lin, E., and Crabtree, R.H. (2015). *J. Organomet. Chem.* 792: 184–189. doi: 10.1016/j.jorganchem.2015.04.015.
18 Preuster, P., Papp, C., and Wasserscheid, P. (2017). *Acc. Chem. Res.* 50 (1): 74–85. doi: 10.1021/acs.accounts.6b00474.
19 Salman, M.S., Rambhujun, N., Pratthana, C. et al. (2022). *Ind. Eng. Chem. Res.* 61 (18): 6067–6105. doi: 10.1021/acs.iecr.1c03970.
20 Shimbayashi, T. and Fujita, K. (2020). *Tetrahedron* 76 (11): 130946. doi: 10.1016/j.tet.2020.130946.
21 Sordakis, K., Tang, C., Vogt, L.K. et al. (2018). *Chem. Rev.* 118 (2): 372–433. doi: 10.1021/acs.chemrev.7b00182.
22 Yadav, V., Sivakumar, G., Gupta, V., and Balaraman, E. (2021). *ACS Catal.* 11 (24): 14712–14726. doi: 10.1021/acscatal.1c03283.
23 Zhang, Y., Wang, J., Zhou, F., and Liu, J. (2021). *Catal. Sci. Technol.* 11 (12): 3990–4007. doi: 10.1039/D1CY00138H.
24 Iguchi, M., Himeda, Y., Manaka, Y., and Kawanami, H. (2016). *ChemSusChem* 9 (19): 2749–2753. doi: 10.1002/cssc.201600697.

25 Fellay, C., Yan, N., Dyson, P.J., and Laurenczy, G. (2009). *Chem. – Eur. J.* 15 (15): 3752–3760. doi: 10.1002/chem.200801824.

26 Papp, G., Ölveti, G., Horváth, H. et al. (2016). *Dalton Trans.* 45 (37): 14516–14519. doi: 10.1039/C6DT01695B.

27 Papp, G., Csorba, J., Laurenczy, G., and Joó, F. (2011). *Angew. Chem. Int. Ed.* 50 (44): 10433–10435. doi: 10.1002/anie.201104951.

28 Graf, E. and Leitner, W. (1992). *J. Chem. Soc. Chem. Commun.* (8): 623–624. doi: 10.1039/C39920000623.

29 Hwang, Y.J., Kwon, Y., Kim, Y. et al. (2020). *ACS Sustain. Chem. Eng.* 8 (26): 9846–9856. doi: 10.1021/acssuschemeng.0c02775.

30 Xu, F., Huang, W., Wang, Y. et al. (2022). *Inorg. Chem. Front.* 9 (14): 3514–3521. doi: 10.1039/D2QI00774F.

31 Stalder, C.J., Chao, S., Summers, D.P., and Wrighton, M.S. (1983). *J. Am. Chem. Soc.* 105 (20): 6318–6320. doi: 10.1021/ja00358a026.

32 Li, Z., Yang, X., Tsumori, N. et al. (2017). *ACS Catal.* 7 (4): 2720–2724. doi: 10.1021/acscatal.7b00053.

33 Shirman, R., Bahuguna, A., and Sasson, Y. (2021). *Int. J. Hydrog. Energy* 46 (73): 36210–36220. doi: 10.1016/j.ijhydene.2021.08.178.

34 Sun, R., Liao, Y., Bai, S.-T. et al. (2021). *Energy Environ. Sci.* 14 (3): 1247–1285. doi: 10.1039/D0EE03575K.

35 Zhou, L., Yao, C., Ma, W. et al. (2021). *J. CO2 Util.* 54: 101769. doi: 10.1016/j.jcou.2021.101769.

36 Bara-Estaún, A., Lyall, C.L., Lowe, J.P. et al. (2022). *Catal. Sci. Technol.* doi: 10.1039/D2CY00312K.

37 Kostera, S., Weber, S., Peruzzini, M. et al. (2021). *Organometallics* 40 (9): 1213–1220. doi: 10.1021/acs.organomet.0c00710.

38 Boddien, A., Mellmann, D., Gärtner, F. et al. (2011). *Science* 333 (6050): 1733–1736. doi: 10.1126/science.1206613.

39 Theodorakopoulos, M., Solakidou, M., Deligiannakis, Y., and Louloudi, M. (2021). *Energies* 14 (2): 481. doi: 10.3390/en14020481.

40 De, S., Udvardy, A., Czégéni, C.E., and Joó, F. (2019). *Coord. Chem. Rev.* 400: 213038. doi: 10.1016/j.ccr.2019.213038.

41 Jantke, D., Pardatscher, L., Drees, M. et al. (2016). *ChemSusChem* 9 (19): 2849–2854. doi: 10.1002/cssc.201600861.

42 Cole-Hamilton, D.J. (2019), *Periodic table: new version warns of elements that are endangered.* https://theconversation.com/periodic-table-new-version-warns-of-elements-that-are-endangered-110377 (accessed 4 July 2023).

43 Das, C., Grover, J., Tannu, et al. (2022). *Dalton Trans.* 51 (21): 8160–8168. doi: 10.1039/D2DT00663D.

44 Federsel, C., Boddien, A., Jackstell, R. et al. (2010). *Angew. Chem. Int. Ed.* 49 (50): 9777–9780. doi: 10.1002/anie.201004263.

45 Wei, D., Sang, R., Sponholz, P. et al. (2022). *Nat. Energy* 7 (5): 438–447. doi: 10.1038/s41560-022-01019-4.

46 Zubar, V., Borghs, J.C., and Rueping, M. (2020). *Org. Lett.* 22 (10): 3974–3978. doi: 10.1021/acs.orglett.0c01273.

47 Xu, R., Chakraborty, S., Yuan, H., and Jones, W.D. (2015). *ACS Catal.* 5 (11): 6350–6354. doi: 10.1021/acscatal.5b02002.

48 Fellay, C., Dyson, P., and Laurenczy, G. (2008). *Angew. Chem. Int. Ed.* 47 (21): 3966–3968. doi: 10.1002/anie.200800320.

49 Vatsa, A. and Padhi, S.K. (2022). *New J. Chem.* 46 (32): 15723–15731. doi: 10.1039/D2NJ03121C.
50 Grubel, K., Jeong, H., Yoon, C.W., and Autrey, T. (2020). *J. Energy Chem.* 41: 216–224. doi: 10.1016/j.jechem.2019.05.016.
51 Elek, J., Nádasdi, L., Papp, G. et al. (2003). *Appl. Catal. Gen.* 255: 59–67. doi: 10.1016/S0926-860X(03)00644-6.
52 Joó, F., Laurenczy, G., Nádasdi, L., and Elek, J. (1999). *Chem. Commun.* (11): 971–972. doi: 10.1039/a902368b.
53 Moret, S., Dyson, P.J., and Laurenczy, G. (2014). *Nat. Commun.* 5 (1): 4017. doi: 10.1038/ncomms5017.
54 Boddien, A., Loges, B., Junge, H., and Beller, M. (2008). *ChemSusChem* 1 (8–9): 751–758. doi: 10.1002/cssc.200800093.
55 Filonenko, G.A., Conley, M.P., Copéret, C. et al. (2013). *ACS Catal.* 3 (11): 2522–2526. doi: 10.1021/cs4006869.
56 Jessop, P.G., Joó, F., and Tai, C.-C. (2004). *Coord. Chem. Rev.* 248 (21–24): 2425–2442. doi: 10.1016/j.ccr.2004.05.019.
57 Majewski, A., Morris, D.J., Kendall, K., and Wills, M. (2010). *ChemSusChem* 3 (4): 431–434. doi: 10.1002/cssc.201000017.
58 Kumar, A., Bhardwaj, R., Mandal, S.K., and Choudhury, J. (2022). *ACS Catal.* 12 (15): 8886–8903. doi: 10.1021/acscatal.2c01982.
59 Tsai, J.C. and Nicholas, K.M. (1992). *J. Am. Chem. Soc.* 114 (13): 5117–5124. doi: 10.1021/ja00039a024.
60 Jens, C.M., Scott, M., Liebergesell, B. et al. (2019). *Adv. Synth. Catal.* 361 (2): 307–316. doi: 10.1002/adsc.201801098.
61 Webber, R., Qadir, M.I., Sola, E. et al. (2020). *Catal. Commun.* 146: 106125. doi: 10.1016/j.catcom.2020.106125.
62 Weilhard, A., Salzmann, K., Navarro, M. et al. (2020). *J. Catal.* 385: 1–9. doi: 10.1016/j.jcat.2020.02.027.
63 Yasaka, Y., Wakai, C., Matubayasi, N., and Nakahara, M. (2010). *J. Phys. Chem. A* 114 (10): 3510–3515. doi: 10.1021/jp908174s.
64 Scott, M., Blas Molinos, B., Westhues, C. et al. (2017). *ChemSusChem* 10 (6): 1085–1093. doi: 10.1002/cssc.201601814.
65 Zhang, Z., Liu, S., Hou, M. et al. (2021). *Green Chem.* 23 (5): 1978–1982. doi: 10.1039/D0GC04233A.
66 Drake, J.L., Manna, C.M., and Byers, J.A. (2013). *Organometallics* 32 (23): 6891–6894. doi: 10.1021/om401057p.
67 Jószai, I. and Joó, F. (2004). *J. Mol. Catal. Chem.* 224 (1–2): 87–91. doi: 10.1016/j.molcata.2004.08.045.
68 Zhao, G. and Joó, F. (2011). *Catal. Commun.* 14 (1): 74–76. doi: 10.1016/j.catcom.2011.07.017.
69 Bulushev, D.A. (2021). *Energies* 14 (5): 1334. doi: 10.3390/en14051334.
70 Coffey, R.S. (1967). *Chem. Commun. Lond.* 18: 923b. doi: 10.1039/c1967000923b.
71 Onishi, N., Kanega, R., Kawanami, H., and Himeda, Y. (2022). *Molecules* 27 (2): 455. doi: 10.3390/molecules27020455.
72 Onishi, N., Laurenczy, G., Beller, M., and Himeda, Y. (2018). *Coord. Chem. Rev.* 373: 317–332. doi: 10.1016/j.ccr.2017.11.021.
73 Onishi, N., Xu, S., Manaka, Y. et al. (2015). *Inorg. Chem.* 54 (11): 5114–5123. doi: 10.1021/ic502904q.

74 Lu, S.-M., Wang, Z., Li, J. et al. (2016). *Green Chem.* 18 (16): 4553–4558. doi: 10.1039/C6GC00856A.
75 Siek, S., Burks, D.B., Gerlach, D.L. et al. (2017). *Organometallics* 36 (6): 1091–1106. doi: 10.1021/acs.organomet.6b00806.
76 Bahuguna, A. and Sasson, Y. (2021). *ChemSusChem* 14 (5): 1258–1283. doi: 10.1002/cssc.202002433.
77 Grubel, K., Su, J., Kothandaraman, J. et al. (2020). *J. Energy Power Technol.* 2 (4). doi: 10.21926/jept.2004016.
78 Russo, D., Calabrese, M., Marotta, R. et al. (2022). *Int. J. Hydrog. Energy.* S0360319922030476. doi: 10.1016/j.ijhydene.2022.07.033.
79 Boddien, A., Gärtner, F., Federsel, C. et al. (2011). *Angew. Chem. Int. Ed.* 50 (28): 6411–6414. doi: 10.1002/anie.201101995.
80 Wiener, H., Sasson, Y., and Blum, J. (1986). *J. Mol. Catal.* 35 (3): 277–284. doi: 10.1016/0304-5102(86)87075-4.
81 Bényei, A.C. and Joó, F. (1990). *J. Mol. Catal.* 58: 151–163.
82 Rebreyend, C., Pidko, E.A., and Filonenko, G.A. (2021). *Green Chem.* 23 (22): 8848–8852. doi: 10.1039/D1GC02246F.
83 Zaidman, B., Wiener, H., and Sasson, Y. (1986). *Int. J. Hydrog. Energy* 11 (5): 341–347. doi: 10.1016/0360-3199(86)90154-0.
84 Wei, D., Sang, R., Moazezbarabadi, A. et al. (2022). *JACS Au* 2 (5): 1020–1031. doi: 10.1021/jacsau.1c00489.
85 Sordakis, K., Dalebrook, A.F., and Laurenczy, G. (2015). *ChemCatChem* 7 (15): 2332–2339. doi: 10.1002/cctc.201500359.
86 Kothandaraman, J., Czaun, M., Goeppert, A. et al. (2015). *ChemSusChem* 8 (8): 1442–1451. doi: 10.1002/cssc.201403458.
87 Federsel, C., Jackstell, R., Boddien, A. et al. (2010). *ChemSusChem* 3 (9): 1048–1050. doi: 10.1002/cssc.201000151.
88 Xin, Z., Zhang, J., Sordakis, K. et al. (2018). *ChemSusChem* 11 (13): 2077–2082. doi: 10.1002/cssc.201800408.
89 Treigerman, Z. and Sasson, Y. (2018). *ACS Omega* 3 (10): 12797–12801. doi: 10.1021/acsomega.8b00599.
90 Treigerman, Z. and Sasson, Y. (2019). *Am. J. Anal. Chem.* 10 (08): 296–315. doi: 10.4236/ajac.2019.108022.
91 Hsu, S.-F., Rommel, S., Eversfield, P. et al. (2014). *Angew. Chem. Int. Ed.* 53 (27): 7074–7078. doi: 10.1002/anie.201310972.
92 Bi, Q.-Y., Lin, J.-D., Liu, Y.-M. et al. (2014). *Angew. Chem. Int. Ed.* 53 (49): 13583–13587. doi: 10.1002/anie.201409500.
93 Laurenczy, G., Joó, F., and Nádasdi, L. (2000). *Inorg. Chem.* 39 (22): 5083–5088. doi: 10.1021/ic000200b.
94 Horváth, H., Kathó, Á., Udvardy, A. et al. (2014). *Organometallics* 33 (22): 6330–6340. doi: 10.1021/om5006148.
95 Horváth, H., Papp, G., Szabolcsi, R. et al. (2015). *ChemSusChem* 8 (18): 3036–3038. doi: 10.1002/cssc.201500808.
96 Horváth, H., Papp, G., Kovács, H. et al. (2019). *Int. J. Hydrog. Energy* 44 (53): 28527–28532. doi: 10.1016/j.ijhydene.2018.12.119.
97 Yamaguchi, R., Ikeda, C., Takahashi, Y., and Fujita, K. (2009). *J. Am. Chem. Soc.* 131 (24): 8410–8412. doi: 10.1021/ja9022623.

98 Vivancos, Á., Beller, M., and Albrecht, M. (2018). *ACS Catal.* 8 (1): 17–21. doi: 10.1021/acscatal.7b03547.

99 Wang, S., Huang, H., Bruneau, C., and Fischmeister, C. (2019). *ChemSusChem* doi: 10.1002/cssc.201900626.

100 Wang, W.-H., Himeda, Y., Muckerman, J.T. et al. (2015). *Chem. Rev.* 115 (23): 12936–12973. doi: 10.1021/acs.chemrev.5b00197.

101 Mejuto, C., Ibáñez-Ibáñez, L., Guisado-Barrios, G., and Mata, J.A. (2022). *ACS Catal.* 6238–6245. doi: 10.1021/acscatal.2c01224.

102 Semwal, S., Kumar, A., and Choudhury, J. (2018). *Catal. Sci. Technol.* 8 (23): 6137–6142. doi: 10.1039/C8CY02069H.

103 Maji, B., Bhandari, A., Bhattacharya, D., and Choudhury, J. (2022). *Organometallics* 41 (13): 1609–1620. doi: 10.1021/acs.organomet.2c00107.

104 Borthakur, I., Kumari, S., and Kundu, S. (2022). *Dalton Trans.* 51 (32): 11987–12020. doi: 10.1039/D2DT01060G.

105 Johnson, T.C., Morris, D.J., and Wills, M. (2010). *Chem. Soc. Rev.* 39 (1): 81–88. doi: 10.1039/B904495G.

106 Orosz, K., Papp, G., Kathó, Á. et al. (2019). *Catalysts* 10 (1): 17. doi: 10.3390/catal10010017.

107 Orosz, K., Udvardy, A., Papp, G. et al. (2023 July 24-29). Synthesis and catalytic activity of Ir(I)-N-heterocyclic carbene-tertiary phosphine complexes in hydrogenation of ketones and dehydrogenation of alcohols. In: *22th Symposium of Homogeneous Catalysis*. Portugal: Lisbon, Book of Abstracts, P22.

108 Schuchmann, K. and Müller, V. (2013). *Science* 342 (6164): 1382–1385. doi: 10.1126/science.1244758.

109 Schwarz, F.M., Moon, J., Oswald, F., and Müller, V. (2022). *Joule* 6 (6): 1304–1319. doi: 10.1016/j.joule.2022.04.020.

110 Volker, M. and Kai, S. (2014 December 24). *Method for storing gaseous hydrogen through producing methanoate (formate)*. EP 2 816 119 A1.

111 Roger, M., Brown, F., Gabrielli, W., and Sargent, F. (2018). *Curr. Biol.* 28 (1): 140–145.e2. doi: 10.1016/j.cub.2017.11.050.

112 Mourato, C., Martins, M., da Silva, S.M., and Pereira, I.A.C. (2017). *Bioresour. Technol.* 235: 149–156. doi: 10.1016/j.biortech.2017.03.091.

38

Low-cost Co and Ni MOFs/CPs as Electrocatalysts for Water Splitting Toward Clean Energy-Technology

Anup Paul[1],*, Biljana Šljukić[2], and Armando J.L. Pombeiro[1,3],*

[1] Centro de Química Estrutura, Instituto Superior Técnico, Universidade de Lisboa, Av. Rovisco Pais, Lisboa, Portugal
[2] Center of Physics and Engineering of Advanced Materials (CeFEMA), Instituto Superior Técnico, Universidade de Lisboa, Av. Rovisco Pais, Lisboa, Portugal
[3] Peoples' Friendship University of Russia (RUDN University), Research Institute of Chemistry, 6 Miklukho-Maklaya Street, Moscow, Russian Federation
* Corresponding authors

38.1 Introduction

The global energy demand is growing rapidly due to economic development, and it is predicted to double in the coming 15 years. However, most energy is currently produced using fossil fuel-based technologies; for instance, c. 25 PWh of electric energy was globally consumed in 2019, 80% of which was produced using fossil fuels: coal, oil, and natural gas [1]. In general more than 85% of primary world energy is produced using fossil fuels [2]. The use of fossil fuels has advantages due to their ease of extraction, processing, and distribution to end users, which grants them a lower cost than alternatives. Nevertheless, fossil fuels are unsustainable as energy sources because they have finite reserves, and although the quantity of fossil fuel reserves is debatable, there is no doubt of their limitation.

The use of fossil fuels also contributes to serious environmental issues, including global warming, sea/ocean-level rising, and climate change, due to the release of carbon dioxide (CO_2) and other greenhouse gases during the combustion of fossil fuels to produce electrical energy. For example, the production of 1 kWh of electric energy by the conversion of chemical energy from fossil fuels can generate as much as 400 g of CO_2. These problems associated with the use of fossil fuels will eventually result in their increased price. However, the production of energy using renewable and sustainable sources is seen as a promising solution for notably decreasing the dependence on fossil fuels; therefore, the development of clean, efficient, and reasonably priced energy technologies has become an imperative. Hydrogen (H_2) is seen as a promising energy carrier or fuel of the future: it has the highest specific energy density and is environmentally friendly, i.e., it is a clean and non-toxic fuel during the use of which no greenhouse gases are generated. The notion of using H_2 as a fuel is as old as 1766, the year of its isolation, and it stands until the present time.

Thus, the question arises: why is hydrogen still not a fuel used worldwide?

The problem is its high price. A new fuel can only be globally used if it has both a low-enough price and a prospect of long-lasting use. However, the price of hydrogen energy (depending on its

Catalysis for a Sustainable Environment: Reactions, Processes and Applied Technologies Volume 3, First Edition. Edited by Armando J. L. Pombeiro, Manas Sutradhar, and Elisabete C. B. A. Alegria.
© 2024 John Wiley & Sons Ltd. Published 2024 by John Wiley & Sons Ltd.

production method) is typically several times higher than the price of energy from a fossil fuel such as natural gas.

This raises the question of *why* hydrogen has such a high price when it is one of the most abundant elements on Earth. The explanation is that hydrogen is not present in the environment in pure form, but rather, in the form of water or hydrocarbons. Consequently, hydrogen gas is a synthetic fuel and needs to be produced. This can be achieved using many different methods, from those using fossil fuels (coal or natural gas) over nuclear energy to renewable energy technologies, such as wind or solar. The current main technologies for producing H_2 gas are fossil fuel-based and release CO_2 (as much as 830 million ton) to the air. Namely, more than 95% of hydrogen for industrial use is produced in this way; specifically, by steam reforming of natural gas (48%), by oil reforming (30%), or by coal gasification (18%) [2]. These processes have a double disadvantage in that they consume fossil fuels and release CO_2. Such hydrogen is called gray hydrogen because, as previously mentioned, its generation is accompanied by the formation of large amount of CO_2–specifically, nine volume parts CO_2 for one part hydrogen. However, a step toward more environmentally friendly and clean fuel is called blue hydrogen. The CO_2 generated during the production of blue hydrogen is captured and is subsequently either used (e.g. in advanced oil recovery) or disposed of (e.g. deep underground).

For hydrogen to be a clean and efficient fuel, its production must fully eliminate the emission of CO_2 and other greenhouse gases and must use exclusively renewable energy sources. This "green" hydrogen can be produced by water electrolysis, i.e., by splitting water into hydrogen and oxygen by passing an electric current through it (Eq. 38.1). The application of electricity generated by renewable sources, such as wind or solar, ensures "zero-carbon" hydrogen.

$$H_2O + energy \rightarrow H_2 + \tfrac{1}{2}O_2 \tag{38.1}$$

Yet only ca. 4–5% of globally produced H_2 is currently obtained by water electrolysis, as green hydrogen fails to fulfil the economic requirement of reasonable price [2, 3]. Green hydrogen has the highest price compared to grey or blue hydrogen, up to ca. $6 per kg–well above the price of grey hydrogen, ca. $0.7 to $2.2 per kg. (It should be kept in mind that these prices are highly dependent on the cost of natural gas or coal.) Consequently, water electrolysis still does not provide an alternative to fossil fuel-based technologies for hydrogen production, but instead only a complementary method used for the production of high-purity hydrogen for special applications. Thus, although hydrogen energy is highly ranked from the Energy-Ecology-Economy (E3) viewpoint, it fails in the aspect of market-competitive price. Some countries are advocating reducing the cost of water electrolysis by selling its by-products (specifically O_2) and heat [4]. Optional methods for the production of hydrogen from water have also been proposed, such as thermochemical and low-voltage water electrolysis or high-temperature steam electrolysis [5, 6].

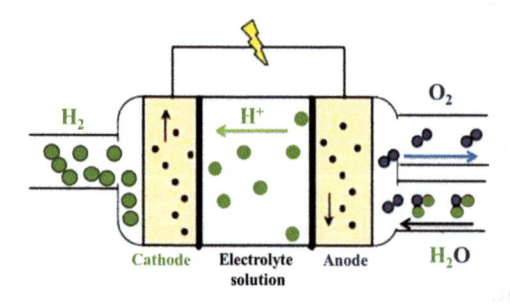

Scheme 38.1 Schematic illustration of the water electrolysis process.

But where does the high cost of the water electrolysis process (Scheme 38.1) come from? It is an outcome of the high overpotential necessary to operate the industrial electrolysis cell, which demands a high consumption of energy. The voltage required to split a water molecule in laboratory conditions is ca. 1.23 V; however, this voltage is higher in real electrolysis cells due to the increase of the overpotentials of the two electrode reactions. Even a small augment of cell voltage will result, according to Ohm's law, in a notable increase of power demand/consumption in large-scale industrial electrolyzers where high currents are used.

As water electrolysis is a rather simple and clean method for producing H_2–its major advantages over fossil fuel-based methods are that H_2 and O_2 are the only abundantly sourced products with zero CO_2 emissions–research has been dedicated to increasing its efficiency and reducing its cost. These studies are mainly focused on the search for novel electrode materials with high electrocatalytic performance (decreasing the overpotentials of electrode reactions), along with high long-term stability and low cost.

Currently, materials for proton exchange membranes and electrolysis devices that use OER and HER catalysis mostly utilize Pt and Ru, noble metal oxides with high costs and low mass activities [7, 8]. As a result, the research and development of oxygen rich catalysts that have remarkable catalytic activity and stability is foreseen to be a way to replace the expensive metals and realize the practical advanced energy systems for fuel cells and electrolyzers. A class of materials called either MOFs (which are porous) or CPs have emerged as highly promising due to their captivating structures and growing range of applications, including gas storage, sensing, catalysis, etc. The use of MOFs has also been expanded to include hydrogen storage, photochemical, and electrochemical energy storage reactions, and in particular, the conversion of water splitting for hydrogen evolution reactions (HER) and oxygen evolution reactions (OER) [9, 10]. MOFs are basically constituted of metal nodes and organic linkers, which represent the active sites for the electrocatalytic OER and HER reactions. They possess a designable structure, large channels, a high surface area, and a very clear chemistry. Furthermore, the chemistry of MOFs/CPs can be tuned to enhance their electrocatalytic properties. Although various transition metal-based MOFs and MOF composites have been reported to show electrocatalytic applications in OER, HER, and overall water splitting reactions, this chapter focuses on selective examples of Co(II) and Ni(II)-based MOFs and composites, which have exhibited noticeable electrocatalytic properties. The future prospects of these MOFs and composites are also discussed.

38.2 Fundamentals of Water Splitting Reactions

As mentioned previously, water electrolysis is the process of splitting water into hydrogen and oxygen by passing an electric current through it. Proton is reduced and hydrogen is evolved at the cathode, with simultaneous oxidation and evolution of oxygen at the anode. The processes are highly pH dependant. The overall H_2 evolution reaction in alkaline media is given by Eq. 38.2 as follows:

$$2H_2O + 2e^- \rightarrow H_2 + 2OH^- \tag{38.2}$$

but the evolution typically occurs as a multi-step process (Eqs. 38.3–38.5) [11]. First is the Volmer step, a primary discharge step that occurs by the coverage of the metal surface by adsorbed protons (Eq. 38.3, where MH_{ads} represents an H atom adsorbed on the metal surface). This step is followed either by the Tafel step (i.e. catalytic recombination of the adsorbed intermediates [MH_{ads}] Eq. 38.4, or by the Heyrovsky step (i.e., electrodesorption of the adsorbed species Eq. 38.5).

$$M + H_2O + e^- \rightarrow MH_{ads} + OH^- \quad \text{(Volmer step)} \tag{38.3}$$

$$MH_{ads} + MH_{ads} \rightarrow H_2 + 2M \quad \text{(Tafel step)} \tag{38.4}$$

$$MH_{ads} + H_2O + e^- \rightarrow M + H_2 + OH^- \quad \text{(Heyrovsky step)} \tag{38.5}$$

OER is also a multi-step process in which oxygen is generated though several proton/electron coupled steps. In alkaline media, hydroxyl groups (OH^-) are oxidized and converted into H_2O and O_2, (Eq. 38.6) along with the formation of several (adsorbed) intermediates (HO_2^- is usually the main intermediate in alkaline media).

$$4OH^- \rightarrow 2H_2O + O_2 + 4e^- \tag{38.6}$$

It should be mentioned that the rate of the water electrolysis process is limited by the reactions of the oxygen electrode, which are considerably slower than the reactions of the hydrogen electrode. This is because the generation of the O_2 molecule requires the transfer of four electrons compared to the transfer of two electrons during the generation of the H_2 molecule. The simultaneous transfer of four electrons within a single step during OER is unlikely to occur. The OER process will be kinetically favorable if it proceeds via several steps with a single electron transferred during each step. The build-up of energy for each step to occur results in sluggish OER kinetics and, consequently, large overpotential.

The overall H_2 evolution reaction in acidic media is given by Eq. (38.7)

$$2H^+ + 2e^- \rightarrow H_2 \tag{38.7}$$

and it is also a multi-step process, where steps are represented by Eqs. 38.8–38.10 [12].

$$H_3O^+ + M + e^- \rightarrow MH_{ads} + H_2O \quad \text{(Volmer step)} \tag{38.8}$$

$$MH_{ads} + MH_{ads} \rightarrow H_2 + 2M \quad \text{(Tafel step)} \tag{38.9}$$

$$MH_{ads} + H_3O^+ + e^- \rightarrow H_2 + M + H_2O \quad \text{(Heyrovsky step)} \tag{38.10}$$

It is worth noting that the exchange current density in acidic media during HER is ca. two times higher than that in alkaline media at commonly investigated electrode materials (such as platinum) [13], most likely due to the shorter Pt-H bond length in the acidic compared to the alkaline media.

During OER in acidic media, two H_2O molecules are oxidized into four protons (H^+) and one O_2 molecule (Eq. 38.11), with hydrogen peroxide (H_2O_2) as the main reaction intermediate.

$$2H_2O \rightarrow 4H^+ + O_2 + 4e^- \tag{38.11}$$

It should also be noted that OER typically does not proceed at the pure metal but at the oxidized metal surface. These oxides can be highly conductive (similarly to pure metals, such as ruthenium, iridium. or platinum oxides), semi-conductive (such as NiO, or oxides of molybdenum and tungsten), or isolators (such as oxides of aluminium, titanium, zirconium, tantalum and niobium). OER proceeds at the highest rate at the highly conductive oxides.

As mentioned, the main downside of using large-scale industrial water electrolysis for the production of H_2 is the high price of the process and the resulting high price of the product itself, with power demand being the major contributor to this price [13]. The stable structure of the H_2O molecule demands a high current density, and thus a high energy input, to break the molecule into H_2 and O_2 at the ambient temperature. The O=O double bond is formed by taking four protons from a water molecule during water oxidation, but the low activity of water oxidation at low potential is the main impediment for water splitting. The theoretical voltage for decomposing a water molecule is 1.23 V; subsequently, the theoretical energy consumption for the production of 1 m^3 of H$_2$ is 2.94 kWh. In practice, however, voltages between 1.65 and 1.70 V are used, and between 1.80 and 2.60 V in industry, due to a higher overpotential and an ohmic drop in the electrolysis cell. For a voltage of 2 V, the energy consumption is as high as 4.78 kWh m^{-3}.

The result is a low net efficiency for the water electrolysis processes: the efficiency calculated as the ratio of the higher heating value of H$_2$ and the input of thermal energy is typically ca. 24% [14]. Thus, an imperative is to optimise the parameters governing water electrolysis to reduce its power consumption and increase its efficiency.

These parameters have been classified into three categories [15]: cell parameters (e.g. cell configuration and geometry: electrodes shape, dimensions, placement, and mutual distance); operating conditions (e.g. current density, cell potential, electrolyte solution composition and electrical resistance, process temperature and pressure, and electrode and membrane materials); and gas bubbles management and external conditions. Industrial electrolyzers typically operate at current densities between 1000 and 3000 A m^{-2} at temperatures of 80 to 90 °C. The choice of the electrolyte–acidic or alkaline– influences the performance of an electrode material. HER is typically more favorable in acidic media, whereas OER is typically more favorable in alkaline media; many materials are unstable in acidic media due to susceptability to oxidation. However, for an efficient water electrolysis process that simultaneously produces hydrogen and oxygen gases, cathodic (HER) and anodic (OER) reactions should be efficient in the same electrolytic solution.

As for the electrode material, nickel (Ni) is the most commonly used in industrial electrolyzers due to its stability and suitable activity. However, a major issue is deactivation of the electrode material; Ni electrodes may still be deactivated due to the formation of Ni hydrides at the electrode surface in the presence of high concentrations of hydrogen. The working electrode determines the reaction rate, depending on its structure, electrical conductivity, degree of wettability, and the accessibility of its surface to an electrolyte. Electrode shape and dimensions also influence efficiency. The increase of electrode surface area creates a larger contact area between the electrode and electrolyte, and thus a higher number of active sites; this results in higher current densities and a greater amount of hydrogen gas evolved.

The parameters for evaluating the electrode material performance toward HER/OER include onset potential, E_{onset}, and overpotential, η, to reach a defined current density j. Onset potential, one of the most important parameters, may be determined as the potential at which the measured current visibly deviates from the background signal, or as the intersection point of the tangent to the baseline with the tangent to the rising current part of the linear sweep voltammetry (LSV) curve. As these lie open to the subjective error of the researcher performing the analysis, the value of potential at 1 mA cm^{-2} or, sometimes, at 10 mA cm^{-2} is considered more reliable and is commonly used [16]. Overpotential is the difference between the potential needed to reach a defined current density and the equilibrium potential. Additional kinetic parameters for the evaluation of the electrode material performance for HER/OER are determined through Tafel analysis. The Tafel coefficient, b–determined by the slope of Tafel plots ($\eta/\log j$)–indicates the rate of change of the current density with the applied overpotential. As mentioned previously, HER proceeds via one of two mechanisms, Volmer-Heyrovsky or Volmer-Tafel [12]. Assuming a charge transfer coefficient α value of 0.5 at the temperature of 25 °C, the Tafel slope will take the value of 120, 40, and 30 mV dec^{-1} for the Volmer, Heyrovsky, and Tafel steps, respectively.

The exchange current density j_0 reflects the rate of electron transfer, which depends primarily on factors such as the electrode material and its composition, structure, and surface roughness. It is also dependant upon the species adsorbed on the electrode surface, electrolyte composition, and temperature.

Finally, the stability and durability of the electrode material are another crucial factor in the practical application of water electrolyzers. The stability is greatly affected by the nature of the electrolyte; as mentioned, most catalysts are very stable in alkaline media but cannot perform well in those that are acidic. Stability is usually assessed by long-term chronopotentiometric measurements (at a constant current density) or by long-term chronoamperometric measurements (at a constant potential).

38.3 MOFs/CPs as Electrocatalysts for Water Splitting Reactions

38.3.1 Co MOFs and Derived Electrocatalysts for OER and HER

In recent years, a great deal of effort has been devoted to developing efficient and economically feasible electrocatalysts for the production of green energy, in anticipation of replacing expensive and relatively rare precious metal catalysts with more efficient and economical alternatives. Cobalt-based compounds, which exhibit notable redox properties, are currently being investigated for their potential as catalysts. Following the discovery of the first Co-based MOF by Yaghi et al. in 1995, many more Co MOFs have been synthesized. In this section we have arrayed some of the Co-based MOFs and their composites that have been utilized as electrocatalysts for OER and HER reactions.

38.3.1.1 Co MOFs and Derived Electrocatalysts for OER

Co MOFs have become increasingly popular as catalysts for OER in recent years; this process is being continuously studied because of the sluggish kinetics of OER at the anodes of energy storage and conversion devices.

Jiang et al. reported the cobalt-citrate MOF UTSA-16 (**1**) prepared by solvothermal method to be highly efficient for electrocatalytic water oxidation [17]. **1** exhibits a very low onset potential of 1.60 V (Figure 38.1a), reaching a current density of 10 mA cm^{-2} at an overpotential of only 77 mV dec^{-1} (Figure 38.1b), and an excellent long-term durability of 7 h (Figure 38.1c) in alkaline media. The excellent activity of **1** toward the catalysis of OER in alkaline media was attributed to the synergetic relationship between its inherent porous structure, high valent cobalt species formed in situ, and existing Co_4O_4 cubane in the structural framework (Figure 38.1d).

In another study, Pang et al. demonstrated the synthesis of ultrathin two-dimensional cobalt–organic framework (Co–MOF) nanosheets labelled as $[Co_2(OH)_2BDC]$ (**2**) (BDC = 1,4-benzenedicarboxylate) for OER [18]. The nanosheets were grown vertically, using polyvinylpyrrolidone (PVP) as an anionic surfactant. In addition to micro-nano Co-MOFs, the authors also synthesised bulk Co-MOFs for comparisons. Atomic force microscopy (AFM), scanning electron microscopy (SEM), and transmission electron microscopy (TEM) were used to analyse the morphologies and microstructures of the synthesised materials. SEM images reveal that the thin nanosheet morphology of 2D Co-MOF possessed a partially stacked morphology, whereas the micro–nano Co-MOFs (**3**) and bulk Co–MOFs (**4**) demonstrated a very large, layered sheet structure (Figure 38.2a-c). The 2D Co-MOF nanosheets were found to be highly electrocatalytically active toward the OER, exhibiting a slope of 74 mVdec^{-1} and a small overpotential of 263 mV under alkaline conditions while maintaining electrochemical stability for 3.3 h (Figure 38.2d-f). Furthermore, the 2D Co–MOF nanosheets demonstrated better OER behaviour than both micro–nano Co–MOFs and

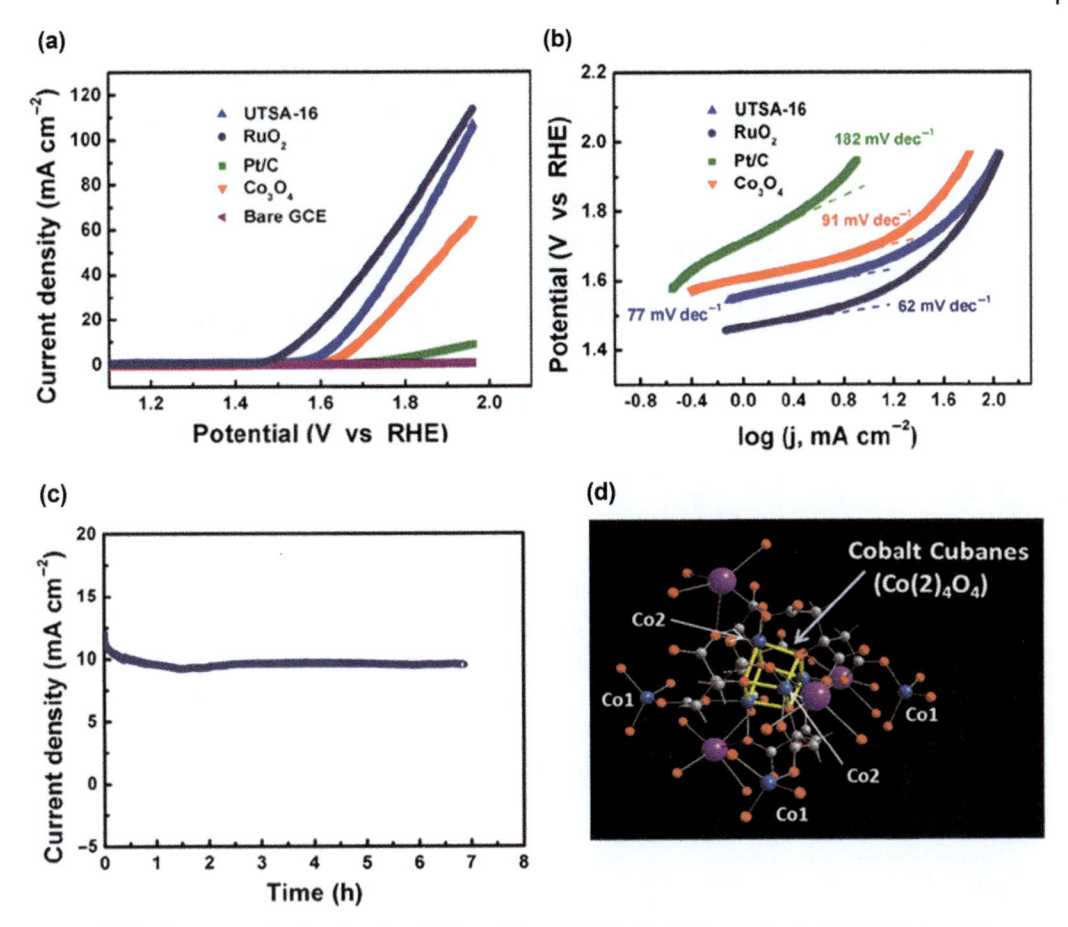

Figure 38.1 Oxygen evolution reaction (OER) activity of UTSA-16 (1). Overpotential (a), Tafel slope (b), stability test (c) and Co_4O_4 cubane in the structural framework of 1. Reproduced and adapted with permission from Ref [17] / American Chemical Society.

bulk Co–MOFs (326 mV), in addition to commercial RuO_2 (360 mV). The excellent electrocatalytic properties of the Co–MOF were attributed primarily to its porous structure, rapid ion transport, and the arrangement of unsaturated Co(II) active sites on its surface.

In addition to pristine MOFs, MOF-derived electrocatalysts, which combine MOFs with conductive nanoparticles (NPs) to enhance the properties of both their parent materials, are also undergoing significant advances [19]. A Co-based MOF (zeolitic imidazolate framework-9 [ZIF-9]) was used for the first time by Yamauchi et al. as a precursor for the preparation of hybrid materials for the OER; they used a two-step thermal conversion method (in an inert atmosphere and in air) to prepare the carbon–cobalt-oxide hybrids Z9-700-250 (**5**), Z9-800-250 (**6**) and Z9-900-250 (**7**) from cobalt-based MOFs (Figure 38.3a) and tested for OER in KOH (0.1 M). Hybrid materials **5** and **6** exhibited lower onset potentials and higher current densities than **7** and Pt/carbon black (Figure 38.3b). Among these hybrid materials, **6** has been found to be an effective catalyst for OER.

In another study, Muhler et al. pyrolyzed a Co-based MOF (ZIF-67) with core-shell embedded in carbon nanotube-grafted (CNT-grafted) N-doped carbon polyhedral in H_2/He atmosphere, followed by oxidative calcination, resulting in $Co@Co_3O_4$ NPs labelled as Co/NC (**8**) and $Co@Co_3O_4$/NC-1

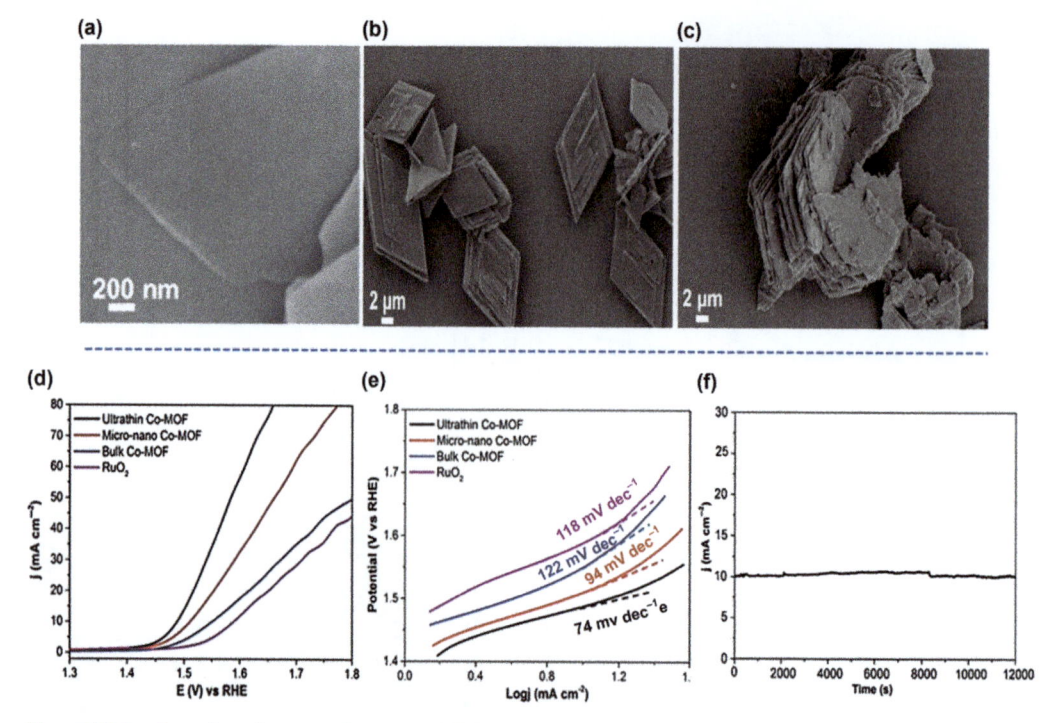

Figure 38.2 Scanning electron microscopy (SEM) images of ultrathin (a), micro–nano (b) and bulk Co–MOFs (c). Overpotential, Tafel plots with slope values of ultrathin, micro–nano and bulk Co–MOFs (d,e). Stability test of ultrathin Co–MOF (f). Reproduced with permission from Ref [18] / Royal Society of Chemistry.

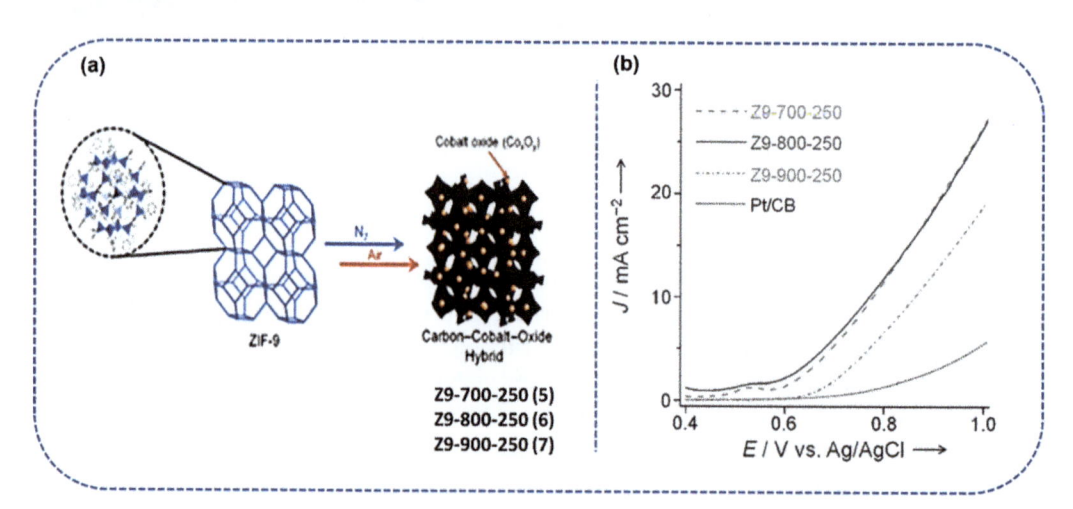

Figure 38.3 Synthesis of carbon–cobalt-oxide hybrids materials 5–7 (a). Onset potentials of 5–7 and Pt/CB (b). Reproduced and adapted with permission from Ref [19] / John Wiley & Sons.

(**9**), respectively (Figure 38.4a) [20]. They also prepared a similar catalyst labelled as Co@Co$_3$O$_4$/NC-2 in O$_2$ flow at 250 °C for six hours. The polyhedral morphology of the catalysts was confirmed through SEM, and the electrocatalytic activity of the catalysts was investigated in O$_2$-saturated 0.1 M KOH solution. In terms of catalytic activity in OER, the catalyst Co@Co$_3$O$_4$/NC-2 displayed the highest activity, reaching a current density of 10 mA cm^{-2} at 1.64 V (Figure 38.4b). This catalyst

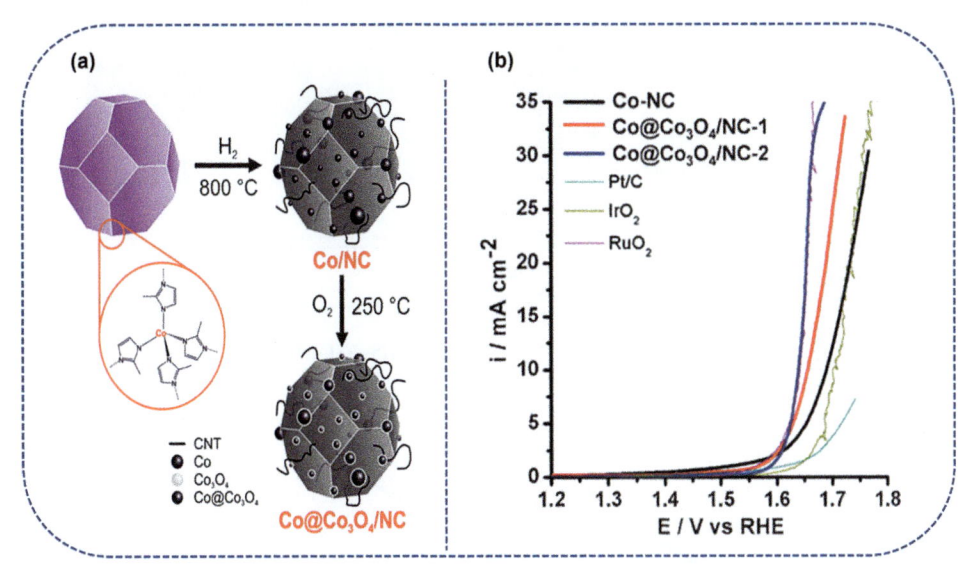

Figure 38.4 Synthesis of Co/NC (**8**) and Co@Co$_3$O$_4$/NC-1 (**9**) (a). Catalytic activity of 9, Pt/C, IrO$_2$, RuO$_2$ recorded in O$_2$-saturated 0.1 M KOH at a scan rate of 5 mVs^{-1} (b). Reproduced and adapted with permission from Ref [20] / John Wiley & Sons.

showed a superior activity to both Pt/C and IrO$_2$, as well as a similar activity to the benchmark OER catalyst RuO$_2$.

The recognition of coordinative unsaturated metal sites (CUMSs), also known as open metal sites (OMSs), has also contributed to the popularity of MOFs [21, 22]. A wide variety of applications have already been established for MOFs with CUMSs; in addition, CUMSs can act as active centres in the catalyst to enhance OER catalysis. Wang et al. reported the synthesis of CUMS-containing electrocatalysts from ZIF-67 [Co(Hmim)$_2$]$_n$ (Hmim = 2-methylimidazole) (**10**) by using dielectric barrier discharge (DBD) plasma (Figure 38.5a) [22]. SEM was used to observe the surface morphologies of ZIF-67 (**10**) and CUMSs-ZIF-67 (**11**) (Figure 38.5b,c). After being exposed to DBD plasma, **10** maintained its dodecahedral structure, while its surface became rougher due to the etching effect of the plasma (Figure 38.5c). **11** showed a different onset potential in OER electrocatalysis than plain **10** (1.58 V vs. reversible hydrogen electrode [RHE]) (Figure 38.5d). Additionally, when **10** and **11** were tested at a current density of 10 mA cm^{-2}, **11** presented a smaller overpotential of 410 mV (1.64 V vs. RHE) and a smaller Tafel slope (185.1 mV dec^{-1}) than pristine **10** (550 mV; 1.78 V vs. RHE; 316.9 mV dec^{-1}); both findings indicate an improved OER catalysis (Figure 38.5e).

In another study, Jiang et al. used a facile microwave-induced plasma engraving technique to finetune the CUMSs of a cobalt-based MOF [Co$_2$(dobdc)]$_n$ (dobdc = 2,5-dihydroxyterephthalate) designated Co-MOF-74 (**12**) with Ar [MOF-74-Ar (**13**)] and H$_2$ [MOF-74-H$_2$ (**14**)] gas, respectively (Figure 38.6a) [23]. A significant improvement in OER activity was observed for the hydrogen plasma engraved **14** without compromising its phase integrity: it exhibited superior OER activity in 0.1 M KOH electrolyte, with a relatively low overpotential of 337 mV, a turnover frequency (TOF) of 0.0219 s^{-1}, and a high mass activity of 54.3 A g^{-1} (Figure 38.6b). Furthermore, the stability of this material was evaluated at a current density of 10 mAcm^{-2} for approximately 10,000 s (Figure 38.6c). As a result of the developed control strategy and the identified correlation of CUMS activity, further microstructure tuning of MOFs can be inspired for OER reactions.

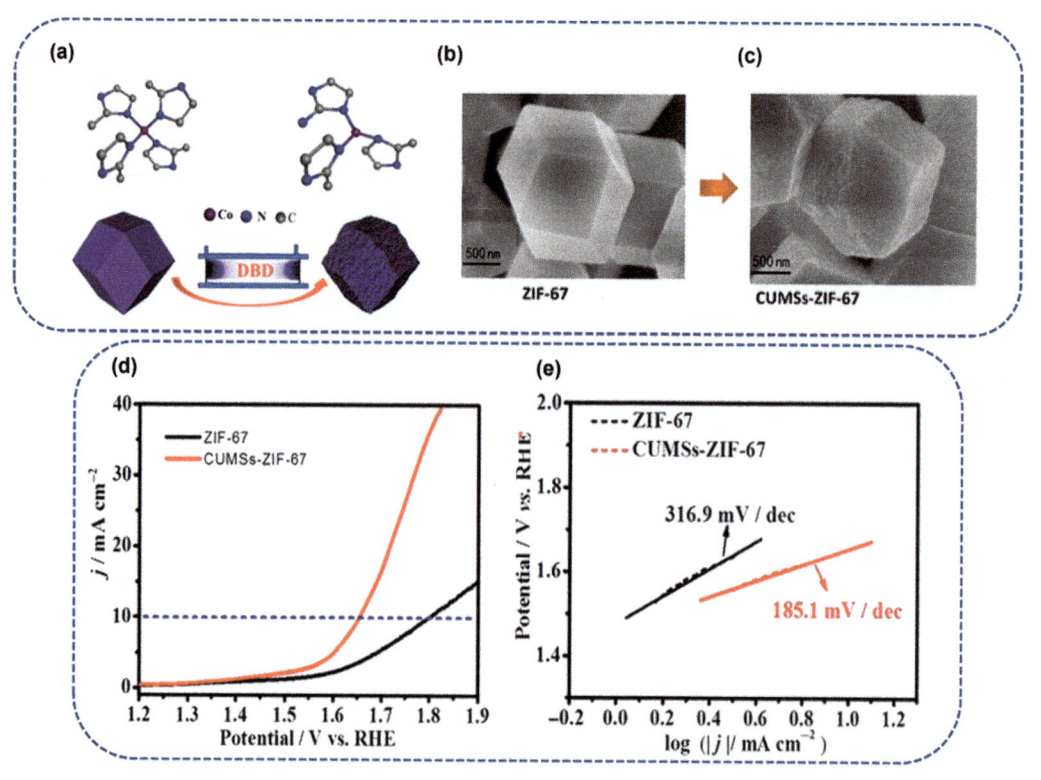

Figure 38.5 Conversion of ZIF-67 (ZIF = zeolitic imidazolate framework) (10) into CUMSs-ZIF-67 (11) (a). Scanning electron microscopy (SEM) image of 10 (b) and 11 (c). Linear sweep voltammetry (LSV) plot of 10 and 11 (d). Tafel slope of 10 (d) and 11 (e) in 0.5 M KOH at a scan rate of 5 mV s^{-1}. Reproduced with permission from Ref [22] / Elsevier.

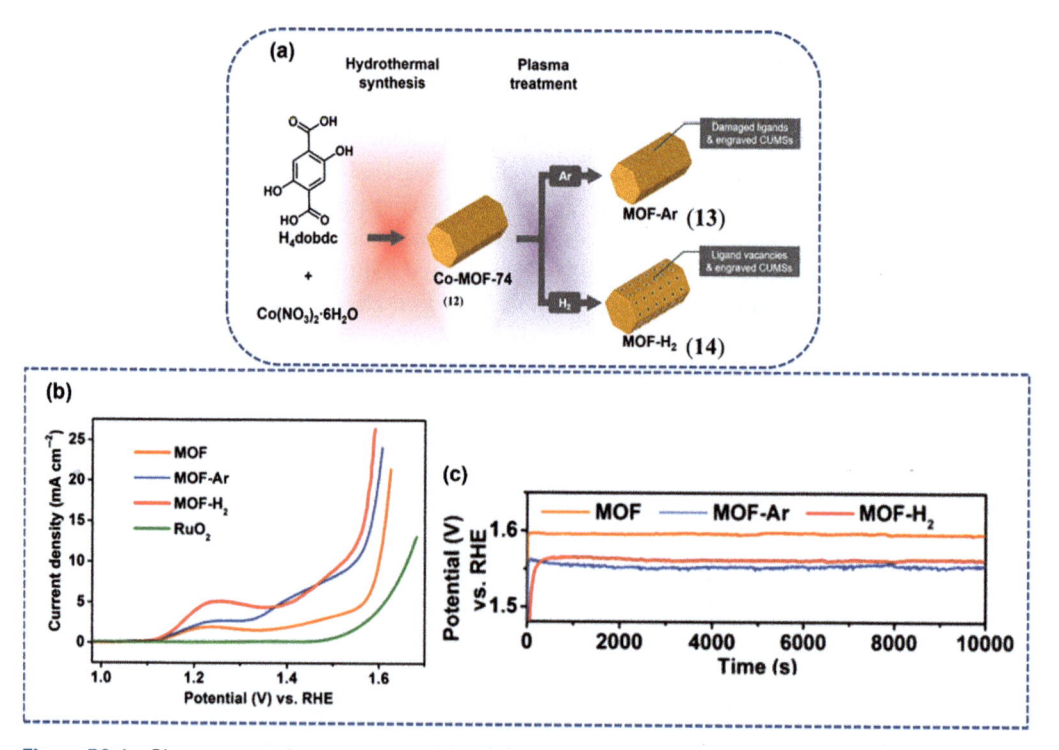

Figure 38.6 Plasma engraving process used for the preparation of catalysts [MOF-Ar (13)] and [MOF-H$_2$ (14)] (a). Linear sweep voltammetry (LSV) plots for Co-MOF-74 (12), MOF-Ar (13), MOF-H$_2$ (14) and commercial RuO$_2$ in 0.1 M KOH (b), and long-term stability test (c). Reproduced and adapted with permission from Ref [23] / American Chemical Society.

Recently, we have reported two isostructural CPs, $[M(L)_2(H_2O)_2]_n$ [M = Co (**15**) and Ni (**16**), L = 4-(pyridin-3-yl-carbamoyl)benzoate], synthesized by the solvothermal reactions of $CoCl_2·6H_2O$ and $Ni(NO_3)_2·6H_2O$ with 4-(pyridin-3-ylcarbamoyl)benzoic acid, respectively [24]. They were tested as electrocatalysts for both OER (Figure 38.7) and HER, showing higher activity for the former. **15** showed a lower OER onset potential of 1.58 V compared to **16** (1.61 V) and reached higher current densities, although the Tafel slope determined for **15** (102 mV dec^{-1}) was higher than that of **16** (72 mV dec^{-1}) (Figure 38.7). The higher catalytic activity of **15** is attributed to lower impedances and faster redox kinetics.

38.3.1.2 Co MOFs and Derived Composites for HER

Co-MOFs and composites are also being tested for their potential application as electrocatalysts in HER reactions. In tests of the HER activity of the previously-mentioned CPs **15** and **16** [24], the Co CP **15** showed a lower onset potential and higher current densities (Figure 38.8). The Tafel slope values of 328 (for **15**) and 252 mV dec^{-1} (**16**) were determined, with the corresponding exchange current densities of 130 and 32 μA cm^{-2}, respectively (Figure 38.8). The HER Tafel slope values were higher than expected, which could result from a hindered electron transfer within the electrode materials (dependent on their structure and morphology, as well as on the presence of hydrides or adsorbed intermediates on the surface).

Zhao et al. reported pristine 3D Co MOF $Co_2(Hpycz)_4·H_2O$ (H_2Pycz = 3-(pyrid-4′-yl)-5-(4″-carbonylphenyl)-1,2,4-triazolyl) (**17**), synthesized under solvothermal conditions, to show HER catalytic properties using a three-electrode electrochemical cell (Figure 38.9a) [25]. With a low onset potential of 101 mV, a small Tafel slope of 121 mV dec^{-1}, an exchange current density of 1.62×10^{-4} A cm^{-2}, and a long-term stability of at least 72 h, the Co(II) MOF catalyst exhibited excellent catalytic performance in HER (Figure 38.9b-d). Although the exact reason behind its excellent HER activity has not been discussed in the article, it has been thought that Co metal ions nodes, which act as active sites of electron transfer, may play a role in this property.

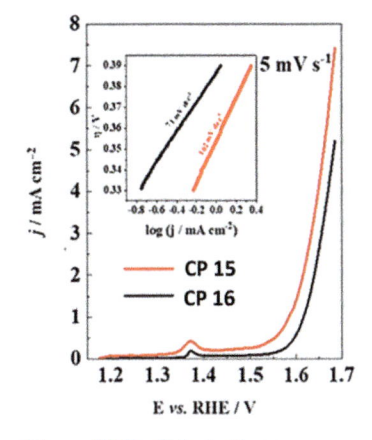

Figure 38.7 Polarization curves of the coordination polymers (CPs) 15 and 16 with the corresponding Tafel plots in the insets for oxygen evolution reaction (OER) [24] / American Chemical Society.

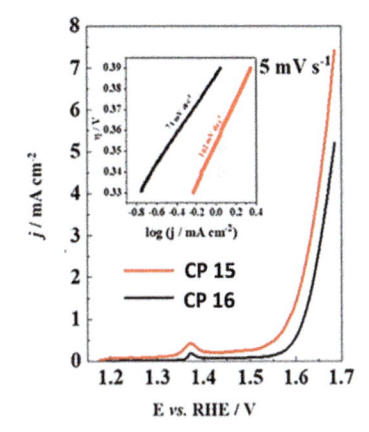

Figure 38.8 Polarization curves of the CPs 15 and 16 with the corresponding Tafel plots in the insets for the hydrogen evolution reaction (HER) [24] / American Chemical Society.

Using a neutral or anionic surfactant such as polyethylene glycol (PEG) or PVP, Bu et al. reported the synthesis of two new polymorphic Co-MOFs, known as CTGU-5 (**18**) and CTGU-6 (**19**), whose crystallization transforms them into pure 2D or 3D networks, respectively (Figure 38.10a) [26]. The two MOFs differ in the way they bind to water molecules. These MOFs were tested on modified glassy carbon (GC) as electrocatalysts: the HER catalytic activity of CTGU-5 (**18**) and CTGU-6 (**19**) resulted in a large current density of 10 mAcm^{-2} at an overpotential of 388 and 425 mV, respectively, as well as positive onset potential of 298 and 349 mV (Figure 38.10b-c) [26]. The 2D CTGU-5 (**18**) has shown a significantly better electrocatalytic activity in HER than the 3D CTGU-6 (**19**). In addition, the authors synthesized composite materials of AB&CTGU-5 (**20**) and AB&CTGU-5 (**21**) by varying the stoichiometric ratio between the MOF and acetylene black (AB); a stoichiometric ratio of 1:4 between AB and CTGU-5 exhibited the lowest Tafel slope (45 mVdec^{-1}) and a high

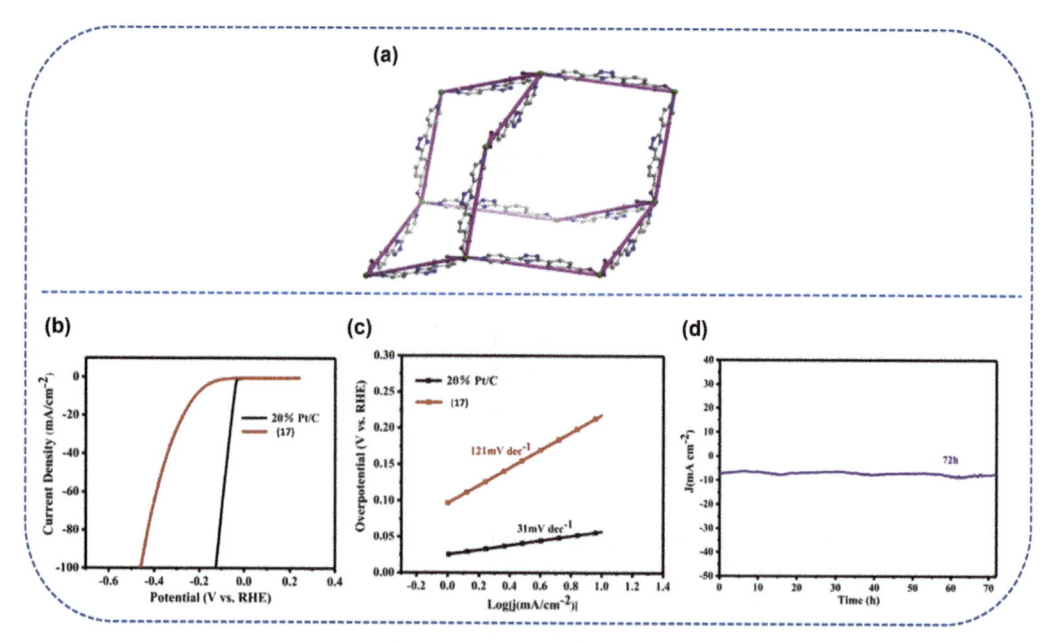

Figure 38.9 Single adamantanoid cage in $Co_2(Hpycz)_4 \cdot H_2O$ (17) (a). Polarization curves for electrocatalysts in 0.5 M H_2SO_4 aqueous solution (b). Tafel plots (c) and long-term stability test (d). Reproduced and adapted with permission from Ref [25] / Elsevier.

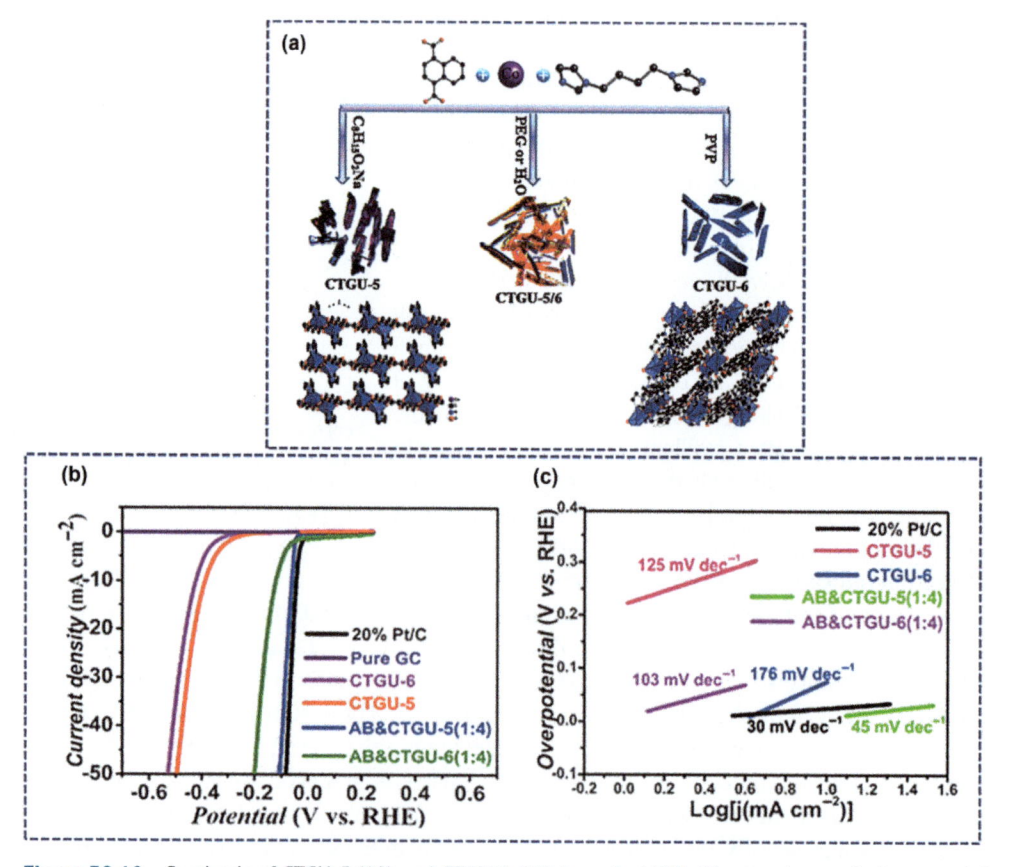

Figure 38.10 Synthesis of CTGU-5 (18) and CTGU-6 (19) from Co-MOFs (Co-metal-organic frameworks). (a) Polarization curves for electrocatalysts 18–21 (b) corresponding Tafel plots (c). Reproduced and adapted with permission from Ref [26] / John Wiley & Sons.

exchange current density of 8.6×10^{-4} A cm^{-2} among all the MOFs (Figure 38.10a). The material can also maintain a highly stable current density at a constant overpotential of 10 mV for at least 96 h. Thus, this study demonstrated for the first time the use of 2D Co(II) MOFs and conducting co-catalysts as efficient and stable electrocatalysts for HER.

38.3.2 Ni MOFs and Derived Composites for OER and HER

MOFs containing redox active Ni(II) and functional organic ligands are also suitable for the design of novel electrocatalysts for OER and HER. A few examples are discussed in this section. Our Ni CP **16** was already mentioned in the sections concerning Co materials (3.1.1 and 3.1.2) for comparison with the related Co CP **15** [24].

38.3.2.1 Ni MOFs and Derived Composites for the OER

2D π-conjugated MOFs synthesized via the bottom-up approach have emerged as a promising type of 2D material [27]. Du et al. were the first to demonstrate the design and successful bottom-up fabrication of a novel noble-metal-free nickel phthalocyanine-based 2D MOF [Ni$_3$(C$_{32}$H$_{16}$N$_{16}$)]$_n$, designated as (NiPc–MOF) (**22**), for highly efficient water oxidation catalysis (Figure 38.11a) [27]. The NiPc–MOF thin film grown on fluorine-doped tin oxide (FTO) gave rise to a blue-black thin film (Figure 38.11b), which displayed an excellent electrocatalytic activity without further pyrolysis or addition of conductive materials; it disclosed a high catalytic OER activity with a very low onset potential (<1.48 V, overpotential < 0.25 V), high mass activity (883.3 A g^{-1}), high TOF value

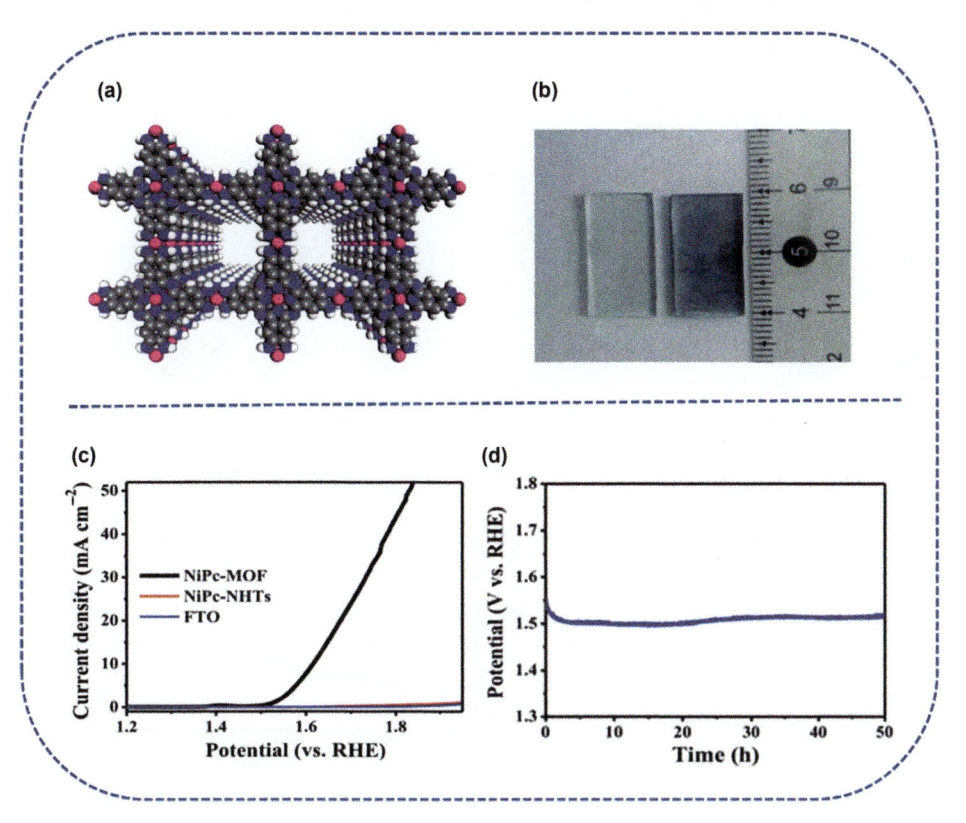

Figure 38.11 Eclipsed AA-stacking mode of 22 (a). NiPc–MOF modified fluorine-doped tin oxide (FTO). (b). Linear sweep voltammetry (LSV) curves for the oxygen evolution reaction (OER) at the scan rate of 10 mV s^1 (c). Stability plot of 22. Reproduced with permission from Ref [27] / Royal Society of Chemistry.

($2.5\ s^{-1}$), and excellent catalytic durability (Figure 38.11c-d). Its excellent OER performance was attributed to its 2D structure and good conductivity. The present study broadens the applications of π-conjugated conductive 2D MOF materials, which may facilitate the development of this class of materials for energy applications.

In another study, Wang et al. reported three Ni-MOFs, designated Ni-MOF-FA (**23**), Ni-MOF-BDC (**24**), and Ni-MOF-BTC (**25**), derived from three different organic linker precursors: formic acid (FA), 1,4-benzenedicarboxylic acid (BDC), and 1,3,5-benzenetricarboxylic acid (BTC) [28]. The number of carboxylates for FA, BDC, and BTC is one, two, and three, respectively; this may lead to different interactions between linkers and Ni ions, resulting in different OER performance of the corresponding Ni-MOFs. The Ni-MOFs comprising different ligands have different shapes and morphologies. The TEM analysis displayed an oblate shape for Ni-MOF-FA (**23**) (Figure 38.12a) with a hole in the center, resembling a donut, while Ni-MOF-BDC (**24**) (Figure 38.12b) and Ni-MOF-BTC (**25**) (Figure 38.12c) displayed rolled-up nanosheets and nanospheres, respectively. During LSV, Ni-MOF-FA (**23**), Ni-MOF-BDC (**24**), and Ni-MOF-BTC (**25**) displayed an overpotential in this order, with Tafel slopes of 97, 156, and 298 mV (Figure 38.12d); this suggests that the OER performance decreases from **23** to **25**, as is supported by the EIS (electrochemical impedance spectroscopy) curves (Figure 38.12e). The OER activity trend of the MOFs was adjudicated on their XPS (x-ray photoelectron spectroscopy) analysis, where O1s spectra of three Ni-MOFs demonstrated peaks at 531.38, 532.11, 532.70, and 533.50 eV attributed to the O=C, Ni-O-H, H_2O, and C-O, respectively. According to the Ni-O-OH pattern, $Ni(OH)_2$ was generated on the surface of Ni-MOFs as a result of the hydrolysis, and the ratios of Ni-O-H for Ni-MOF-FA, Ni-MOF-BDC, and Ni-MOF-BTC were 38%, 27%, and 18%, respectively, in agreement with the variation trend of

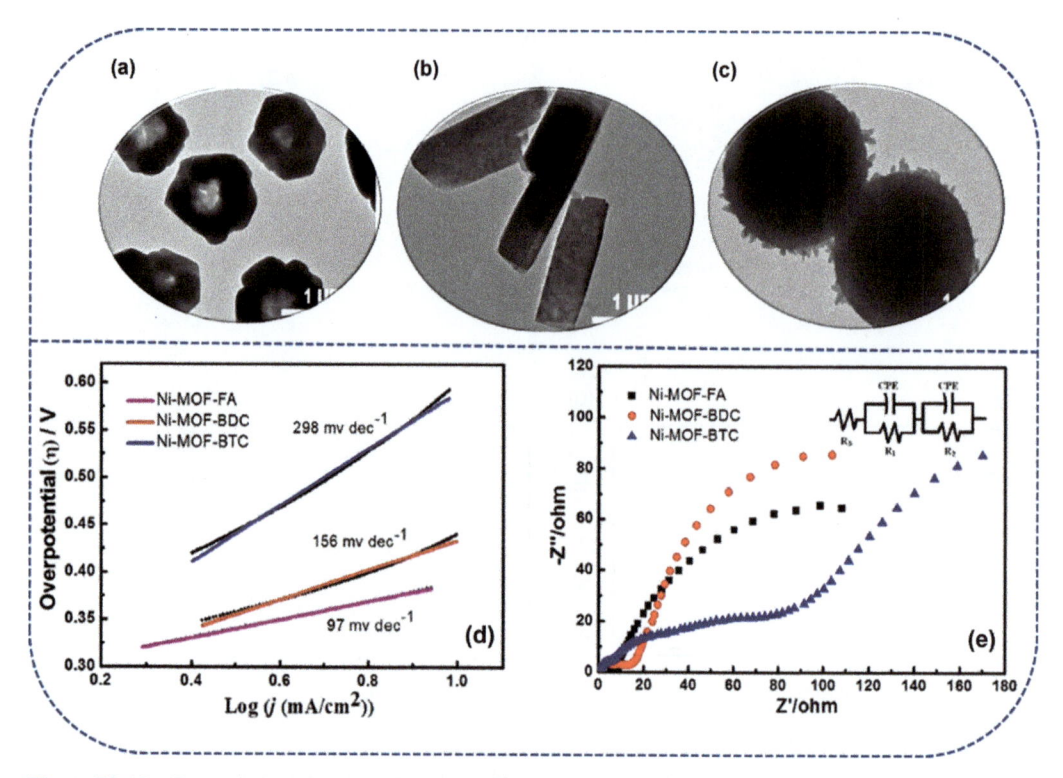

Figure 38.12 Transmission electron microscopy (TEM) images of 23 (a), 24 (b), and 25 (c). Tafel slopes of 23–25 for OER (d) and the electrochemical impedance spectroscopy (EIS) curves of 23–25 (e). Reproduced with permission from Ref [28] / Elsevier.

Ni-MOF structure. It was observed that the degree of hydrolysis of Ni-MOFs was reduced with an increase in carboxylic groups, which contributes to a smaller ratio of Ni-O-Hs and, therefore, to a lower OER activity within the Ni-MOFs.

More recently, Sun et al. reported highly stable and efficient Ni-MOF electrocatalysts for OER in alkaline media [29]. They used 4,4′-biphenyl dicarboxylic acid as an organic ligand precursor to develop a Ni-MOF nanosheet array on nickel foam (NF), labelled Ni-MOF/NF (**26**). SEM images indicate the formation of an array of interconnected Ni-MOF nanosheets on the surface of the Ni foam (Figure 38.13a-c). The OER catalytic study of **26** was performed in a 1 M KOH solution. In addition, for comparative purposes, commercial RuO_2 on NF (RuO_2/NF) and bare NF were investigated under the same conditions while Ni-MOF powder was immobilized on NF with the same loading as a control catalyst (denoted as NiMOF powder/NF). The LSV analysis indicated a strong OER performance for the 3D Ni-MOF/NF, which drives a current density of 20 mA cm^{-2} at a low overpotential of 350 mV, comparable to RuO_2/NF (Figure 38.13d). It also displayed a strong long-term electrochemical stability (Figure 38.13e) for at least 24 hours and achieved a high TOF of 0.24 mol O_2 per s at an overpotential of 400 mV. Its remarkable OER performance was attributed to a larger surface area with exposed metal active sites.

Ashiq et al. reported novel Ni(II) MOF composites–namely, carbon dots (CDs), MOF (**27**), composite (CDs@MOF) (**28**), calcinated MOF (**29**), and calcinated composite (CDs@MOF) (**30**)— thencompared them with bare nickel foam (BNF) [30]. SEM was used to analyze the surface morphology and particle size of these novel composites. The synthesized MOF $[Ni_2(BDC)_2(DABCO)]_n$ (**27**) (BDC = 1,4-benzenedicarboxylate; DABCO = 1,4-diazabicyclo[2.2.2]octane) showed a rod-like structure with varying particle sizes, whereas a slight change in the morphology and structure of the MOF occurred after composite formation (Figure 38.14a-c). Calcinated products showed a significant change in their morphology after calcination at 400 °C due to Ni-MOF derived carbon

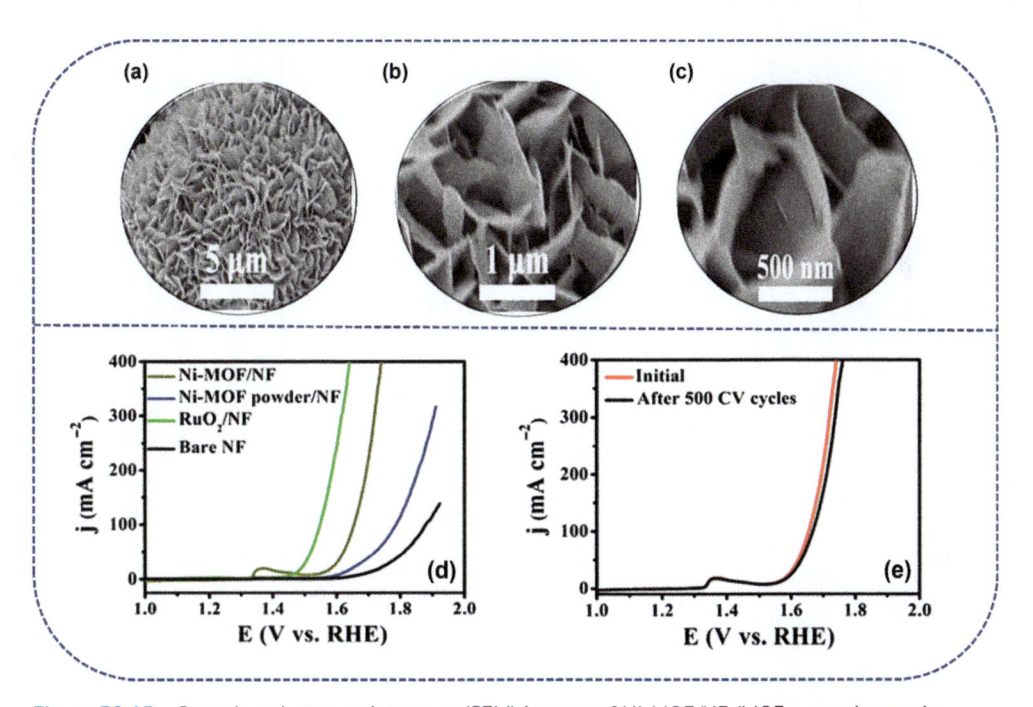

Figure 38.13 Scanning electron microscopy (SEM) images of Ni-MOF/NF (MOF = metal-organic frameworks; NF = nickel foam) (26) under different magnifications (a-c). Linear sweep voltammetry (LSV) curves of Ni-MOF/NF, Ni-MOF powder/NF, RuO_2/NF, and bare NF (d). Stability test for 26 (e). Reproduced with permission from Ref [29] / Royal Society of Chemistry.

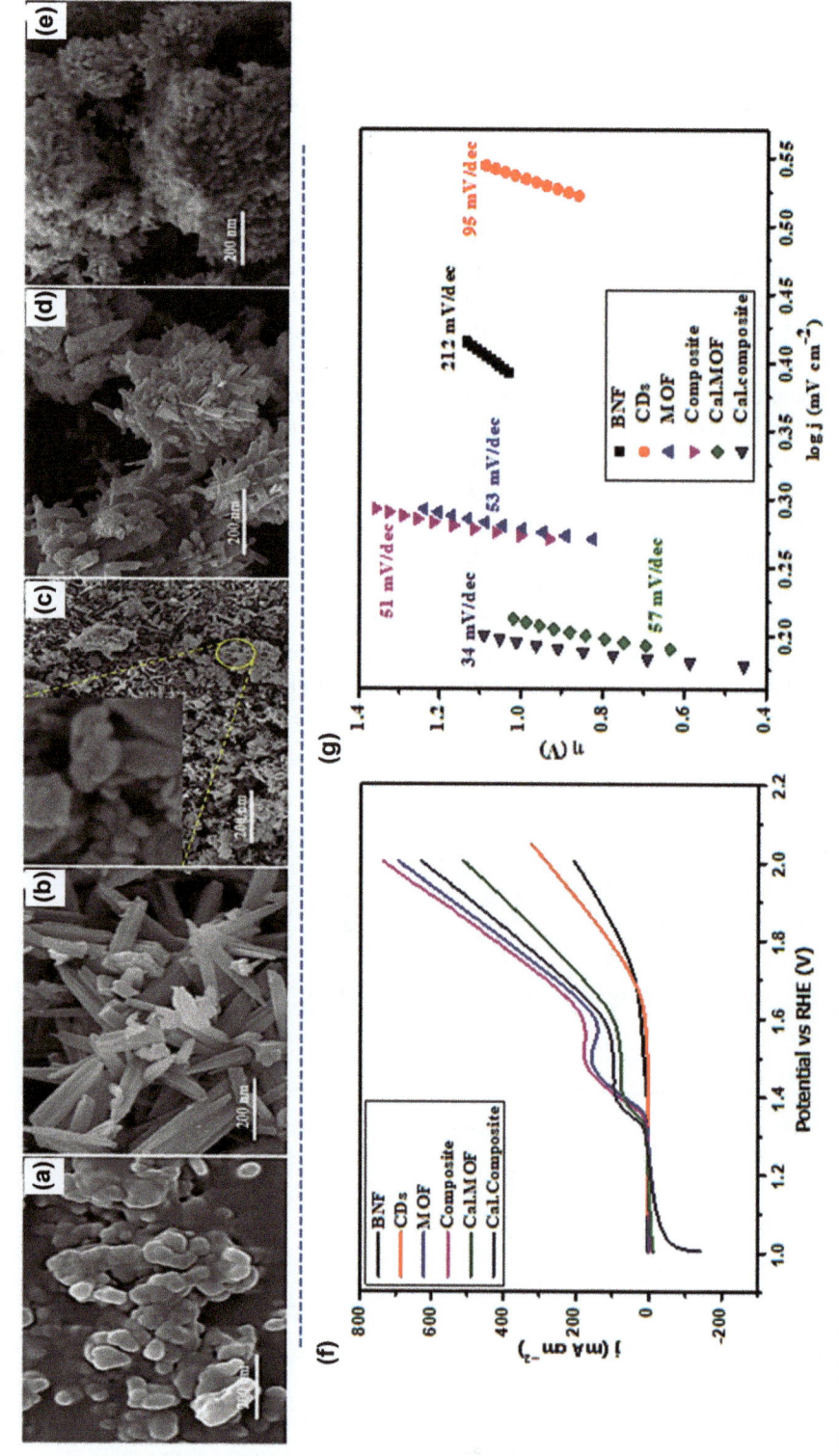

Figure 38.14 Scanning electron microscopy (SEM) images of carbon dots (CDs), metal-organic frameworks (27), composite (CDs@MOF) (29), and calcinated composite (CDs@MOF) (30) / from Taylor and Francis / CC BY 4.0 (a-e). Linear sweep voltammetry (LSV) polarization curves of bare nickel foam (BNF), CDs, MOF, composite, calcinated MOF, and calcinated composite obtained in 1 M KOH solution at a scan rate of 5 mV s^{-1} at room temperature (f). Tafel plots and slope values of BNF, CDs, MOF, composite, calcinated MOF, and calcinated composite (g).

composites (Figure 38.14d-e). These materials were investigated for their electrochemical properties in 1 M KOH solution; the calcinated composite material displayed a current density of 10 mA cm^{-2} at only 194 mV and an ideal Tafel slope of 34 mV dec^{-1}, which suggested a quick and easy electron–proton transfer during the OER (Figure 38.14f-g). The remarkable OER activity shown by the calcinated composites recommends an opportunity to design and synthesise desirable composite materials for OER.

38.3.2.2 Ni MOFs and Derived Composites for HER

Li et al. reported the pyrolysis of the Ni-based MOF [Ni(BDC)(ted)]$_n$ (BDC = 1,4-benzenedicarboxylate; ted = triethylenediamine) in Ar/NH$_3$ to yield Ni NPs with surface nitridation and thin carbon coating layers labelled as Ni–Ar (**31**), Ni–0.2NH$_3$ (**32**), Ni–0.4NH$_3$ (**33**), Ni–0.2H$_2$ (**34**), and Ni–Ar–0.2NH$_3$ (**35**) [31]. These Ni NPs were used as electrocatalysts for HER (Figure 38.15a). With a current density of 20 mA cm^{-2}, the surface-modified Ni NPs Ni–0.2NH$_3$ (**32**) exhibited a very low overpotential of only 88 mV. The catalyst exhibited a mixed Volmer and Heyrovsky mechanism based on the Tafel slope (71 and 83 mV dec^{-1}) (Figure 38.15b-c); it also showed a good stability for c. 12 hours (Figure 38.15d). By using surface modifications, the authors were able to develop a cheap, improved, and effective electrocatalyst for HER.

Jiang et al. employed a Ni–BTC MOF (**36**) as an effective sacrificial template for producing NiS$_2$ hollow microspheres (NiS$_2$ HMSs) (**37**) by a simple direct sulfidation process with thioacetamide (TAA) under microwave irradiation [32]. A significant change in microstructural morphology was observed during sulfidation, followed by phase transformations (Figure 38.16a-b). The NiS$_2$ HMSs electrocatalytic HER performance was then evaluated using a typical three-electrode system in 0.1 M KOH solution. With an onset potential close to zero, Pt/C displayed the expected HER activity,

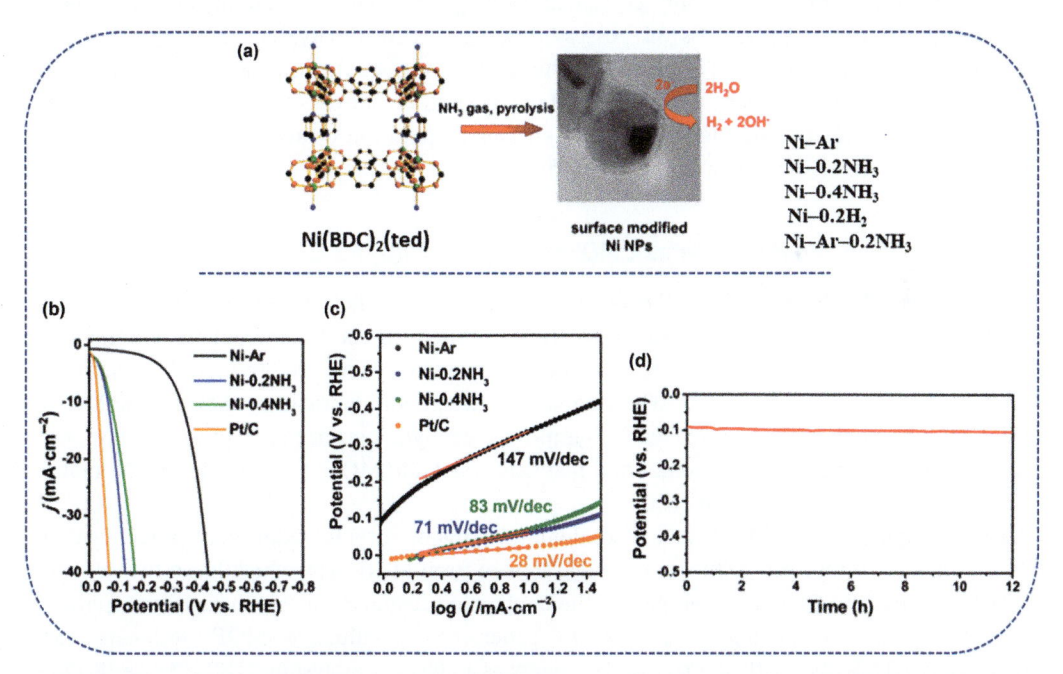

Figure 38.15 Surface modification on a Ni-based MOF in Ar/NH$_3$ atmosphere: Ni–Ar (31), Ni–0.2NH$_3$ (32), Ni–0.4NH$_3$ (33), Ni–0.2H$_2$ (34), and Ni–Ar–0.2NH$_3$ (35) (a). Linear scanning voltammograms (b) and Tafel plots (c) of the MOF-derived catalysts 32–36 in 1 M KOH solution. Reproduced with permission from Ref [31] / Royal Society of Chemistry.

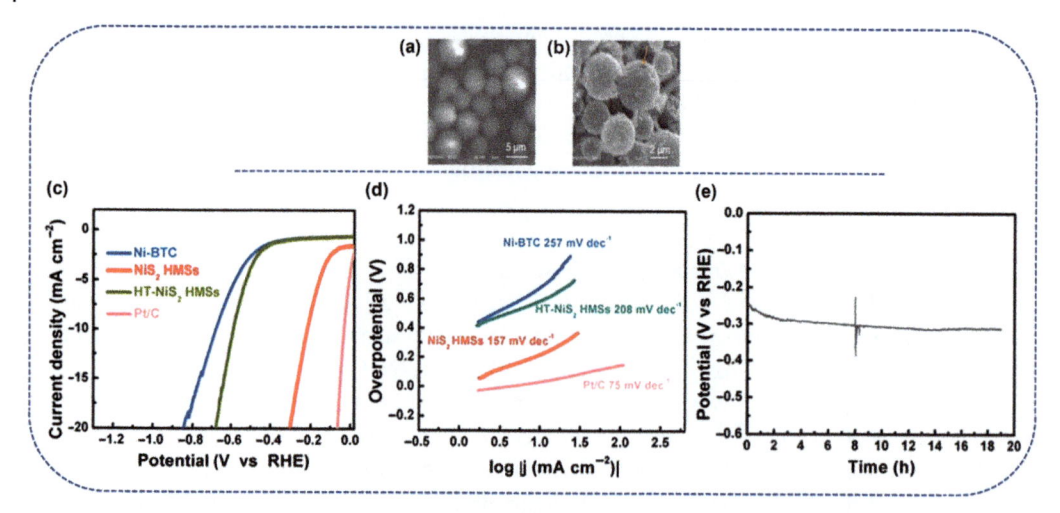

Figure 38.16 Scanning electron microscopy (SEM) images of Ni–1,3,5-benzenetricarboxylic acid (Ni-BTC) (36) and NiS_2 hollow microspheres (HMSs) (37) (a,b). Polarization curves © and corresponding Tafel plots of the Ni–BTC (36), NiS_2HMSs (37), HT-NiS_2 HMSs (HT = hydrothermal) (38), and Pt/C reference at a scan rate of 5 mV s^{-1} in 1.0 M KOH solution (d). Chronopotentiometric durability test for NiS_2 HMSs (37) at a constant current density of 10 mA c© (e). Reproduced with permission from Ref [32] / Royal Society of Chemistry.

whereas the Ni–BTC MOF (36) demonstrated poor HER activity. However, NiS_2 HMSs (37) displayed a remarkable HER activity—as indicated by its lower Tafel slope (157 mV dec^{-1})—with a current density of 10 mA cm^{-2} at only 219 mV and a long-term operating stability of 19 h (Figure 38.16c; Figure 38.16d-e). NiS_2 HMSs (36) possessed a greater HER activity than HT-NiS_2 HMSs (38) synthesized by hydrothermal (HT) method without anion surface modification, and NiS_2 HMSs (37) proved to be more efficient than electrocatalysts based on nickel sulfide. Thus, Jiang et al. demonstrated an effective method for the design and preparation of advanced electrocatalysts based on surface anion-rich metal sulfides for efficient HER.

38.3.3 Polyhomo and Heterometallic MOFs of Co(II) or Ni(II) for OER and HER

Co and Ni-based MOFs have received extensive attention due to their dispersed metal sites, high activity, and stability. MOFs are considered potential non-noble metal-based catalysts for OER and HER due to their intrinsic high surface area and high porosity. More importantly, the structural flexibility of most Co and Ni-based MOFs enables the design and fabrication of bimetallic MOFs with different second metals, which leads to improved catalytic performance. In this section, we illustrate a few examples of polyhomo- and polyheterometallic MOFs that use Co and Ni metal ions as electrocatalysts for OER or HER.

Heptaazatriphenylene (HATN) is a tris(bidentate) polyheterocyclic ligand with a conjugated planar structure. It interacts with several metal ions to form M_3.HATN, which has a two-coordinate (M–N2) moiety. Chen et al. employed the conjugated ligand hexaiminohexaazatrinaphthalene (HAHATN), an analog of HATN, to generate a conductive MOF designated as $Ni_3(Ni_3 \cdot HAHATN)_2$ (39), with an extra M–N_2 moiety as an electrocatalyst for HER (Figure 38.17a) [33]. SEM analyses of $Ni_3(Ni_3.HAHATN)_2$ (39) show that it consists of thin-layered nanosheets with abundant wrinkles in a petaloid morphology. These nanosheets interact to create a hierarchical porous structure, where pores have a size in the range of dozens of nanometers. The anomalous porous morphology effectively boosts electroactivity by exposing active centers and

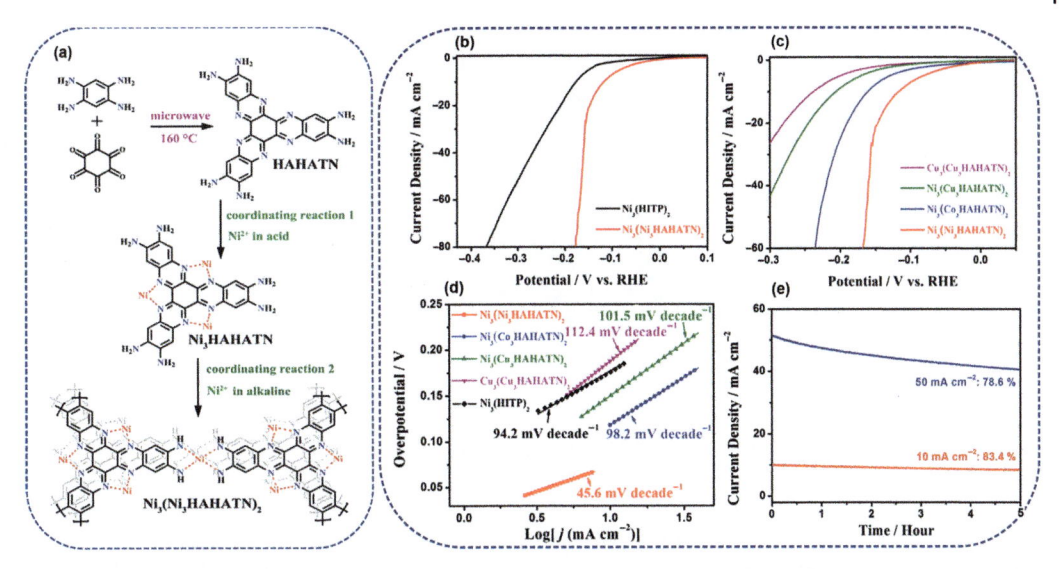

Figure 38.17 Synthesis of $Ni_3(Ni_3\cdot HAHATN)_2$ (HATN = heptaazatriphenylene) (39) (a). Polarization curves and corresponding Tafel plots of $Ni_3(Ni_3\cdot HAHATN)_2$ (39), $Ni_3(HITP)_2$ (HITP = hexaiminotriphenylene) (40), and various $M2_3(M1_3\cdot HAHATN)_2$ samples (b-d). Time-dependent HER current density curves for $Ni_3(Ni_3\cdot HAHATN)_2$ sample at 10 and 50 mA©$^{-2}$ (e). Reproduced and adapted with permission from Ref [33] / John Wiley & Sons / CC BY 4.0.

accelerating mass diffusion. These conductive $Ni_3(Ni_3\cdot HAHAT)_2$ nanosheets exhibited outstanding HER performances in alkaline solution, with a low overpotential of 115 mV at 10 mA cm^{-2}, a small Tafel slope of 45.6 mV dec^{-1}, and excellent electrocatalytic stability, which was even found to be better than Co(III) or Cu(II) metal ions analogs (Figure 38.17b-e). Furthermore, the $Ni_3(Ni_3\cdot HAHATN)_2$ (**39**) nanosheets exhibited remarkably enhanced activity for HER in comparison to $Ni_3(HITP)_2$ (HITP = hexaiminotriphenylene) (**40**) conductive MOFs.

In another study, Tang's et al. reported a NiCo bimetallic–organic framework labelled NiCo-UM—Ns (**41**)–synthesized from a mixed solution of Ni^{2+}, Co^{2+}, and benzenedicarboxylic a—d (BDC)–that exhibited high electrocatalytic activity toward OER in an alkaline medium (Figure 38.18a) [34]. The ultrathin NiCo bimetal–organic framework nanosheets **41** on glassy-carbon electrodes achieved a current density of 10 mA cm^{-2} at an overpotential of only 250 mV. This was compared with CoUMOFNs, Ni-UMOFNs, and bulk NiCo-MOF electrodes in order to demonstrate its **41** remarkable OER activity (Figure 38.18b-c). Its Tafel slope of only 42 mV dec^{-1} further confirms its excellent electrocatalytic activity toward OER, which is driven by the presence of coordinatively unsaturated metal atoms (the primary active centers in the electrocatalytic OER process).

In a different study, Zhao et al. fabricated ultrathin Fe/Ni-based MOF nanosheets that demonstrated a high electrocatalytic performance toward OER (Figure 38.19a) [35]. The material had high macroporosity, with a pore size between 200 and 400 μm as shown by SEM image (Figure 38.19b). The electrocatalytic performance of NiFe-MOF for OER was tested in 0.1 M KOH electrolyte in a typical three-electrode system. At a current density of 10 mA cm^{-2}, LSV plot of NiFe-MOF (**42**) revealed an overpotential of 240 mV, which is significantly lower than that of Ni-MOF (**43**) (without Fe, 296 mV), Fe-MOF/NF (**44**) (without Ni, 354 mV), Ni foam (without MOF, 370 mV), NiFe-MOF powder loaded on NF (**45**) (denoted as bulk NiFe-MOF, 318 mV), calcined NiFe-MOF (**46**) (at 650 C for 6 h in N$_2$, without molecular NiFe sites, 336 mV), and NiFe-MOF grown on a GC (**47**) (Figure 38.19c). Its superior OER performance was also confirmed by its smaller Tafel slope derived from LSVs (34 mV dec^{-1}) compared to other controlled samples and IrO$_2$ (43 mV dec^{-1}) (Figure 38.19d). Moreover, the catalyst demonstrated excellent stability in a 20 h chronoamperometric test at 1.5 V

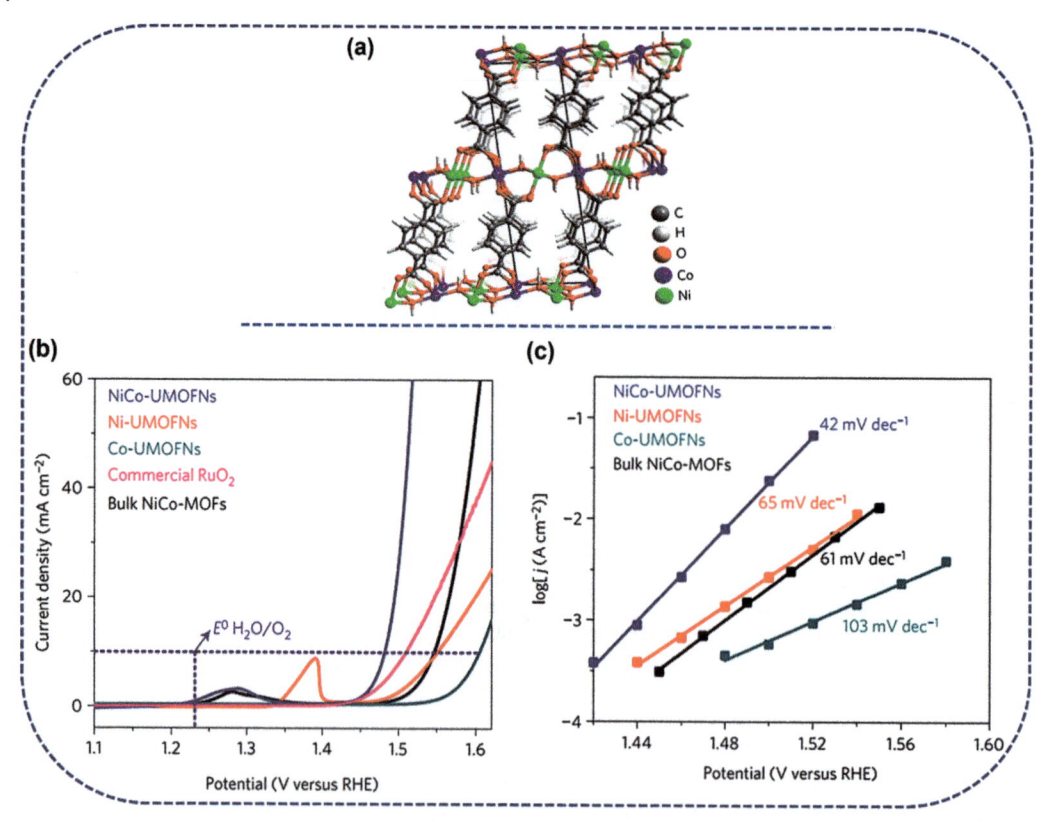

Figure 38.18 Crystal structure of NiCo-UMOFNs (**41**) (a). Polarization curves of NiCo-UMOFNs (**41**), Ni-UMOFNs, Co-UMOFNs, RuO$_2$, and bulk NiCo-MOFs in O$_2$-saturated 1 M KOH solution at a scan rate of 5 mV s^{-1} (b). Tafel slopes of NiCo-UMOFNs, Ni-UMOFNs, Co-UMOFNs, and bulk NiCo-MOFs in O$_2$-saturated 1 M KOH solution (c). Reproduced and adapted with permission from Ref [34] / Springer Nature.

(Figure 38.19e). The authors thus demonstrated an effective method of fabricating ultrathin nanosheets of a 2D MOF to be used as an electrocatalyst.

Su et al. reported a low-cost electrocatalyst NiMo$_2$C@C (**48**), which was synthesised via the calcination of a NiMo-MOF ([NiMo(bpp)$_2$O$_4$]·0.5H$_2$O, bpp = 1,3-bis(4-pyridyl)propane) that was finally coated with a graphene layer to obtain the desired composite (Figure 38.20a) [36]. The complete encapsulation of Ni/Mo$_2$C NPs into the graphene shell was proven by SEM analysis (Figure 38.20b). In addition, they synthesised a NiMo$_2$C@C catalyst mixture (**49**) by a pyrolysis process using Ni(NO$_3$)$_2$·6H$_2$O, Na$_2$MoO$_4$·2H$_2$O, and bpp ligands in a ratio of 1:1:2 corresponding to the constitution of the NiMo-MOF. After HCl leaching and carburization at 700 °C for two hours, the NiMo$_2$C@C catalyst mixture exhibited the same components as the NiMo$_2$C@C catalyst–as examined by PXRD curves, HRTEM, EDS, and XPS analysis. The composites were tested as electrocatalysts for HER in both acidic and basic media. To reach a current density of 10 mA cm^{-2}, MOF composites **48** and **49** displayed an overpotential of only 169 mV and 345 mV, respectively, indicating the superior performance of the former (Figure 38.20c). This outcome was validated by the corresponding Tafel slopes (Figure 38.20d-e). The favorable performance of the composite in HER can be attributed to the synergistic effect of Mo$_2$C and Ni, in addition to the homogeneous coating of graphene and the mesoporous structure that favors charge transfer in the HER. Based on these designed MOFs, the authors have provided a protocol for the fabrication of nanostructured hybrids from transition metal carbides and graphene.

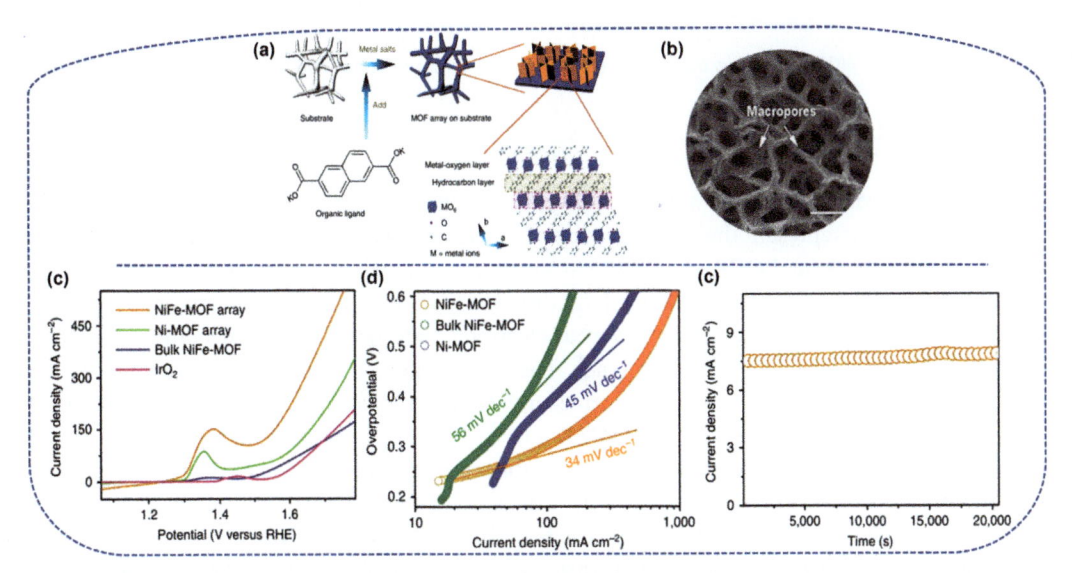

Figure 38.19 Synthetic strategy for the synthesis of mertal-organic frameworks (MOFs) nanosheets (a). Scanning electron microscopy (SEM) image of NiFe-MOF (42) nanosheet (b). Linear sweep voltammetry (LSV) plots obtained with NiFe-MOF (42), nickel-based metal-organic framework (Ni-MOF), bulk NiFe-MOF, and IrO_2 for OER at 10 mV s^{-1} in 0.1 M KOH [35] / from Springer Nature / CC BY 4.0 (c). Tafel plots obtained with NiFe-MOF, Ni MOF, and bulk NiFe-MOF (d). Chronoamperometric testing of NiFe-MOF for 20,000 s at 1.42 V (vs. reversible hydrogen electrode [RHE]) in 0.1 M KOH (e).

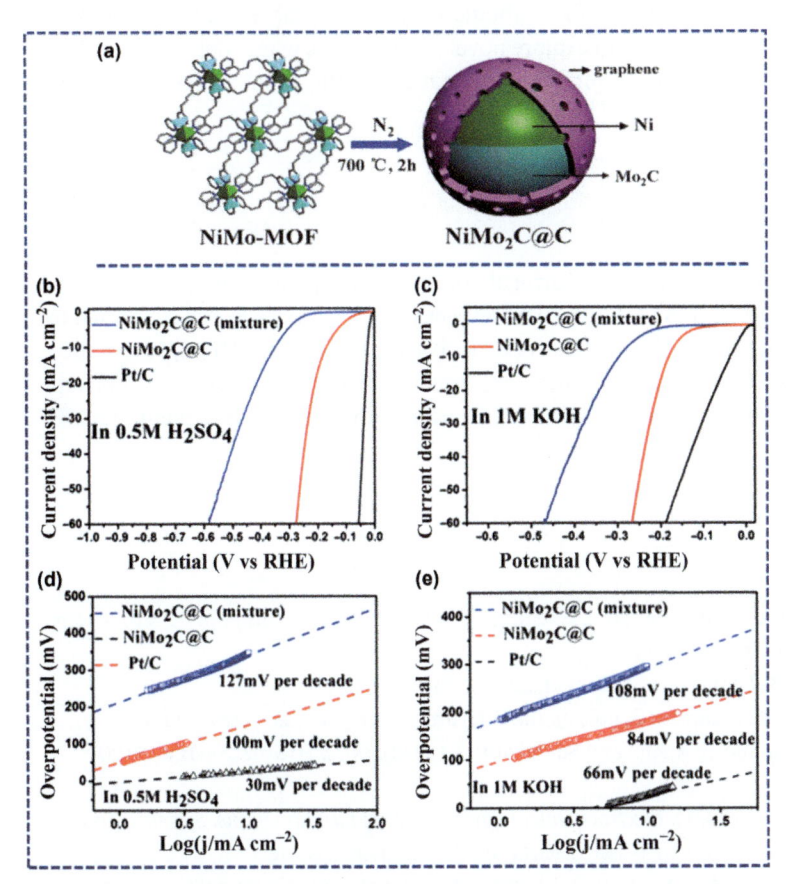

Figure 38.20 Strategy for the synthesis of NiMo$_2$C@C (48) (a). Polarization curves at 5 mV s^{-1} and the corresponding Tafel plots in 0.5 M H$_2$SO$_4$ and 1 M KOH for the composite NiMo$_2$C@C (48) (Figure 38.19b-d) and the composite mixture NiMo$_2$C@C (49) (Figure 38.19c-e), respectively. Reproduced and adapted with permission from Ref [36] / Royal Society of Chemistry.

38.4 Conclusions

This chapter concerns the use of water electrolysis to produce green hydrogen as a future main energy vector. We first present the fundamentals of the water electrolysis process (i.e. of HER proceeding at the cathode and OER proceeding at the anode of a water electrolyzer). The complex mechanisms of these two reactions are briefly discussed, along with the parameters governing their performance. Special attention is given to electrode materials and electrocatalysts used to optimize the water electrolyzers' performance. The main issue is that the most commonly used electrocatalysts are currently based on scarce and expensive platinum-group metals (PGM), namely Pt/C for HER and RuO_2 or IrO_2 for OER.

We draw attention to non-PGM electrocatalysts for both HER and OER, specifically to Ni and Co-containing MOFs (abundant and low-cost transition metals), their composites, and MOF-derived electrocatalysts. MOFs have the advantages of tailorability, structural diversity, functionality, and porosity, making them suitable for various applications. MOFs were shown to be suitable supports or precursors of functional materials for water electrolysis applications, exhibiting improved physicochemical properties and performance compared to their parent components.

The presented results on the high activity and long-term durability of Ni- and Co-MOFs and MOF-derived composites for OER and HER confirm their suitability as novel materials for water electrolyzers. There are some remaining issues to address that are related to MOFs' electrocatalytic activity, including MOFs' low conductivity: focus should be put on scalable synthesis involving mild conditions. It would also be valuable to explore novel compositions where density functional theory (DFT) calculations could be used for their screening and the rational design of novel materials.

Acknowledgements

A.P. and B.Š. are grateful to the FCT and IST, Portugal, for financial support through DL/57/2017 (Contract no. IST-ID/197/2019 and IST-ID/156-2018, respectively). This work is also supported by the Fundação para a Ciência e Tecnologia (FCT), Portugal, projects UIDB/00100/2020, UIDP/00100/2020, and LA/P/0056/2020 of Centro de Química Estrutural. It has been further supported by the RUDN University Strategic Academic Leadership Program (recipient A.J.L.P., preparation).

References

1 Fossil fuels account for the largest share of U.S. energy production and consumption (2020). *Indep. Stat. Anal. U.S. Energy Informtion Adm.*
2 Santos, D.M.F., Šljukić, B., Sequeira, C.A.C. et al. (2013). *Energy* 50 (1): 486–492.
3 Maggio, G., Nicita, A., and Squadrito, G. (2019). *Int. J. Hydrog. Energy* 44 (23): 11371–11384.
4 Nicita, A., Maggio, G., Andaloro, A.P.F., and Squadrito, G. (2020). *Int. J. Hydrog. Energy* 45 (20): 11395–11408.
5 Smolimka, T. and Garche, J. (2021). *Electrochemical Power Sources: Fundamentals, Systems, and Applications, Hydrogen Production by Water Electrolysis*. Amsterdam: Elsevier.
6 Scipioni, A., Manzardo, A., and Ren, J. (2017). *Hydrogen Economy Supply Chain, Life Cycle Analysis and Energy Transition for Sustainability*. Amsterdam: Elsevier.

7 Li, C. and Baek, J.B. (2020). *ACS Omega* 5 (1): 31–40.
8 Reier, T., Oezaslan, M., and Strasser, P. (2012). *ACS Catal.* 2 (8): 1765–1772.
9 Nemiwal, M., Gosu, V., Zhang, T.C., and Kumar, D. (2021). *Int. J. Hydrog. Energy* 46 (17): 10216–10238.
10 Lu, X.F., Xia, B.Y., Zang, S.Q., and Lou, X.W. (2020). *Angew. Chemie - Int. Ed.* 59 (12): 4634–4650.
11 Šljukic, B., Santos, D.M.F., Vujkovic, M. et al. (2016). *ChemSusChem.* 9 (10): 1200–1208.
12 Šljukić, B., Vujković, M., Amaral, L. et al. (2015). *J. Mater. Chem. A* 3 (30): 15505–15512.
13 Dubouis, N. and Grimaud, A. (2019). *Chem. Sci.* 10 (40): 9165–9181.
14 Pinsky, R., Sabharwall, P., Hartvigsen, J., and O'Brien, J. (2020). *Prog. Nucl. Energy* 123 (March): 103317.
15 Zeng, K. and Zhang, D. (2010). *Prog. Energy Combust. Sci.* 36 (3): 307–326.
16 Bandal, H.A., Jadhav, A.R., Tamboli, A.H., and Kim, H. (2017). *Electrochim. Acta* 249: 253–262.
17 Jiang, J., Huang, L., Liu, X., and Ai, L. (2017). *ACS Appl. Mater. Interfaces* 9 (8): 7193–7201.
18 Xu, Y., Li, B., Zheng, S. et al. (2018). *J. Mater. Chem. A* 6 (44): 22070–22076.
19 Chaikittisilp, W., Torad, N.L., Li, C. et al. (2014). *Chem. - A Eur. J.* 20 (15): 4217–4221.
20 Aijaz, A., Masa, J., Rösler, C. et al. (2016). *Angew. Chemie - Int. Ed.* 55 (12): 4087–4091.
21 Wu, Y., Li, Y., Gao, J., and Zhang, Q. (2021). *SusMat.* 1 (1): 66–87.
22 Tao, L., Lin, C.Y., Dou, S. et al. (2017). *Nano Energy* 41 (July): 417–425.
23 Jiang, Z., Ge, L., Zhuang, L. et al. (2019). *ACS Appl. Mater. Interfaces* 11 (47): 44300–44307.
24 Paul, A., Upadhyay, K.K., Backović, G. et al. (2020). *Inorg. Chem.* 59 (22): 16301–16318.
25 Zhou, Y.C., Dong, W.W., Jiang, M.Y. et al. (2019). *J. Solid State Chem.* 279: 120929.
26 Wu, Y.P., Zhou, W., Zhao, J. et al. (2017). *Angew. Chemie - Int. Ed.* 56 (42): 13001–13005.
27 Jia, H., Yao, Y., Zhao, J. et al. (2018). *J. Mater. Chem. A* 6 (3): 1188–1195.
28 Li, X., Fan, M., Wei, D. et al. (2020). *Electrochim. Acta* 354: 136682.
29 Meng, C., Cao, Y., Luo, Y. et al. (2021). *Inorg. Chem. Front.* 8 (12): 3007–3011.
30 Yousaf ur Rehman, M., Hussain, D., Abbas, S. et al. (2021). *J. Taibah Univ. Sci.* 15 (1): 637–648.
31 Wang, T., Zhou, Q., Wang, X. et al. (2015). *J. Mater. Chem. A* 3 (32): 16435–16439.
32 Tian, T., Huang, L., Ai, L., and Jiang, J. (2017). *J. Mater. Chem. A* 5 (39): 20985–20992.
33 Huang, H., Zhao, Y., Bai, Y. et al. (2020). *Adv. Sci.* 7 (9): 1–9.
34 Zhao, S., Wang, Y., Dong, J. et al. (2016). *Nat. Energy* 1 (12): 1–10.
35 Duan, J., Chen, S., and Zhao, C. (2017). *Nat. Commun.* 8: 1–7.
36 Li, X., Yang, L., Su, T. et al. (2017). *J. Mater. Chem. A* 5 (10): 5000–5006.

Index

a

Absorption edge 722, 726
Acetic acid 2, 25–26, 29–32, 48, 375, 567, 587, 656, 839
Acetic acid from CH_4 and CO_2 21, 31, 56, 61
Acetins 220
Acetobacterium woodii 839
Acetogenic bacteria 839
Acetophenone 171–172, 544–545, 580, 839
Activated carbons 209, 214, 218, 487, 492, 504, 732, 762
Active sites 31, 59, 94, 105, 121, 148–149, 154, 168, 209–211, 215–217, 229–232, 237, 239–240, 247, 252, 268–271, 429–431, 441, 582, 596, 759, 777, 851
Acylation 474, 577–578, 580, 586, 588, 605
2-acyl imidazoles 710
ADMET 659–660, 663, 666–670, 671
Adsorption 106, 123, 143, 147–148, 167, 239, 247, 253, 268–269, 275, 283, 285, 589, 595, 604, 722, 743, 762–763
Advanced oxidation processes (AOPs) 3, 165, 170, 248, 720
Advanced reduction processes 248, 288–290
Ag(I)-NHC compounds 837
Ag-/Cu-catalyzed asymmetric [3+2] cycloaddition 621
Agglomeration 216, 259, 482, 501, 725, 765
 see also Sinterisation
α-haloketones 701
Alcohol-based hydrogen economy 837
Aldehydes 173, 301–302, 304, 308, 319, 435, 469, 482, 529, 530, 541, 641, 651, 691, 711

Aldimines 700
Aldol reaction 614, 640, 642, 809
Aleglitazar 612–619
Alkane dehydrogenation (ODH & DDH) 231–240
Alken-1-ol 330–332
α-alkylation 691–692, 711
Alkylhydroperoxide (ROOH) 201, 515–516
Allylic alkylation 707
Amination and deamination reactions 311, 435–436, 813
Amino acids 310–311, 648, 703
Amino-cinchona organocatalysts 651–653
2-aminoethanol 838
Aminolysis 344–345
Ammonium carbamate 118–119, 124, 126, 132
Ammonium enolates 702
Anion exchange 832
Aqua regia 481
Aqueous phase and biphasic catalysis 85, 393, 524, 568, 583, 744, 749, 811, 830, 837
Aromatics 74, 150, 153, 752
α-aryl esters 703
Asymmetric tandem reactions 439–440
Azaarene 696, 698
Azomethine ylide-based [3+2] cycloaddition 619

b

Ball-milling 212–213, 218
Band structure 723
Bandgap 720
Bazarov reaction 118

Benajamin list   657
Benzene   153, 162–167
Benzotetramisole (Birmann catalyst)   655
Benzoyl peroxide (BPO)   186
Bi- and tri-metallic Au catalysts   487
Bifunctional catalyst   59, 217–218, 220, 266, 429, 654, 812
Bio-based   2, 38, 43
Bio-based substrates   37, 43, 50
Biocatalysts   20, 826, 840
Bioelectrochemical systems (BESs)   19
Biofuels   10, 133, 248, 390, 402–403
Biogas upgrading process   21
Biomass   8, 13–14, 44, 80, 214, 219, 261, 341, 360, 659
Biomass conversion   14, 216, 262
Biomass valorization   3, 13–16, 214, 216, 247–249, 253, 261–262, 291
Biomass-derived sugar valorisation   214, 219
Biomimetic and bioinspired catalysis   369, 375
Biomimetic chemistry   369
Biorefinery   248, 261–262, 264–265, 291
1,2-bis(diphenylphosphino)methane   831
(5,6-bis(5-methoxythiophen-2-yl)pyrazine-2,3-dicarbonitrile)(DPZ)   696–698
Bisoxazoline   703, 707–708
bmim, 1-butyl-3-methylimidazole-2-ylidene   836
Bnmim, 1-benzyl-3-methylimidazole-2-ylidene   838
Boron mediated diastereoselective aldol reaction   614
Brønsted acid   214–215, 696, 700, 711
Brunauer–Emmett–Teller theory (BET)   165
BTEX (Benzene, Toluene, Ethylbenzene and Xylenes)   228

C

C1 molecules and building blocks   32, 785–786, 807, 810, 815
$Ca(HCO_3)_2$, hydrogenation of   830
Calcination   85–87, 482–483, 487–488, 493, 495, 497, 499, 501–502, 504, 584, 723
Carbazole   694
Carbene transfer reactivity   445–457

Carbohydrates   44–46, 210, 220, 304, 732, 821
Carbon
  gels   210, 212
  materials   3, 189, 200, 209, 211–212, 214–216, 250–251, 253–254, 256
  nanofibers   544
  nanotubes   175, 209, 213, 215, 217, 227, 247, 316, 732
  neutral   820, 822
  xerogels   200, 210, 212, 215, 504
Carbon Capture and Utilization (CCU)   16, 26, 32, 35
Carbon circular economy (CCE)   16
Carbon dioxide (CO_2)   20, 26–27, 30–32, 36, 45–47, 55, 61, 76, 101, 117, 120, 128, 165, 757, 815, 819–820, 847
  activation   29, 57–58, 64, 119, 123, 263, 416–420
  capture   35, 45, 117, 124, 131–133, 395
  as CO surrogate   26, 28
  hydrogenation   55–61, 68, 77, 79, 101, 103–106, 549, 822, 830, 833–834
  methanation   56–57, 61, 63–64, 68
  utilization   36, 58
Carbon monoxide (CO)   28, 30, 57, 119–120, 133, 143, 312, 492, 797
Carbon nanotube-glucose composite catalysts   775
Carbonaceous materials   55, 172, 219, 256, 274, 283, 286–287, 721, 732, 735
Carbon-based catalysts   3, 209–220, 225–240, 262
Carbonization   210, 212, 215, 220, 228, 258, 261, 279, 290
Carbonylation   25, 28, 31, 127–130, 787, 793, 808
Carbonylation reactions   787
Cascade   14, 18, 216–219, 429, 433, 462, 466, 469, 793
Cascade reaction   18, 216, 217, 262, 429, 466, 698, 711, 788
Catalysis   4, 7, 10–18, 21, 25, 73–108, 141–157, 240, 389–404, 429, 481–505, 611–635
Catalysts for oxidative reactions   167
Catalytic membranes   210
Catalytic oxidation   144, 150, 155, 161–168, 170, 266, 441, 502, 518, 524, 769, 771, 777, 778

Catalytic reactor   103, 640, 644, 646, 655, 760
Catalytic synthesis   119, 122, 127, 134
Catechols   15
Cathode   19, 20, 120, 248, 254, 277–282, 420, 684, 849, 868
C-C bond formation and C-H activation   92, 370, 412, 462, 465, 567, 692, 694, 696, 705, 793–794, 797, 808
Cellobiose   215–217
Cellulose   13, 44–46, 214–220, 262
Cesium formate   836
C-H activation   370, 383, 412, 462, 465, 705, 786, 793, 794, 797, 808
CH_4 20, 21, 27, 29, 31, 55–58, 61, 68, 89, 104, 106, 118, 259, 340, 341, 549, 785, 822
Charge carriers   287, 722, 723, 724, 726, 735, 736
Chars   219, 283
Chemical catalysts   13
Chemical industry   8, 10–13, 37, 50, 58, 248, 288, 429, 611
Chemical recycling   339, 340, 342–344, 347, 356, 363, 663, 671
Chemical vapour deposition   734, 736
Chemistry of Fischer Tropsch   75
   alpha value   80, 83
   metals   76
   commercial catalysts   83
Chiral phosphoric acid (CPA)   699
Chlorinated VOCs   155
Cinchona alkaloids   651, 690, 702
Cinchona-picolinamide organocatalysts   651
Cinchona-squaramide organocatalysts   652
Circular economy   1, 16, 340
Claus reaction   490
Clean fuels   2
CO oxidation   490–499
CO_2 see Carbon dioxide
CO_2 conversion, Power-to-Gas   61
CO_2 methanation, Synthetic natural gas   55
CO_2 reduction by microbial electrosynthesis   19, 26, 31
CO_2 reduction reaction   259
Cobalt (Co) catalysts   82, 90, 189, 572
Cobalt Fischer Tropsch catalysts   82
   design   402
   composition   83
   preparation   85

regeneration   100
   life-cycle   100
   materials circularity   100
   oxidative regeneration   101
   water tolerance   88
Co-factors   15
Colloidal gold   482
Composite materials   163, 220, 247, 857
Condensation   431, 434, 660, 661
Condensation polymerization   332, 659, 660, 661, 671
Conduction band   144, 720
Conjugate addition reaction   708
conPET (consecutive photoinduced electron transfer)   681, 683
Coordination of water to metal in catalytic intermediates   554
Copolymerization   325, 327, 328, 330, 664
Copper   191, 472, 475
Copper catalysts   191, 472, 475, 105
CO-surrogates   788
Covid-19 pandemic   1, 4, 10, 55
Cp*, pentamethylcyclopentadienyl   814, 837
Cross-coupling reactions   546, 559, 560, 570, 574
Cross-dehydrogenative coupling (CDC)   692
C-X formation (X = C, N, O, Metal)   410
Cyanoalkylation   692, 705
Cyclic (Alkyl)(Amino)Carbene (CAAC), polymerized (PCAAC)   826
Cyclic carbonates   37, 44, 46
Cyclic olefin   325, 327, 330
Cyclic ureas   130, 132
Cyclization reactions   468, 793
Cycloaddition   474
[2+2] cycloaddition   687–688, 708
[3+2] cycloaddition   468–469, 619, 621–622, 708
[8+2] cycloaddition reaction   656
Cyclobutane   708, 790
Cyclohexane   41, 181–203
Cyclohexanol   181–182, 187, 189, 190, 192, 195, 198, 200–202
Cyclohexanone   181–182, 185, 187–192, 195, 197–198, 200–202, 548, 641
Cyclohexylhydroperoxide (CyOOH)   182, 201
Cyclopropyl ketones   708

d

Danoprevir   623
DBU, 1,8-diazabicyclo(5.4.0)undec-7-ene   40, 126, 132, 656, 751
Deacetalization-Henry reaction   433, 442
Deacetalization–Knoevenagel condensation   431, 432, 433, 442
Deactivation   489, 501, 502
Decarboxylation   614, 694, 698, 705
Defect density   722
Dehalogenation   549, 688, 698
Dehydration   58, 118, 124, 127, 262, 785
Dehydrogenation of aqueous formate   827, 836
Deliverable hydrogen capacity   840
Density Functional Theory   121, 253, 358, 412, 539, 807, 868
Depolymerisation   45, 339, 348, 350, 354, 356, 670
Desulfurization   743, 746, 757, 758, 761, 767
Diamines   129–132, 351–356
Diazocarbenes, Metal-carbene complexes   455
Dienes   413, 520, 708, 801
Diffuse Reflectance Infrared Fourier Transform (DRIFT)   163
Dihydropyridines   471, 707
1,2-dimethyl-3-uracilyl-4-imidazolylidene   837
Diols   47, 661
Dioxygen   187, 370, 384, 698, 700
Direct CO_2 conversion over iron   101
  effect of conditions   103
  stability against oxidation   80
The direct conversion of cellulose into polyols   216
Dispersion   482, 483, 501
Disulfuvibrio desulfuricans   112
DMSO   351, 356, 468, 470, 812, 828
Doping   722
Durability   489

e

Earth abundant metal based catalysts   369, 370
Edaglitazone   613
Electric energy   820, 847
Electrocatalysis   120, 247
Electro-catalytic incorporation of CO_2 17
Electrolysis   683
Electron donor acceptor (EDA) complex   711
Electron-hole pair   720
emim, 1-ethyl-3-methylimidazole-2-ylidene   836
Enamine catalysis   692, 702
Enantioselectivity   301, 304, 462, 475
Endangered element   827
Energy storage   4, 61, 820
Energy transfer (ET)   687
Energy-related reactions   291
Environmental catalysis   481
Environmental protection   3, 209
Enzymatic   18, 31
Enzymes   13–18, 217, 370, 523–524, 529, 532, 617, 628–629, 814, 821
Epichlorohydrin   42
Epoxidation–ring-opening of epoxide   436
Epoxides   37, 39
Esherichia coli   839
Ethylbenzene   171
Ethylene copolymer   332
Ethylene urea   130, 131, 133, 335
Excitation   721

f

Fatty acids   42
Fenton-like processes   272
Fermi level   723
Ferrimagnetic   725
Fine chemicals synthesis   320
Fischer Tropsch, selectivity   91
  activation conditions   98
  promotion   4, 64, 108, 149, 233, 538
  role of pore diameter   96
Flow-chemistry   640
Fluorescent   732
Formate-bicarbonate equilibrium   831
Formation of radicals in reactions in presence of water in homogeneous systems   556
Formic acid dehydrogenation   824
Formic acid, pK_a 827
Formic acid   60, 394, 770, 801, 824, 827–829, 831–832, 834
  properties   820
  synthesis   828
Fossil fuels   259, 261, 421, 499
Fossil-C   7, 8
Free radical   144, 172, 191–192, 213, 237, 375, 700, 703
Friedel-Crafts   597
FFT   494

Fuel cell (FC) 822
Functional groups 43, 211
Functionalization 370
Future perspectives for Fischer-Tropsch 106
   decentralized PtL plants 107
   green chemicals 108
   greener catalysts 108
   sustainable design 107

g

Gas hourly space velocity (GHSV) 162, 163
Gas-phase oxidation processes 225
George Olah renewable methanol plant 27, 822
GHG 7, 74
Glucose 14–15, 45, 47, 214–216, 219, 262, 264, 283, 544
Glycolysis 344
Glycosidic bonds 214
Gold 481
   Nanoparticle size 482, 483, 484, 485, 486, 487, 488, 489, 492, 494, 495, 497, 501, 503, 504, 505
   Nanoparticle properties 487
      activity 487
      durability 489
      poison resistance 490
      selectivity 488
   Oxidation state 481, 488, 494, 497, 499, 501, 503
   Preparation methods 482
      Co-Precipitation (CP) 483
      Deposition Precipitation (DP) 484
      Impregnation (IMP) and Double Impregnation (DIM) 483
      Ion-Exchange 485
      Liquid-phase Reductive Deposition (LPRD) 484
      Photochemical deposition (PD) 485
      Sol-immobilisation (COL) 482
      Ultrasonication (US) 486
      Vapour-Phase and Grafting 486
Gold catalysts 200
Graphene 19, 28, 210, 236, 238, 251, 260, 269, 273, 279, 726, 763, 772, 775, 866
Graphitic carbon nitride (GCN) 826
Gravimetric hydrogen density (capacity) 823
Green 730
   catalysis 732
   chemistry 737
   synthesis 730

h

H_2S selective oxidation to elemental sulfur 226
Half-titanocene catalyst 330
Hantzsch ester 696
Heteroatoms 211, 215, 235, 239
Heterodimerization 708
Heterogeneous 719, 720
Heterogeneous catalysis 209, 262, 313, 482, 768
Heterogeneous catalyst surface modification with water molecules 537
Heterojunction 722
Heterometallic complexes 409
Hierarchical catalysts 585
High-resolution transmission electron microscopy (HRTEM) 165, 494
High temperature Fischer Tropsch 74, 79
   Iron catalysts 79
   plants 80
   carbon number distribution 76
Homogeneous 18, 60, 64, 85, 153, 162, 175, 313, 473, 537, 556, 749, 823, 826
Homogeneous catalysis 40, 420, 424, 523–524, 532, 787, 822–823, 830, 836
Hot spots 723
HRTEM 493, 494
see also TEM
Hybrid catalysis 11, 13, 16
Hybrid materials 256, 283
Hydrochars 219
Hydroformylation 545, 555
Hydrogen 257, 399
   economy 837
   evolution 826
   generation 830
   properties 835
   storage 835
Hydrogen dependent CO_2 reductase enzyme (HDCR) 839
Hydrogen evolution reaction 257, 849
Hydrogen peroxide (H_2O_2) 166, 402, 532, 550, 850
Hydrogenation 216, 217–220
Hydrogenation and dehydrogenation reactions 403, 826, 835
Hydrogen-atom transfer (HAT) catalysis 692
Hydrogenolysis 25, 262, 265, 291, 357–358, 759

Hydrolysis   214–219, 342–343, 361, 539, 548–549, 620–623, 628, 692, 793, 800, 860–861
Hydrolytic oxidation   217
Hydroperoxyl radical   166, 267
Hydrothermal carbonization   210, 220, 252
2′-hydroxychalcones   708
5-hydroxymethylfurfural   14–15, 539
Hydroxyl radical   144, 149, 165–166, 187, 201, 212–213, 282
Hypervalent iodine reagent   370, 694

i

Idasanutlin   612, 619, 635
Iminium ion   690, 694, 696, 711–712
α-iminyl radical cation   690, 694
Immobilization   200, 257, 267–268, 285, 313–317, 317, 639–640, 643, 651, 657, 773, 776–778, 725
Immobilized organocatalysts   651, 653, 657
Impregnation   27, 58, 66, 85–86, 91, 107, 163–165, 171, 200, 217, 226, 228, 254–255, 263–266, 283, 289, 483, 496, 496, 765, 775
Influence of hydrogen bonds on catalytic reactions   538, 539, 556
Interaction of support with gold   488, 492, 493, 494, 497, 499, 500–502
Interface   27, 58, 122, 253, 492, 495, 499, 537–538, 721, 723, 725, 730
Ionic liquid(s)   41, 45, 133, 188, 196, 198, 202, 400, 580, 641–672, 744, 751, 828
Ionophore   839
Ipatasertib   612, 626, 635
[IrCl(cod)(NHC)]   836, 838–839
[{IrCl$_2$(Cp*)}$_2$]   837
[Ir(cod)(NHC)(mtppms)]   836, 838
Ir-dipyridilamine   836
[IrH$_2$Cl(PPh$_3$)$_3$]   828
Iron catalysts   76, 79–80, 102–105, 186, 746
β-isocupreidine   702
Isothiourea   655–656, 702
Isothiourea Catalysts   655

k

KA oil   181–182, 190
β-ketocarbonyl compounds   694
Ketoreductase promoted asymmetric reduction of a prochiral ketone   626
Ketyl radical   696–697, 710

l

Late-stage functionalization   712
Lewis acid   32, 43, 126, 130, 153, 263–364, 348, 394, 397, 401, 412, 420, 433–434, 441, 515, 518–520, 575, 577, 581–591, 600, 688, 703, 707
Limonene
Limonene oxide   39–40
Liquid hydrogen   73, 403, 825–826
Liquid organic hydrogen carriers (LOHC)   822–823
Low temperature Fischer Tropsch   79
   cobalt catalysts   13, 28, 57, 73
   carbon number distribution   76
   plants   2, 75, 402, 500, 539
Low-temperature Reforming   399
Luminescence resonance energy transfer   735
Lysine, potassium salt (LysK)   830, 834

m

Magnetocatalysis   2
Mars and Van Krevelen mechanism   162
m-Chloroperoxybenzoic acid (mCPBA)   186, 190, 202, 372
McMillan catalysts (Imidazolidinones)   647
MCR   461
m-Dinitrobenzene (m-DNB)   696
Mechanism   21, 38
Mechanism, Oxidation   375
Mechanochemical treatment   164
Meinwald rearrangement–Knoevenagel condensation   434, 435, 442
Mesoionic triazolylidene ligands   185, 837
Mesoporous carbons   210, 217, 227, 229, 262
Mesoporous materials   57, 150, 153, 210, 220, 589, 773
Metal nanoparticles   167, 216, 547, 723, 826
Metal oxides   63, 142, 150, 153, 165, 167, 175, 214, 247–251, 262, 267, 482, 483, 487, 488–505, 722, 725
Metal-free carbon catalysts   3, 212
Metallaphotoredox catalysis   688, 702, 703
Metallocene catalyst   323, 329, 330
Metal-Organic Frameworks for fuels treatment   42, 762, 765, 858, 860
Metal-support interaction   29, 56, 64, 67, 492, 502, 834

Metathesis polymerization   4, 670
Methanol
  dehydrogenation   59, 399, 811, 822
  economy   27, 391, 822, 837
Methanol from CO_2 and H_2 26
2-methyl-quinoline (2-MeQ)   837
2-methyl-1,2,3,4-tetrahydroquinoline
  (TH-2-MeQ)   836–837
Methylation reactions   799
Michael-addition   620, 622, 641, 651, 653, 708
Microorganisms   12–14, 18–20, 31, 38, 839
Microwave irradiation   169, 202, 585, 744, 863
Minisci-type reaction   698
Mixed oxides   164, 500, 502, 503
mmim, 1,3-dimethylimidazole-2-ylidene   838
Molybdenum catalysts   198
Monensin   839, 861
mtppms-Na   826
mtppts-Na_3 829–830
myrcene   38, 333–335

n

$Na_2[Ir(cod)(NHC)(mtppts)]$ 836
Nanocatalysis   254
Nanocomposite   123, 149, 202, 251, 254–255, 268, 272–273, 275–276, 286, 723, 726, 731–732, 763, 765
Nanoparticles   3, 27, 77, 121, 163, 168, 481–498, 501–505, 725, 765, 828, 853
Nanostructured carbon materials   3, 209, 254
Nature-based   720–721, 732, 737
N-heterocycles, hydrogenation/
  dehydrogenation   462, 468, 822–823, 829, 831, 836, 837
N-heterocyclic carbene (NHC)   563
N-hydroxyphthalimide (NHP) ester   703
Nickel (Ni) catalysts   570
Nitrilase promoted chiral resolution of nitriles   628, 813
Nitrogen oxides (NOx)   2, 7, 117
Nobel prize 2021   639
Nobel prizes of chemistry   639, 657
Non-biodegradable   339, 720
Non-thermal plasma (NTP)   161
NOx   see nitrogen oxides
NTP technology   162
n-type   282, 723, 726

o

O_2 and H_2 production   4, 61, 81, 107, 118, 423, 822
Olefin metathesis   336, 356, 526
Olefin polymerization   323, 334
"On water" reactions   537
Ordered mesoporous carbons   210
Organic photoredox catalysis   675–676, 681, 684
Organocatalysis   639, 640, 645, 657, 690
Oxidation   3, 104, 122, 142, 144, 146, 148, 157, 161, 167, 225, 240, 414
Oxidation processes   225, 240, 248, 266, 414, 752
Oxidation-esterification   437, 442
Oxidation-hemiacetal reaction   438
Oxidative removal of VOCs   163
Oxides   see metal oxides
2-oxindole   705
Oxyfunctionalization   181, 202
Oxygen evolution reaction   248–250, 254, 849
Oxygen reduction reaction   249
Oxygen vacancies   29, 63, 123, 130, 149, 162, 253, 289, 491, 492, 494, 540, 726
Oxygenated VOCs   150, 154
Ozonation   144, 146, 148–149, 150–151, 155, 157, 212, 266, 728
Ozone   8, 141–145, 153, 154–157, 189, 482, 768
Ozone catalytic oxidation (OZCO)   166

p

$P(CH_2CH_2PPh_2)_3$ 826, 828
Palladium (Pd) catalysts   398, 567, 569–570, 661, 707, 801, 834
Participation of water in formation of catalytic intermediates   554
(iPr)2PCH2CH2P(iPr)2   826
Pd nanoclusters   826
Pd/C   15, 45, 358, 398, 543, 826, 830
Pd-catalyzed cyclocarbonylation   791, 795
Peptides   574, 650
Perennial energy sources   8, 10
Perfluoroalkylation   702, 710
Perimeter   492, 495, 499
Peroxidative oxidation   169, 170, 189–190, 202
Phase composition   723
Phenyl-benzylimine   837
1-phenylethanol   838–839
Phosphine ligands   304, 517, 520–522, 560, 561, 568, 707, 826

Phosphorus ligands 301, 303, 564
Photocatalysis 719, 720
Photocatalytic oxidation (PCO) 166
Photocatalytic reduction of CO_2 28
Photocatalytic tandem reactions 440
Photochemical reaction 490, 687
Photodegradation 166, 283–287, 722
Photo-Giese reaction 708
Photooxidation 441, 696, 700, 732
Photoredox catalysis 675, 681, 687–690, 693, 696, 700, 707, 712
Photosensitizer 19, 172, 291, 420, 445, 489, 702, 707–708, 710
Photosynthetic biohybrid system 21
Photothermal 721
pH-responsive ligands 829
Phtotocatalysis 122, 144, 165, 281–283, 440, 677, 683, 684, 720, 732, 737
Phytomolecules 730
Pincer
  chemistry 389
  complexes 3, 348, 389, 390–403, 830–832
  compounds 389
Piperazine-2,5-dione (glycine anhydride) 838
Plant oil 659
Plasmonic 730
Plastic 339–340
Platform chemicals 216, 220, 262
Platform molecule(s) 14, 18, 21, 216
Platinum 515, 516, 532, 850
Pollution control 141
Polycondensation 660–663
Polyester 332, 342, 345–347, 659–663, 662–664, 670, 671
Polymer 10, 36, 108, 117, 133, 182, 584, 661
Polyolefin 323, 335, 340, 357, 417
Polyphenolic compounds 730
Porous adsorptive materials 163, 235, 772, 778
Post-treatment 487, 602
Potassium formate 828
Preferential oxidation of CO in presence of H_2 496
Pre-treatment 165, 175, 218, 237, 262, 482, 487, 489, 501
Process Mass Index (PMI) 611, 612, 616, 619, 623, 626, 634, 635, 684

Proline 639, 640, 641, 642, 646, 650
Promoting effect of water in hydrogenation and hydroformylation 301–320, 545–549, 555–556, 794–797, 826
Propargylamines 133, 462, 463, 464, 465
Propiolic acid 694
Proton exchange membrane (PEM) 291, 489, 496, 497, 822, 833, 849
Proton-coupled electron transfer (PCET) 122, 420, 696
PROX see Preferential oxidation of CO in the presence of H_2
Pta (1,3,5-triaza-7-phosphaadamantane, 9) 828
2-pyrazinecarboxylic acid (PCA) 182–183
Pyrolysis 239, 254, 255, 256, 260, 263, 265, 266, 279, 285, 288, 340, 341, 345, 356, 363, 772, 859, 863, 866
Pyrroles 462, 466
Pyrrolidine organocatalysts 645–646
Pyrroloindolines 700

q

QSPR analysis 599, 601, 602
Quantum confinement effect 732
Quinoline 290, 466, 696, 698, 748, 793, 794, 809, 837
Quinolones 698, 708, 711
Quinoxaline 837

r

Radical conjugate addition 694
Radical reactions 570, 689
Reactions efficiently catalysed 490
Reactive oxygen species (ROS) 162, 165, 166, 725
Reactive species 212, 272, 546, 720, 746, 752
Reactors for Fischer-Tropsch 81
  fixed bed reactor 27, 81, 82, 89, 94, 100, 105, 230, 760
  microchannel reactor 107
  microstructured reactor 81
  slurry reactor 82, 90
Recycling 27, 30, 32, 35, 42, 55, 106, 127, 182, 189, 197–198, 201–202, 216, 264, 268–270, 279, 289, 323, 335
Redox reaction 15, 278, 675, 688, 723
Redox-neutral reaction 689

Reduced graphitic oxide (rGO)   210, 218, 250–251, 253, 261, 269, 272–273, 286, 731, 834
Reduction   2, 18, 28, 35, 37, 56, 64, 88, 94, 100, 107, 165, 168, 175, 187, 248, 274, 288–289, 345, 394, 501, 794, 826
Reductive amination   311, 435, 436, 565, 800
Renewable-C   11, 15, 214
Reusability   725
[RhCl(mtppms-Na)$_3$]   830
[RuCl$_2$(PPh$_3$)$_3$]   812, 826
Rhenium catalysts   199
Roche Environmental Awareness in Chemical Technology (REACT) Award   612
Ru- / Ir-catalyzed asymmetric hydrogenation of alpha,beta-unsaturated acids   616
[Ru(2-Me-allyl)$_2$(cod)]   831
Ru(II)-PNP pincer complex   831
Ru-catalyzed ring closing metathesis (RCM)   623, 625, 626
[{RuCl$_2$(benzene)}$_2$]   831
[{RuCl$_2$(mtppms-Na)$_2$}$_2$]   826, 835–836
[RuHCl(CO)(PNN)]   838
[Ru(H$_2$O)$_6$](tos)$_2$ 828
Ru-MACHO   393, 394, 395, 398, 399, 400, 832

s

Sandmeyer Award   635
Saturation magnetization   268, 269, 726
Scanning electron microscopy (SEM)   165, 852
Schottky contact   723, 730
Secondary metabolites   730
Selective isotope labelling   121, 810, 814
Selective oxidation of CO in the presence of H$_2$ *see* Preferential oxidation of CO in the presence of H$_2$
Semiconductor   121, 166, 282, 720, 722–726, 737
Semipinacol rearrangement   698
Silica-based materials for clean energy   493
Single electron molecular orbital (SOMO) activation   690
Single-electron transfer (SET)   676, 688
Sintering *see* Sinterisation
Sinterisation   482, 483, 488, 495
Size *see* Gold, Nanoparticle size
Sodium formate   832
Solar-driven   721, 737
Solubility of alkali formates   830

Solvent-free   40, 42, 184, 190, 195, 274, 358, 438, 466, 542, 580, 750
Solvolysis   342, 363
Solvothermal method   164, 268, 279, 280, 852
Sorbitol   218, 219, 220
Speciation   559, 564, 567, 568, 569, 570, 572, 574
Specific activity   15, 253, 590
Spillover   27, 83, 105, 218
1,1′-spirobiindane-7,7′-diol (SPINOL)   697–698
Stereoselectivity   44, 369, 371, 381, 409, 646, 649, 656, 696
Storage   487
Sugar alcohols   218
Sulfur oxides (SOx)   2, 743
Superoxide anion radical   698, 720
Superoxide radicals   149, 166, 282, 774
Support   482–495, 497–505
  active supports   493
  inactive supports   488
  inert supports   *see* inactive supports
Surface chemistry   210–215, 217, 218, 247, 262, 771
Surface groups   210, 220, 238, 771
Surfactants   210, 523–527, 530, 532, 583–585, 594, 745
Sustainable   7, 17, 32, 35, 36, 37, 38, 44, 46, 50, 55, 73, 74, 81, 101, 107, 108, 117, 119, 162, 164, 181, 190, 202, 203, 209, 214, 220, 225, 226, 231, 232, 240, 247, 248, 253, 257, 261, 262, 265, 291, 292
Sustainable Aviation Fuels   73
  climate impact   10
  pathways for production   74, 81, 106
Sustainable catalytic processes   248
Sustainable coupling reactions   559
Sustainable organic synthesis   785
Sustainable processes   220, 389, 403, 612, 635
Synergistic effects   215, 216, 414, 488, 493, 764, 767

t

Tammann temperature   489
Tandem reactions   31, 220, 264, 429, 430, 431, 433, 434, 435, 438, 439, 440, 442
TBHP   170, 172, 175, 182, 184, 186, 187, 190, 192, 195, 200, 201, 439, 745, 768, 769, 771, 772, 773
TEM   483, 486, 488 *see also* HRTEM

Temperature-programmed desorption   210
Templated carbons   209, 210
Terephthalic acid   25, 162, 172, 173, 175, 342, 358, 661
Terpenes   38, 39, 40, 307, 324, 332, 335
Tert-butyl hydroperoxide (TBHP)   182, 186, 439, 745, 768
Tetrabutylammonium decatungstate (TBADT)   705
(Tetracarboxylate)-M_2 complexes of M=Rh, Pd, Ag, and Co, Ni, Cu   447, 454–455
Tetrahydroisoquinoline (THIQ)   696, 702
Tetramethylpiperidine 1-oxyl radical (TEMPO)   700
Textural properties   63, 66, 211, 220, 235, 289, 765, 768, 770
Texture   209, 210, 217, 235, 585
Thiocyanato alcohol   705
Thioxanthones   711
Toluene   150–153, 167–170
Total oxidation of volatile organic compounds (VOCs)   502
Transesterification   220, 349, 660, 664
Transfer-hydrogenation   800, 810, 812, 813, 815
Transition metal   *see* Transition metal catalysis
Transition metal catalysis   334, 518, 689, 705
Transition metal catalyst   182, 323, 327, 330, 335, 515, 810
Transition metal oxides   56, 131, 142, 150, 153, 167, 251, 268, 488–489
1,2,3-triazole   641–642
Triazolylidene-ligated Ir-NHC complexes   836
5-trifluoromethylpyridonate   837
Triphenylphosphine (PPh_3)   568
Triphos   389, 831
Tryptamines   700

u

UN Paris Agreement on Climate Change   1
UN Sustainable Development Goals (SDGs)   1
United Nations (UN) Sustainable Development Goals (SDGs)   *see* UN Sustainable Development Goals (SDGs)

Upconversion   735
Urea   117–134
   derivatives   118–134
   synthesis   117–134
Urea-hydrogen peroxide adduct (UHP)   186

v

Vacuum ultraviolet (VUV)   166
Valance band   723
Valorization of real biomass-waste   69, 247, 262
Vanadium catalysts   182
Vegetable oils   42–44
Vinylcyclohexene oxide   41
Vinylpyridine   696–698
2-vinylquinolines   698
Visible light   441, 687–712
VOC(s)   482, 490, 502–505
Volatile organic compounds (VOCs)   *see* VOC(s)
Volumetric hydrogen density (capacity)   823

w

Wastewater treatment   247–292, 719–737
Water   42, 64, 77, 89, 104, 421, 522–524, 540, 546, 552, 555
Water electrolysis   10, 27, 55, 61, 82, 849, 851, 868
Water-gas shift   482, 490, 497, 499–502
Wet Air Oxidation   212, 220
Wet Oxidation   212
Whole cell biocatalysis   839–840
Wittig reaction   617
Wohler reaction   117–118

x

Xanthenes   692
Xanthone   707, 711
XXII International Symposium on Homogeneous Catalysis (ISHC)   xxvi
Xylene   172

z

Zeolites   57, 64–68, 577–605, 763–765, 768–769, 775–776